T0143399

Circuit Design: Anticipate, Analyze, Exploit Variations

Statistical Methods and Optimization

RIVER PUBLISHERS SERIES IN CIRCUITS AND SYSTEMS

Series Editors

MASSIMO ALIOTO
National University of Singapore
Singapore

KOFI MAKINWA
Delft University of Technology
The Netherlands

DENNIS SYLVESTER
University of Michigan
USA

Indexing: All books published in this series are submitted to Thomson Reuters Book Citation Index (BkCI), CrossRef and to Google Scholar.

The "River Publishers Series in Circuits & Systems" is a series of comprehensive academic and professional books which focus on theory and applications of Circuit and Systems. This includes analog and digital integrated circuits, memory technologies, system-on-chip and processor design. The series also includes books on electronic design automation and design methodology, as well as computer aided design tools.

Books published in the series include research monographs, edited volumes, handbooks and textbooks. The books provide professionals, researchers, educators, and advanced students in the field with an invaluable insight into the latest research and developments.

Topics covered in the series include, but are by no means restricted to the following:

- Analog Integrated Circuits
- Digital Integrated Circuits
- Data Converters
- Processor Architecures
- System-on-Chip
- Memory Design
- Electronic Design Automation

For a list of other books in this series, visit www.riverpublishers.com

Circuit Design: Anticipate, Analyze, Exploit Variations
Statistical Methods and Optimization

Stephan Weber

Germany

Cândido Duarte

Universidade do Porto
Portugal

River Publishers

Published, sold and distributed by:
River Publishers
Alsbjergvej 10
9260 Gistrup
Denmark

River Publishers
Lange Geer 44
2611 PW Delft
The Netherlands

Tel.: +45369953197
www.riverpublishers.com

ISBN: 978-87-93379-75-6 (Hardback)
    978-87-93379-76-3 (Ebook)

©2017 River Publishers

Contents

PART I: Engineering, Circuit Design, Flow and Methods

PART II: Basic Statistical Design Techniques

Foreword

Fueled by the continuous downscaling of transistors in the CMOS VLSI technology, integrated electronic circuits are the cornerstone of most applications and devices we use in our daily lives today. From the radio and TV to our smartphone, from our car to the dishwasher, from our laptop to the heart rate monitor we use while jogging, just to name a few. Electronics add intelligence and controllability to objects, and wireless connectivity adds on top mobility and universal connectedness across the globe. But while digital integrated systems are mostly designed in a highly automated manner at higher abstraction levels starting from some form of language-based description, analog and mixed-signal electronic circuits are still today mostly designed at block and circuit level and with little or no automation, be it of course not without individual CAD tools.

Such electronic circuit design is by far not an easy task, not easy to carry out and not easy to learn. The main reasons are the many complex and often conflicting relationships between design variables and circuit performances, the underdetermined nature of the design problem at hand, and the impact of internal and external variations. The many degrees of freedom in the design and the sheer high-dimensional complexity challenge our human brain. As a result, designing a circuit that is optimal in some desired sense and that is guaranteed to meet the targeted requirements and specifications under all fabrication and operational circumstances is not at all trivial. But although it may look as an art to some, it's a skill that actually can be mastered by many by combining experience with analytic insight and systematic design methodology.

While various widely used analog "circuit design" textbooks essentially restrain themselves to presenting and analyzing circuit schematics, this book goes way beyond that and actually presents design from a pragmatic design viewpoint: it describes a systematic variation-aware approach to interactively design fully functional circuits with the help of advanced CAD tools. While EDA tool research is progressing continuously in academia and industry, also for analog and mixed-signal circuits, fully automatic synthesis of analog circuits in general is probably not to be expected soon as commercial offering. Yet, powerful CAD tools for simulation/analysis as well as for optimization have been developed over the last decades. In combination with a systematic design strategy and the power of today's computers, these can be used to successfully crack the nut of designing a functional electronic circuit.

This book will show you how, and does that in a practical way with several design examples. Large focus is on dealing with the impact of variations due to the fabrication process as well as through the environment. Using advanced statistics and going beyond classical corner and Gaussian analysis, the impact of these variations is analyzed and circuits are optimized for robustness and maximum yield. Though the CAD techniques used are state-of-the-art commercial offerings, the authors are no blind tool believers, and clearly pinpoint the limitations of the tools used. As always, the computer only gives you what you ask for.

Considering its unique focus on a systematic variation-aware and tool-supported approach for designing electronic circuits, this book is recommended reading for practicing design engineers, electronic design engineering students as well as EDA developers. As such, the book illustrates in a pragmatic way that circuit design in today's world is so much more than magic, it is magic that everyone can master!

Prof. Georges G. E. Gielen
University of Leuven, Belgium

Preface

Writing a book, you probably start with an outline, with collecting ideas, with interesting chapters, etc. and if you write an introduction, you often end up too many things, in something too *long*. So here is the two page, shortest possible, introduction and preface. In the subsiding (long) chapter just read further about "why" and "how", about motivation, background information, challenges, trends, etc. Then we really move over to real design, design techniques, to their problems and to new techniques!

People are fascinated by beauty, which has often aspects of maximum simplicity, and endless complexity: nature, arts, sports, technics, philosophy, electronic circuits, and math! The authors really like circuits, but this book is not so much about circuits itself; we focus on *techniques*, on connecting these parts well: circuits and *design techniques* and math; and we will find beautiful and ugly things. To some degree designers also love ugly things and need to find workarounds, and sometimes, but not always, you again end up in something elegant, something yours.

What is the status? Electronic devices are very complex and tricky, but very cheap. This is because essentially they are just *printed*, like this book! The manufacturing of chips is complex, but amazingly efficient and cheap, per device or function even extremely cheap. The biggest invention in design itself is the use of computers and software for *simulations*. So in modern designs people work – usually together in a team – on *virtual* prototypes. And at some point they need to become confident, that the product would also work in reality. This is a challenging task, and many variables have an impact whether a design (component, chip, board, system) fails or succeeds.

Circuit simulation is mathematically something quite special, solving a set of nonlinear equations, like $f(x) = 0$. The real simulation breakthrough was in the 1980s, so what is *beyond* pure simulation?

Dealing with variables and function helps to translate a problem into math and algorithms, but unfortunately we have to deal with many variables, and with functions we simply do not know so well. Usually special *combinations* of variables are critical (like low supply voltage, high temperature, heavy load capacitance, etc.); and having many variables, means having even more combinations of them, so that at some point also simulation time matters or the full simulation is becoming even unaffordable (like taking years). Imagine you have a budget to run realistically n = 100,000 simulations, "which" simulations you should execute to achieve a certain goal, like "Find the most critical parameter combinations regarding all performances".

On top of such verification tasks, designers want to find the "best" circuit topology and the according component values to make sure that the design works even under these difficult conditions. This is minimization of errors or optimization; and mathematically it is basically the *minimization* of functions.

Dealing with all such problems is possible with modern design software, and this has not only a combinatorial aspect, but also a statistical one. The latter is just because many variable variations are not completely known, having a statistical nature (e.g., production tolerances). Often design techniques are directly related to statistics, like "Verify that simulated production yield is above 99% for all valid environmental parameter combinations" or "Make sure that the standard deviation of the offset voltage is below 10 mV".

Having software for difficult statistical and optimization problems is quite a second "revolution" which has taken place *now* in the industry. Mathematically the step from simulation $f(x) = 0$ to function minimization (e.g., via $f'(x) = 0$) is not so large, indeed there are many similarities, and also statistical methods often pick up optimization algorithms. These beautiful things in math help a bit to find consistency regarding the topics and new methods we address in the book.

Not at all, this means that experienced designers have to through away old "manual" techniques. Actually the opposite is true. Many times we will see that basic math results fit very well to the designer's intuition and how they act, how we anticipate problems. However, often indeed we can further improve amazingly. Two nice basic examples are corner analysis and Monte-Carlo. To some degree these are "optimum" techniques already, but still modern math offers further enhancements like adaptive worst-case corner finders and low-discrepancy sampling. A third example is design tweaking: Often it is done manually by parameter sweeps, but clearly more efficient optimization techniques exist. Not only in math libs, but applicable to complex state-of-the-art chip designs.

That is about 2016, so why many people have not heard about this "revolution", and what we can expect further? Important for researchers, for young students, for investors! Often "time is ready" for certain innovations, so actually many people presented the "first" telephone or "first" electric light bumb. Gary Nagel's SPICE (Simulation program with integrated circuit emphasis) was extremely successful; and his software picked up many ideas which are available right now at this time, for the math part it was e.g., having sparse matric solvers. SPICE was also easy to use, and the available models were good enough, so everybody learned it, and the success was so large, that even some bugs become a feature (even a reference). The "second revolution" has not yet lead to a kind of "standard" program which designers have to learn, so there is more to read than a manual. At the moment, there is even a "war" on "high-sigma methods"; read why we think it is more a little banter, because also the person in front of the computer screen makes a big impact. However, what is needed for design, the "second revolution" little pieces, and the bigger ones, they are all *present* already and *will remain*! This books is for experienced circuit designers who want to dive deeper, but also for young designers and student in circuits and software. Some will become the next famous personalities in the big industry around electronics.

We have met people who are already working on the next "big thing" in engineering, and what could be this, *technically*? Designers follow a *strategy*, e.g., you need to solve the most critical problems first. However, sometimes it happens that something unexpected occurs, and e.g., other effects become important. So you need to change your strategy. You can also do so in math and computer programs. "Learning" or "decision making" is maybe a bit too much, but creating adaptive, more flexible algorithms is clearly a hot topic since years in research, and it will find a way into real design software too, enabling a true third revolution.

However, now, in this book, there is enough to say for transferring the "second revolution's" techniques to designers, giving a survey, a field guide with many circuit-related examples. Keep yourself up-to-date! We go beyond simulation, go for optimization and statistics; use them in circuit design! This often means getting answers to urgent but difficult design questions, e.g., regarding sensitivities, nonlinearities, design risks quantification, etc. And because already pure optimization, pure advanced statistics are sometimes a bit difficult, we focus on the real interesting, the real circuit-related stuff. This is unique and why we hope to find many readers.

Generic versus commercial. We present many examples and results, and we also add few screenshots from commercial environments to be authentic. This does <u>not</u> imply <u>any</u> judgments, e.g. on tool quality or preferences; and in some cases we unfortunately haven't received the permission to publish our results.

Sometimes the use of screenshots limits unfortunately the printing or display quality a bit (e.g. compared to vector graphics), even if the shot from the tools *itself* is perfect, but we still feel showing that a *direct* tool output *is* available is sometimes important.

Of course, some presented algorithms are almost brand new and will find application in hopefully near-future EDA tools. Often such new methods itself are not that much "present" in the user interfaces; they "are" just a *click* to enable a certain setting.

The real great advantage of modern design environments e.g. against automation by batch files, is that many integrated *debugging* features are available. If e.g. something gets wrong, you can often see it marked red in a table, often with hyperlink to the source of problem, or with a context-sensitive menu providing further options, like cross-probing to schematic, getting a plot, creating a simulation subset related only to the critical parameter combination, etc.

No Fear about Math

For the first time, we deliver circuit design methodology, math, and[1] tools in a well-aligned package. We try to do this according to the golden rule that no scientific method can replace common sense. Actually, anyone *has* a certain gut feeling e.g., on probabilities—like that p is equal to 1/6 for a rolling dice. We want to extend this to treat more difficult problems, like for yield cases with more extreme numbers, or problems with many more variables, including non-uniform or non-normal random variables. We will clarify about judgments on *rare* events and present techniques to treat them correctly.

Besides the book and e-book, there are also two real-time applications, which give you many more insights and a true "feeling" for both optimization and statistics. Complex circuit design environments are simply not designed to give the user a direct quick feedback on anything! Circuit simulations are *time-consuming*, so learning on circuit examples purely based on SPICE in an electronic design environment takes simply too long, and in addition, the graphical tools are often too limited, just because they are optimized for design purposes; and not for algorithm comparisons.

---

[1]In scientific papers, it is not common to highlight or underline important parts. We will not follow this ill convention, because it makes the understanding harder.

Of course, also the electronic design automation (EDA) R&D departments do not only use the EDA products; of course they use *dedicated* math and statistical software (such as Matlab® or R©). On the other hand, *as much as possible*, we will stick to real design-related examples, because they are simply more convincing compared to simplified, non-circuit simulation-related spreadsheet examples.

Note: Searching for EDA and related keywords gives you often already quite many hits, but CSE (computational science and engineering) is a more general term.

A few words upfront to the math in our book: We try to keep it as simple as possible, but not simpler, because as an engineer you should not rely too much on hope; often you need hard numbers! Most designers are fully aware of the fact that Monte-Carlo results have some randomness, often even some systematic inaccuracies, but *how* large these errors are and *which* techniques are best suited to minimize them is a key question. We focus on concepts, prerequisites, intuition, pictures, examples, and rules of thumb truly linked to circuit design, not on proving mathematical theorems. *Logic merely sanctions the conquests of the intuition (J. Hadamard)! So skilled intuition is our main target.*

On the other hand, it is important to know what is really proven and what only a meaningful assumption is! Marketing material is usually not good in this aspect, and even some mathematical techniques have fancy, a bit too fancy, names: like *maximum likelihood*—can there be something better? or *confidence intervals*—you do not trust them?

In quite many cases, we have to look to the real details; too many scientific articles are not suited as true field guide: They may mention also difficult cases, but *why* problems occur is too often unclear. Usually, this is exactly the most interesting part: If a circuit does not work, you may choose another, but a better idea is to repair, modify, and extend it! As designer in hardware or software, you simply have to do so and to learn.

Engineers are luckily often pragmatic, and you need to be, because you need to apply methods for solving non-trivial problems. People usually transfer things they learned on simple cases to more difficult problems, but sometimes really something essential changes if you move from a one-dimensional problem to two dimensions or even to n dimensions. Examples are helpful, but sometimes also misleading. Even experts can fall into this trap, and simplifying concepts that work well in some cases (like assuming normally distributed data or replacing true parameters by sample estimates, or using standard confidence interval formulas) can fail completely in other cases, other theoretical cases, but also in real circuit design! This is because analog circuit design can be very complex and highly nonlinear! Concepts working well in one field may fail completely in others, and this triggers the creation of more advanced theories; even when having such an advanced theory or method—like maximum-likelihood estimation (MLE) or bootstrap—people may be overenthusiastic, and limitations may be discovered later—we will give examples for this. Just one is Latin hypercube sampling (LHS), which has been implemented in many simulators for ten years or more. The idea is good, and initial benchmarks had shown a significant speed-up. However, for complex circuit designs, the advantage disappears too often—and there are better methods. This book is <u>not</u> at all on benchmarking different algorithms or tools from different vendors, but of course we <u>do</u> core algorithm comparisons; usually on difficult but still easy-to-understand cases—often these are also representative. Unfortunately, there is no single example test case on which you can show everything—analog examples are often not as easy to scale as digital ones. So, sometimes we also use mathematical examples on which you can indeed prove certain algorithm characteristics, like quadratic convergence of an optimizer.

In fact, this is not at all a mathematical book covering statistics and optimization *in general*. Also we assume that the reader knows about key circuits like amplifiers or filters, but the circuit topology is usually not the primary interest. We need to focus on techniques that <u>work for circuit design</u>, so you will not find detailed discussions on ANOVA, hypotheses, Runge–Kutta integration or linear simplex optimization, yet references will be provided for the readers who need to revisit those concepts in more detail. We show also methods dealing with non-statistical variables, because they are of course important for circuit design and we will see many similarities, e.g., when talking about coverage, sensitivity and correlations.

The authors have both a strong circuit and a mathematical background; remember from school and university that statistics and matrices can be a very boring topic. In this book and by the inclusion of auxiliary software, we want to demonstrate the opposite. Also we always want to relate good <u>manual</u> design techniques to the math background. Learning something challenging is often difficult, but <u>not</u> if you really have to solve your own (hopefully) fascinating highly motivating problems! For instance, what is a designer doing if he/she follows a certain design strategy and how can a design software act in a similar way? Tools can remove boring work from the designer, so that he/she can focus on more fruitful topics, like on exploring different circuit or system topologies and for preparing your design for a perfect design review. The great thing with computers is that you can often verify amazing things with little setup effort, and you can also use software not only to your direct advantage of solving a specific problem but also for becoming a better, more experienced engineer! For this reason, we do not split the book content in a theory part and examples; instead, we always want to apply algorithms immediately on interesting realistic—often only slightly simplified—design tasks. You should start with such basic understanding to build up more understanding and to be able to explain what the general phenomena are. Actually, the source of all great math is the special case, the concrete example; it is even frequent that every instance of a concept of seemingly generality is in essence the same as a small and concrete special case.

Educated application is one key for success, because there is no "best" algorithm in general for complex tasks like circuit yield optimization. The better the selected method fits to your problem structure, and the better you set convergence parameters and starting points, the shorter the way to success. Custom IC design, statistical analysis, and optimization are difficult topics, so several highly automated and efficient methods have been already used very successfully for more *special* topics like digital circuits (here you can focus on timing and leakage) or memory circuits (here the focus is on read and write capabilities), and here, we may use at least partially analytic expressions instead of running full circuit simulations [Jiajing Wang]. Also in "SPICE model fitting", you can find very efficient algorithms, because you optimize all the time very *specific* models with *fixed* parameter sets to fit for measured data (having also a fixed *structure*, like IV and CV data) and using usually the least-square criteria. Use such methods *if* possible, but here in the book on analog variation-aware design, we typically need more *flexibility* and will describe more generally applicable methods and tools.

In our examples, we need to make a balance: Real ASIC designs are often very complex, and a big mix of techniques is required to solve all problems, so we usually have to apply some simplifications to be able to directly apply and demonstrate new techniques. For instance, the way an optimizer takes in the parameter space is hard to visualize for more than two variables, but luckily almost 99% of the optimization problems can be fully explained with 2-variable examples! Actually, the real advantages of an advanced optimizer become often much more prominent if you apply it to more complex problems, like with more than ten variables. In addition, we wish that the reader is really able to follow and apply the key ideas—the limited book size and often intellectual property (IP) issues unfortunately prevent us to showcase very complex examples. However, when e.g., looking at an optimizer log file, the user can see that state-of-the art tools really use the described algorithms, and he/she can observe in them the same problems as we demonstrate in our shorter examples. The main difference is only that in big problems, you typically have to fight against multiple problems and the runtimes are much longer!

Hi! This is us, the authors! This is a book, and learning from books is often a bit more difficult, because in long sentences it might be not clear where the key point is. In teaching, you can *stress* words, but in scientific articles, it is not usual to use underlines. So we do not follow this tradition! If we write, e.g., ... almost accurate ..., it could be quite a difference if the focus is on *accurate* or on *almost* or even on both, and if both in which sense? The context for us is usually circuit design, but sometimes it is indeed up to you to decide whether 99% accuracy (or yield) is good or not. Also in medicine, you do not need always a high confidence level, and in urgent cases, better give a heart massage! Modern statistical algorithm come of course with internal accuracy checks, but also in a circuit simulator, the checks may be not done in the way *you* need. In a transient simulation, you may want 0.1 dB accuracy for your FFT, but that is seldom in sync with simulator settings like *reltol*. That is the reason, why engineers do the work, and actually change the world with chips, smartphones, and software—not that many designers, maybe many in electrical engineering, quite many in IC design, some in EDA, and quite a few in statistics and optimization. Luckily, these techniques are a hot field also in other areas.

Front-End Design Flow

To follow this premise of circuit-oriented concept, let us start with a little sketch on an analog design. "Analog design" can have different meanings—our focus is analog design in its *widest* sense, including RF design and mixed-signal and custom digital designs. In a single book, you cannot describe each aspect in full detail from a bare semiconductor wafer to the final tested product. So we focus on what is often called design "front-end" flow, the part that includes defining the circuit in a schematic entry, setting up a simulation testbench, verifying the circuit behavior and checking specifications, plus tweaking the circuit component values; so in our methodological context, "front-end" has not the same meaning as in RF (radio frequency) or sensor design. A very typical approach in analog design is shown in Figure 1.

Product definition
specs, marketing, costs, volume

Feasibility study
general system concept, critical blocks

Design
system partitioning, **block design**

Production
fabrication, test equipment, analysis

Redesign
block redesigns

Production
fabrication, analysis

Front-end design
topology, testbenches, calculations,   **Primary focus**
simulation, sweeps, corners,
Monte Carlo, sensitivity, optimization,
find worst case, verify yield, review

Back-end design
floorplan, package, placement, routing
LVS, DRC, parasitic extraction

Postlayout verification
Simulations, final block review

Figure 1   General IC design steps.

Imagine an RF receiver has to be designed, and it might be part of a bigger transceiver chip or even of a full system-on-chip (SoC) containing also bigger digital parts, a PLL synthesizer, ADCs, DACs, etc.

- A system designer (team) has decided on the concept; for example, based on experience and system specs (like IEEE802.11 for WLAN), it has made a certain system partitioning into sub-blocks such as low-noise amplifier (LNA), mixer, variable-gain amplifiers, several filters, an demodulator.

- Next, the system gets refined further, e.g., by creating a detailed frequency plan; with system simulators, MATLAB® and spreadsheets, the system designer will also create a level plan giving an overview of the ranges for wanted signals, interferers, noise, etc. Step by step, the designer can hopefully obtain a consistent and realistic set of sub-block specs (such as gain, area, and power consumption) derived from the top-level system specs. We need to make sure that the system works even in the presence of many known imperfections, such as noise, filter tolerances, group delay ripple, jitter, mismatch, and different kinds of nonlinearities (compression, intermodulation, etc.).
- At some point, a circuit designer has to work on "his" block and related sub-blocks, e.g., an operational amplifier (op-amp) being part of a bigger filter. The block designer either will take an existing circuit or may compose a new op-amp, e.g., based on the input and output voltage range, the supply voltage, gain, speed and noise requirements, etc. For the latter, he would execute several hand calculations for the different component values in the circuit (e.g., based on power budget, g_m, on-resistance, and slew rate). In both cases, he/she needs to verify the circuit behavior in a set of testbenches under all the different environmental conditions. For instance, he/she would do sweeps on temperature, supply voltages, load resistance, load capacitance, etc.
- Usually, some design weaknesses will become visible, like we consume more power than allowed, so the circuit has to be modified, e.g., till the variations are acceptable (like in sync with hand calculations) and till the design is fulfilling all specifications (being "in-spec").
- Of course, we should also perform simulations together with neighboring blocks, like the usual bias generators—if available. At the very end we could end up in a top-level simulation, which is often a mixed-signal simulation (mix of transistor-level blocks and behavioral descriptions).
- Besides simple one-parameter sweeps, we may also do 2D sweeps to account for correlations or pick certain critical corners (combinations of parameters) directly for debugging and design tweaks. In those 2D sweeps or critical corners, the design is often more stressed than in simple sweeps, so tweaking becomes harder and maybe we need to extend the circuit a bit or even need another circuit topology.
- To account for offset voltage, we may do a Monte Carlo (MC) analysis and get a certain standard deviation for it. With corners and MC, we could check the design quite well, and we can calculate the overall most extreme behavior; for example, we can add the offset errors from systematical and statistical imperfections (e.g., PSRR and random offset).

- All in all, at some point we might be completely in spec with enough margin and create a layout ("Back-end" part in Figure 1). From the layout, we can calculate wiring parasitics and we can include them in our simulations. If our initial design margins were large enough, also the postlayout verification would indicate that our design is really "ready for production" (tape-out). Hopefully, we included all important aspects to our verification, so that indeed also the production samples work as expected.

This "simple" flow shows that analog design can be quite efficient. We have many well-established standard techniques, but also some key questions arise immediately; we will address them in our book:

- Imagine your circuit simulations show good results in a corner run and also in a short MC run: *How much does this mean with respect to yield? Is it really ready for tape-out?*
  This has many aspects, and we will guide you (e.g., on MC count) and show you common pitfalls (like assuming normality). We will explain how to get most out of your simulation results, more than just sigma and a histogram!

- Imagine your specifications are very difficult and hard to achieve. *What should be the next steps? Which strategy should you follow?*
  We will create a systematic flow from design start to sign-off verification. We give you an overview on the methods and explain which ones are suited and how to set them up. This also includes optimization techniques (Chapter 8); we tell you when they are useful.

- This leads to the often asked question: *What is the best <u>way</u> to make a design?*
  "Define best!" is probably a good answer. "Best" in the sense that it would <u>always</u> work would lead to a big and very complex flow! "Best" in the sense of <u>efficiency</u> would lead to a flow based on a lot of experience. So "overall best" is for sure a mix of techniques, which makes it sometimes difficult to see a systematic behind the flow. On the other hand, we will show that even when you use methods with strange names (like "high-yield estimation"), you are often still using something close to the techniques you have applied manually from time to time! Generally, there is <u>no</u> need for big changes in your habits.

Actually, one of the nicest questions I ever heard as consultant was this: *How much do I need to overdesign <u>in Cadence</u> to make it working <u>in reality</u>?*

Cadence® is indeed a synonym for custom IC design tools, so technically it would be better to ask for "overdesign in *simulation*"! As analog designer, you know the answer of most questions is "it depends." If you know your laboratory measurement equipment has an uncertainty of 0.5 dB, then you can use this as <u>safety margin</u> and specify it in a datasheet. However, it is quite some work to find out the margins for all the other effects like temperature drift, aging, circuit tolerances, modeling, Monte Carlo variances, and simulator inaccuracies! Ultimately, the whole concept of margins is very efficient but leads to severe errors if applied in a too simplistic way. We will tell about the risks and <u>extend</u> the concept!

> **Is overdesign a problem?** Regarding verification, it is no problem, but for being competitive it could be—especially when talking about bigger key blocks and/or critical subsystems. Actually, it is quite hard to find out whether you are overdesigning or not. You have to run the circuit under extreme but still realistic conditions! Finding those is one key task in variation-aware design, and optimizing the circuit to meet the specs and/or improve on yield is a second key technique. Both are closely related, e.g., in both sensitivities, and search algorithms are important; the more you read and learn about it, the more links you will find, and the more intuition you will get, the better you can design by problem anticipation.

Actually, our op-amp and filter example is just one simple example, and often each described step (or at least one) is far more difficult, especially when designing critical blocks or using the newest silicon technologies. So Figure 2 shows some major trends, and many of them can make circuit design unfortunately more difficult.

One trend among many others [Graeb ITS] in modern IC design is that already the blocks become more and more complex, e.g., to enable multimode operation. In our amplifier/filter example, one may decide to put several stages together—into a kind of subsystem, often including bias circuits and calibration blocks. So the meaning of "block" design can be a bit fuzzy.

Design and Verification

In IC design, many individuals and teams are involved, and they often have different opinions on what should be improved. For example, behavioral modeling is a high-priority problem too, as variability. For both, the benefit

Technology	Moore's law difficult to maintain, fin-FETs, stacked-die, 3D integration, SiP, double patterning, more leakage, MEMS, photonics
Economics	growing mask costs, less tape-outs, fabless companies, big teams, mobile, price pressure, time-to-market, IoT, tool costs
Signals	more digital, more mixed-signal, more sensors, multiple domains
Complexity	SoC, bigger subsystems and sub-blocks, more I/Os, more software
Difficulty	more impacts from statistical variations, more corners, more trade-offs, more complex models, higher frequencies & BW, layout counts, lower V_{DD}
Algorithms	more adaptive, high capacity, fastSPICE simulation, optimization, statistical techniques, higher automation
Design	more awareness of layout-dependent effects, parasitics, statistical variations, shorten iteration loops, avoid redesigns, IP reuse, fast time-to-market, $\Sigma\Delta$ techniques, real-number modeling, ensure quality

Figure 2   Trends in IC design.

is larger, the earlier you use them. In software the creation of a "design" is quite easy (e.g., due to availability of powerful libraries), but testing can be difficult and very complex; and in digital design the situation is often similar, because reliable standard cells and synthesis tools are already available. So in these communities many people say *only* verification is a problem, we have a "verification gap". This is not completely true for analog and custom designs! Verification is a difficult topic too, as almost all kinds of designs become more and more complex and harder to test. So nowadays, there is pretty much talk about the "verification gap" or "productivity gap." Best don't let you bother and just take the best ideas coming up.

When digital designers talk about verification problems, they typically have complexity in mind, like how to find input test vectors leading to a bug. However, it does typically not mean that they cannot run the testbench with that test vector anymore. However, in analog, mixed-signal and RF, indeed, e.g., a single postlayout simulation pushes even a 1 TByte RAM multicore compute server to the limits! However, there is seldom a verification problem in the sense that you do not *know* what to simulate! On the other hand, for sure also digital designers push their tools to the limits (especially when thinking of HW-SW co-design). Also "new" digital techniques like randomized tests and coverage-drive verification can make highly sense in context of mixed-signal or even full analog designs.

In analog, there is also a true "automation gap" in topology selection and initial circuit design. Here, a mix of techniques is required. No single commercial tool offers here "push button" solutions—quite opposite to simulation and verification. However, some of these things are also the fun part of analog design in general, so why automate? A tool needs to be really good to convince in that domain! On the other hand, luckily, there <u>are</u> advanced design tools plus powerful compute servers that still allow efficient design plus giving the engineer a great cockpit for steering the design, implementation, and the verification tasks. Further improvements like tool support for parasitic and layout awareness are in trend, giving not only speed but more insights, thus also making the designer better.

We are also not sure whether there is really something like a "productivity gap" or if people need a new smartphone every six month. Software is a big lever to improve your individual productivity, like just designing more transistors (or lines of code) per day, but design *difficulties* are hard to quantify. Terms like "transistors per hour" or "mm^2 chip area per week" are only very rough measures in both front-end design and layout. You can spend weeks on an LNA or months on an RF PA having few transistors only and you might be highly productive, because the person at the competition fails on this problem. Especially in the field of statistical methods and optimization techniques, a breakthrough is observable; so any engineer working in that area should have an interest to become familiar with them. They help you to identify the critical parameters in your design and quantify and minimize their impacts, till you are confident enough to tape-out.

The biggest step forward in analog design in the last 30 years was clearly the introduction of standard circuit simulators, allowing an almost virtual verification, instead of pure breadbording and pure hand calculations. Also some discussions in engineering are also quite old: If there would be "only" a sign-off verification problem, people may say, why not going for a fast production lot, wait four weeks and measure everything—*only* the *silicon* tells the (full) truth!? The big problem with such approach is that the debugging will not be much easier that way, and actually, it is far better to use advanced verification tools, modeling techniques, etc., already from the <u>beginning</u>, for the design phase, not only for sign-off verifications. This way you can improve on product quality, design understanding, reuse IP, and reduce need for costly iterations (especially long, time-consuming iteration loops). In addition, even the silicon is not the truth, e.g., in RF or high-performance ADC design packaging, external matching networks, supply bypassing, etc., can have a big impact! In addition, the more complex a design is, the more virtual the design and the verification of it will be, naturally, just to reduce risks and costs.

Verification can be indeed treated quite well with systematic mathematical methods, but we feel that this is a too narrow approach, because whether a design is good or not can be often decided much earlier, so it is better to avoid problems early, instead of finding them in a final verification phase. Plan for this as early as possible. Explain how difficult things should be verified in a verification plan, best in conjunction with the target datasheet. Iteration is always required, but keep at least iteration loops *short* and improve your understanding, and the understanding in the whole team.

Therefore, looking at the whole front-end flow as a pure verification problem is a trap! Solving the "verification gap" is often much easier if you do not split the flow into parts, better take a more unified approach and include manual best practices for circuit design. A good way for solving complex problems is to anticipate the "disturbing" influences on the design, to analyze them; often the results from techniques already lead to solutions—step by step, starting with the most urgent problems.

Let us pick up an op-amp example again: At the beginning, the design is almost for sure not working as desired, at least not in the whole operating range. You have to pamper up your baby design and make it work under DC nominal conditions, making it work under wider conditions (over temperature by deciding on a certain bias concept, or for constant performance over a wider supply range by adding cascodes, etc.) and also for your typical input signals, till you end up in a good and robust circuit that works also under the most difficult allowed conditions. In such a flow, the design and your knowledge about it grow in parallel with the verification! This way verification or complexity becomes much less of a problem, and more design insight is a further valuable output.

In digital or software design, there is a trend to split the parts of design and verification, i.e., different people do it. This solves the problem that a designer may "love" his "baby" too much. Actually, the idea that a verification engineer should "hate" the design is good and can lead to better verification, but for analog designs, it should be only a complementary concept. In critical cases better create several "baby designs" and let the best "survive"; once you have made them detailed enough for a fair comparison. Design is often a fight of ideas! Often also a split between front-end design and layout—usually seen in bigger companies—is critical, e.g., misunderstandings on priorities or constraints can lead to unnecessary extra work or just bad layouts.

Layout aspects (or more general the "physical implementation") are important, and the reader is highly encouraged to also read books on technology, layout, and device modeling! Many of these topics will be referenced in our design flow: Modeling is a limiting effect on verification accuracy, and the layout part (unavoidable wiring parasitics and "bad" layouts, respectively) is often a reason for design iterations—unfortunately.

Note: In this book, we concentrate mostly on <u>front-end</u> design not on physical implementation aspects like layout and packaging (or ESD and latch-up), but for sure you also need to include these extra-effects (plus sometimes others like aging), and often, they are as essential as supply voltage or temperature effects. So you should also include them, <u>as soon as possible</u>—not only in final verifications. For some more details, read Chapter 10.

For sure, also simple bad design practices can let design projects fail or at least delay, e.g., last-minute changes without careful re-verification, creating different block versions but not making clear which one to use, hoping that someone else has already verified something, not applying manual checks (often automated run decks for DRC, ESD, latch-up, etc., are incomplete), using models in extreme regions (breakdown, ultralow currents, etc.), or typical interface problems, e.g., between blocks from different departments. Actually, communication is a key part for success. Partly, it can be also supported by EDA tools, e.g., via constraints. Table 1 presents some reasons for re-designs which have been appeared over the years to the authors. Note, that the table focused on first, almost immediate redesigns; in later stages, after a more detailed analysis, problems introduced by variability are much more typical.

At some point, it might be impossible to simulate the design, like a postlayout simulation (maybe still excluding substrate, self-heating effects, aging, ESD, latch-up, etc.) will take a month. In such cases, it would be better to go for production, but being sure to apply at least careful manual checks and calculations. On this, do not underestimate the human factor: We have seen design fails and need for a redesign, just because designers claim "I checked this in Monte Carlo," but due to a small bug in the design kit, it was simply completely impossible to run MC. So the designer was lying so and so.

You may think why everything is so difficult? It should be not a matter of complexity, and it is more on following an important rule that says:

> *"Know what you want, and use what you have. Make no bullshit, do not rely on hope or magic. Double-check your assumptions. There are no stupid questions. <u>Be an engineer!</u>"*

Table 1 Few reasons for chip redesigns

Block	Chip	Class	Comments
Logic part	RF receiver	Last minute change	No full re-verification applied after change. Solution: FIB applied, logic changed in redesign
Bandgap	RF power amplifier	Modeling	Substrate transferred the RF output signal to sensitive bandgap net Solution: Non-Miller-cap frequency compensation of the bandgap, better shielding of pads, FIB and metal redesign
Substrate contacts	RF PA control IC with negative supply	Layout	Contacts connected to gnd instead of V_{EE}. Solution: Layout modified in metal redesign
$g_m C$ filter	IR remote control	Variability	Only corner simulation done, no analysis for mismatch Solution: Larger transistors in bias part
Buck converter	Hearing aid chip	LVS	Incorrect LVS setup for self-created metal capacitors. So a short circuit was not detected. Solution: LVS run deck improved, via causing the short removed
RF power amp	RF power amp	Modeling	Amplifier had strong oscillation tendency, modeling for package, substrate, etc. too simplistic. Solution: Modeling improvements and concept change to a narrow-band amplifier
Memory	SoC	Variability	Circuit function correct, but yield too low. Solution: Yield improvement redesign on critical parts.

This book wants to show and explain further advanced methods—beyond simulation and verification—that work for creating analog designs. When we talk about "analog", we include also mixed-signal, high-speed, or custom digital and RF designs—so all kinds of designs where typically a highly flexible design strategy that is required. Not always mathematical and automated techniques can be directly applied, especially for the art part of circuit design, but for sure the described techniques can help good designers and students to solve more difficult problems and to speed-up the design process in several ways.

We feel also non-math experts, but engineers should <u>know</u> what they are doing when using certain electronic design automation (EDA) software. Users need be able to select and set up a suitable method, because (like circuit simulators) also the new tools come with many options. This book will give you guidance and an overview and should enable the reader to get in touch also with more advanced original papers and more specialized mathematical literatures. Those are clearly recommended if the reader wants to get highly detailed information or creates his own algorithms. This marks also the point where we need to stop in this book and we want to focus on the really important concepts, algorithms, and tools.

The book is written for engineering students and engineers who have already some background in standard techniques like just creating a little design and simulating it, and knowing about technology variations. So we address simulation techniques themselves only in aspects relevant to variations, statistical methods and optimization. We hope we can motivate you, as the reader, to think more about the topics we cover, and you should be able to understand, able to work, and probably also able to tell, to talk, and (at some point) able to teach.

Actually, too often designers end up in SPICE monkeying, hoping the simulator makes the design job. Sure, many things base heavily on models and simulation, but at least you should do it in an efficient way, doing the really required simulations, with some realistic margin and just some acceptable overhead for confidence and understanding.

Often EDA vendors are asked: How much does this tool help to increase the productivity? In some cases, you can indeed quantify this, but we feel—even in the book—we partially focus too much on such a simplified view on engineering. For example, reducing the number of simulations can lead to a measurable speed-up and reduced costs, but actually other aspects are of equal importance, like being able to manage complexity of new system architectures or advanced process nodes, and getting confidence and understanding of your design.

Organization of this Book

The topics we treat are both relevant to scientific researchers, EDA software engineers, electrical engineering students, and circuit designers. You do not need to be an expert in everything; we point you in each chapter also to the literature available for further reading—to more basic and sometimes also to more advanced references. In the appendix, you can find an ordered list of all references, and we think it is a good mix for getting a deeper introduction and also for digging into the details.

Of course, unfortunately, we have to deal with many abbreviations and special terms, and we collect them in a big list—many are very common, and you should know them.

There is no eat-all-or-nothing, the book has a modular structure (Figure 3): The advanced designer may skip the introduction and the chapter on manual design methods and could jump to the advanced statistical techniques; or the

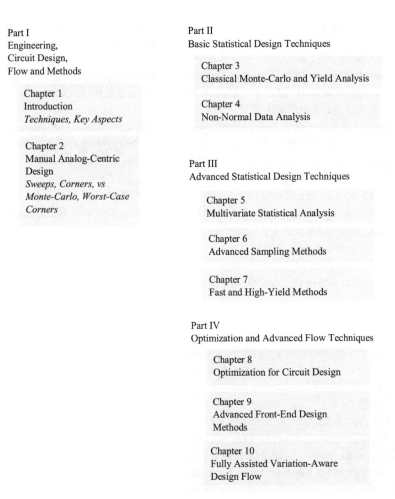

Figure 3   Design techniques and book chapters.

statistics expert could directly go to the optimization chapter, but we think if
you read the book page by page, you will not miss clear guidance. Topics that
are interesting but not key for further reading are delivered in separate boxes.

We always start with the problem descriptions and discuss different methods to solve them, starting from one of the most straightforward approaches
and then moving to more advanced techniques that typically are more accurate
or more efficient in difficult cases. The examples we start with are typically
quite short because with too complex ones, we would lose the focus too
much. Therefore, we shifted several more complex examples into separate
subchapters. Of course, also our initial examples are often not so simple in
some aspects, e.g., for circuit design, it makes little sense to talk about "linear"
optimization, so starting with a quadratic function is quite native, and even
such an example can be simple like $x^2 + y^2$ (because here you can optimize
both parameters independently and even the simplest optimizer will work very
fine) or more difficult like $x^2 - x \cdot y + 100\ y^2$ (here, we have correlations and
highly different sensitivities).

The book starts with two chapters on circuit *design* and design *flow*, not so
much on circuits. We explain the most important design targets and measures,
like yield, corners, and worst-case and Monte Carlo analyses.

In Chapters 3–7 we discuss statistical methods in detail, starting with
Monte Carlo and basic result analysis for yield, sample variance, confidence
intervals, etc. Next, we extend the method in Chapter 4 to non-normal data
analysis, which includes a generalized process capability index C_{GPK}. All
these analyses are possible for data obtained both from tests in the foundry or
laboratory or from MC simulations, but the simulator has really <u>access</u> to all
statistical variables in your design, so you can also do *multivariate* analysis
to obtain correlations or to fit complex models to MC data. This is the next
step, before we also address MC speed-up techniques and non-MC statistical
techniques like the worst-case distance method, which is often much more
efficient for high-yield verification. Chapter 7 is a long chapter, just because
there is no one-size-fits-all method, almost each method you can break, at least
in special difficult test case—and you should understand why.

Chapter 8 is explaining optimization methods. One key part in setting up an
optimization is the goal and constraint definition, because often optimizing the
circuit under typical conditions is not enough, and at best this could be used as a
starting point. Instead, you should really inspect the most critical conditions,
and for this, you have to treat both statistical variables and the ranges for
your operating conditions like supply and temperature, and then you have to
optimize on this. The difficulties and the benefits of such overall optimization

approach are discussed in Chapter 9, together with further advanced frond-end design topics. To some degree, we leave here the area where most design environments have fully built-in solutions; so some scripting or manual interaction is required to perform very advanced tasks. Actually, assembling worst-case analysis and optimization is a key part for obtaining a complete and automated or at least a highly assisted variation-aware design flow.

The overall flow aspects, including layout, design-supporting tools, and IP management, are described in Chapter 10. Variation-aware design is not at all only about process variations, mismatch, temperature, and supply effects, and the physical implementation can change the circuit performances sometimes much more! This means layout effects, such as wiring parasitics, may be substrate couplings and nowadays even the neighborhood of each critical transistor matters, so you should include them as soon as possible. Our summary (Chapter 11) includes also an outlook for upcoming further advanced design methods and environments.

Real-Time Apps and Auxiliary Material

Please regard the software as an *integral* part of the book! This does not mean that anything is really missing without these pieces of software, but to really understand what the circuit design is, you have to do circuit design; to learn what statistics can do and what not, you should look at statistical data, at best from circuits!

In some way, EDA design environments are already quite good for learning! The only pity is that circuit simulations can be very time-consuming, so it is almost impossible to get a "feeling", e.g., for how large the variations from one MC run to another could be, just because already *one* "meaningful" MC run can take a day!

How can you learn driving a curve with a speed of 1 mph? You cannot, you need real-time speed to really get the feeling; and for statistics it is very similar. The human eye and the brain has different mechanism to interpret what we see; there is a static and a dynamic part. It is best to use both.

Both apps (see Figure 4) run under Microsoft Windows®, and a detailed documentation is delivered as separate pdf files. With both programs, you can make your own experiments on optimization and statistics, to learn interactively and from many examples just much <u>faster</u> than is possible within a design environment, and to some degree, also more details can be provided.

In the book, usually at the end of each chapter, we provide a collection of questions and answers, and the two apps can support you to answer them.

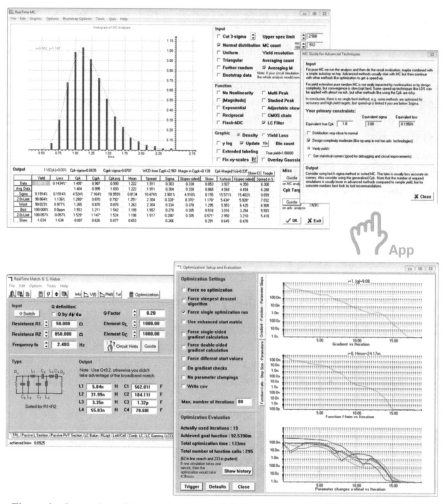

Figure 4 Screenshot of the two learning and design programs RealTime MC and Match.

When you could use one of the apps for solving problems, we often add a little green hand in the book. The newest app versions are available under http://www.riverpublishers.com

It could also make sense for more complex experiments to use your IC design environment, and we mark this and connections to further small design tools with a blue hand. At the River Web page you can download supporting material.

In addition, we prepared some statistical experiments with Excel®. These you can also download at River Publisher.

In the last chapter, we talk about layout effects and intellectual property (IP), and the creation of it. For that further design programs are available and provided by the River Web page!

These books we highly recommend as a refresher on circuits, math, and statistics:

- Willy. M. C. Sansen, Analog Design Essentials, Springer US, 2007.
- Tony C. Carusone, David A. Johns, Kenneth W. Martin, Analog Integrated Circuit Design, John Wiley, 2012.
- R. E. Walpole, R. H. Myers, S. L. Myers, K. Ye, *Probability & Statistics for Engineers & Scientists*, 9th Edition, Prentice Hall, 2012.

In the different chapters, we point you of course to further, more specific references. Note that especially for mathematical topics, there is a big amount of free and great material available in the Internet, and in the appendix we also give links to such *highly valuable* material. There you can also find links to the homepages of several EDA vendors. Many of them offer a forum and advanced material as download.

Software tools, Math, Statistics and Numerics? Doing math by dealing with data and programming are the best approach to become an expert. In our book we present many spreadsheet examples, not because such tools are best, but because such tools can display data quite well, and the user can see each step applied to the data. Of course, at some point you can push a spreadsheet to the limits, like analyzing 10 million data points; here true programming languages are best, like C, FreePascal or Java. Also in these many libraries for charts and math are available, it is just some work to put all together. The most elegant way for doing math is using math environments like Matlab® or R©. Matlab has the advantage that even a graphical simulation environment is available. R is perfect for all who need access to the true state-of-the-art in math and statistics. Both software packages are quite easy to learn, some things are even easier than in Excel®!

Acknowledgements

Attention is the highest of all skills and virtues.
—Johann Wolfgang von Goethe

We want to thank all people giving us attention, which includes of course you, the reader.

The trigger for writing this book came up in the preparation phase for the ESSCIRC 2015 Conference in Graz, but the authors have their ideas in their minds since many years—at least when not overloaded with everyday work. So we want to thank Mr. de Jong a lot for all his work around this book and his valuable hints and always quick help.

Special thanks for valuable discussions with Frank Pronath and Michael Schenkel (MunEDA). We included also some screenshots from MunEDA's WiCkeD^{TM} design environment. Similarly, we got also very helpful feedback, clarifications and often additional material from EDA vendors, colleagues and many authors of books, research reports and journal articles, which we included to our set of references!

We would like to thank those who reviewed the book and provided extensive and valuable suggestions, especially Kris Breen, Dennis Silvester and Frank Wiedmann, but also to the whole River team, and of our colleagues. Actually, the first two chapters were quite difficult to write for us, so when it comes to math, in Chapter 3, it was much easier to address also difficult topics, just because math is beauty, in harmony and contrast. Beside this, often, there is also a big effort behind a small screenshot or even one numerical entry in a table!

BTW, *feedback* is one of the greatest inventions ever, and it is not only for amplifier design or control systems! Actually, most algorithms we present use it in a clever way, as designers do. A pure Monte-Carlo analysis is like a simple single transistor stage *without* feedback: Set up it up and run it, nothing more, and just accept the result or not. We will also present what is possible if a simulator gets driven by its own results.

XL   *Acknowledgements*

As mentioned, you can skip chapters or read selected chapters in any sequence, if you are interested only in certain topics. Or vice versa, you can get more material at the River webpage, or give us feedback by e-mail, for suggestions, program enhancements, removing bugs, etc.

Finally, and again, thank you for picking up this book or e-book. We hope that you find it useful and valuable in your own work.

List of Figures

List of Tables

List of Glossary and Abbreviations

analog design   We use this term in a wider sense, i.e., for the
                     design of non-standard electrical circuits, which
                     include also e.g., RF and custom digital blocks,
                     ADC's, DAC's, etc.

Bandgap reference   Circuit to generate a stable reference voltage

Bayesian statistics In this probability is described as lack of
                     knowledge, not primarily as frequency of
                     occurrence. Created by Thomas Bayes
                     (1761–1701)

Bootstrap   Method to generate artificial random data from
                     existing data, e.g., using resampling with
                     replacement

bias   Difference between the expected value of the
                     estimator and the true parameter value

cdf   Cumulated distribution function, integral of pdf

empirical cdf   Stair-case shaped cdf directly according to the
                     sampled data

estimate   Result of an estimator, like the value of the
                     sample mean

estimator   A rule for calculating an estimate of a given
                     quantity based on observed data, e.g., the sample
                     mean

Gaussian mix   Distribution created by adding samples from
                     different normal Gaussian distributions.
                     Example: 10% of the samples come from $N(0,1)$
                     and 90% from $N(3,1)$. This leads to a non-normal
                     distribution, having two modes.

Halton set   A low-discrepancy set based on prime numbers.
                     Nowadays better methods exist, e.g. according to
                     Sobol or Faure.

ISO	International Organization for Standardization, e.g. ISO 9000 quality management norm or ISO 26262 on functional safety, requirement tracking, etc.
LDS	Low-discrepancy set or sequence, designed to give good space-filling.
LHS	Latin hypercube set. Near-random sampling method giving a variance reduction in MC estimates.
mode	Location for the highest density of a distribution. For a normal Gaussian distribution the median, the mean and the mode are identical.
model error	Error resulting from the selection of a wrong model
multimodal, unimodal	Distribution with multiple modes or only one mode.
Newton method	Method to solve optimization problems efficiently
norm	Non-negative functional measure for the length of a vector $\mathbf{x}$. Has to be linear scalable and needs to fulfill the triangle inequality. Example: p-norm $L_p = (\Sigma |x|^p)^{1/p}$; with $p = 2$ we get the Euclidian norm.
normal distribution	Bell-shaped distribution, pdf follows the e^{-x^2} law, often also called Gaussian distribution, according to Carl Friedrich Gauss (1777–1855)
Op-amp	Operational amplifier, universal voltage amplifier
pdf	Probability density function
quasi-Newton method	Newton method without direct calculation for the Hessian matrix
Sampling error	Error resulting from making inferences about the total population based on a sample from that population
ADC	Analog-to-digital converter
API	Application programming interface
ASIC	Application-specific IC
BE	Back-end, physical implementation like layout and packaging

BFGS	Broyden–Fletcher–Goldfarb–Shanno, a quasi-Newton optimizer
BSIM	Berkeley short-channel IGFET model
BW	Bandwidth
CAD	Computer-aided design
CI	Confidence interval, depending on confidence level, CI = [LCB,UCB]
CLT	Central limit theorem
CSE	Computational science and engineering
CV	Capacitance-voltage
DACE	Design and analysis of computer experiments, like DOE, but for computer simulations instead of experiments
dB	Decibel (logarithmic measure, e.g., $A_v = 20 \log(V_{out}/V_{in})$
DfM, DfY	Design for manufacturing/yield, e.g., layout improvement tools to increase the manufacturability of a chip
DFP	Davidon–Fletcher–Powell, name for a certain quasi-Newton optimization scheme
DOE	Design of experiments, methodology to efficiently set up experiments for finding the relation between inputs and outputs of a system
DRC	Design-rule-check
EDA	Electronic design automation
EM	Electro migration, aging in wires
FIB	Focussed-Ion-Beam, technique to modify the metallization on chips
FOM	Figure of merit
FE	Front-end, schematic design including block partitioning, topology development, and element sizing
GUI	Graphical user interface
HCI	Hot carrier injection
HDMR	High-dimensional model representations
HYE	High-yield estimation, yield estimation with methods suited for $Y > 0.998$
IC	Integrated circuit, circuit on a chip, typically in silicon
I/O	Input/output
IP	Intellectual property
IS	Importance sampling
ISO	International Organization for Standardization, e.g., ISO 9000 quality management norm

IT	Information technology
IP3	Intermodulation point of third order
IV	Current-Voltage
LCB	Lower confidence bound
LDE	Layout dependent effect
LNA	Low-noise amplifier
LSL	Lower specification limit
LSB	Least significant bit
LVS	Layout vs. schematic, check for comparing both netlists
MC	Monte Carlo
MC-MM	Monte Carlo with mismatch variables only
MDO	Multi-disciplinary design optimization
MLE	Maximum-likelihood estimation, method for parameter estimation
MSB	Most-significant bit
NBTI	Negative bias temperature instability
NF	Noise figure
NMOS	N-channel MOSFET
MOSFET	Metal–oxide–semiconductor field-effect transistor
NR	Newton–Raphson to solve nonlinear equations
OTA	Operational transconductance amplifier
PA	Power amplifier
PCA	Principal component analysis
PCM	Process control measurement
PMOS	P-channel MOSFET
PDK	Process development kit (usually for IC design in a fix technology and a certain design environment)
PSRR	Power supply rejection ratio (usually in dB)
PTAT	Proportional to absolute temperature
PVT analysis	Corner analysis for process, supply voltage, and temperature, usually also for further variables like load impedances and clock frequencies
QMC	Quasi Monte Carlo, MC using not true random numbers
R&D	Research and development
RF	Radio frequency
rms	Root mean square
RTL	Register-transfer level

RSM	Response surface modeling
SBO	Surrogate-based optimization, optimization based on a meta-model instead of direct simulations
SKILL®	Programming language, Lisp-like, used in Cadence tools
SoC	System-on-Chip
SPICE	Simulation program with integrated circuit emphasis
Tcl	Tool command language, an Open-Source scripting language
USL	Upper specification limit
UCB	Upper confidence bound
VGA	Variable-gain amplifier
WPE	Well-proximity effects
WCD	Worst-case distance
WCC	Worst-case corner
WID	Within-die variations (mismatch)

List of Mathematical Symbols

A	Amplification factor, e.g., A_V for voltage gain
C_{in}	Input capacitance
G	Gradient vector
g_m	Mutual transconductance, e.g., of a transistor or OTA
H	Hessian matrix
I_D	Drain current
μ	Charge carrier mobility or mean (average) value
n	Count or number
N	Normal distribution
S_{ij}	Scattering parameter (ij as port index)
V	Variance
V_{GS}	Gate-to-source voltage
V_{DD}	Positive power supply voltage
V_{TO}	Threshold voltage
σ	Standard deviation
μ	Mean

PART I

Engineering, Circuit Design, Flow and Methods

1

Introduction: What Makes an Engineer a Good Designer?

We present key measures, elements and problems in design. We discuss the efforts for design and verification tasks; and the inputs and outputs.

In this Chapter 1 we describe the problems of being a design engineer from a still quite general perspective, so many things appear in similar ways also in other fields (like car design). Of course, we have circuit design in our mind, and we describe the typical manual IC-specific design style in Chapter 2.

Design and circuit design is a fascinating topic, and it is a science and also a kind of art—for many amateurs and professionals. There are systematic approaches and there are physical foundations, but usually there is also something "special", especially when designing integrated circuits for high-performance areas like high-speed, high-power, or radio frequencies, but also smaller PCB (printed circuit board) designs, e.g., you often have to minimize the number of components with some "tricks." This is because—almost by definition and in opposite to digital design—there are *many* more things that matter (not only speed, area, and power consumption) and analog circuits are inherently much less error-tolerant.

Analog design is quite an art, because you need creativity to find the right compromises to fulfill many specifications, written and non-written ones. Due to more and more stringent requirements on minimizing size, power consumption, and costs (of course), designs moved in 50 years from the classical simple twenty-transistor op-amp to highly complex, multimode, mixed-signal circuits with billions of transistors (actually memory design is also very close to analog design). On the other hand, the basics have not changed much! By far, not all problems are related to complexity, and small circuits can be most tricky—a small amateur PCB design can fail for similar reasons than a high-end smartphone. It is very tough to be prepared for everything that could go wrong or just varies by nature, like transistor length and width, threshold voltages, load impedances, and gate oxide thickness. Of course, such complex designs are done by very experienced design teams, but if something gets wrong, it is indeed very often due to such variations and/or complexity (e.g., in interfaces and states).

We hope to show that also almost all numerical algorithms are based on "common sense" ("gesunder Menschenverstand")—and also dealing with them, improving them, and even applying them is something creative and fascinating! Common sense fails seldom, just some training is required, and some clarifications.

> **Styling versus Design.** In German these are foreign words, so often both terms are misused. Adding a fancy chrome spoiler to a car, which is no race car, is styling. So something between taste, bad taste and art. Doing it because you need it for a perfect driving behavior is design. Design is closer to science, real construction, problem solving, but of course there is often still pretty much freedom. Also designers have personality, and style; and the solutions from different designers may reflect this. However, to a high degree it is indeed usually possible to clearly state, what is really required or an even optimum solution, and which parts are nice to have. Usually circuits contain not much styling elements, but incorporate quite some art.

What are the key techniques every student, engineer, and designer should know and apply? Of course, learning about circuit design is good, and doing it is even better, but can we be more specific? Two sentences I remember from my professors as a student were as follows:

The most important skill to learn at a university is to learn how to learn

In IC design you can create almost everything, the problem is always making a reliable design that also works under varying parameters.

The first statement was to some degree a clear disappointment when I heard it because I want to learn much more, but in our ever-changing environment, this is clearly an important point. It just takes some time to pick that up.

The second quote was quite a surprise, because the technology in 1988 was by far not as advanced as it is today (e.g., most ICs use single metal layer routing, and the bipolar processes had lateral pnp with very low speed and gain)—and I did not have that much experience in how much tolerances can really make your life difficult! As an amateur designer, soldering for an audio amplifier or AM transmitter, you are typically done when the circuit is just running, but that is not the case when giving the circuit to someone else! Often optimism lacks in information.

Both sentences are very important, because as an engineer you make important and costly decisions for your company, and overlooking something can happen easily. Here, professionals are even under more pressure, because you rely much more on virtual techniques; any simulation usually can <u>only</u> answer the questions you <u>prepare</u> for—like "Is the amplifier stable?" In a laboratory the amplifier circuit might just oscillate when you turn on the supply—and you can immediately see it with an oscilloscope. However, in IC design a dedicated testbench is needed, and often different options are available, so a simple question like proving stability, can become quite difficult, especially if you want to go to the limits (like achieving also a large gain and high efficiency).

Just entering a design in a schematic editor and simulating it for a kind of virtual verification is possible since 1970s for professionals (when the circuit simulator SPICE becomes popular) and for amateurs since 1980s (PSpice® came up, running on PCs). In digital design, the flow progress in the following years was amazing: Essentially, nowadays you can create circuits and even whole digital systems with millions of transistors from software because clever programs can synthesize the whole hardware in a given technology, based on few core libraries featuring the basic logic cells.

A Very Short History of Digital Synthesis. In the 1970s, digital designers made designs in quite similar way as analog designers. So they drew schematics with little standard cell sub-blocks (like NAND gates, flip-flops, ALUs, and counters). In the 1980s, for simulation purposes and more compact design description, co-called behavioral languages like VHDL and Verilog have been created. This allowed more complex designs and better documentation. In 1994, Synopsis® created a synthesizing program that has been very quickly adopted by the industry. Recent developments are e.g., regarding verification with new languages like "e" and System Verilog.

Unfortunately, analog synthesis is much harder, because the way from a unit element like a transistor or resistor to a whole block like a programmable amplifier or even to its layout is much longer. Also languages like Verilog-A are not very powerful, and better ones have found no wide application and have no industry standard.

However, in analog, RF and mixed-signal automatic synthesis failed mostly— at least commercially, although for some special areas like DAC or filter design, some kind of synthesis makes indeed sense. On the other hand, further techniques have arrived in real commercial tools and enabled engineers to do things that analog designers dreamed for years. One is statistical design, and the other is optimization—and doing it not only on small academic examples, but also on real professional often highly nonlinear circuit designs!

People working in one area like EDA tools or circuit design can often learn a lot from other fields, even from topics faraway like biology, stock pricing, weather forecast, disaster prediction, or insurances. For instance, lot of attention in many of these fields is on advanced Monte Carlo techniques, whereas for most electrical engineers MC or corner analysis has *not changed much* over the last 30 years!

Not all techniques presented in this book are brand new. For instance, historically optimization is *not* a new topic, and some of the most important algorithms have been developed in the late 1960s and have been applied to electronic design in the 1970s, e.g., for passive RF filter design. However, the option to use such advanced techniques was only present for few academic institutes, and no user-friendly software was really available. One of the earliest commercial successful optimization tools came up in the 1980s and was Super-Compact^{TM}, able to simulate and optimize linear RF circuits. We used it intensively for transistor modeling and for wide-band amplifier design. Generally optimization has found widest use in modeling, e.g., for semiconductor devices—but until now not really much for true complex analog circuits, so we will discuss when optimization makes sense, and how users can make circuits easier to optimize.

A situation perfect for learning is if something gets *wrong*! Of course you can learn from "best practice" examples, but often the real difference between two algorithms becomes visible if something becomes more difficult, so that one fails, whereas the other methods may still work! That is a good starting point for more thinking and for innovations—of course not only for circuit designers, but also for CAD researchers, and for CAD managers. Unfortunately, in EDA environments and in real design situations you have seldom much time to inspect all difficulties regarding algorithms in detail. So in case of problems clear guidance is needed.

A last motivation for reading this book should be this: We all carry around in our head rules and guidelines that give us a sense of intention. The topological map of a big foreign metro will seem "obscure" to a casual visitor, but a *resident* must understand its structure and some details to enable daily travel by memory. Similar to this, engineers have to deal with numerous equations, tools, models, etc., and they must be sorted in one's mind for everyday work. For instance, you do not need to have knowledge about Bessel's function directly in your mind, but should have a feeling for V_{BE} and its behavior versus temperature, versus current, etc., or you should know the meaningful range for current densities in your used technology. Such issues must be at your fingertips, and beyond that, they must be integrated into the *instinctive* fabric—that *is your core being*. You will not get very far on the metro if you need to consult the map each day as you travel to work! Engineers frequently have to make journeys to places far from familiar landmarks. Returning from such, we can return to the challenges of daily work with a new perspective, a little better equipped to examine problems under a brighter light. Do not wait to be told what to do: Do it anyway. Do it soon. Indeed, statistics can be as interesting as circuit design! Optimization has an even closer relationship to design and can be very helpful too. As often in engineering, there are many better ways than "try the same but harder"!

1.1 Key Problems in Circuit Design

In a design project, engineers have to deal with many variables and we have to treat them in a systematic way. Intuitively, you do it mostly, but sometimes confusions can arise. So let us introduce some common simple notations and conventions. In our book, we mark vectors in **bold** face, and we use bold uppercase characters for matrices (like **H** for the Hessian matrix). For random variables, we follow the convention of using uppercase characters (like X). As performance functions, we use f (or $\boldsymbol{f}$ for multiple performances). For names like resistor instance R_2, we use the normal font, but for variables (usually real or complex numbers), we use italics like $A = g_{\mathrm{m}} R_{\mathrm{L}}$. (the m stands for mutual and the L for load—both are names, so <u>no</u> variables).

Note: Sometimes it is hard to say whether a threshold voltage or the supply is fixed or a variable, so we follow our convention when it really matters for understanding—like in flowcharts or equations—but not as slaves. On top of the mentioned conventions, often just all symbols are written in italic style. However, in the context of circuit design these are not always a good ideas.

Example: For an optimization, we usually need to vary multiple parameters x_i, and we can handle this easier by putting them into a vector $\boldsymbol{x}$, e.g., $\boldsymbol{x} = (R_1, C_1, R_2)^{\mathrm{T}}$ (T stands for transpose, turning the "horizontal" vector into a vertical one. This is sometimes needed for matrix calculations).

A designer has to manage many kinds of parameters $\boldsymbol{x}$ which impact circuit performances $\boldsymbol{y} = \boldsymbol{f}(\boldsymbol{x})$:

- Design parameters $\boldsymbol{x}_{\mathrm{D}}$: They are controllable to the designer, so can be set dedicatedly. We assume they will not change during production, only in the design phase. Examples are the value of resistor R_1, the number of resistor segments in parallel in R_1, the capacitance of a capacitor C_1, and the width or the number of fingers of transistor N_2, and on top of these *nominal* design values, there can be of course variations from process or mismatch!

- Statistical parameters $\boldsymbol{x}_{\mathrm{S}}$: The resistor R_1 may have a nominal value of 1 kΩ set by R_{sheet}, length, and width, but in production you may observe statistical variations. Usually, mathematical models are available defining, e.g., the standard deviation of R_{sheet}. Elements can vary, e.g., due to <u>global</u> statistical variations (like from wafer to wafer), but also even two resistors constructed in the same way and on the same wafer may have different values—within-die variation (WID)—due to the so-called <u>mismatch</u>, so $\boldsymbol{x}_{\mathrm{S}} = (\boldsymbol{x}_{\mathrm{P}}, \boldsymbol{x}_{\mathrm{MM}})$. Often the designer can hardly influence global variations, but mismatch can usually be reduced by increasing the device area. Even for a perfect layout, you have to accept a certain mismatch, unfortunately.

- Conditions, constraints, environmental or operational parameters, range parameters x_R: like temperature supply voltage, load resistance, etc. – usually defined as parameter ranges. Also operating modes like temperature, over-current or over-voltage shutdown, power-down, low-gain mode, etc. might be part of x_R.

Parameters, Variables, Constants. Often there is quite some confusion about what is what! It actually depends on the *context* and the *analysis you apply*. Surely, ε_0 of vacuum is a physical *constant*, but for other materials ε_r it might be a function of temperature or a statistical *variable* even! Another example is this: We can treat statistics this way, that we assume there is an ideal model from which we get random samples, e.g., in the background there is a normal Gaussian distribution having a fix well-defined mean μ and standard deviation σ. If we take a sample (via measurements or simulations), we can calculate the mean of the sample data, and it might be different from the ideal mean value, just due to chance. So is μ a variable? Having a fix model in mind, it would be no real variable. And what about the mean from the sample? A specific sample is just a sample, it is as it is: once it is, it might be also regarded as a (specific) constant set! Looks strange, but actually this is the way we follow if we do a parameter *estimation* e.g., via maximum likelihood method (ML). Here we take the data is given, so fix. And we search for the model parameters (like μ and σ) which fit best to the data, so we *treat* the parameters as *variables*. Although later we interpret them as fix, e.g., when using the model in a Monte-Carlo analyses.

In many cases it makes also sense to differentiate between (global) variables (like sheet resistance) and e.g., instance-specific parameters (like length of transistor #3 or its threshold variation against the ideal value), but also this is a convention which is usually not followed strictly.

In IC design, there is quite a clear trend that the number of all kind of parameters increases, i.e., design becomes more complex and also the models (Figure 1.1).

In addition, also the impact of variations tends to increase, e.g., an IR drop of 100 mV matters much more in modern low-voltage designs than in older technologies. Also the changes of threshold voltages (e.g., from statistics and temperature) in *relation* to supply or the absolute thresholds become

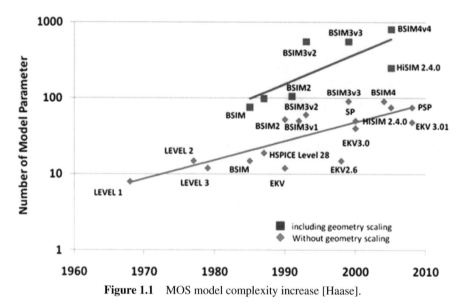

Figure 1.1   MOS model complexity increase [Haase].

more critical (Figure 1.2)—besides several other problems like increased local variations and layout-dependent effects (LDE).

Note: Unfortunately Figure 1.2 shows no units on the y-axis (we just found none!), but the intention is to show that the speed improves by using modern process nodes. This together with smaller area, lower costs and lower power consumption is the major benefit of new technologies. However, unfortunately also the *relative* performance *spread* becomes larger, so harder to manage. Of course, the exact values depend also on technology features, devices sizes, supply voltage tolerances, temperature, etc. What is also not easy to show in

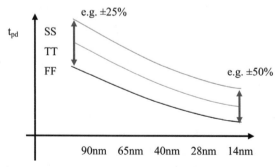

Figure 1.2   Increasing process corner spread on CMOS speed.

a picture, is that e.g., due to clever self-adaptive circuits the spread becomes manageable. And there are tools which can address the problems of variations accurately and efficiently.

In some design environments, and especially in older process development kits (PDK) and model cards, global statistical variations are usually treated not as statistical parameters, but only with fix sets of process <u>corners</u>, like FastMOS (=best-case speed), SlowMOS (worst-case speed), MaxR, MinC, and SlowNMOSFastPMOS (lowest threshold) or just nominal. This is a simplification, because "slow" can only be the worst-case in dedicated terms, like with respect to propagation delay time for a certain class of circuits like CMOS logic but not for other circuits or other measures (like bandwidth and phase margin). The major advantage of process corners is that the designer can directly pick them, simulate, and get at least an approximated worst or best case. In many design environments, you also have different setups available, so you can <u>decide</u> whether you want to treat process variations as corners or via MC. Best use both methods for understanding and efficiency. Actually, also classical logic design was already done in a variation-aware sense, but it excluded statistical variations almost completely.

More Statistical Methods? In principle, we can treat statistical variables x_S with combinatorial methods—which makes sense with discrete random variables, like coins—or we may use Monte Carlo. Actually, there are good attempts to use statistical techniques also for range parameters (corners) x_R or for design variables x_D. The idea of *randomized verification* for corners is quite clever in cases where the number of *directed* tests would be huge, like in big digital or software systems! Random methods for design variables can make sense for difficult optimization problems to achieve global convergence. In the near future, more and more statistical methods will come up—also in analog design.

1.1.1 Brute-Force Design—No Way!

If you want to address the general problem of "design" mathematically and want to describe it in high detail, we would have to deal with <u>all</u> performances f collected in vector $\boldsymbol{f}$ as a function of <u>all</u> variables $\boldsymbol{x} = \boldsymbol{x}_D, \boldsymbol{x}_S, \boldsymbol{x}_R)^T$.

Note: This "art of design" is actually only a subtasks, although a very important and time-consuming one. Usually, there is a kind of exploration phase upfront, which consists also of testing different circuits and composing/extending circuits. And afterward, there is also a longer sign-off phase, and there the focus is on verification (x_D almost fix). However, often there is no clear separation, neither in project time, nor in the tools; there are many overlaps and iterations. This is almost a characteristic for analog design (Figure 1.3).

Unfortunately, the performance function $f(x)$ can be extremely complex. In a clever testbench, we might be able to get all f with a single circuit simulation like a transient analysis driving the circuit to all modes, but even then we can typically only cover one <u>single</u> point $f(x)$ (also called sample) of that function; already this can take a minute or an hour. As circuit simulation is often the most time-consuming (automated) part of the design, overall efficiency can be often measured in *many* simulations needed to achieve the targets. In fact, simulators are quite complex and have dozens of analyses and hundreds of options, whereas the classical methods on top—like parameter sweeps or Monte Carlo—have little *internal* runtime and a simpler setup.

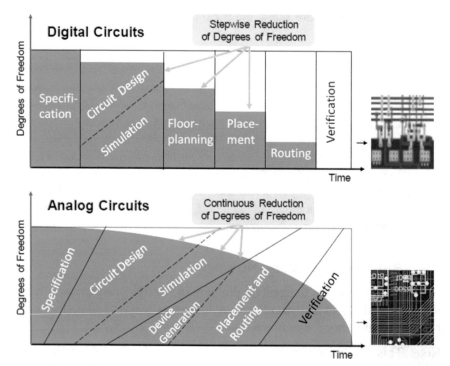

Figure 1.3   Degree of freedom in digital and analog flows [Scheible2015].

The major difference is probably that designers are very familiar with simulator options, because settings like *gmin, reltol,* or *maxstep* can be directly linked to electrical measures, so the other tools on top often come sometimes with something you are less familiar with.

To get a feeling for the circuit, designers usually apply many hand calculations and do many sweeps. To cover nonlinearities accurately enough, the sweeps should be dense enough. Especially temperature behavior is often nonlinear, so you would set up a sweep with 10–100 points. Also for the supply voltage, it is often good to hit the transition, when problem starts to appear, accurately enough. Such sweeps are perfect for understanding, but pure sweeps of *one* parameter at a time do <u>not</u> often show well the complete behavior, because of correlations, or mathematical due to *mixed* terms like $x_1 \cdot x_2$ (here the impact of x_1 on f depends on x_2).

Example #1: CMOS logic delay usually increases with temperature due to lower mobility μ. However, at low supplies V_{DD}, this effect can change because the negative TC of the threshold V_{TO} starts to become more important, and at very low supplies (like for hearing aid applications), also the overall TC might be negative, instead of positive! So the usually helpful picture of increasing delay versus temperature gets wrong, just because delay, temperature, and supply are highly correlated and nonlinear. A one-parameter sweep can be captured in a vector for input and output values, but two-parameter correlations need to be captured in a matrix. If we look to 5 discrete values for both parameters, we end up in $5 \cdot 5 = 5^2$ combinations.

Unfortunately, even if you would run all 25 two-parameter combinations, you might still miss some critical cases, because more than two parameters also can form such correlation group! And we do not know exactly which parameter correlations to treat.

Example #2: If we would like to inspect all combinations in our design (like an op-amp), we would have to treat 20 design parameters, 100 statistical parameters, and 5 operational range parameters. For each parameter, we may want to run 5 values, so to get a full picture, we end up in all-in-all $5^{20} \cdot 5^{100} \cdot 5^5$ combinations to simulate. Even if one simulation takes only a second to get all f in $\boldsymbol{f}$, we would end up in a simulation time of more than 7E79 years. Doing this and looking at all results, we would have the guarantee to find the best design values for the given circuit topology, and its behavior under all conditions. For pure verification (i.e., for fix $\boldsymbol{x}_D$), we would only have to cover the two last parts, so we need $5^{100} \cdot 5^5$ simulations or 7E65 years and even brute-force <u>verification</u> (without exploiting any assumptions on the design) is almost impossible.

The biggest part in our example is the statistical part taking 5^{100} points, so using a dedicated statistical technique like Monte-Carlo can already give some speed-up! We will do so and will also discuss the risks. However, even if we have a clever statistical method, we would have to run it for all the range parameter combinations and for design also at each design point. What about numbers? For verification using the sample yield being in the order of 3σ or approximately 99.8%, we need rougly 3,000 MC points for 95% confidence; so we still end up in $3,000 \cdot 5^5$ simulations or 3.5 months for *pure verification*. Only such exhaustive or brute-force methods would really give a kind of guarantee for any arbitrary complex and nonlinear design. As this is hopelessly inefficient, we need better methods which really exploit the <u>structure</u> of f by finding in which variables we have high sensitivities, strong nonlinearities, and correlations. This way we can avoid "uninteresting" simulations providing us almost redundant results. We need to compose a clever search strategy that leads us quickly to the design limits. Luckily, this is possible because many circuit design problems are similar.

Of course any such efficient design strategy has both parts which can be applied in general (like doing sweeps) but also adaptive parts (like we need to find out which variables are important and form a group with strong impact on a certain output f). Usually, the variables with the highest nonlinearity cause most pain, e.g., temperature characteristics are often difficult, but even more extreme cases can occur. For instance, you may want no monotony errors in a DAC, but to check this, you may really need to simulate each bit, because such errors may take place anywhere. In such cases, best create a *dedicated* testbench, maybe one with autostop if we have found a monotony error or using an algorithm which starts at a place with the highest fail probability (e.g., around half-input, when the MSB would toggle).

A Very Short History of Statistics and Numerics. Using statistical methods to invest on card games and coin flipping is very old, but in opposite to other mathematical areas like geometry, statistics as science is quite young! For instance, a clear judgment why least-square techniques should be used for fits, and when not, was just given in 1921 by R. A. Fisher. The way statistics are often taught based on the axioms of Kolmogorov dates to 1933! The correct confidence interval method for the mean of a normal distribution was given in 1908 by W. S. Gosset, under pseudonym "Student"! Of course bigger breakthroughs are done by C. F. Gauss in the nineteenth century, e.g., he solved many difficult physical problems by applying least squares, problems on which, e.g., Leonhard Euler still failed. The central limit theorem has a longer history starting in the eighteenth century, but proof has taken time as well. Monte Carlo techniques came up in 1940s when numerical computers came up more and more. First quasi-Newton optimization algorithms have been invented in the late 1950s. Bootstrap techniques have been created in the late 1970s. The popular latin hypercube sampling method has been described in 1979 by McKay. Advanced worst-case distance methods are even newer. Matrices are an elegant method to collect numbers and equations, and the term came up in 1850 by J. J. Sylvester.

1.2 Engineering Techniques

Engineers are discoverers, hunters and gatherers, seldom dancers, or actors. A first key technique—and maybe even the most important one—is knowing what you want to do and being able to apply your knowledge.

1.2.1 Ground Work and Anticipation

The circuit behavior is usually defined by physical relations like the Ohm's law or the transfer characteristic of a MOSFET or an amplifier. For a block, this usually ends up in a set of equations like the total gain is $A_{\text{tot}} = \prod A_{\text{stage}}$ with $A_{\text{stage}} = g_{\text{m}} R_{\text{L}}$. In nonlinear cases, such equations might be hard to solve for obtaining the element values, so often simplifications are needed, e.g., based on Taylor series. In an ideal op-amp-based amplifier (having an infinite open-loop gain), the (closed-loop) voltage gain is defined by the feedback resistor ratio, like $A_{\text{stage}} = -R_2/R_1$ (Figure 1.4).

Obviously, a design is more robust if it relies on *ratios* instead of *absolute* values, but sometimes it is not so clear, e.g., because the loop gain might be not as high as desired, so that on top of the (resistor) ratio mismatch error, other effects could be present, and even dominating—"bad luck." Also "good luck" is possible, e.g., you may find a clever bias concept to make g_{m} proportional to $1/R_{\text{L}}$ to *cancel* out the absolute variations even in a simple transistor amplifier stage. Via hand calculations, you can typically obtain only some start values for the circuit elements, e.g., for those inside the amplifier or for the RC values of a filter, and finding the really best-suited values requires

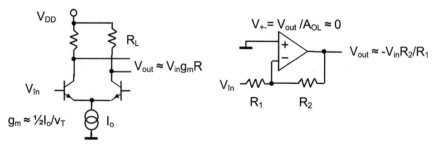

Figure 1.4   Typical transistor amplifier stage and op-amp as inverting voltage amplifier.

some tweaking and resimulations or even multiple prototypes and redesigns, respectively.

1.2.2 Iterative Refinement

Our example clearly shows that iterative refinement is a key technique too: System design may start with simple budget sheets, often entered in spreadsheet programs. At some point, you want to include more effects and running quick simple simulations, e.g., in Mathworks® MATLAB®. Later, in a real circuit design environment, you can switch part by part to more complex models—based on Verilog-A—or to transistor-level circuits which also take loading effects into account. Last you create full layouts, extract parasitic elements, and run really time-consuming sign-off performance verifications, whereas the functional verification and the testbench creation are usually done at a much higher abstraction level. At the end, you can decide whether the design is good enough to make tape-out, i.e., creating an expensive mask set and fabricating the design.

Of course modeling is a key part of the design process and is partly done by modeling experts. Modeling is very helpful in testbench creation, in debugging, and also in the specification phase because having a testbench with models is a kind of "executable specification." It is very helpful for circuit implementation to see how each block should act in the system context, like what are the input signals and the desired outputs. Such "executable specs" help a lot regarding team communication and give also a good status overview. Ultimately, this gives high confidence already in *early* design stages, because it allows to have always something that works and can be demonstrated. All these points are often even more important than the simple simulation speed-up you may get with simpler models compared to transistor-level simulations. For this reason, start the modeling early in a project. Read a bit more about modeling in Section 1.3.2.

1.2.3 Composition in Design

Besides refinement, also composition is important: building of complex systems or blocks by simpler elements. Analog circuit design is a bit like Lego®, and digital is even almost 100% Lego! For instance, you may start directly with a known op-amp circuit topology and optimize just the parameter values, or you may construct a new op-amp:

- Decide on the input stage type according to the input common-mode voltage range (for ground-sensing op-amps, you could use a PMOS input, but no NMOS, and for rail-to-rail signals, you typically need both types or a level shifter) and bandwidth requirements.
- Decide on the number of stages to fulfill the overall gain requirements.
- Choose an output stage based on drive requirements, technology limitations, output voltage range, etc.
- Further decisions could be related to use either a simple class A concept or more power-efficient class AB stages (or even switched-mode amplifiers).

Construction often comes with decisions, and these might be tough to make, because you have to work out each solution to some degree till you are able to make decisions. Decisions are much harder to automate than pure parameter refinements! And analog designers use a lot of different Lego keystones—some are small like a differential pair, and others are complex like a PLL or ADC.

Of course, design tweaking and composition methods are usually in competition, but can also complement well. For instance, if you design a second-order LC lowpass, you know you can get 40 dB/dec, so a certain attenuation for the fifth harmonic. However, in reality, the elements have self-resonances, and with good luck, you can exploit the series inductance of your SMD capacitor and get a much better damping for HD_5! In this case, an optimizer might have found a similar solution, but designer's knowledge could outperform any optimizer in such simple case—but often not in more complex case.

1.2.3.1 Construction vs. optimization

Exploiting the problem structure is usually the key for design efficiency. Optimizers can *partly* act in this way, because they follow a certain *strategy* which can be mathematically even quite optimal (see Chapter 8). In this book, we address *parameter* optimization based on a fix circuit topology, because we want to talk about methods that work in commercial EDA tools.

In several academic papers, true synthesis techniques for analog have also been reported. Usually, these are based on a circuit library and optimization and construction techniques. Construction techniques, which are often rule or knowledge-based, are important too and can be more efficient regarding the number of simulations than pure optimization.

Note that there is, in principle, no clear difference between optimization for component parameters only and optimization which would include topology optimization. You could just give all of your circuits an integer number, and let the optimizer optimize on both this integer and the usual component parameters! However, this is (by far) not the best way, because it just does not exploits the problem structure and nonlinear mixed real integer optimization but is very difficult, thus creating a big burden for the optimizer!

As mentioned, construction is often regarded as an alternative to optimization, e.g., you could try to code [Berkely] your design strategy from spec to circuit for each circuit type—like two-stage op-amp with Miller compensation, NMOS input, folded cascade stage, and PMOS class A output—in a script (e.g., in a programming language like Perl or SKILL®), maybe even including the layout. Unfortunately, such scripts are obviously much harder to create and usually quite limited (at least without optimization), e.g., regarding the specs, you can address as input, and maintenance is a problem too. In addition, it is not easy to make such scripts technology-independent—although interesting approaches at least exist, e.g., by doing the sizing according to $g_\mathrm{m}/I_\mathrm{D}$ technique and by the inclusion of optimization or lookup tables [Iskander2013] (Table 1.1).

It is an interesting question if such fully automated methods will be available in "analog", would they be really well *adopted* by designers? And what about competing methods which may focus on more design *insight*? One current prominent example is the mismatch contribution analysis (see Chapter 5)! Essentially, the whole idea of "awareness" is based not only on "automation" but mainly on avoiding long iteration loops and for getting more insights: for variations, for parasitics, for layout-dependent effects, electromigration, etc.! Also tools like IP management systems have strong user-specific aspect: Any IP system is only as good as the users and administrators are in structuring and maintaining it. Analog designs will probably never be as "simple" as logic design.

Already in existing environments, many companies have made clever extensions to let the designers work in a convenient way, like offering property editors not only for editing but also with immediate feedback for design

Table 1.1 Construction versus optimization-supported flow

	Construction-Based Design	Optimization-Supported Design
Topology definition	Typically in a script	By designer in schematic
Parameters to design	Defined in script	Defined by designer, e.g., supported via contribution analysis
Rules for sizing	Defined in script, e.g., according to g_m/I_D method and circuit-specific calculations	To fulfill block performance, support by sizing rules and many other ones (see Chapter 2)
Flexibility	Limited, need script changes	High
Speed	High, because circuit-specific calculations can highly avoid SPICE simulations	Low
Suitability for high-performance designs	Limited, especially if you want to avoid SPICE simulations	Yes

parameters, like for transistors you get immediately after entering width W and length L also a value for the threshold voltage standard deviation based on the process matching constants or for capacitors by setting W and L you will get not only the capacitance but e.g., also the parasitic substrate capacitance and the parasitic series resistor. Implementing more (like giving a layout preview, or displaying key performances like f_T, f_{res}, Q factor, S_{11}, MAG, noise density, and maximum allowed current—whatever makes sense) is often no big thing. Information at your fingertips is often just work—or a talk with your CAD team! In modern design environments, most customers use only roughly 65% of all tool features they buy for and individuals often even less, so training and continuous improvement is essential.

1.2.4 Team Work and Divide-and-Conquer

In bigger projects, many engineers work together. Usually, some experienced system designers decide on system specs and system partitioning. Once the system topology is defined, we can derive block specifications; however, there is some flexibility in doing that like you can obtain an overall amplification of 1,000 by using either 1 or 2 or 3 amplifiers in a chain. This limits the application of the classical "divide-and-conquer" approach—as maybe number one general design technique. Besides its limitations, in general, this approach is *extremely* successful in chip design because it enables working on many blocks *in parallel*. Each of the blocks will then be designed in quite a similar way, using similar tools. Also in block design, we often apply divide-and-conquer, e.g., when we split the verification into corner runs, and for treating mismatch we use MC.

1.2.5 Automation and Tools

Automation is an important method as well. Designers love their circuit "babies", and they love and hate tools. Designers hate repeating uninteresting tasks, and usually, it pays out to automate things. I remember the old days where you can just run a single simulation, look to the waveforms, take some notes, and tweak the design. After some time, you inspected the more critical corners on temperature T and supply voltage and made a little table on a sheet of paper. Of course, a circuit is work in progress, so you changed it a bit later, and so the table went out of date and becomes quickly inconsistent! Already using a simulator was some kind of automation, and also setting up a clever testbench is a key part of your everyday work. Nowadays, you can also easily automate your result evaluation interactively with a "few" mouse clicks, with built-in calculator and assistances—sent by heaven. This makes also the application of more advanced techniques much easier, like Monte Carlo analysis.

Automation is not only to support lazy people or just to enable design. Being efficient, creating affordable products, and fighting not only for the best but also for economic products are must for engineers. So automation is a strong driver to reduce costs and becoming more and more important, because the risk for failure is always present—redesigns are becoming even more expensive in modern technologies (e.g., due to increasing mask costs), and unexpected redesigns are one major reason for missing the design-in time window.

You may anticipate many problems—maybe a dozen—like a chess player, but surely at some point in design (still many), things may get wrong. Then, you become a hunter for bugs and you have to debug and improve. In this case, you typically do not know completely what is happening, but you should have a working hypothesis and create tests to check it. In theory, there is no difference between theory and practice, and in practice, there is. This is usually because in "real" problems, we often just have a mix of problems.

In a design project, progress means removing the unknowns step-by-step or at least quantifying their range and minimizing their impacts till you are confident enough that you can tape-out. This comes not only from simulation and verification plans! You should understand *what* is causing the limited PSSR of your circuit, and with analytical methods like small-signal equivalent circuits, you can calculate your expectations and verify them using power supply sweeps. The same you can often do for other parameters like temperature as well as for other design metrics like offset voltage. Later you will check for critical corner combinations like low-V_{DD} and slow technology corner, best with a sweep on temperature on top. Or you will set up an MC analysis to check for the production yield. And if your yield target is high, you may switch to special high-yield estimation techniques; and over-all analog design bases also heavily on experience. For luck all many tools do not only provide automation, but many can bring also much insight to the designer. So it is not only "tool speed-up versus costs" that matters.

Last but not least: I remember, in a big tool demo provided by our leading experts at the end, this question came up:

> *OK, we saw that great demo, but what else do you still need to do?*
> *Can we use it already?*

The answer was nice too:

> *Next step is to enable designers that they can do what I have shown!*

Indeed, in making real EDA tools, this aspect is important as well, because if something is difficult to use or confusing, analog designers will not use it. This also points out well that education and training are indeed key points in becoming and staying a good engineer! In fact, some techniques like Monte Carlo are quite old, but still there is a lot of confusion in MC result interpretation! So from time to time, engineers should stop in following the usual habits and focus on things which may look boring or confusing at first glance.

1.2.6 Re-Use in Designs

A last "last but not least" might be "do not reinvent the wheel." If you already know a good solution from experience, then it is often best to reuse it and to focus on other problems. Why applying optimization on a low-performance circuit with known design strategy and well-defined construction steps? Luckily, a big part of design work can be already simplified by pure reuse, e.g., using existing Verilog-A models or testbenches for standard circuits like bandgap, op-amp, ADCs, or voltage regulators. Also analog designs can reuse some circuit blocks like digital gates and flip-flops, or you may use layout macros for differential pairs or current mirrors, etc. Further examples of reuse are the use of macro compilers (for scalable memory blocks, etc.), sharing verification templates in the team which collect known critical corner conditions. In Chapter 10 we will further describe IP and re-use techniques.

1.2.7 Summary

The sweetspot of EDA tools is usually accuracy (like being able to treat very detailed and complex models), and capacity (doing calculations fast). However, tools are <u>not</u> very good in following most of these *manual* approaches (Figure 1.5). For instance, only slowly advanced partitioning techniques are available in EDA tools, usually for becoming even faster and to enable application to extremely complex systems. Simple examples are Fast-SPICE simulators and parameter screening techniques in an optimizer for calculating worst-case distances.

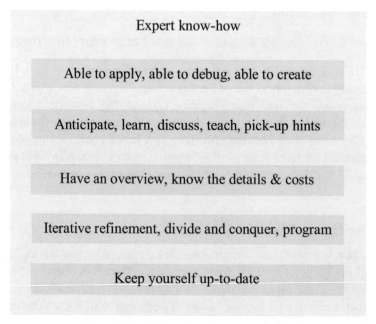

Figure 1.5   Engineering core competences.

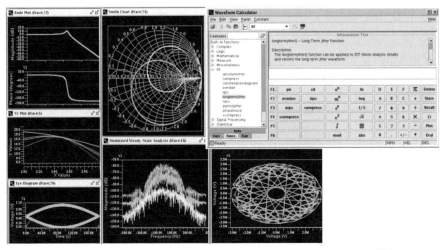

Figure 1.6 Simulation plots quickly created from a special calculator (EZWave^{TM}, Courtesy Mentor Graphics).

Usually the <u>decision</u> about te next design step and which circuit should be used is up to the designer, not to the tool. At best, an optimizer can optimize multiple predefined circuits in parallel, and then it can hand out the best solution found. Only in academic research, true circuit topology optimization indeed exists already. Debugging of circuits and construction are still almost beyond the scope of EDA tools, but of course all the software is also designed to highly support these tasks. For instance, special calculators are available to derive standard circuit performance measures (like 3dB-bandwidth or 10%–90% risetime, and much more) quickly from simulation data (Figure 1.6, not shown is the comfortable graphical stimuli editor). So, since roughly 1985 IC design is a clever mix of manual and semi-automated techniques.

All over the world, engineers have made tools to support you in solving problems and these use the same engineering techniques as described. Often you just have to read the software manuals or ask the EDA vendor for a product update presentation.

1.3 Key Elements and Aspects in Circuit Design

Let us now take a look to further elements in design, specific to circuit and IC design. A native starting point is of course a datasheet.

1.3.1 Datasheets, Conditions, and Trade-Offs

The datasheet is the key document for any electrical device, as target datasheet is often the base for future products and discussions. Here, you promise certain functionality and characteristics. The circuit simulations during the design phase help to find the best-suited circuit and also allow a virtual verification based on simulation models, but for this we need clear guidance for efficient work. Usually, the datasheet reports both the typical performance and the guaranteed minimum performance; and it defines a bunch of testbenches. Often a performance can be defined in different ways, e.g., in terms of power in Watt or in dBm. Usually, the designer set up tests up in a convenient way, e.g., fitting to measurement equipment and to get numbers easy to handle. The latter is also important for numerical algorithms, e.g., the period of an oscillator could be infinite, just in case that the oscillator does not work. To avoid infinite numbers, better use the oscillator frequency $f = 1/T$. Of course, terms of "pass" versus "fail" and for the yield, the unit does not matter at all, but for other kind of data analysis or for optimization, it does!

Of course, a circuit should not only work at nominal conditions but also provide correct operation in a certain *range* of important environmental parameters such as temperature, supply voltage, and load capacitance. In older environments often designers spend many hours to collect simulation data and to create spec *compliance* tables for reviews and for documentation, but since several years this is a feature provided automatically (in the user interface and e.g., as HTML as CSV file) in most EDA environments (Figure 1.7). With context-sensitive menus or additional buttons also many more options are available, such as backannotations to schematic, plotting window access, sorting and filtering features, log file access, selection of a subset of corners for debugging, automatic datasheet generation, etc.

Often there is confusion about which performances are required under which conditions; a small change can have a big impact on whether the design is easy to create or almost impossible! For instance, a small change in the input voltage range of a DC–DC converter could impact the whole topology (buck vs boost vs buck-boost) and pin-out. Clarify these points *early* and *explicitly* in a verification plan, e.g., as appendix to the target datasheet.

When designing a product, you have to make many trade-offs, e.g., you can make a product cheaper using a simple process technology (like pure digital CMOS process), but this can make the design (much) more difficult, because older processes offer usually only moderate bandwidths.

	Variable	Nominal	Spec	Min	Max	Pass/Fail	PVT0	PVT1	PVT2	PVT3	PVT4	PVT5
	Vcc	1.5					1.7		1.9		1.7	
	gpdk090.scs	NN					SS				FF	
	Temperature	27					-40	100	-40	100	-40	100
Testbench	Output	Value	Spec	Min	Max	Pass/Fail	Value	Value	Value	Value	Value	Value
Test1	BW05dB	8.25M	>5M	7.94M	16.5M	pass	11.95M	8.519M	16.55M	10.38M	9.889M	7.936M
	BW1dB	12.5M		12.2M	94.8M	n.a.	19.52M	13.76M	94.84M	71.29M	15.04M	12.24M
	BW3dB	101.8M	>20M	77.9M	104M	pass	110.4M	82.84M	103.9M	77.87M	103.4M	85.95M
	BWfromtr	19.0M		17.5M	45.1M	n.a.	37.88M	41.78M	44.28M	45.1M	23.4M	17.54M
	HD2	66.2	>60	48.35	71.7	4 fails	71.71	66.42	67.39	62.11	57.61	52.22
	HD3	91.58	>70	73.4	110.9	pass	100.2	100.8	110.9	88.35	82.84	75.91
	Icc	143u	(100u;1m)	141.8u	167u	pass	144.8u	141.8u	150.1u	145.3u	158.1u	151.9u
	Vinoise(1M)	606n	<1u	265.3n	817n	6 fails	460.2n	686.6n	513.2n	715.7n	265.3n	770.7n
	LoopGain DC	50.83	>50	29.9	55.4	4 fails	55.41	50.21	49.92	44.65	40.54	34.98
	Peaking	971m	<1	0	4.57	4 fails	3.449	4.572	3.812	4.411	125.5f	661.1m
	PM	73.9	>70	65.3	73.9	pass	69.59	70.22	65.34	67.03	72.64	72.47
	PSRR(10M)	4.698	>0	3.69	6.21	pass	6.132	3.921	6.211	3.859	5.404	3.692
	PSRR(DC)	46.57	>30	28.9	49.1	1 fail	49.15	45.28	45.11	40.94	38.62	33.67
	t05percent	50.12n		36.2n	74.9n	n.a.	46.54n	68.64n	49.69n	74.94n	39.21n	56.44n
	t1percent	44.9n	<40n	31.2n	60.8n	fail	31.24n	57.33n	38.03n	60.76n	35.65n	41.92n
	trise	18.4n	<30n	7.76n	19.96n	pass	9.24n	8.377n	7.904n	7.76n	14.96n	19.96n
	Voffset	1.22m	(-25m;25m)	1.18m	7.48m	pass	1.178m	2.383m	1.868m	3.426m	2.229m	4.24m
	Summary	1 fail				fail	1 fail	2 fails	3 fails	4 fails	2 fails	3 fails

Figure 1.7 Typical compliance table for a corner analysis (first 6 corners only).

For a given technology, you can often select a high-speed parallel architecture, but that usually consumes more power and occupies a larger chip area. In many cases, some overdesign with respect to performance is possible, so that you will be on the safer side (e.g., on noise, offset, and distortion), but for critical blocks, this is harder and leads usually to the use of more power consumption and chip area, so your design will not be competitive! Underdesign is risky, maybe you can still keep the specs, but the yield may drop significantly or you will be out-of-spec and need a redesign.

1.3.1.1 Trade-off examples

In analog, mixed-signal, or RF, there are generally many compromises, requiring much experience and careful well-organized work is required. Figure 1.8 shows the major trade-offs, but there can be even more (like costs and stability) or some need to be split up (like distortion into odd and even order, or speed into rise time, fall time, delay, bandwidth, and settling time).

Note: The green connections in Figure 1.8b show which performances have *positive* correlation (like unity-gain frequency and power), and the red ones show negative correlations (like phase margin versus gain). But look up, also the positive correlations often compete, because it also matters whether we have upper or lower spec limits.

Trade-off examples:

- Low noise is often a key requirement and is often directly related to bias currents (so power) and device area (especially for flicker noise).
- Also linearity and output range are related to bias currents and of course also to supply voltages.
- Low offset voltages (and good DC accuracy in general) require large area devices to minimize mismatch, but this increases the chip area, and it also leads to speed restrictions (or increasing power).
- Often high DC accuracy and low distortion come in sync, but at higher frequencies, they can also compete due to reduced loop gains.
- If you want a certain output impedance (often required for RF circuits), it may give severe restrictions on the supply voltage or your impedance transformation networks, which unfortunately need some area, limit the bandwidth, and reduce efficiency.

In Chapter 2, we pick up the trade-off topic when discussing the typical manual design flow and transistor sizing.

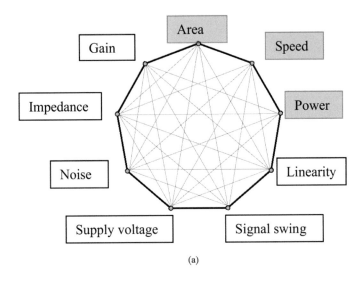

(a)

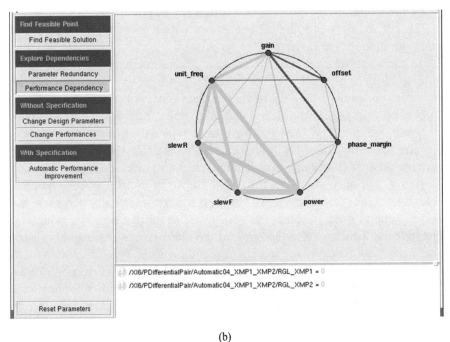

(b)

Figure 1.8 Typical design trade-offs (red=digital) and circuit-specific tool output (Courtesy of MunEDA, red=fighting specs).

Testing versus "Guaranteed by Design" Chips would be very expensive if really everything would be *tested* in hardware under all environmental conditions. Therefore, production tests are usually done only at room temperature and some critical conditions. For this reason, most datasheets are split, e.g., in a part describing the performance at 25°C with usually quite tight tolerances and parts for the characteristics over a wider range of T, V_{DD}, R_L, etc. Detailed tests under these wider conditions are usually done only from time to time, in laboratory, not during production. This way many specs are not really guaranteed by 100% testing, but "by design." Only for very expensive components, it is affordable to really perform a near-100% production test, like for military or spacecraft applications.

1.3.1.2 Datasheet contents

Datasheet for commercial products could also serve well as a reference for blocks on an integrated circuit. Let us do so by inspecting the datasheet of a commercial high-performance operational amplifier (excerpt, Courtesy of Texas Instruments, for the complete information go to http://www.ti.com/lit/ds/symlink/opa1612.pdf).

In the same way we can also create a design documentation e.g., of an op-amp block in an ASIC. For instance, we can look to several commercial examples, or we may use a datasheet template generator (see Figure 1.9).

An official target datasheet is usually quite complete from the pure customer viewpoint (at least you have to convince the customer), but some key characteristics for yourself are usually missing like yield and worst-case corners. Also it is usually not defining chip area, block shapes, bonding diagrams, and second-order effects like substrate noise. A real complete datasheet in the "IC-design sense" is good for documentation purposes, but also to support other designers in your team, to make a designer review or just to learn. This way also the reuse of the block can be made much easier. Some specs are typically not interesting for customers, but very important to know internally. Actually, for the customers, maybe the guaranteed minimum performance matters, just to fulfill system specs, but in other applications, it may matter if your performance variations, e.g., in an ADC, are due to temperature or supply voltage or due to mismatch, and for pure ADC design, maybe just the total variation matters. However, if you want later to reuse the design for a multichannel or IQ ADC application, the mismatch is usually more critical, compared to temperature effects. For such reason, documentation can

TEXAS
INSTRUMENTS

OPA1611, OPA1612

SBOS450C – JULY 2009 – REVISED AUGUST 2014

OPA161x SoundPlus™ High-Performance, Bipolar-Input Audio Operational Amplifiers

1 Features

- Superior Sound Quality
- Ultralow Noise: 1.1 nV/√Hz at 1 kHz
- Ultralow Distortion:
  0.000015% at 1 kHz
- High Slew Rate: 27 V/µs
- Wide Bandwidth: 40 MHz (G = +1)
- High Open-Loop Gain: 130 dB
- Unity Gain Stable
- Low Quiescent Current:
  3.6 mA per Channel
- Rail-to-Rail Output
- Wide Supply Range: ±2.25 V to ±18 V
- Single and Dual Versions Available

2 Applications

- Professional Audio Equipment
- Microphone Preamplifiers
- Analog and Digital Mixing Consoles
- Broadcast Studio Equipment
- Audio Test And Measurement
- High-End A/V Receivers

3 Description

The OPA1611 (single) and OPA1612 (dual) bipolar-input operational amplifiers achieve very low 1.1-nV/√Hz noise density with an ultralow distortion of 0.000015% at 1 kHz. The OPA1611 and OPA1612 offer rail-to-rail output swing to within 600 mV with a 2-kΩ load, which increases headroom and maximizes dynamic range. These devices also have a high output drive capability of ±30 mA.

These devices operate over a very wide supply range of ±2.25 V to ±18 V, on only 3.6 mA of supply current per channel. The OPA1611 and OPA1612 op amps are unity-gain stable and provide excellent dynamic behavior over a wide range of load conditions.

The dual version features completely independent circuitry for lowest crosstalk and freedom from interactions between channels, even when overdriven or overloaded.

Both the OPA1611 and OPA1612 are available in SOIC-8 packages and the OPA1612 is available in SON-8. These devices are specified from –40°C to +85°C.

Device Information[1]

PART NUMBER	PACKAGE	BODY SIZE (NOM)
OPA1611	SOIC (8)	4.90 mm × 3.91 mm
OPA1612	SOIC (8)	4.90 mm × 3.91 mm
	SON (8)	3.00 mm × 3.00 mm

(1) For all available packages, see the orderable addendum at the end of the datasheet.

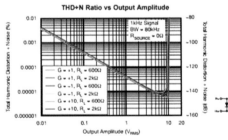

THD+N Ratio vs Output Amplitude

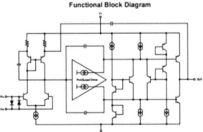

Functional Block Diagram

An IMPORTANT NOTICE at the end of this data sheet addresses availability, warranty, changes, use in safety-critical applications, intellectual property matters and other important disclaimers. PRODUCTION DATA.

OPA1611, OPA1612

SBOS450C –JULY 2009–REVISED AUGUST 2014

www.ti.com

6 Specifications

6.1 Absolute Maximum Ratings

over operating free-air temperature range (unless otherwise noted)[1]

		MIN	MAX	UNIT
Supply voltage	$V_S = (V+) - (V-)$		40	V
Input voltage		(V–) – 0.5	(V+) + 0.5	V
Input current (all pins except power-supply pins)			±10	mA
Output short-circuit[2]		Continuous		
Operating temperature	(T_A)	–55	+125	°C
Junction temperature	(T_J)		200	°C

(1) Stresses beyond those listed under *Absolute Maximum Ratings* may cause permanent damage to the device. These are stress ratings only, which do not imply functional operation of the device at these or any other conditions beyond those indicated under *Recommended Operating Conditions*. Exposure to absolute-maximum-rated conditions for extended periods may affect device reliability.
(2) Short-circuit to V_S / 2 (ground in symmetrical dual supply setups), one amplifier per package.

6.2 Handling Ratings

			MIN	MAX	UNIT
T_{stg}	Storage temperature range		–65	+150	°C
$V_{(ESD)}$	Electrostatic discharge	Human body model (HBM), per ANSI/ESDA/JEDEC JS-001, all pins[1]	–3000	3000	V
		Charged device model (CDM), per JEDEC specification JESD22-C101, all pins[2]	–1000	1000	
		Machine model (MM)	–200	200	

(1) JEDEC document JEP155 states that 500-V HBM allows safe manufacturing with a standard ESD control process.
(2) JEDEC document JEP157 states that 250-V CDM allows safe manufacturing with a standard ESD control process.

6.3 Recommended Operating Conditions

over operating free-air temperature range (unless otherwise noted)

	MIN	NOM	MAX	UNIT
Supply voltage (V+ – V–)	4.5 (±2.25)		36 (±18)	V
Specified temperature	–40		+85	°C

6.4 Electrical Characteristics: $V_S = \pm 2.25$ V to ± 18 V

At $T_A = +25°C$ and $R_L = 2$ kΩ, unless otherwise noted. $V_{CM} = V_{OUT}$ = midsupply, unless otherwise noted.

	PARAMETER	TEST CONDITIONS	MIN	TYP	MAX	UNIT
AUDIO PERFORMANCE						
THD+N	Total harmonic distortion + noise	G = +1, f = 1 kHz, V_O = 3 V_{RMS}		0.000015%		
				-136		dB
IMD	Intermodulation distortion	SMPTE/DIN two-tone, 4:1 (60 Hz and 7 kHz), G = +1, V_O = 3 V_{RMS}		0.000015%		
				-136		dB
		DIM 30 (3-kHz square wave and 15-kHz sine wave), G = +1, V_O = 3 V_{RMS}		0.000012%		
				-138		dB
		CCIF twin-tone (19 kHz and 20 kHz), G = +1, V_O = 3 V_{RMS}		0.000008%		
				-142		dB
FREQUENCY RESPONSE						
GBW	Gain-bandwidth product	G = 100		80		MHz
		G = 1		40		MHz
SR	Slew rate	G = -1		27		V/μs
	Full-power bandwidth [1]	V_O = 1 V_{PP}		4		MHz
	Overload recovery time	G = -10		500		ns
	Channel separation (dual)	f = 1 kHz		-130		dB
NOISE						
	Input voltage noise	f = 20 Hz to 20 kHz		1.2		μV$_{PP}$
e_n	Input voltage noise density [2]	f = 10 Hz		2		nV/√Hz
		f = 100 Hz		1.5		nV/√Hz
		f = 1 kHz		1.1	1.5	nV/√Hz
I_n	Input current noise density	f = 10 Hz		3		pA/√Hz
		f = 1 kHz		1.7		pA/√Hz
OFFSET VOLTAGE						
V_{OS}	Input offset voltage	$V_S = \pm 15$ V		±100	±500	μV
dV_{OS}/dT	V_{OS} over temperature [2]	$T_A = -40°C$ to +85°C		1	4	μV/°C
PSRR	Power-supply rejection ratio	$V_S = \pm 2.25$ V to ±18 V		0.1	1	μV/V
INPUT BIAS CURRENT						
I_B	Input bias current	V_{CM} = 0 V		±60	±250	nA
		VCM = 0 V, DRG package only		±60	±300	nA
	I_B over temperature [2]	$T_A = -40°C$ to +85°C			350	nA
I_{OS}	Input offset current	V_{CM} = 0 V		±25	±175	nA
INPUT VOLTAGE RANGE						
V_{CM}	Common-mode voltage range		(V–) + 2		(V+) – 2	V
CMRR	Common-mode rejection ratio	(V–) + 2 V ≤ V_{CM} ≤ (V+) – 2 V	110	120		dB
INPUT IMPEDANCE						
	Differential			20k ‖ 8		Ω ‖ pF
	Common-mode			10^9 ‖ 2		Ω ‖ pF

(1) Full-power bandwidth = SR / (2π × V_P), where SR = slew rate.
(2) Specified by design and characterization.

	PARAMETER	TEST CONDITIONS	MIN	TYP	MAX	UNIT
OPEN-LOOP GAIN						
A_{OL}	Open-loop voltage gain	(V–) + 0.2 V ≤ V_O ≤ (V+) – 0.2 V, R_L = 10 kΩ	114	130		dB
		(V–) + 0.6 V ≤ V_O ≤ (V+) – 0.6 V, R_L = 2 kΩ	110	114		dB
OUTPUT						
V_{OUT}	Voltage output	R_L = 10 kΩ, A_{OL} ≥ 114 dB	(V–) + 0.2		(V+) – 0.2	V
		R_L = 2 kΩ, A_{OL} ≥ 110 dB	(V–) + 0.6		(V+) – 0.6	V
I_{OUT}	Output current			See Figure 27		mA
Z_O	Open-loop output impedance			See Figure 28		Ω
I_{SC}	Short-circuit current			+55		mA
				-62		mA
C_{LOAD}	Capacitive load drive			See Typical Characteristics		pF
POWER SUPPLY						
V_S	Specified voltage		±2.25		±18	V
I_Q	Quiescent current (per channel)	I_{OUT} = 0 A		3.6	4.5	mA
	I_Q over Temperature [3]	$T_A = -40°C$ to +85°C			5.5	mA
TEMPERATURE RANGE						
	Specified range		-40		+85	°C
	Operating range		-55		+125	°C
θ_{JA}	Thermal resistance, SOIC-8			150		°C/W

(3) Specified by design and characterization.

INSTRUMENTS

www.ti.com

OPA1611, OPA1612

SBOS450C – JULY 2009 – REVISED AUGUST 2014

6.5 Typical Characteristics

At T_A = +25°C, V_S = ±15 V, and R_L = 2 kΩ, unless otherwise noted.

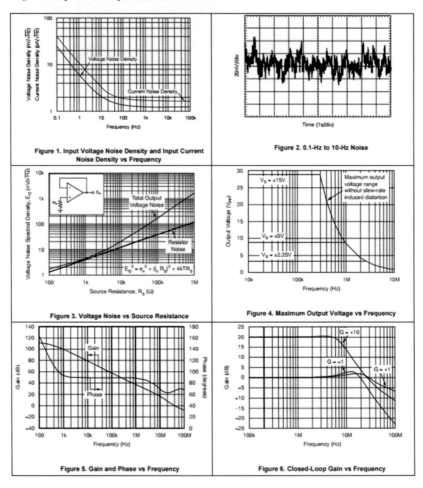

Figure 1. Input Voltage Noise Density and Input Current Noise Density vs Frequency

Figure 2. 0.1-Hz to 10-Hz Noise

Figure 3. Voltage Noise vs Source Resistance

Figure 4. Maximum Output Voltage vs Frequency

Figure 5. Gain and Phase vs Frequency

Figure 6. Closed-Loop Gain vs Frequency

OPA1611, OPA1612

SBOS450C – JULY 2009 – REVISED AUGUST 2014

www.ti.com

Typical Characteristics (continued)

At T_A = +25°C, V_S = ±15 V, and R_L = 2 kΩ, unless otherwise noted.

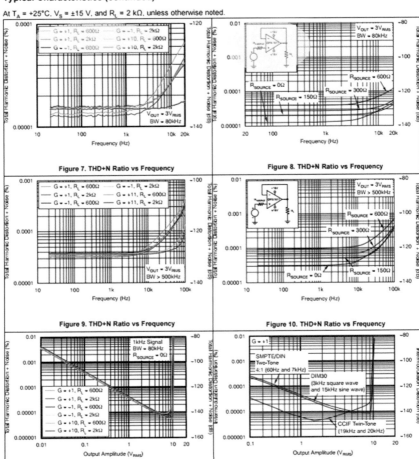

Figure 7. THD+N Ratio vs Frequency

Figure 8. THD+N Ratio vs Frequency

Figure 9. THD+N Ratio vs Frequency

Figure 10. THD+N Ratio vs Frequency

Figure 11. THD+N Ratio vs Output Amplitude

Figure 12. Intermodulation Distortion vs Output Amplitude

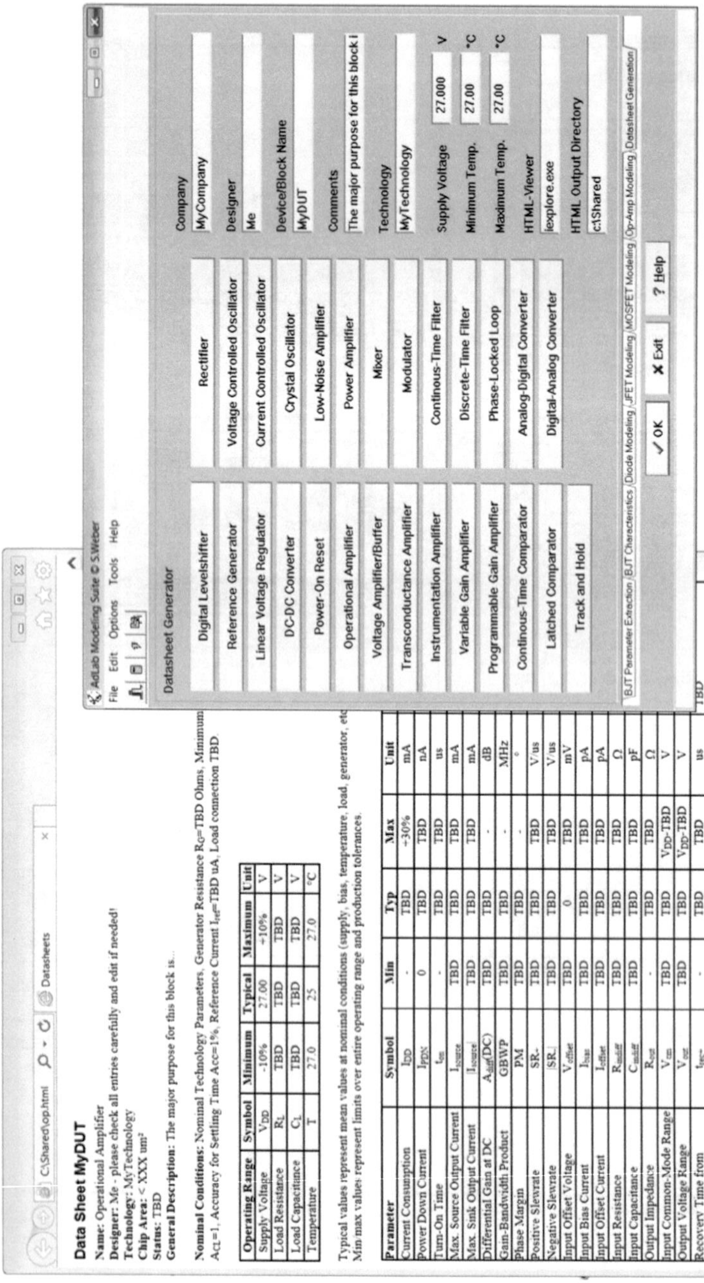

Figure 1.9   Freeware datasheet generator and its HTML output.

be hardly too good! Also it is convenient to see *where* improvements make sense, e.g., where you have already followed best practices and reached the state-of-the-art.

Often not all specs are fully confirmed by the customer, or you may want to add internal specifications, for documentation purposes or to avoid design iterations. For instance, in a system, only the overall offset or noise figure may matter, but to understand the design, it could be also interesting to know about the offset voltage generated in each amplifier stage. In addition, you may want to limit layout-depending effects on offsets. Or you have a spec on bandwidth, and by anticipating critical nets, you may want to limit the parasitics at several internal nets.

1.3.2 Modeling Is Key

This is not a book about modeling or about simulation, but very often a project failure is due to bad or even "lack" of modeling. Actually, if you do "nothing", assume "no model", then you typically implicitly assume a too ideal model, just a bad model. Even if you have no good model(s) e.g., for device mismatch or package inductance, it is a stupid idea to assume no mismatch or no inductance! It is indeed a good method to start with something almost ideal, but then also check the design with the use of realistic models; do it soon, and step by step.

Already when started using simulation techniques in the 1960s, many things rely on modeling (Table 1.2), and of course also for hand calculations you would use models. Actually, mathematically any function can be interpreted as a model, there might be a strong physical background, like for structural models, or even no direct meaning at all, just a fit. In this chapter, we focus more on the first type of models, but for some design methods like corner or sensitivity investigations also pure mathematical models, pure response models have their benefits.

Luckily, the device models have been improved a lot over the years, and partially, you can trust them more than measurements. On the other hand, *more and more* things rely on modeling, not only the simulator results, but also the way the outputs vary, e.g., in a Monte Carlo analysis. So not only accurate IV + CV and noise modeling is essential, but also accurate *statistical* modeling! Luckily, also the MC models have been improved a lot over the years. In fact, the more physically based the model is, the easier the statistical modeling will be, e.g., in the simple old bipolar Gummel–Poon model,

Table 1.2   Different model type for circuit design

Type	Tool	Comment
Device models	Circuit simulator	e.g., classical SPICE models, built-in to the simulator
Statistical models	Circuit simulator e.g., for Monte-Carlo simulation, dedicated statistical tool	Describing the parameter variations of the device models
Behavioral models	Circuit simulator	e.g., Verilog-A models for blocks to get a speed-up over transistor-level simulations or to test ideas or system performance quickly and without having a full implementation
Auxiliary models	e.g., for substrate, package, parasitics, aging, etc.	e.g., SPICE subcircuits

the parameter "IS" is quite difficult to model because it is *not* related to a single physical property! The opposite is true e.g., for the oxide thickness of a MOSFET—here, we can expect much less impacts and correlations with other parameters like doping concentration, bandgap voltage, or sheet resistances. For this reason, the accuracy of most models found in modern PDKs is quite good, although for sure some deviations to reality exist. For instance, often a uniform, normal, or lognormal distribution is assumed. Often this fits to a simplified physical theory, but frequently it is only a meaningful or just acceptable fit to measurement data.

The foundries monitor the process continuously by making process control measurements (PCM). The results will be double-checked in simulation (Figure 1.10 from [Pieper2008]) by just using the same testbenches as in the fab, e.g., on sheet resistance, capacitances, saturation currents, and small circuits like ring oscillators. Based on PCM results the foundry can make sure that only good wafers will be delivered to the customer. Often it is good to be in tight contact with the technologists. I remember in a new process the fab had problems with the current gain β of the new vertical pnp device, but luckily our new circuit was robust enough to work accurate even with a very low β. So instead of throwing away the wafers, we were able to deliver our customer.

Variations may come for different physical reasons, so process variations in general can be classified as random and non-random (e.g., temperature or age),

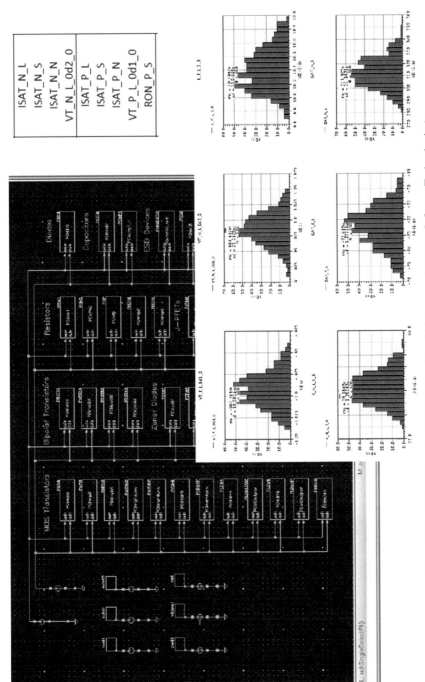

Figure 1.10 Typical foundry PCM testbench and histograms (courtesy Infineon Technologies).

and statistical effects are usually split into intra-die and inter-die variations. For circuit designers, this level of classification is usually enough and models for this are usually available from the foundry in the process development kit PDK. Actually, for quality investigations, also a deeper split into lot-to-lot, wafer-to-wafer, and die-to-die variations makes sense. Statistical variations might be independent or correlated (Figure 1.11).

Usually, an additive law is assumed for parameters:

$$p = p_0 + p_{\text{process}} + p_{\text{mismatch}} \tag{1.1}$$

where p_0 is the nominal value of the parameter (but it might be a function of temperature), p_{process} models the global variations and is shared among all instances on your chip, and p_{mismatch} is intra-die variation specific to instance. Physically, the mismatch depends on the distance between the instances and also on layout details, but as in front-end design, the layout is often not yet defined and these details are typically ignored. They are also not that large *if* you follow good layout practices, like having the same orientation for devices which should match well.

Typically, process variations on threshold voltage V_{TO} are not much depending on device sizes and are often larger than mismatch variations, but the latter become larger for smaller devices. Knowing this, designers can create quite accurate circuits <u>if</u> they can manage that <u>global</u> process variations <u>cancel</u> out! This is done in structures like differential pairs or current mirrors, so that in these now the mismatch dominates. Another key technique to reduce variations is calibration, e.g., one time (in production test), dynamically (e.g., switching to a calibration mode), or sometimes even in the background.

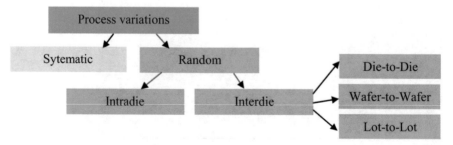

Figure 1.11   Typical classification for circuit design.

Note: For some parameters like leakage currents, often a multiplicative law is used and lognormal statistical parameters. This makes an analysis typically not more difficult; it actually even helps because the voltage across a PN junction follows typically a logarithmic law so that it would show a normal Gaussian distribution because the exponential function in the lognormal distribution and the logarithmic junction behavior would cancel each other! For external components, it is usually more realistic to assume a uniform distribution, not a Gaussian, although sometimes discrete elements have also very strange distributions, e.g., if you buy $\pm5\%$ SMD components, it is not unlikely that the $\pm1\%$ samples are sorted out and sold for a higher price!

Designers should check <u>all</u> models carefully, because sometimes one kind of resistor or transistor is only "better" (e.g., on mismatch or temperature coefficient) due to bad and too simple modeling! Usually, "special" things like noise, mismatch, or breakdown are not treated well in seldom used or special components (like native or low-V_{TO} transistors or coils). A further problem is often that more extreme devices like very small or very big ones are modeled not as good as typical devices. One example for this could be mismatch modeling, and often the simple $\sqrt{A}$-law is assumed and implemented in the model files (Figure 1.12), more complex, more accurate models are usually only available in advanced technologies like 28 nm CMOS or lower (although of course highly advanced models could be also created for older technologies).

Discrete versus Chip Design. In principle there is no big difference between a discrete design, e.g., using SMD components, and IC design, but if you really exploit the advantages of each you can end up in many differences. If you need high accuracy elements, you can choose e.g., discrete components with tighter tolerances, spending a bit more money for critical parts. In IC design you have to live with quite large process variations and some area and component-type dependent mismatch. So for discrete designs Equation (1.1) becomes easier: We have no real process *tracking* for element tolerances, but of course we *have* absolute tolerances causing also mismatch, e.g., between two SMD resistors. In discrete designs, the manufacturer usually guarantees a certain maximum tolerance (like $\pm5\%$), whereas e.g., in typical IC technologies we have e.g., $\pm15\%$ from technology and $\pm1\%$ from mismatch (being usually differential normal distributed). IC designers can often build very good pair stages, whereas in discrete designs two packaged transistors (or resistors, capacitors, etc.) never match very well. Of course discrete designers have much more freedom regarding element choice, e.g., we can choose a dual-transistor or a transistor array, or even a full op-amp in an 8-pin package. In some aspects IC designers are much more limited, e.g., on-chip inductors cannot really compete with SMD coils on Q-factor or current handling capability. It is also hard to create IC technologies which have both very small transistors (for optimum logic and memory implementation with lowest costs) and e.g., power elements (e.g., able to handle 40 Volts or more). At some point a very universal IC technology would become too expensive, so that e.g., a multi-chip system make much more sense, often also regarding design time, flexibility, time-to-market, etc. Another aspect is design methodology: of course discrete designs are quite easy to breadboard, but in IC design intensive simulations are almost a MUST for verifications.

```
library mos090
section stat_mis
parameters
+ pvt_mc=0
+ pu0_mc=0
+ plw_mc=0

statistics {
 mismatch {
  vary pvt_mc  dist=gauss std=1/1
  vary pu0_mc  dist=gauss std=1/1
  vary plw_mc dist=gauss std=1/1
  } }
endsection stat_mis

.....

inline subckt pmos1v   (D G S B)
parameters l=0.1u w=10u M=1 nrd=s1v_hdif_pe/w nrs=s1v_hdif_pe/w  as=1p ad=1p ps=1u pd=1u
+ varvt =  .0029   // 1 sigma Vt mismatch variation in unit of v-um
+ geo_fac =  0.7071 / sqrt(l*w*M*1e12) mm_delvt = varvt * geo_fac * pvt_mc
+ mm_mu0 = 1-(.005 * geo_fac * pu0_mc ) mm_dl= 2e-03  * geo_fac * l * plw_mc
.....
```

Figure 1.12   Transistor model card (typical older process, part for mismatch modeling marked bold).

Many outcomes rely on modeling, so often some small extra-margin might be included for this (like make wires wider than needed according to EM and IR drop requirements or let circuit work to 20% higher clock frequency)—this helps a bit to be prepared for the unknown.

As mentioned, also circuits can be modeled, for example, we may create simplified equation-based models and use these for early simulations and planning on system behavior. In this book, we do not focus on modeling, but using a modeling language clearly helps a designer to solve his problems efficiently. Luckily, model reuse is often easier than circuit design reuse! For instance, the same model might be used for an LNA, PA, or just any amplifier. And even if you need a very complex and accurate model, you may still end up in a single LNA model, and it can represent transistor-level models of many kinds and many technologies. One advantage is that optimization with such models is much faster, because less parameters are involved; you can directly optimize on key parameters like gain, NF, and IP3, which is much easier then tweaking the element values to achieve the desired performance. Figure 1.13 show a Verilog-A model of a voltage reference, also here we can define e.g., the noise level directly as parameter, without changing other parameters.

1.3.3 Design, Debugging, and Tools

A designer should have clear opinions on what he wants to achieve and how. Coming to that point requires of course some discussions and experience, but then there are still quite many things that could go wrong. One interesting aspect in tools is that often they are useful for much more than only one specific task—if you know them well.

Designers do experiments, collect data, and decide for further experiments based on the results of the previous experiments. In circuit design, statistics play a role, and also in math, such approach is known, the so-called design of experiments (DOE). DOE covers techniques like parameter sweeps, corner analysis, and Monte Carlo, but of course a big part of design is also intuition and problem anticipation.

Good debugging capabilities (in laboratory and on computer) are very essential, and using iterative refinement and divide and conquer helps a lot because often the error is easiest to identify if you are at the transition from something that works to something that does not work. Actually the word "engineer" comes from the Latin word *ingeniator*, meaning a keen-witted artificer. In the circuit tweaking phase, the designer learns a lot about the circuits, the system, the testbenches, and the technology by doing many parameter sweeps. If you make the sweeps extreme enough, you always have to debug

```
`include "../../veriloga.inc"

// Author: Stephan Weber, Munich
// Version/Origin: New model
// Status : Initial model [/] In work [X] Fully qualified [/]
// Description : Bandgap block
// Limitations : TC are only present in parametric sweep, not in DC sweep
// Testbench Schematic : test_bg


module bandgap(en, out,VCC);
input (* integer inh_conn_prop_name="vcc" ;
          integer inh_conn_def_value= "\\vcc! " ; *) VCC ;

input en; output out; electrical out, en, VCC;

parameter real vref = 1.2 from [0:inf);
parameter real rout = 1k  from (0:inf);
parameter real vnoise = 10e-9  from (0:inf);
parameter real fc = 1k from [0:1G];       // voltage noise flicker corner frequency
parameter real fc2 = 10M from (0:inf); // voltage noise roll-off frequency
parameter real trise=1u from (0:inf);
parameter real tfall=0.1u from (0:inf);
parameter real tondel=2u  from [0:inf); // tdoff is 0s
parameter real vthres=1.5;
parameter real psrr=80 from [20:inf);   // at DC
parameter real fpsrr=100k from (0:inf); // roll-off frequency for psrr
parameter real tcvout=0 from [-10m:10m]; // TC in V/K
parameter real tc2vout=-0.12u from [-1u:1u]; // Square law in V^2/K
parameter real icc=100u from (0:inf);
parameter real ileak=1n from (0:inf);
parameter integer fullhints=1 from [0:1];

electrical  noise, noisef, psr;

real       Vout, VrefT, tdel,Cnoise,Cpsr,rPSRR, Vnom, i, v, pwr;
integer    enstate;

 analog function real set_vout;
   input state, vactive; integer state; real vactive;
   begin
          if(state>0) set_vout = vactive;
          else set_vout=0;
   end
 endfunction
 analog function real set_delay;
   input state, tondel; integer state; real tondel;
   begin
          if(state>0) set_delay = tondel;
          else set_delay=0;
   end
 endfunction

 analog begin

  @(initial_step) begin
         i=icc;
         v=V(VCC);
         pwr=i*v;

      VrefT=vref+($temperature - `TNOM)*tcvout+pow($temperature - `TNOM,2)*tc2vout;
      Vnom=`vcc_min/2+`vcc_max/2;
      enstate=(V(en)> vthres) && (V(VCC)>=`vcc_min);
      Vout=set_vout(enstate,VrefT);
      tdel=set_delay(enstate,tondel);

      // Check voltages at initial step:
      if (V(VCC)<`vcc_min) $strobe("%M: Warning! Supply TOO LOW in BG at initial step");
      if (V(VCC) >`vcc_max) $strobe("%M: Warning! Supply TOO HIGH in BG at initial step");
      // Noise roll-off
      Cnoise=1/(2*`PI*fc2);
      // psrr
    rPSRR=pow(10,-psrr/20);
    Cpsr=rPSRR/(2*`PI*fpsrr);
     if(fullhints) $strobe("%M: Temperature influence vref-Vrefnom %f5.3",VrefT-vref);
   end
```

```
@ (cross(V(en)- vthres, 1)) begin
enstate=V(VCC)>=`vcc_min;
Vout=set_vout(enstate,VrefT);
tdel=set_delay(enstate,tondel);
end;
@ (cross(V(VCC)- `vcc_min, 1)) begin
enstate=V(en)>=vthres;
if(enstate) $strobe("%M: BG turned on by Vcc>Vccmin and EN=H.");
Vout=set_vout(enstate,VrefT);
tdel=set_delay(enstate,tondel);
end;

@ (cross(V(en)- vthres, -1)) begin
enstate=0;
Vout=set_vout(enstate,VrefT);
tdel=set_delay(enstate,tondel);
end;

@ (cross(V(VCC)- `vcc_min, -1)) begin
$strobe("%M: Warning! Supply TOO LOW in BG.");
enstate=!`reset_on_vccmin;
Vout=set_vout(enstate,VrefT);
tdel=set_delay(enstate,tondel);
end;

@ (cross(V(VCC)- `vcc_max, 1)) $strobe("%M: Warning! Supply TOO LARGE in BG");

 V(noise) <+ white_noise(vnoise*vnoise, "BG noise")+flicker_noise(vnoise*vnoise*fc,1, "BG 1/f noise");
// RC lowpass filter for noise roll-off
`CAPG(noisef,Cnoise);
`RES(noise,noisef,1);
// RC highpass filter for psrr
`CAP(VCC,psr,Cpsr/1M);
`RESG(psr,1M);
V(out) <+ I(out)*rout + transition(Vout, tdel, trise, tfall) + V(noisef) + (V(VCC)-Vnom)*rPSRR + V(psr);
I(VCC) <+ V(VCC)*`gmin + (icc-ileak)*transition(Vout/VrefT, tdel, trise, tfall) + ileak;
$pwr(I(VCC)*V(VCC));

end

endmodule
```

Figure 1.13 Verilog-A model for a bandgap reference cell.

something! Of course, sometimes, you also have to debug not only circuits, e.g., you may need to check whether this is a model problem or a simulator accuracy problem. For this, inspect log files and tighten the simulator accuracy.

Modeling is also perfect for debugging, e.g., the "assumption" that the gain is lower due to package inductance by 1 dB is often meaningful, but of course it is much better to include the package to your testbench. This way your setup reflects the idea directly (even if you forget the assumption) and even much more accurately!

Mistakes can be costly, so to be able to make decisions for difficult problems, you need high trust. A good technique is "always double-check." Do not rely too much on thinking or "obvious" things: Imagine there is a design problem, and you measure ten samples in laboratory. Maybe the variations are not large, but to conclude that the samples behave like in a nominal simulation is risky. If your samples are from one production lot only, you may have significant process deviations on top of the usual mismatch. So it still can make sense to run, e.g., a short Monte Carlo process and mismatch analysis to clarify the performance problem; first get an overview before making conclusions!

Also tools like simulators double-check things internally, e.g., by applying multiple convergence criteria, and often you can double-check further by making a "golden run" using very tight accuracy settings (like *reltol* = 1 *e*-9). Sometimes this is difficult (e.g., due to convergence problems, maybe caused by floating nets, and high-Q elements) or time-consuming, so your deep expertise is required, like treating not only *reltol*, but also tweak more advanced options (like *maxstep*, *minstep*, the integration method or whatever).

In the advanced techniques described in the book, it is absolutely the same, e.g., just run MC twice with different settings or inspect the reported confidence intervals and inspect the log files in detail.

In statistics and optimization, there are luckily only a few icy places where you need to look up carefully, probably confidence intervals are one (Chapter 3.5) and we will tell you! A good method is usually doing an analysis in a different way, e.g., checking transient results against what you expect from AC behavior or double-check yield calculated from sample yield and process capability index C_{PK} (Section 3.6.2).

Often you have to decide which to trust more—and that depends on many things—e.g., phase margin PM gives you a number to quantify stability, but a single number cannot fully represent all kind of instabilities in a nonlinear system, so double-check with transient analysis, S-parameters, manual calculations, waveform inspections, etc.—exploit what you have; and try to get what is missing.

The good thing is that tool problems are often related to circuit problems! So most designers apply such techniques anyway to some degree and extend it hopefully. You should never really stop: Some outputs of analysis are for sure almost trivial and check for what you directly want to verify, like that a unity-gain buffer really reproduces the input signal—easy to check in a transient analysis. The more experience you have, the more you can do: Check also overshoot and distortions, and look maybe to the differential input voltage to check whether the loop gain is high enough forcing a low difference. Check the recovery behavior: Is your circuit coming back quickly to correct operation in case of overdrive? It is hard to be aware upfront of everything, e.g., the filter cutoff frequency sensitivity to RC elements should be one to one (like 10% in R gives a shift of 10% in frequency), but in high-Q filters, this will change even depending on topology. Also the sensitivity to other parameters is not always easy to predict, e.g., because op-amp loop gain might not be really large anymore at the frequency of interest.

Maybe some problems remain, and you have to discuss them with an expert and have to train yourself further. Do not *only* learn from your own mistakes.

Almost all kinds of tools have a direct obvious output, like a histogram from a Monte Carlo analysis or the performance variations in a corner analysis or parameter sweep. However, there is often much more, unfortunately sometimes "hidden" in log files or menus. If a design is bad, you may want an optimization, but often other methods are more efficient: Many advanced simulators feature an analysis to check the impact of mismatch (or transistor parameters) to DC operating point. The result of such analysis could be a ranking list of the instances causing the biggest performance changes and the total variation, e.g., in the output voltage of a reference generator. A designer can do a lot with that information: If, for example, the top 4 transistors dominate the mismatch and we make their area $4\times$ bigger and we can expect an improvement of the overall mismatch by almost $2\times$! So we can improve directly without using an optimizer!

Also automated optimizers and high-yield estimation (HYE) methods provide much more benefits than just improving the circuit or verifying the yield—in addition, you get valuable design information for your under-standing and for more efficient work. We will tell you because this way advanced designers have often even much *more* benefits from advanced tools than less experienced ones. For instance, sensitivity analysis results are available as a "by-product" of more advanced analysis like worst-case corner search or just a Monte Carlo analysis. In opposite to simulator built-in analysis, those have often the advantage of higher flexibility, like being not limited to DC or AC behavior, but valid for any kind of output (like noise figure (NF), total harmonic distortion (THD), or third-order intercept point (IP3)).

To some degree, statistical analyses are often not done because statistics is so interesting, but also because it is one important piece for enabling sensitivity-driven design. But watch out, and this is not for free, e.g., it is quite easy to calculate the sensitivity of a certain performance metric with respect to a certain transistor width (like W_1), but this does not mean that you as a designer can do really much with it, because in a low-offset differential pair, nobody would usually change the width W of only one of the two transistors forming the pair! What you really need is the sensitivity to W_1 being in synch with W_2 and that is no netlist information available to the simulator. Also many simulators can provide sensitivities to many transistor parameters, but you as a designer cannot really change the technology or just one individual transistor parameter like mobility or current gain β without compromising others (there is, e.g., a trade-off between β and early voltage V_{EA}). As a designer, you have to formulate your questions in testbenches—and this is always some significant work. In addition, there is also much design work *beyond* pure sensitivity in general, because the sensitivity cannot catch circuit modifications like adding a buffer chain or a cascade or a capacitor for more frequency compensation flexibility.

In conclusion? In this book, we give you guidance on many methods, and sometimes the very basic methods—like simply simulating really all variable combinations or doing a huge Monte Carlo analysis—are extremely inefficient, so maybe we condemn them too often and we stress the disadvantages too much in the readers mind? We are fully aware that sometimes only almost-brute-force methods are completely foolproof, and indeed, you can always construct test cases, where "too" clever methods would fail! Running all combinations allows to find the worst-case safely, but an automated search can be much faster. On the other hand, having really the results from all combinations would also offer to find the best one, which is not of much help for verification, but is indeed helpful for starting laboratory investigations or for keeping your design alive and improving it further.

Murphy's Law versus RTFM? Besides doing the setup and decision making, designers are also challenged by tool bugs and limitations, sometimes. For a transistor-level simulator, hundreds of options may solve problems, or cause them. Luckily, most statistical or optimization techniques feature much less options, e.g., for Monte-Carlo, you may need to decide what do you want to save, which random seed you want, and how many run points, but not much more! Also more advanced methods do not need a big setup luckily, and mostly, this is because—even and especially the most advanced—algorithms came with a lot of internal automatically adjusted options. With the options available directly in the user interface, you typically set a certain compromise between speed and accuracy, e.g., by setting stopping criteria.

However, of course something could go wrong like an optimization gives no progress or a high-yield estimation algorithm is not able to provide an accurate solution. In these cases, read the log files carefully, try to follow the hints, and read the fantastic manual. Actually, 20 years ago, software documentation was sometimes horrible, but nowadays the products are quite mature and well-documented, featuring many examples, screenshots, and even demo databases or videos. Ask the service team, and often you are not the first one having this problem. If something gets really wrong, double-check not only for the direct tool output message window, but look <u>also</u> to the general log window, to messages in the Unix shell, etc. Best go back to an easier testbench till you get something that works. From that, you can extend the setup again, step by step, till you narrow down the problem and locate it—like problem happens with a certain version or is only present in a certain device or type of distribution, etc. Of course, there is a general problem in documentation: A manual is typically focused on explaining all the different features, so it is often not really solution-oriented! However, often there is further material like app notes, videos and white papers giving more background explanations and examples.

Also the simple methods may come with further options, e.g., the overhead in running all combinations can be used to derive internal error limits, which might be not available to that level in highly advanced methods which would really only do the absolute minimum number of necessary simulations.

There is no free lunch! "Greedy" methods can fail, often in a quite spectacular way! We will get some examples later. On the other hand, the design challenges are often so large that indeed, brute-force methods would be far too inefficient (like we would need more than one million simulations to run) and too simple basic techniques (like doing only one-dimensional sweeps and ignoring all correlations) would become highly inaccurate. Then, mixed approaches and iterative techniques become attractive, but still it is good to know what their benefits are and how they mitigate the remaining risks.

Trust and Error Limit. Tools often report error limits, this gives trust. And actually a pure point estimate is not enough, you should really have also an error estimate. "Error Estimate", seldom you can get more; and such error estimates can have different quality! You would be in a perfect situation, if someone gives you a true guarantee, like $900\,\Omega < R < 1\,k\Omega$. If you buy an SMD component, you get almost such hard limits. Unfortunately, in statistics you typically have only statistical "limits", like "The chance that R is larger than $1k\Omega$ is below 0.1%". These kind of limits are better than nothing, but not as good as hard limits. In addition, in many cases estimations depend on model assumptions, but it is hard to say if a a model is really valid or not. There is a gray area! So not only the estimate (e.g., on yield) but also the method to estimate its tolerance might be not 100% correct. Try to understand how the error estimation works in the specific algorithm. Using Taylor series is one general method, but often you can hardly know how many terms you need to include for error calculations; and usually error estimation works only with low risk for interpolations, not for extrapolations.

Check if model assumptions are meaningful, do not misuse special methods, e.g., check by eye inspection if the data is Gaussian if you use confidence intervals on the mean based on the Student-t distribution. Of course, multiple errors can be present, and they may add up significantly. Here double-checking is best. If someone is promising that a certain method is "trustable" and "verifiable", he often promises too much (at least in a mathematical or legal sense), or he forgets to mention the prerequisites.

1.3.4 Simulation Aspects

Of course you need to be able to simulate your circuits, usually on transistor level to design your circuits with computer support. This is standard since 1980s. For complex analysis, the runtime could unfortunately cause problems; usually more in design automation and for advanced analysis, then e.g., for pure interactive manual design and debugging plus waveform inspection. Therefore, e.g., advanced statistical methods need to be efficient regarding the number of simulations to get a certain output, like sensitivity or the standard deviation of your circuit performances.

These aspects are quite obvious, like a 10-corner simulation may take $10\times$ more times than a 100-corner simulation. The good thing is that also most advanced methods have still quite a moderate internal runtime, but one aspect is often overlooked—accuracy! For instance, it can be already challenging to get accurate enough transient results, e.g., for a DFT output with low-noise floor or to get the overshoot really accurately. For instance, reading out the maximum output voltage $\max(V_{out})$ can be impact by tiny spikes or small shifts in the simulation steps the simulator takes. Usually, designers can manage such problems for their verifications, with careful testbench setup, but for some advanced analysis, you need indeed truly a higher accuracy. This is mainly the case for comparisons, like for gradient calculations by finite differences for an optimization. Having a too large numerical noise could prevent to find the best circuit solution, could prevent getting accurate sensitivity results.

Later we will see that optimization is also one step for finding worst-case distances (WCD), so also some advanced and very useful statistical techniques need really a good testbench setup.

Not only transient analyses are critical regarding accuracy, but also an AC analysis can create numerical noise (even if the related DC solution is already very accurate), e.g., by using too few frequency points and when looking to characteristics like bandwidth, peak frequency, deepness of a notch, or filter passband ripple! Therefore, always inspect your results manually with a waveform viewer and read out the performances manually; also double-check the accuracy setting (like tighten the error limits and double the number of points and check how much the results change).

All in all, quite a big part of engineers' work is spent on making good testbenches, universal testbenches, and tests for debugging, up to a full optimization setup which really captures all performances correctly and reliably.

1.3.5 Total Yield and Partial Yield

The sample yield is easy to calculate as the number of good samples n_{pass} divided by the total sample count n. You can calculate it not only for each specification, but also for all specifications together. In both cases, yield is only well defined if you have enough pass and fail samples to guarantee a "stable" statistic!

Of course, changing one design parameter like resistor R_1 may improve the regarding performance A but may make performance B worse; not only performance matters, and with respect to costs, the production yield has similar importance as performance itself.

Note: The real production yield is also impacted by layout defects like broken vias. Here, in the book we focus on what the front-end designer can do to improve the yield. The term "design for yield" or "design for manufacturing" (DFM) can be used in different ways. In layout, yield improvements are possible too, e.g., by avoiding single vias, following more rigid design rules, etc. Also note that in statistics and in production, the term "sample" often has slightly different meanings: In production, a sample is usually a single piece, but in statistics, also a certain set of samples (or several MC points) are regarded as sample. Note, because also such sets of random samples are random samples, not fully representing the whole statistic.

To simulate the production yield of our design (defined by x_D) in a computer, we can mimic the fabrication and its production tolerances (defined by x_S) with a Monte Carlo simulation. For instance, we can generate a set of n = 1,000 designs, each having different statistical parameters. To be in spec, we need to run many simulations on each design to cover all *combinations* of operating (range) parameters x_R like temperature, load resistance, and supply. If we want to verify at least three values for each of the r range parameters, we need to do three-corner simulations. For realistic designs, this may lead quickly to >250-corner simulations, to be executed on *each* MC sample (coming from our virtual production), so overall to >250,000 simulations. This is a simple but very time-consuming way to check the design. If you want to improve your design on yield with given performance specs you even need to tweak your design and the step with >250,000 simulations is required for *each* individual design, which ends-up in a very slow over-all progress!

On the other hand, truly only this extremely exhaustive flow has no systematic errors. In general, the Monte-Carlo simulation effort for design is given by:

$$\#\text{simulation} = \#\text{design combinations to inspect} \cdot \#\text{tests/simulations}$$

$$\text{per test} \cdot \#\text{corners}^{\text{sweep-points for each corner}} \cdot \text{ MC points} \qquad (1.2)$$

Note: If you as a designer make a very clever testbench setup, you might be able to treat multiple corners already in <u>one</u> simulation. This is often done for important parameters, like doing a DC sweep on temperature or supply voltage! However, "too clever" testbenches are often harder to manage, to extend or less handy for debugging.

Mathematically (see Chapter 3), the overall yield is defined as volume integral over the product of the indicator function and the joint pdf. The indicator function gives a 1 in the pass (or acceptability) region (the region where all performances are in-spec) and 0 in the fail regions.

Even if we exclude the condition parameters, it is typically a very difficult and highly nonlinear function of a huge number of statistical variables; the more performances we have to check, the more difficult the spec-to-failure boarder (and the yield integral) will look like. Later we will give you some pictures and equations.

Do you hate buzz words? Like "new," "breakthrough,," "brute-force MC"? You are right, e.g., why taking a stupid brute-force method as "reference"? In this case, however, there are indeed good reasons to do so. One is that using fancy adaptive or empirical methods as reference would lead to very fuzzy comparisons. Also there is often no better standard way, and new adaptive methods are often simply not available to all authors! In addition, e.g., a full-factorial analysis is a brute-force "stupid" method, but it leads to very well-defined measures: If the number of variables is given and the individual values, you can directly calculate the total number of all combinations and your simulation effort. Also for MC, something like this is possible. On top, you can also often quantify the remaining inaccuracies of such methods. Often the user has to decide for the setup of two different more advanced statistical methods, which might be hard to understand and unfortunately not really well documented. In this case, go one step back and inspect MC as reference; then relate both advanced algorithm against MC for the manifold aspects. Whenever possible, we try to give also references to manual best practices for design, but there is unfortunately no "gold standard" for more advanced methods including those based on design experience, intuition, and common sense!

When checking production samples against the spec limits, we can calculate the yield; each performance leads to a certain *partial* yield. Only if a sample is in-spec for all performances, we can ship that sample to the customer as a good sample. The total or overall yield is lower or equal to the lowest partial yield. It is well known that the partial yield for spec1 and spec2 can be 50%, but the overall yield might be 0 to 50%—depending on correlations. Typically, we have both "fighting" specs (often bandwidth or rise time versus phase margin—see Figure 1.14), where we have almost to add the yield losses and almost redundant specs (like bandwidth and rise time), where we can almost just use the minimum partial yield. So the "compromise" of assuming no correlation, giving 25% is often not so unrealistic, luckily.

Of course, for too difficult and too many competing specs, the design becomes completely infeasible! Luckily, for high yields (and you typically aim for this), the total yield uncertainty from correlation relaxes a lot: e.g., $Y_1 = Y_2 = 99.8\%$ can lead to $Y_{\text{tot}} = 99.6\% \ldots 99.8\%$ which is often an acceptable accuracy. In such cases, the non-correlated case (99.6004%) is anyway very close to the worst-case. In Chapter 5, we will address the difficulty of performance correlations in more detail.

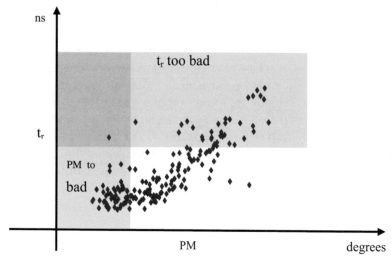

Figure 1.14   Yield for two fighting performances (phase margin and rise time).

Big chips should be still fabricated with a high yield like 90%, but to guarantee this, *each block* needs to have a much higher yield like 99.9%. If we have replicated blocks in our design, like digital standard cells, memory cells, or subcells of a high-resolution DAC or ADC, we need even much higher block yields, which often cannot be verified efficiently with standard MC methods (Figure 1.14).

As you can see, dealing with yield numbers can be a bit difficult, especially if we want to address yield yields, like 99.999%. For this reason, it is very common to express the yield in terms of sigma for a yield-equivalent normal Gaussian distribution.

One problem is unfortunately that sometimes we have single-sided spec and sometimes double-sided ones, and for a single spec placed at 3sigma, the equivalent yield would be app. 99.85%, but for a double-sided spec ± 3 sigma, we would have two times the loss, so $Y = 99.7\%$. The latter number is used a bit more frequently, but most real specs are single-sided, e.g., for the yield or for PSRR, you are only interested in avoiding production samples with a too low yield or PSRR! Instead of using "sigma", you can also use the C_{PK}, and we will discuss it in detail in Chapter 3. The C_{PK} is only valid for normal Gaussian data, and in Chapter 4, we extend the idea and explain the generalized C_{PK}.

Figure 1.15 shows the pdf of a normal distribution with readouts for yield. If the sigma of a design is fix, then one good way to improve on yield is to

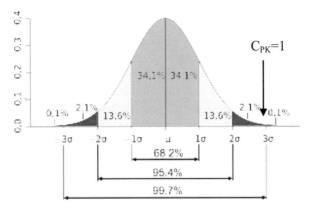

Figure 1.15 Yield parts for a normal Gaussian distribution.

Table 1.3 Yield in terms of sigma and C_{PK}

Spec Setting* or Yield in Sigma	C_{PK}	Rule of Thumb	Single-Sided Yield Loss	Double-Sided Yield Loss
0 sigma	0	50% fails	50%	100%
1 sigma	0.33	about 1 failure in 6	15.9%	31.8%
2 sigma	0.67	about 1 failure in 50	2.3%	4.6%
3 sigma	1	about 1 in 700	0.14%	0.27%
4 sigma	1.33	1 in 30 thousand	0.003%	0.006%
5 sigma	1.67	1/3 in a million	290 ppb	590 ppb
6 sigma	2	1 in a billion	1 ppb	2 ppb

*Distance of spec to mean for using a normal Gaussian distribution.

"center" the design. This way you can minimize the total loss according to upper and lower spec limits; better have a balance than too much loss on one of both spec limits. Another way is of course to try to make the design more robust and to reduce the sigma, so yield optimization is more than only design centering (Table 1.3).

How many sigmas do you need? Please start to like "sigma", it allows dealing with less extreme numbers, and it provides you a better feeling for statistics. If your plan a high-volume production, a good chip-level yield makes life (e.g., testing) much easier (and cheaper). So maybe $Y = 95\%$ is a realistic target, maybe even 99%. However, if your design contains 1,000,000 memory cells or more, then we need for each cell a real high yield, easily six sigma. For blocks which are placed only once on the chip the situation is more relaxed, but still a chip can contain easily a hundred blocks, and already one block fail can lead to a bad chip sample; so each block should still have a yield in the order of 4 to 5 sigma. So to some degree we have applications where high-yield verification is a must, and also cases where low-yield methods like Monte-Carlo fit well. However, as shown, also the intermediate region is important, and already for 4 sigma Monte-Carlo might be impractical (read the chapter on confidence intervals and yield verification), at least if your circuit requires time-consuming simulations.

A further important questions is also how accurate your MC estimates (for yield, mean, standard deviation, etc.) should be. Usually it does not matter so much if the standard deviation of your offset voltage is 5 mV or 6 mV, so sometimes 20% error in terms of sigma might be still acceptable. However, for an ADC or DAC too large mismatch can quickly cause severe errors like missing codes or non-monotonic behavior. In pure Monte-Carlo analysis all estimates have a certain tolerance; and tighter tolerances require more simulation effort. Find more details in Chapter 3.

1.3.6 Robust Designs

You are typically happy if your design is in specification over the full operating region. But how to achieve it? By far the best way is to make the design robust by construction and not to rely on pure simulation and verification techniques!

In analog circuit, we represent signals directly by physical natures (I, V, C, etc.), so they are much more sensitive to manufacturing process and environmental parameter than digital circuits. Design robustness requires the systematic elimination (or at least minimization) of sensitivities to all those parameters. This is only possible by careful choice of the circuit and system architecture, circuit topology, and very careful implementation. This is time-consuming and requires accurate device modeling, and good understanding for the circuit operation and the technology behind. Many problems need to be <u>anticipated</u>, so that a timely project execution and verification are feasible.

A big trend in making analog circuits robust is using clever mixed-signal techniques, e.g., $\Sigma\Delta$ ADC and PLLs. Those were only a first step and a lot of innovations can be further expected, because pure analog techniques tend to become more difficult or just too expensive compared to clever mixed techniques.

In analog circuits, some old tricks may not work so well anymore, e.g., due to reduced supply voltages, or you may need to use a technology optimized for digital, giving less flexibility as an analog-oriented BiCMOS process. Often innovations are coming from both: new restrictions and new opportunities. In Chapter 2, we will give several examples. More complex examples and an excellent overview on ADC design (but not only this) can be found in [Murmann]. Performance gains in circuits are not only coming from CMOS scaling (triggered by the down-sizing of transistor dimensions in new technologies), but also coming from great innovations and surprising concepts. Sometimes the improvement is *not* in making a *better* op-amp but just using *no* op-amp anymore (like replacing them by oscillators or comparators or charge multipliers).

Some of the general techniques for yield improvement are visualized in Figure 1.16; (a) shows a non-optimized design which is "in spec" at nominal conditions, but it fails on performance f_2 at the worst-case corner. In (b), we accepted the variations, but we improved overall performance, and this might be difficult but often possible by spending more area or current (assuming a big spec margin here). In (c), we reduced the variations in performance f_2, but it

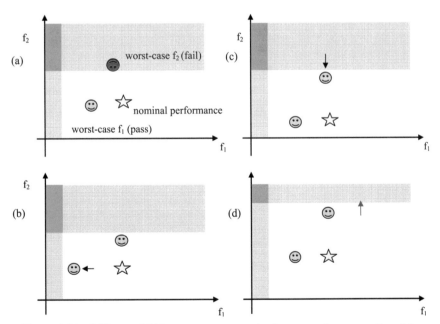

Figure 1.16 Different yield improvement strategies for two performances f_1 and f_2.

often comes with sacrificing other performances or the performances spread in other performances! What is also often possible is to ask for a spec relaxation (Figure 1.16(d)). The ideal case of reducing the *variations* in general (so that it is <u>not</u> needed to make the nominal performance extremely good) is usually also quite difficult to realize; e.g., you may need more chip area to reduce mismatch variations. Sometimes there is no other solution then just spending more area or current, often this is the case for critical parts, and e.g., we may need to compensate the area increase by using smaller transistors in less critical parts. Later we will give further examples, and one solution is of course just to try another circuit variant.

Note: Worst-case (WC) refers not only to environmental conditions, also to statistical variations. A nice circuit example is a Butterworth filter, having a *maximum flat* passband gain. If we design a Butterworth g_mC filter at nominal conditions and process corner (NN) it can happen, that e.g., far too many Monte-Carlo samples have a large undesired filter ripple. So if we really need a flat response for almost all samples and conditions, we actually should design our filter this way that also these extreme MC samples—also being a kind of worst-case corner – are in spec. And this is usually only possible by limiting the mismatch impacts and by reducing the filter Q factors. So at the WC we would get a Butterworth behavior (quite high Q factors) and at nominal we get a filter closer to a Bessel filter (quite low Q factors). Having an eye on nominal performance for understanding *and* on WC for being in spec is the perfect method for achieving robust designs efficiently. Doing this we could see *early* enough that our filter is almost impossible to design, and we may need indeed another circuit, e.g., and to increase the filter order.

1.4 Design Flow Inputs and Outputs

Some elements in the custom IC design flow we already mentioned, beside schematic, specifications, testbenches, layouts, etc., there are also many other documents important for you and your customers, like a guarantee for a certain life time or a limited number of bad devices in the delivery.

Especially for reuse purposes, a (much) more detailed design-oriented datasheet—more a real design documentation—is usually desirable. In addition, make a presentation to your colleagues, and describe well the circuit and its tricks (Table 1.4).

Table 1.4   Custom IC elements

What	Requirements	Created by	Comment
Datasheet	Product idea, electrical and mechanical specifications	Designer, marketing, customer	
Design documentation	Datasheet plus additional information (e.g., on tricky parts, on sensitivities)	Designer	
Process developments kit PDK	Featuring component libraries, simulation models, layout cells, run decks to check design rules, etc.	Foundry	Is technology-specific, and number of rules and complexity increases more and more
System topology	Datasheet	System designer	e.g., checked with Excel® and MATLAB®
Floorplan	Datasheet, chip size estimate, pin positions, block size estimations	Lead designer, lead layouter	
Schematics	Inputs and outputs, circuit function	Designers, using a schematic entry	For circuits and testbenches
Netlists	Schematic	Automatic, triggered by designer	Usually in SPICE format or a similar one
Postlayout netlists	Layout, LVS results	Automatic, triggered by layouter	Tools offer also table outputs, backannotation of parasitics into schematic, etc.
Layouts	Schematic, layout hints, e.g., in OA format	Layouter or designers, using a layout editor	Hints can be provided verbally, as comments or as constraints
Bond plan	Package and die drawing	Lead designer	To be send to fab
LVS report	Schematic and layout, LVS run decks to extract devices	Layouter	To make sure that what you layouted is fit to schematic
DRC report	DRC run deck, layouts	Layouter	To check that design can be manufactured
GDS	Layout	Layouter	Defining the coordinates for all elements to be created at each layer

(*Continued*))

Table 1.4   (Continued)

What	Requirements	Created by	Comment
Evaluation board	Laboratory test hardware (and software)	Designers and application engineer	
Test circuit/program	Production test hardware and software	Designers and test engineers	Test is usually related to production tests, but verification is usually referred as part of design
Quality plan/report	Checklists, etc.	Designers and quality engineers, etc.	This is clearly a team effort. Often it is required to follow certain norms like ISO 9000 on quality management
Third-party IP	Usually, you get only a minimum on documentation on files, like GDS, SPICE netlist, and datasheet	IP vendor, foundry, etc.	Often used for digital standard cells, memories, IO cells, etc., but can be also a major part like ADC, PLL, DDR3 interface, etc.

Figure 1.17 is giving a picture for different design and analysis methods according to the different variable types. For circuit simulations, all three types matter, whereas some other techniques like DRC and LVS run are usually only done for a fix design defined by x_D. Note, that the complete space $x = (x_R, x_S, x_D)^T$ can be huge and the performance functions f depend on all three types, so doing a special analysis, like a corner run is capturing only a little subspace, which might be not fully representative. So mathematically, doing only these basic analysis is working without really having the eyes fully open. Actually doing only isolated analyses in *one* kind of variable, would be mathematically only acceptable if the circuit would be have according to Equation (1.3).

$$f(x) = f_R(x_R) + f_S(x_S) + f_D(x_D) \qquad (1.3)$$

In this case e.g., the sensitivities $\delta f/\delta x_D$ would not depend on x_S and x_R, but this is clearly unrealistic.

Let us see in Chapter 2 how open the eyes are in a typical manual IC design flow.

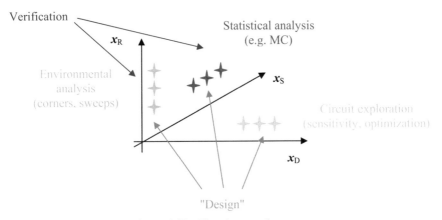

Figure 1.17 Flow input and outputs.

What is in a PDK? To make real designs to be manufactured, you need library elements, These represent the chosen technology, and coming usually from your foundry. So in modern complex technologies, a process development kit contains quite a bunch of material. From the technology library, like cmos90rf, you can pick cells (actually the symbol e.g., for a certain NMOS transistor, a certain resistor type, etc.) and create your circuit blocks in a schematic entry. To run simulations, the process development kit also includes simulator models, like Gummel-Poon models for bipolar transistors.

Also layout views are part of the PDK, and such layout cells are usually *parametrizable*, because in opposite to a SMD transistor, chip designers can e.g., choose the width and length of their elements in a quite large range to optimize circuit characteristics. Such layout cells are called programmable cells (pcells), and each contains the geometric construction statements for the different chip layers to form a specific component (like a high-voltage transistor).

Also available are e.g., rule decks. Using them we can make sure that the design becomes really manufacturable, e.g., all designed element need to be separated by at least a certain minimum distance, to avoid e.g., problems with short circuits, leakage, etc.

The tools picking up the PDK content are typically coming from an EDA vendor. Some required tools are even available for free, like the original old SPICE simulator. However, usually commercial implementation offer more features (like special analysis types) or higher performance (like faster matrix solution, parallel processing, etc.), plus service (not only around the tools itself, but e.g., also regarding design IP, methodology, hosting, etc.).

1.5 Questions and Answers

1. How complete should a datasheet be?
   In the style of official datasheets, there are huge variations, some provide only the absolute minimum, and some are readable like a book on learning circuit design! For good examples, look to the old datasheets of OP-07, the famous PMI low-noise high-precision operational amplifier or to newer products of leading manufacturers. Often the devil is in the details, e.g., some performance specs require an accurate testbench description. For instance, the distortion might be small as inverting amplifier, but much larger in unity-gain configuration due to common-mode distortion. In addition, load and frequency will have a significant impact.

2. Assume the error in your yield calculation is 0.3σ, and what is the error in yield loss?
   The relationship is highly nonlinear, e.g., $2.3 \times loss$ error at 3σ, $6 \times$ at 6σ.

3. Could it happen that in a full MC analysis the sigma of a reference voltage is $2\times$ smaller than from a production?
   This can happen as it also can happen that two results of an MC analysis are not identical! For instance check whether at least the MC simulation confidence interval hit what you get in production. In addition, the production in one fab and few lots (see Figure 1.18) might not show all the allowed tolerances, which you may see over the whole product lifetime or at other fabs using the same process technology! Usually, the limits a fab has to guarantee also come with some margin, and bad wafers will be thrown away, so often process parameter distributions look Gaussian, but with cuts or like multiple narrow Gaussian distributions shifted against each other. Of course, you should design for high long-term yield!

4. Look at the Texas Instruments op-amp datasheet, and how many specs are included? Is this typical? Compare to Figure 1.6!

There are more than twenty specs, which are quite a lot for a simple block with seven pins. Few important characteristics like saturation voltages, and recovery times are not included.

5. Discuss when a design is "good"!
   This is an important question, because only when we can formulate this, we could think of a true design automation.

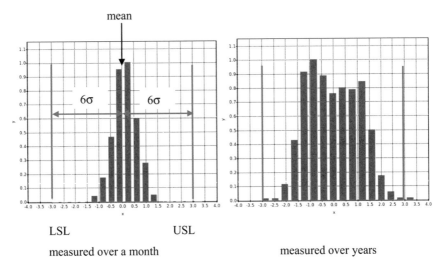

Figure 1.18 Typical short-term and long-term distributions in a fab [Pieper2008].

2

Manual Analog-Centric Design Style(s)

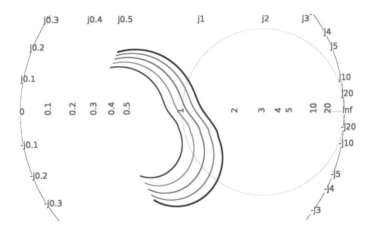

We collect now the most important best practices for efficient design and verification, the typical analog *flow*. We introduce briefly the concepts of specification margins, corner simulations, worst-case and Monte-Carlo. Corner and MC simulations are two standard approaches to discover the design behaviors, and both are simple in essence, because you just run a certain *fix* scenario, a *fix* "design" of experiments (DOE). Both methods act like a simple signal chain, without feedback; you as user have to inspect the results, and you have to decide on further simulations, if needed. To succeed in analog design a lot of experience and anticipation is required, but also the more systematical you work, the higher the chance for being in time and making a successful tape-out.

In this chapter, we also introduce the concept of worst-case corners (WCC), and discuss how to find them. Basically this is a simple task: Just run the simulations and look which ones are most critical. However, actually here we could apply also a more "mathematical" approach, like trying to find how the corner variables influence the corner results. This is (multivariate) modeling in terms of (performance) functions. Note, as in this chapter the focus is more on manual best-practices, we discuss here on the basic math, and later when multivariate techniques are required for more difficult statistical techniques we pick up the topic in more details (Chapter 5).

We end with a little "benchmark" on "men versus machine"! Sometimes men wins, sometimes it is really good to have clever adaptive methods, because worst-case search is a highly nonlinear problem; and with brute-force methods this task would be often very time-consuming.

So this chapter is also important as a starting point, and because we discover some ideas to improve the flow and to compose a more assisted overall flow. An important outcome for a making a design is *learning* about design, e.g., you should understand why a design fails; in fact, even the yield is not the only thing you need to care about, e.g., you may sell non-perfect devices for reduced temperature range or reduced clock frequency. To showcase the impact of variability we present at the end of the chapter a small CMOS RF PA circuit, this also opens a series which we have named "Design with Pictures"—math, pictorial design techniques, and circuits, going slightly beyond purely introductionary examples.

One can learn a lot from looking to other fields of engineering, math, or science in general. The problem is usually that project time is limited, and designers have to focus too much on everyday problems and cannot care so much about flow problems, unfortunately. Actually, in the old days, designers typically focussed and spent probably more time on circuit functionality, for analysis and deep understanding, besides pure verification. During the design phase, designers <u>collect</u> a lot of know-how and invent new circuits and system solutions. The collection is step by step, design, testbenches, and verification coverage grow <u>in parallel,</u> and trust in your design comes also step by step, not in a final sign-off verification (this is maybe true only for LVS).

Let us pick up again the op-amp design example and similar ones:
A first step is usually picking a meaningful circuit topology (often with partially ideal sources and elements—resistor, capacitors, etc.) and making an initial testbench (e.g., for DC behavior). The circuit selection is an essential step, but it usually does not stop early (only in case of hard IP reuse). Often many refinements are needed, like outputs have to be extended, cascodes have to be added for high PSRR, and maybe you need protection circuits. Sometimes also little concept changes are needed because the full requirements have been available too late, and one concept might be if you need only one block (like a bias block), but if you need multiple ones, another implementation might be preferable. Often we can decide early how trustable and robust a design is, like

a CMOS Schmitt trigger is usually much more technology dependent (e.g., thresholds differ significantly for SF vs. FS corner, the 1st letter stand for the NMOS corner, the 2nd one for the PMOS, so SF means the combination of slow NMOS and fast PMOS) than one based on a differential pair (voltage accuracy depends almost only on mismatch and of course reference voltage accuracy). However, exactly how many differences exist depends on many things like spec ranges, technology, temperature, and supply range—and on additional circuit tricks like calibration or replica parts. In this circuit "finding" phase, a lot of interactive work and many simulations are required, so tool speed and usability often matter much more than automation. If done effectively, this "SPICE monkeying" phase is not so bad and you can learn a lot. Typically, if something goes wrong, you even learn more, e.g., you may be able to exclude bad solutions or improve existing ones for your specific application and for effects that might be not important in an older application of the predecessor block.

For Further Reading:
There are many good books available on analog design. As mentioned, we focus an analog in a wide sense, so we do not cover in much detail digital or mixed signal aspects, layout topics or modeling aspects, but in the following list some top references also on these topics are included. Have fun reading them!

- Willy, M. C. Sansen, *Analog Design Essentials*, Springer US, 2007.
- R. A. Pease, *Troubleshooting Analog Circuits*, Butterworth-Heinemann, 1991. R. A. Hastings, *The Art of Analog Layout*, Pearson Prentice Hall, 2006.
- J. Chen, M. Henrie, M. F. Mar, M. Nizic, *Mixed-Signal Methodology Guide*, Cadence Design Systems, 2012.
- W. H. Press, *Numerical Recipes in C: The Art of Scientific Computing*, Cambridge University Press, 1992.
- H. E. Graeb, *Analog Design Centering and Sizing*, Springer Netherlands, 2007.
- H. Graeb, *ITRS 2011 Analog EDA Challenges and Approaches*, in Design, Automation Test in Europe Conference Exhibition (DATE'2012), Dresden, Mar 2012, pp. 1150–1155.
- W. K. Chen, *Computer Aided Design and Design Automation*, CRC Press, 2009.
- R. Spence, R. S. Soin, *Tolerance Design of Electronic Circuits*, Imperial College Press, 1997.

2.1 Biasing and Transistor Sizing

Unfortunately, not all variations can be minimized easily, e.g., a single transistor amplifier stage will have a certain specific temperature variation for gain, output power, noise, and distortion. For obtaining a constant transconductance g_m and gain, usually proportional-to-absolute-temperature (PTAT) biasing is well suited (Figure 2.1), but (for sure) this lowers the intermodulation point IP3 and the slew rate at low temperatures. The latter can be stabilized with a constant bias current instead of PTAT, but this can e.g., lead to severe phase margin problems at low temperatures in feedback circuits—although feedback has certainly many benefits. For instance, negative feedback can make performances generally more stable (usually at the cost of larger noise and lower gain). Another method is using class-AB circuits (which are unfortunately more complex and often have limited common-mode voltage ranges); these can offer more constant g_m over the input voltage, and more current drive capabilities.

Besides the bias concept, also the transistor intrinsic behavior is a key factor, and not only one measure like transconductance g_m matters. The first decision is usually almost trivial and is on the operating region like off-state, ohmic region, or saturation region. In the later, for a given transistor current I_D (or

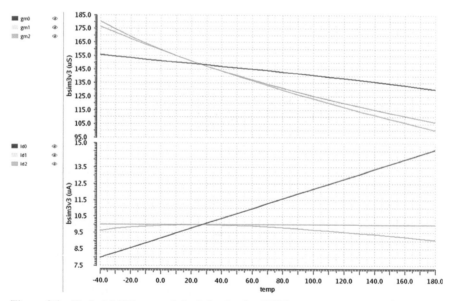

Figure 2.1   Typical MOS g_m and I_D behavior for PTAT, constant current, and constant–voltage biasing.

I_C)—e.g., set according to experience, total current budget, noise level, slew rate, and output power—we can influence many kinds of measures in different ways, just by *sizing* the transistor width W and length L differently [Binkley]:

1. g_m-based sizing makes sense for the MOS saturation region and requires mainly a certain W/L ratio. Maximum g_m requires minimum length L, but that could be non-optimum regarding matching, flicker noise, or output conductance g_{DS}.

2. To make a circuit functional, maybe the V_{Dsat} is most important. Again, the W/L ratio matters mainly.

3. White-noise-based sizing requires a large transconductance and thus a certain minimum current and large W/L for voltage-amplifying stages, whereas for current sources you should not use too short transistors due to their bad matching and larger current noise.

4. Flicker-noise-based sizing requires usually larger gate areas than pure white-noise-based sizing. Also it could be that PMOS transistors are significantly better than NMOS in some technologies.

5. C_{in}-based sizing matters if you use MOS transistors as capacitors, but also in charge amplifiers, the optimum noise performance is reached if the transistor input capacitance is roughly equal to the generator capacitance.

6. Matching-driven sizing might be needed for low-offset amplifiers and requires a certain minimum area $A = W \cdot L$.

7. I_{Dsat}-based sizing makes sense in logic circuits and also in bandgap start-up transistors. W/L matters most. When looking to the maximum transistor current not only the silicon part transistor performance matters, for high-current applications also reliability, metallization and vias needs to be checked carefully.

8. R_{on}-based sizing is useful for switches and depends also on W/L. For lowest R_{on}, we need minimum L. Larger L makes sense if a certain matching accuracy is needed and is mainly useful for non-switching applications (like variable-gain amplifiers).

9. f_T-based sizing makes sense for many high-speed circuits. Typically, this requires a certain minimum current density, low gate capacitance, and short transistors. Unfortunately, it comes with bad DC characteristics like low voltage gain per stage and large mismatch. Of course in an op-amp, the stages should have a f_T beyond the GBWP—though you typically do not exactly know how much, because this depends also on layout parasitics.

10. TC-based sizing is often used in discrete designs or when temperature stability is of high priority. The idea is that in a MOSFET, there is a

certain V_{GS} in which the TC of I_D is zero, because the effects of V_{TO} and mobility cancel. Also a stable on-resistance can be obtained. The only pity is that this point is corner dependent and it does not lead to constant g_m; in addition, the area might be quite large, which matters most in power circuits.

11. In rare cases, you might be able to even tweak on technology parameters (like epitaxi thickness or substrate doping), but there will be still many compromises. For instance, high speed comes for sure with lower breakdown voltage for given semiconductor material (like silicon or GaAs). Also a BJT with high current gain β will have low early voltage V_{EA}.

Overall (and this list is not complete), no single method alone fits! For uncritical transistors, you may only need few seconds to select I_D, W, and L, but for critical ones, you need a mix of methods and many hours plus a full inspection of the full block performance (often even including Monte Carlo and corners). So overall, you have to check almost all the listed characteristics. For true RF circuits, also other electrical parameters (such as f_{max} and k-factor) and the layout matters, like number of fingers and metallization. Only really unimportant "near-digital" transistors can be regarded as uncritical; and we can use any small $L \geq L_{min}$ and any meaningful width. For something in between RF and simple digital, designers often follow a little flowchart that takes step by step at least the major measures like I_D, g_m, and f_T into account (see, e.g., [Sansen2]).

Besides the pure *sizing*, also a careful *type* selection is needed for all components like resistors (like poly versus well) and capacitors (like MIM caps versus MOS caps, with big differences according to substrate parasitics, quality factor Q, linearity, and breakdown voltage) and of course transistors (low V_{TO} versus high-V_{TO}, deep N-well, thick gate versus thin gate according to maximum terminal voltages, etc.). Even if we would follow all the mentioned rules, it may still happen that also other rules like on electromigration dictate us a change, e.g., increasing the transistor width to a certain minimum width like 100 um, although for electrical performance maybe 50 um would be better.

Note: For many circuits, figure of merits (FOM) are available, which describe the power-performance trade-off, like for ADCs, PAs, or VCOs. Check how close your circuit is to the best-designed circuits in similar technology to get a feeling how difficult your design and transistor sizing will be. Of course often such FOMs are only related to the core circuit and exclude bias generators, voltage reference generation, additional buffers, etc. In addition, you need some margins for production tolerances, temperature variations, etc.

Knowing about the circuit details and dependencies is of course still the key competence, and luckily with the support of computers, it is bit easier to obtain all the performances than with breadboarding. So when a designer wants to understand e.g., the region of stability for an op-amp or how the operating point for a transistor is defined by the bias circuit, he/she can still make it based on equations and datasheet plots, etc. And of course also doing parameter sweeps and minimizing TC, improving PSRR, etc. are important steps to make a design robust. Many problems have to be solved, and often graphical methods are very helpful (e.g., the load-line method or the Smith chart). They can help a lot to understand circuits, but later we will also see that the same is true for many advanced numerical methods. In this context and in our book, "manual" design should not mean "without computer" but using techniques already available in older EDA environments, i.e., design by the use of a schematic entry and a simulator, being able to do a corner analysis and Monte Carlo—but not "more."

There is no single most efficient best flow applicable to all kind of blocks. And one consequence of applying a mix of techniques and dealing with difficult problems is that almost always some iteration is required.

Good judgement is the result of experience. Experience is the result of bad judgement! By experience and working in a systematic way you can usually avoid too much stupid brute-force verification and too much trial-and-error. There are several examples which show that manual design can outperform computer simulations—and a clever mix is often the best. For instance, computer programs usually work completely numerical, whereas designers can apply analytical hand calculations to find at least an approximate solution (which is often good enough). This typically leads also to deeper design understanding, e.g., regarding sensitivities. For a computer, sometimes even the simplest things may become time-consuming: A designer often knows from symmetry reasons that the sensitivity to two parameters is the same, or he/she knows that the sensitivity on differential gain to many bias components is very low because they only have an influence on common-mode signals or that certain sensitivities are even zero because related transistors are not active in the current operation mode.

Such insights also lead to efficient design strategies. One example is that you typically first need to make sure that your analog circuit is having the correct DC operating point and e.g., achieving a certain gain, before thinking about other characteristics such as noise performance, speed, and distortion. So circuit design is often "pampering up" a circuit step by step, whereas a pure verification engineer could be already happy with finding one condition in which the design breaks.

In design, this translates usually to solving the most urgent problems first, before solving second-order problems. Of course, what is major and what is second order is not always so easy to know upfront and requires some experiments and experience. For instance, if reverse isolation is critical for your amplifier, you should consider using a cascode stage, but that might be harder to implement at low supply voltages. So you often need to inspect both variants: normal common-source stage versus cascode stage. Once you "pampered up" your design, the next step should be testing it under more difficult situations, like check whether it still works at extreme temperatures, or at minimum or maximum load and supply voltage. In this phase, designers do a lot of parameter sweeps and typically tweak their design further.

2.2 Specification Margin Approach: Fast but Risky

As explained, the brute-force simulation effort to maximize the overall yield is usually far too huge. One way to divide and conquer and for better design understanding is to focus on partial yields, because often the designer has an idea what to change if the power consumption is too large, what else is to do for better bandwidth, etc. Looking only to the overall yield would mean ignoring important information! In fact, also just looking to the partial yields is still a method with big waste of information, because looking only for yield means that we would act as a 1-bit ADC, just because when we calculate the sample yield we only check for pass or fail, but ignore e.g., the information on *how much* we fail! Let us go back to our op-amp example in more detail and analyze what could go wrong if we follow a very fast approach.

An approach with really huge speed-up would be doing just a nominal analysis and <u>tightening</u> the specs (see Figure 2.2). If you know from previous designs, circuit topology, reading the technology documentation, or swept simulations that sheet resistance is your major impact on supply current, and you know it is varying by $\pm 15\%$, you should tighten the current consumption spec by 15%! To get some <u>safety margin,</u> you may use 20% (so 80% of the original spec) to also include second-order effects like TC of the resistance and reference voltage variations. This safety margin can be called specification margin or performance margin, because it is related to the worst-case performance and the spec limit. For comparisons, it is often good to define it not as absolute measures, but in percent.

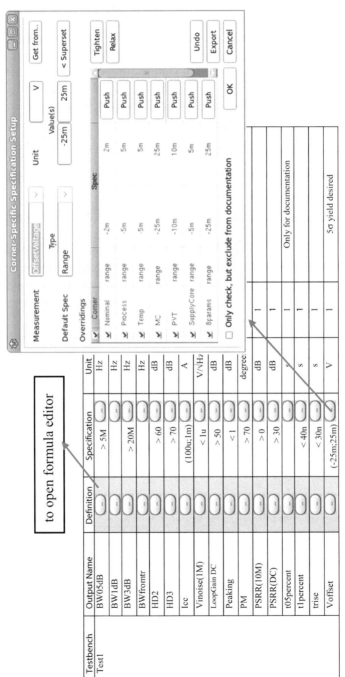

Figure 2.2 Flexible corner setting and typical design environment functions.

In our op-amp example, mentioned a similar margin approach when simply adding the worst-cases from the MC analysis and the corner analysis. In MC, the performance delta from average performance to the WC corner can be often expressed in terms of standard deviation sigma σ, and later we pick up the margin method when discussing the process capability index C_{PK}.

Also the structure of many datasheets supports a margin approach, like defining a tight spec at $27°C$ and a relaxed spec for full temperature range. Also for PCB designs the developer starts typically with spec margin methods, just because the tolerances are often well-documented, and executing sweeps on temperature, supply, etc. is quite time-consuming. However, there are also problems with specification margin approaches:

- You need to determine the design sensitivities quite carefully, either by hand calculations or by simulations.
- The approach usually only works fine with almost linear relationships and if no strong correlations (so-called mixed terms) are present.
- If many effects are important, they can add up too much, like close to $\pm100\%$. In such cases, the margin approach is too conservative, but in other cases, it is often too optimistic!
- You cannot directly debug the design at the point of spec violation!
- The margin approach might be suitable to *center* a design on specs, but making the design really better and reducing the performance spreads is difficult. Here, and for asymmetric tolerances a corner approach can be much better.

In conclusion, the margin approach works fine only for few performances like current consumption or maybe bandwidth or noise figure, but seldom for difficult specs like phase margin, settling time, or IP3. So you should use it mainly in the planning phase or in the starting phase of circuit design, but not for careful verifications. For instance, an RF designer may know from experience that gain is typically 1 dB worse than simulated without layout parasitics. Here, a good way to go is to improve the modeling, e.g., to include at least expected or hand-calculated wiring, package, and substrate parasitics! Maybe this degrades the gain by 0.7 dB—so that you can safely reduce the "fear" design margin to 0.3 dB to be protected against the "unknown."

One big advantage of modeling enhancement is better debugging, and another is that you can also improve on other unwanted effects, e.g., the parasitics may also cause stability problems or can cause cross talk.

How to treat tolerances? This is a big problem in the spec margin approach, and doing it wrong, it could fail. However, unfortunately doing it right often ends up in over-pessimism! If we simply do sweeps and add the magnitudes of the ranges, like $\sum |\Delta y_i|$, it only looks as if we would do a worst-case analysis! This is because we typically do the sweeps of one parameter with the other parameters kept fix, like at nominal. However, it could easily happen that putting such a fix parameter to another value ends up in a bigger Δy_i! In conclusion, this is actually <u>no</u> WC method. If a sensitivity analysis is simple or if the sensitivity S is just known quite well, e.g., current is PTAT or TC of a BJT V_{BE} is app. -1.8 mV/K, then we could also estimate Δy as $S\Delta x$. However, again we would need to find the <u>maximum</u> sensitivity! Actually, doing it this way, the whole approach tends to become both more complex and also often far too pessimistic, unfortunately. So better use the spec margin approach if you are allowed to overdesign (like spending quite a lot of area and current) and if your design is not too nonlinear. Unfortunately, we have seen that even for a classical linear circuit like an op-amp, OFAT can fail for WC finding for that reasons.

For statistical variables, the classical approach is to add quadratic, so add the variances $V = \sigma^2$ and then take the square root. This leads to quite a realistic error propagation. It should be mentioned that the approach is correct if there are <u>no</u> correlations and if you really only aim for standard deviations. For real worst-cases, it is only suited if you have pure Gaussian distributions. Statements like "beyond $\mu + 6\sigma$ there are only 1ppb samples" are critical, because if you do not have found a 6σ sample by simulation, you actually make a risky extrapolations [Schmid]. Beside all these problems, it is also extremely important to clearly state, what is meant when e.g., writing $\pm 10\%$. Clearify if it is a hard limit or only ± 1sigma!

Besides all the criticism in the concept of design margin, it is a good starting point. For instance, one outcome from a swept analysis is the sensitivity, but another one could be the point in which the design starts to "break." Understanding the design behavior in this "breaking" point (e.g., caused by saturation or breakdown) helps usually a lot in finding where and how to improve the circuit! Figure 2.3 gives a design example; for the sketched transistor stack, we need $V_{DD} > V_{GS} + 2V_{Dsat}$, but you need to do a similar analysis also for many other stacks, like to obtain the range for the input common-mode voltage and for the output voltage.

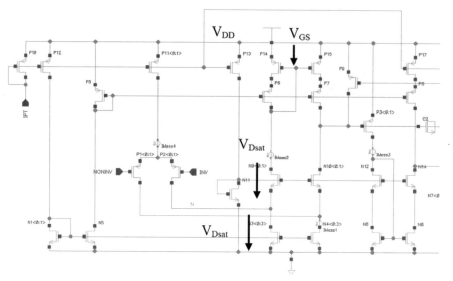

Figure 2.3   Typical transistor stack in an analog circuit to show the worst-case on saturation.

In conclusion, this method of doing parameter sweeps till the design really breaks should apply intensively in early in design stages, but not only. Of course, as extension, you can also apply *combined* sweeps or execute the sweeps around an expected worst-case like temperature sweep at Low V_{DD}+SS corner for saturation-critical specs.

Note: Shifting the break points more and more to the true wanted spec ranges (and beyond) is also a method to make difficult optimizations *feasible*! It acts like g_{min} or source ramping in the DC analysis by a circuit simulator! In difficult circuits and DC simulations, it may happen that the circuit equations are too hard to solve for supply $V_{DD} = 3$ V, so the idea is to start with a simpler problem, like finding the circuit solution for a smaller value (like 1 V or even 0 V; in these the nonlinearities are often lower) and then increasing V_{DD} till we reach the full value. Also in the laboratory do not directly apply the full supply voltage immediately after building the prototype.

If the simple but also very efficient spec margin strategy can be successful depends on the design itself, e.g., phase margin PM is usually uncritical for one-stage op-amps, but might be difficult topic for multistage amps. Such performances can never be treated by spec margins, here we really need to simulate additional critical corner cases!

Also a mixed approach can be created for reducing the set of worst-case corners, the "stretched parameter" method: Often *different* parameters can change the circuit operation in a very *similar* way like a decrease of 100 mV in V_{DD} or $\Delta T = -100°C$, or switching from FF process corner to SS, could have a similar impact on saturation effects and minimum possible supply voltage! So instead of combining all sweeps (giving unfortunately many combinations!), we can also pick one variable only and make its range more extreme! In automotive designs, often the temperature effect is the strongest one, so we can make it e.g., 25 K wider and cover this way also smaller effects like threshold voltages & V_{DD} tolerance! Often you know that V_{TO} changes over process by 100 mV and the TC is -1.5 mV/K, so (for a given circuit) you can directly "translate" it to a change in temperature. This way the designer has still a good overview and can focus on the most important problems, e.g., debugging the circuit at the breakpoint (now in a simple temperature sweep). Problems can appear if there are multiple failure mechanisms, so again this method is typically used in earlier design stages.

In conclusion, if we would know about the set of "worst-case conditions," i.e., the combination giving the worst performance *within* allowed (valid) parameter ranges, we could easily check against the specification <u>directly</u> (without margin) and we could also <u>speed-up</u> design, verification, and maybe also the production test dramatically, with <u>much less risk</u> than in using the specification margin approach! We have to pay a very little price: It cannot be as fast as an approach purely based on design margins, but it can treat quite strong nonlinearities and correlations.

A general compromise for design is also to do design tweaks not at typical conditions but already at an expected worst-case corner, like min V_{DD}, WC load for stability and total harmonic distortion THD, and highest clock frequency. This way the performance margins and the errors in estimating them becomes smaller. Of course, at some point like when moving to layout using the real worst-case corners is <u>much</u> better.

2.3 The Worst-Case Approach

Checking the design at the most extreme parameter combinations is intuitively a good verification method as it is (much) less risky than spec margin approaches. Finding the worst-cases is also a method to speed-up the whole design flow against the exhaustive method with running full corners and full MC. Figure 2.4 shows such flow in alignment with sensitivity-driven design.

The testbench setup can be done in the same way as in any flow, like the brute-force flow. Such testbenches run in a circuit simulator, and typically, an automated calculator tool can extract key performances (like bandwidth, rise

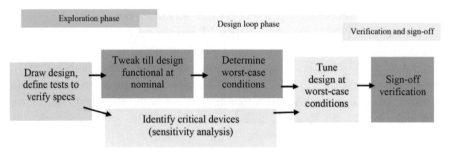

Figure 2.4   Flowchart for efficient circuit design.

time, and phase margin) from the simulation raw data (like voltage signals versus frequency or time usually saved in a binary format).

Finding the critical conditions (represented by $x_S \cdot x_R$) is helpful for both spec checking and debugging. To improve the design, we usually have to identify which parameters x_D are influencing and critical and tune them. Typically, the designer knows quite well about the *major* sensitivities, but often only qualitatively, and also surprises are possible, caused by unwanted resonances, "dirty" circuit tricks, etc. Once the designer is happy, he can proceed with more detailed sign-off simulations, layout, etc.

Looking only to the set of deterministic parameters x_R: The worst-case is given by the worst <u>performance</u> f defined by a certain value <u>combination</u> of <u>environmental</u> parameters x_R, each being in its *valid* parameter range, whereas the design parameters x_D are fix. Note that we do not need exact specifications, and the decision whether an upper or a lower spec limit would be set is enough.

This way finding the worst-case (WC) is possible with a simple grid-based approach and is similar to parameter search or an optimization. Often the worst-case occurs at the most extreme parameter settings, but not always (e.g., for a bandgap output voltage, having a quadratic behavior). We can expect at least one WC combination for each performance, and of course sometimes WC for different specs appears at the same condition (at least approximately). Due to correlations, also the overall worst-case is often <u>not</u> simply the combination of the individual worst-cases obtained from individual parameter sweeps!

If we want to include statistical parameters x_S, the definition of worst-case becomes more difficult; because parameters following a normal Gaussian distribution do not have a finite "allowed" range, they may vary (theoretically) from $-\infty$ to ∞! What we can do is to assign a certain minimum yield (like 99.85%) to our "worst-case." And this way we can calculate also the worst-case statistical parameter sets.

The major problem is that many older design environments have no support for this, and unlike normal environmental range parameters x_R, designers have no way to even set the statistical parameters directly—even if we know the worst-case combination on x_S.

> **Interval Analysis.** There would be also other mathematical ways to treat that problem, e.g., the so-called interval analysis assumes finite ranges for parameters and tries to calculate from that the variations of circuit performances. However, unfortunately, this approach tends to be far too conservative, leading to strong much overdesign, and it is hard to apply for nonlinear circuits, and many variables, or even statistical variables.

Looking to the overall variations in a design, the performance f, e.g., DC gain, delay, or bandwidth, is usually quite amazingly good at nominal conditions and without mismatch. However, process variations cause significant degradations, like 20%. To reach the worst-case across all corners (Figure 2.5), you have to accumulate also all environmental variations like supply voltage and temperature (classical PVT analysis). For CMOS logic, the individual variations are usually in a similar range. In analog design, it depends highly on circuit type, but with a few tricks, the supply changes can often be reduced quite a lot compared to logic design. On the other hand, quite some effort is needed to make analog functions stable against mismatch, e.g., because in modern technologies you often need to operate at quite small supply voltages, so that the ratio between offsets and V_{DD} is not so small as one might expect.

Setup design, testbenches, performance evaluations, and specs

Define parameters ranges, like $x_{R1} = x_{R1min}...x_{R1max}$.
Select technology corners (like SS, SF, FS, TT, FF)

Search for worst-case performances and corners giving the largest spec violations

If needed tweak, your circuit parameter x_D till specs met even at WC corners

Figure 2.5 Design based on worst-case corners.

2.4 Worst-Case Corner Finding

Solving the problem about the worst-case is a key point in design, and the other one is finding ways to improve the design efficiently. Modern design environments offer several methods *beyond* pure corner analysis and Monte Carlo (next subchapter), and the worst-case corner creation is a good starting point to inspect advanced *automated* design techniques. Based on manual techniques we compose some worst-case search strategies, and let them run against standard methods, and adaptive methods in modern EDA environments.

Let us start with observations from typical design situations. Often it is possible to make a design running well at nominal conditions, and often even for all points in a *sweep*, like design is in-spec for both a temperature and a supply sweep. However, this does *not* guarantee that the design works also if we combine the environmental variations. It is also not sure that the worst-case is already given by the individual worst-case for each variable! Designers address this difficulty usually by not only performing single parameter sweeps but also running multidimensional sweeps and the so-called corner simulations. A corner is a set of parameter values, usually including environmental variables but often also technology parameters $R_{sheet} = (minR_{sheet}, nomR_{sheet}, maxR_{sheet})$.

The key problems are unfortunately also quite manifold, e.g., there is no single worst-case like the combination of maximum temperature, minimum supply, and maximum load capacitance—almost for sure it is different for each key performance f! Also it can easily happen that the worst-case conditions change if you change the design, which means that design tweaking and verification at worst-case conditions are tasks that *influence* each other!

So how designers typically solve that problem of finding the worst-case set of corners and variables? Let us do it now at least for the *deterministic* variables like temperature T, supply voltage V_{DD}, and load resistance R_L. We also include technology corners like slowNMOSfastPMOS (SF) and fastNMOSfastPMOS (FF), and such process corners are usually predefined by the foundry and cover some worst-case combinations (e.g., regarding CMOS speed, important to avoid timing violations in digital circuits). The user can access them usually by setting a string variable. In opposite to the typical range variables x_R (like T or V_{DD}), they are *discrete*, and often there is no real ordering or ranking on the process corner string variables (e.g., FS vs. SF).

If we have only one variable, then the WC search problem is the problem of a (absolute) minimum or maximum search regarding the performance function

f under inspection. And basic math shows that such extreme value is either at the end points of an interval or at a point with zero derivative—if the relation between parameter x and performance f is continuously differentiable. If the variable x is discrete like a (binary) bus value (e.g., with eight values or for process corners) and if the relationship is highly nonlinear, then we would indeed need to run a full sweep of all values and sort the results to find the most extreme values. However, if f is fully linear, then the WC will be always at an extreme corner, and never somewhere in the middle. For nonlinear cases, we should have at least 3 levels for the variable! The more values we take, the more time we need, but a benefit would be that this way some error estimations become possible.

A further problem is that we usually have to deal with multiple operational range parameters $x_R = (x_{R1}, x_{R2}...x_{Rn})$. A simplified overall worst-case finding algorithm would be to sweep each of the n parameters (so-called factors) alone, look for the individual worst-cases, and then just combine the individual ones to compose the overall worst-case set. This method is called OFAT, one factor at time, and it needs $2n + 1$ combinations to simulate (if we use three values for each variable of the n variables: minimum, nominal, and maximum). As mentioned, unfortunately there is (by far) no guarantee that we will really find the true overall WC with this technique; even if you use more than 3 levels! If the circuit behavior is highly nonlinear and has strong parameter interactions, then OFAT can easily fail! Experiences show that OFAT typically can only find perhaps roughly 50% of the true WC combinations (or e.g. with other combinations, like all extreme pairs, as done in Table 2.1). So more robust algorithms are desirable, especially if you need to treat many corner variables. Of course, almost all these advanced methods unfortunately require some *more* variable combinations to run compared to OFAT. The exhaustive brute-force approach is just running <u>all</u> combinations and is called <u>full-factorial</u> method. It guarantees ending up in the true worst-case (if the sweeps are dense enough), but it is not efficient at all. Typically, you switch back to this only in small corner sets, fast-running testbenches, for a golden sign-off run, if you have a bad feeling regarding some aspects of the design or if you have anyway enough time (e.g., executing an overnight run).

Is WC finding an optimization problem? Mathematically a clear yes, because both are minimization or maximization! Just the goals and variables differ compared to a circuit optimization.

On the other hand, there are quite many differences in the problem characteristics and on how designers treat corners and circuit optimization.

The optimum value of a resistor or a capacitor is often somewhere in the middle of the interval of the possible values, whereas the WC circuit behavior appears much more often at the interval ends, like at maximum temperature or minimum V_{DD}. In both cases, there are of course also counter examples, but these are less likely. For instance, the maximum bandgap voltage occurs usually somewhere at moderate temperatures, or also $L = L_{min}$ is often an optimum value, when only looking for speed.

In addition, it is quite native to treat corners with a quite raw grid, whereas for optimization, e.g., on filters, few percent changes in a parameter are very critical. So often designers just check for three variable values in a corner analysis, but do really dense sweeps for parameter optimizations. A further difference is that the number of corner variables is quite fix and often all combinations have really to be verified, but for optimization the designer can simply decide whether he/she wants to optimize it or not.

Therefore, also numerically the optimizers for both tasks differ a lot, e.g., WC corner search is focused on quite few discrete variables (like 3 to 10) and on the global extremum for each performance f_i. Actually, in both applications finding the *global* extremum is desirable, but in WC search finding it is even more important.

Circuit optimization, is usually more challenging in other aspects, because we need to treat *all* the different spec-relevant performances *simultaneously*, and often there are much more variables, more correlations, and quite *flat* goal functions (small gradients make the optimum usually harder to find). So, often already finding *some* improvement in overall performance, just finding a better trade-off is enough. More in Chapter 8.

Actually the situation is a bit strange: If f is linear, than OFAT would surely find the WC with linear rising effort $n_{OFAT} = 2 \cdot n + 1$. However, for the 3-level full factorial corner set (in math, any such combination is a so-called "design") we need $n_{3full-fac} = 3^n$ (+1 point for including the nominal case). So people invented also further fix or adaptive algorithms (see Table 2.1, showing a $3\times$ speed-up of an automatic method against full-factorial [Weber2015]). Those algorithms are typically regarding speed and accuracy essentially in between OFAT and full-factorial (see Figure 2.5).

One example of such a fix algorithm is named 2-level full-factorial, and it uses all possible parameter combinations, but only the most extreme value

Table 2.1 Automatic worst-case corner search for an op-amp [Weber2015]

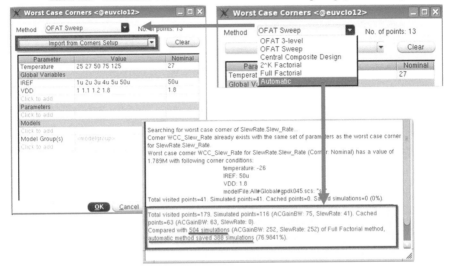

(so we skip the combinations with nominal values). For linear functions, this would still work fine, but *not* for significant second-order nonlinearity (because here the WC is often somewhere in the middle of the interval). Another classical method is called central-composite CC, It is combining the OFAT sweeps e.g., with a 2-level full factorial (or e.g. with other combinations, like all extreme pairs, as done in Table 2.1). Later we will also discuss even more advanced methods, but for now let us check how we can improve the fastest method, OFAT.

2.4.1 Worst-Case Corner Example and Heuristics

As mentioned, the simple, intuitive OFAT method can easily fail for circuits with strong nonlinearities and/or strong parameter interactions. For complex designs or already for an op-amp, this is <u>not</u> always easy to anticipate from circuit understanding. Luckily, there is one nice example showing some difficulties immediately: Already for a basic CMOS inverter and looking for its delay, you can see one major reason why OFAT fails surprisingly often.

The major corners for inverters are process corners (SS, NN, FF, FS, SF), temperature T, and supply voltage V_{DD}. In most of our cases, designers are intuitively right: Beside maximum load capacitance, the usual WC combination on speed is *maximum* temperature (due to mobility degradation), *minimum* V_{DD}, and of course the *slow* corner SS. However, if V_{DDmin} is really very small,

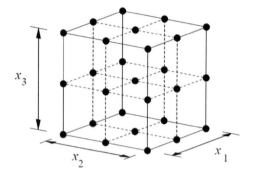

a) 3-level full-factorial n=3^s=27

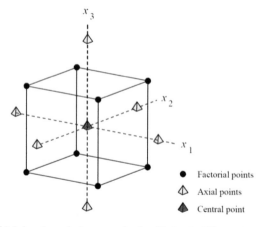

● Factorial points
△ Axial points
▲ Central point

b) 3-level central composite (n=2^s+2s+1=15)

Figure 2.6   Visualization of different frequently used corner methods (s = 3 dimensions).

like for hearing aid applications, it could happen that the gate-overdrive V_{DD}-V_{TO}, and thus the speed, becomes very small, especially at <u>low</u> temperatures! This is because $|V_{TO}|$ has a negative TC, close to the famous –2 mV/K.

Actually, the *same* circuit could have the WC corner combination <u>either</u> at T_{min} (as shown in Figure 2.7) or at T_{max} (as usual for large V_{DD}—so when the upper blue curve is not of interest) just the *ranges* for V_{DD} and T matter mostly (and a bit also the transistor sizes and the technology)!

So let us inspect the more difficult case of a very low minimum V_{DD} and how OFAT would act on this scenario:

Plain OFAT would make three sweeps around the *nominal* values, and we would find that the individual worst-cases are at SS and minimum V_{DD},

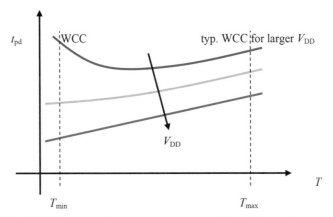

Figure 2.7 CMOS inverter delay versus temperature T for different V_{DD} at SS corner.

but on temperature OFAT would still get maximum T (look at the *middle* green curve of Figure 2.7)! So the *composed* overall WC by OFAT is simply wrong, and only two parameters (V_{DD} and process) are correctly treated! The OFAT-WC on T is wrong due to nonlinearities (in his case mostly quadratic) and correlations (mixed functional terms)! What bothers also is that the WC combination could indeed "jump" in circuit designs, even if the design is robust and meaningful. This is also the case for many other circuits besides CMOS inverters. What helps is that the "jump" is *not* very critical regarding performance f (the y-value might be still quite similar even between T_{min} and T_{max}). To some extent, such difficulties could also happen in other tasks like finding the worst-case for statistical corners or during an optimization.

One question is of course:

> *How would a clever designer address this problem?*

And another one is:

> *Are there clever mathematical algorithms with both: proven effici-ency <u>and</u> reliability?*

Which options to improve, e.g., the setup, do we have? EDA tools have usually only access to information about the design if potentially time-consuming simulations have been done, and if the results are available in suitable form. The designer has also his experience and know-how, e.g., from remembering the problems in older often similar circuit designs or from reading the technology manual. He/she can influence the analysis setup and can exploit this by making an educated guess for the worst-case corners, like:

- Worst-case for leakage current is almost for sure at fast technology corner and high temperatures and supply voltages.
- Lower worst-case on RC filter cutoff frequency is usually given at slow-resistor and slow-capacitor corner.
- Worst-case on CMOS gate delay is at slow MOS corner and minimum supply, and at least usually at maximum temperature. However, we have seen that there is an exception for ultralow supply voltages, because here also minimum temperature can be critical! So just check both of them, which is still better than running all PVT combinations.
- Analog circuits suffer from saturation mostly at minimum supply and slow MOS corner. Again the most critical temperature is often harder to predict (e.g., depending on bias scheme and transistor sizes).
- Of course static performances such as leakage current or DC gain are not impacted by corners from load capacitance or package inductance.
- Worst-case on noise is usually at high temperature and sensitivities on supply or load are usually small.

Following these assumptions may come with some risks, and by far, not in all cases, you can completely determine the full WC combination by experience, but often you can reduce the simulation effort significantly, e.g., ending up in a short sweep instead of running many combinations. Table 2.2. gives some more examples on typical worst-case corners; you should inspect them early in the design phase.

2.4.2 Advanced OFAT Methods

There are many options to improve searches, e.g., exploring the space adaptively or exploiting *a-priori* knowledge. Just making an educated guess for the WC corners and simply taking them is (very) risky. However, we can also try to find out more general concepts of taking designer know-how into account. One simple method for combining *a-priori* knowledge with a mathematical algorithm is this: A WC corner search is typically faster if you have a good starting point. So OFAT around the nominal conditions has usually a bigger chance to fail than doing OFAT around the "expected *a priori* estimate" worst-case! For the ultra-low supply CMOS inverter, plain OFAT can *fail* as we have already seen, but let us inspect what will happen if we "expect" the WC is at $minV_{DD}$, process at SS, and e.g., maximum T (which is wrong for the ultra-low V_{DD} application!). We would indeed find in the sweeps around this "expected WC" the correct over-all WC! The formerly critical sweep on T would be

Table 2.2 Typical overall WC corner combinations for some circuit classes

Circuit	Performance	Range Parameters	Process/ Statistical Parameters	Comments
CMOS inverter	Delay	$minV_{DD}$, maxT or minT	SS	Might be difficult to find by standard OFAT
	Leakage current	$maxV_{DD}$, maxT	FF	Easy to find or anticipate
	Dynamic current	$maxV_{DD}$, minT	FF	Easy to find or anticipate
	Threshold	Depends on application	SF, FS	Easy to overlook the mixed corners
Op-amp	Speed	lowBias	SS	
	Stability		FF	Often load impedance is critical too, but it is hard to anticipate the WC. Low temperatures critical for constant current biasing.
	Supply current	highT, highBias	FF	
	Offset	highT		Often the offset increases with T
	DC gain	highT		g_m drops via T, and with PTAT bias you often get no 100% compensation
	Noise	highT	SS	FF might be critical regarding current noise
	Saturation	$minV_{DD}$	SS	WC for T is hard to predict
Wide-band amplifiers (resistive load)	DC gain	highT	SS + lowR	Due to $A = g_m R$. In CML gates, this might be critical too
	BW	lowBias	SS, highC, highR	Quite obvious, but stability problems can cause deviations
	Distortion	lowT, $lowV_{DD}$	SS, lowT, lowR	The WC may differ for different distortion mechnisms.
LNA	Supply current	highT	lowR	If PTAT biasing
	Gain	$minV_{DD}$, highT	SS	
	IP3, P1dB	$minV_{DD}$		For PTAT bias lowT gives often bad linearity
	NF	maxT	SS	
	Stability	$maxV_{DD}$		Hard to predict further

now at $\min V_{DD}$ and SS (instead of $\text{nom}V_{DD}$ and TT), and we would nicely find the *correct* worst-case temperature, with even *no more* simulations to execute.

Another very fast OFAT variant would be to do OFAT, but to start in the direction of the expected WC; if the expected WC is indeed worse than the starting point (like nominal conditions), there is usually no need to run the opposite OFAT points, because the later would at least typically give the best case, not the worst-case! Such "OFAT with shortcuts" would give a speed between pure guessing and normal OFAT, but it would be not *much* more accurate than OFAT, and so it would be still risky (e.g., if strong second-order nonlinearities are present).

A good variant that *reduces* the OFAT risk is doing a "preordered stepwise OFAT." Let us try to solve the CMOS inverter WC problems manually: As mentioned, the delay WC on V_{DD} and process is almost trivial (even OFAT would have no problems!), but the temperature characteristic is (quite often) more difficult, and here, two effects (TC of mobility and TC threshold voltage) can "reverse" the overall sensitivity; so we have a sign change of the TC, resulting in a strong quadratic behavior. We can modify OFAT further: we should <u>start</u> with low-sensitivity and almost linearly behaving variables, because here the WC is almost trivial, i.e. we should run a sweep on V_{DD}, finding min V_{DD} as WC. Then, we should <u>take</u> this setting and run the sweep on process corners, finding SS as most critical. <u>Last</u> we should sweep T with $V_{DD} = V_{DDmin}$ and process corner set to SS. We will find the correct WC on T and also the overall WC as desired!

The consequence for the general case is that OFAT is <u>more</u> successful if we do it <u>stepwise</u>, with using the currently found WC combination, and we should sweep difficult variables in late sweeps!

Actually, the stepwise OFAT is exactly what designers do in the circuit construction phase, and they just follow the golden rule of only changing one parameter at the time; if the change was good (i.e., we moved indeed a bit to the worst-case), then keep it, think a bit, and consider the next improvement step! It would be usually no good idea to consider the next improvement, but not taking already the first improvement!

Note: The resulting algorithm works like a very simple coordinate-based optimizer, and it would be only a local optimization algorithm. Keep this only in mind, we discuss optimization in Chapter 8 in much more detail.

A small disadvantage of these enhanced OFAT methods is that the setup effort is a bit higher, because the user needs to decide *upfront* which variables are critical (large sensitivities, high nonlinearity, strong correlations) and/or *what* the expected WC corner combinations are. In addition, if we use "too" clever shortcuts, we may take too much risk. An extreme speed-up example would be making a step Δx and using it immediately, just if it makes indeed the behavior worse. Of course, it is more fool-proof to double-check with an opposite step like $-\Delta x$, although it might be "waste" of time in quite many cases. Better avoid risky assumptions in circuit verification. In statistics, one example for a large-risk method is using the C_{PK} without checking for normality. Also in corner analysis there are several well-known critical cases:

- Actually, it is quite typical that temperature is often the most critical parameter, especially for circuits which should work over a big temperature range (like automotive designs). This is also because circuit designers apply usually some "tricks" to make the circuit robust against temperature changes, e.g., by implementing a clever bias scheme (bandgaps, PTAT current generators, replica circuits, resistors with opposite TC, etc.). This often ends up in nonlinear behavior, like the well-known quadratic bandgap voltage characteristic.
- Another difficult parameter might be load impedance, e.g., an op-amp might be (quite) stable for both very small and very big load capacitances, but truly unstable for moderately large capacitances, and this characteristic might be also related to other parameters like load current or the ESR of the output capacitor.
- A third example could be the gain setting parameter in a variable-gain amplifier VGA; if the circuit is tricky, maybe the WC (e.g., on third-order intermodulation point IP3) is at an intermediate position where you may not expect it. Such difficult variables should be treated with quite dense sweeps. For an ordered OFAT such variables should be treated in late sweeps, because they decide mostly on the overall WC, whereas the more linear or less sensitive variables should be set earlier to the WC search.

Of course, if the designer would know that his design is indeed brand-new or difficult, he/she would go indeed for a full-factorial analysis, and by sorting the results table, you can manually find all WC combinations correctly. Unfortunately, for many corner variables, this is not efficient and *much* slower than OFAT.

2.4.3 Advanced Fitting Methods and Adaptive Search Methods

Fix sampling schemes can provide only a certain compromise between speed and accuracy, and one way to improve was to include *a-priori* know-how.

There are some similarities to fully *adaptive* algorithms; e.g., we could just run OFAT and use the obtained information as knowledge to decide on the next sampling points to be simulated. Actually our *stepwise* OFAT is already adaptive, but we can do even better, and such advanced adaptive methods have been implemented very successfully in several custom-design environments already.

Figure 2.8 gives an overview on different worst-case corner search methods. Note that we just show the *typical* behaviors. The enhanced OFAT methods are not shown there to provide a better overview: they could maintain or even exceed the OFAT speed, in addition, due to the *a-priori* know-how reflected in the setup, the *risk* compared to pure OFAT can be even reduced significantly (*if* the setup is done well).

All methods have in common that designers can get rid of boring repetitive work! You need specs, and you just have to define the variables and their values (like in a normal corner analysis) and press the run button!

Automatic methods usually offer a very *good* compromise between speed and accuracy. They usually start with the OFAT method to identify the most important variables; and then more detailed search and space exploration methods are used, e.g., with the inclusion of correlations on these more important variables. Internally, a kind of model-based search algorithm is performed, e.g., from OFAT results, we can create a linear model.

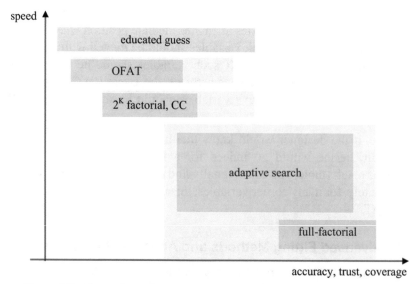

Figure 2.8   Comparison of some important worst-case corner finding methods.

If you think of the next better method beyond linear interpolation and extrapolation; then probably splines are a good choice. Splines also act piece-wise, so if we make a local change (e.g., adding new simulation points), only a local recalculation is needed. Splines also "behave" better than high-order polynomials; the later tend to generate severe "oscillations" (even much more than cubic splines, see Figure 2.10a). However, interestingly also somewhat even better methods are available! Indeed, splines very popular and easy to calculate for 1D to 3D cases, but at *higher* dimensions (so more corner variables) several positive characteristics get <u>lost</u> [Neamtu2001]. Look at Figure 2.9 (from [Erikson 2012]) for a difficult 2D application of so-called thin-plate splines. Here the "overshooting" problem is so severe that the spline fit is not bijective anymore.

With splines we assume that a low-order polynomials fits well to data, and we e.g., force certain continuity characteristics (like continuity in f and first derivative f'). This way different splines can be defined, like cubic splines or Akima splines. Even smoothing splines, which can fit to noisy, statistical data, can be used. However, there is actually little *theoretical* foundation for the use of splines, and on several functions we can outperform splines. At the truly *simulated* points (and using a good model, e.g., splines, high-order polynomials [Daems2002] or so-called Gaussian process models (GPM) [Jones2001, McConaghy]), we can make the fitting error ε arbitrary small, but in the "simulation gaps", any model is surely less accurate. So having not only a fit, but also an error indication (a kind of confidence interval, see Chapter 3) would be a clear advantage.

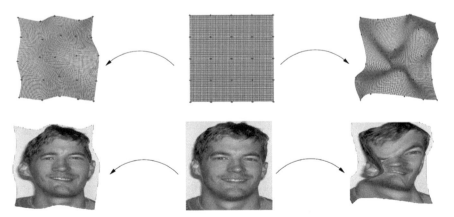

Figure 2.9   Bijective and non-bijective result from spline fitting [Erikson 2012].

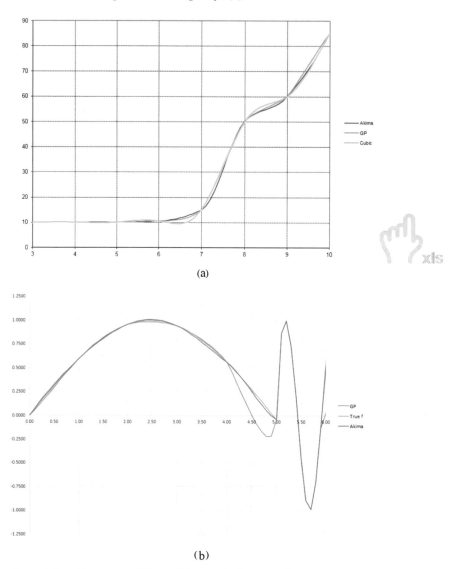

(a)

(b)

Figure 2.10   Comparison different fitting methods on a nonlinear ramp function (data sampled with $\Delta x = 1$) and for a sine function of two different frequencies (10 samples per period).

Note: WC search has aspects of optimization, interpolation and model fitting, but actually also to *learning*! The available simulation points are a kind of training for our model generation, and when we run further simulations we actually test the model, and our learning success.

Splines use low-order polynoms, but what is a Gaussian process model GPM? GPMs use no piece-wise approach, but we try to maintain the advantages of acting quite local at least to some degree. And this near-local behavior, using the Gaussian bell-shape function e^{-x^2}, allows also fits to almost any data. GPM are also very flexible in other aspects, e.g., there is no need that the sampling points are on a certain grid; even random placements would be allowed.

$$f_{GPM}(\mathbf{X}) = a(\mathbf{X}) + \sum b_i \cdot \exp(-|x - x_i|/c_i)^2 \qquad (2.1)$$

Note: In the statistical part of the book we come back to Gaussian distributions in more detail. Here this formula is currently not much more than just a mathematical attempt, just an approach: We start with it, and are just happy that it works quite fine, like splines. For now there is not much to understand, which will become quite different in statistics!

The 2nd term in Equation (2.1) is the Gaussian part, and because the Gaussian pdf diminishes for large Δx, we can make the local approximation at x_i very good, without impacting much the fit at the other sampling points, and we generate no oscillations (as high-order polynomial would do). For more flexibility often the exponent is not set to p = 2 but is allowed to be fitted within 1 and 2. The function e^{-jxj} is a peaky function; and this will also allow fits to true peaks; this would be much harder for splines. Also the first part $m(\mathbf{X})$ might be different in different GPMs, e.g., it could be a constant or e.g., a linear function. One nice GPM feature is that these little Gaussian peaks look a bit like a landscape with little mountains. And knowing the height h at one point x_i obviously helps us to make estimations for the height in the neighborhood; and the correlation between (x_i) to $h(x_i + \Delta x)$ is obviously the higher the smaller $|\Delta x|$ is. And this is also the case for our Gaussian functions. Interestingly, the whole idea has been created for "earth modeling", and these and similar methods are called *kriging*.

One simple model fit verification method is cross-validation by splitting the data into two equal parts and double check the estimates ($f_{fit1}(\mathbf{X}_i)$ versus $f_{fit2}(\mathbf{X}_i)$). An often used variant is "leave-out-one": We can fit a model to n points, and then we compare the model predictions from this full model to models just using one point less. This way an error estimation is possible, actually for all fitting methods. With such an error indication, e.g., an adaptive auto-stop for the WC corner search can be implemented, so that the worst-case combinations can be found in a reliable *and* quite fast way. Gaussian process models are nowadays often preferred, because for statistical applications they have a certain foundation, and offer basic built-in error estimation capabilities. The latter is desirable, because cross-validation and "leave-out-one" can be time-consuming and surprisingly inaccurate (see also Chapter 6 on bootstrap).

However, look up! How can *statistics* help in computer *corner* simulations? In WC corner search we have the situation, that indeed our simulated points are fully reproducible, so there are almost <u>no</u> further (statistical) errors on top (in opposite to lab measurements)! Here the GPM advantage regarding better error estimation would not matter so much! However, in the corner simulation *gaps* there is a "risk", a model error, which we should limit; and that makes the good GPM error estimation also for corner analysis useful, although such error estimate is not the same as the true error.

Actually the situation is tricky: In a normal least-square fit (like linear regression, see Chapter 5), we have a unknown errors, and we assume that typically the error is everywhere (independent on **X**), with the same *distribution* and *standard deviation*, but here in corner performance modeling, we have an error $\varepsilon = \varepsilon(\mathbf{X})$, which is zero for all $\mathbf{X} = \mathbf{X}_i$ (ith simulated point = sampling value).

So to some degree it is often a matter of taste to use GPM or splines, in many cases the results will be very similar (see Figure 2.11). Just having a solid concept with GPM, and at least the option to include statistical errors, is a certain advantage. For instance, you can see in Figure 2.10b that the GPM fit is not perfect, actually the Akima spline works better here. This is because the hyper parameter p is set to 1.5 as in the other examples, but due to the sharp edge (at x = 5, when the sine wave frequency is increased by 10×), e.g., p = 1.1 would give a much better fit. Using GPM <u>and</u> solid estimation methods such "mistakes" are very rare, and actually not necessary. In addition, our 1D example is not 100% representative for complex situations; in many dimensions all methods become more difficult, but spline fitting degrade faster regarding calculation time and accuracy than GPMs! That is an important factor when you need to treat 3 to 20 corner variables.

It is often said that GPM come *without* prerequisites, but actually this is <u>not</u> completely true, usually some internal so-called hyper parameters have to be tweak for an optimum fit (in advanced GPMs it is not only the exponent p). However, indeed the GPM assumptions are very mild ones, i.e. just having (tweaked) Gaussian error distributions; so there are usually applicable with very low risks.

With splines we more or less let the data "speak", and we "only" *interpolate* when we talk about range parameters x_R. In this aspect GPM is very similar, and therefore both work quite well in the circuit design context! This is no big surprise, because many similar algorithms have been even developed

Samples

x	f(x)
0.00	0.00
1.00	0.62
2.00	0.97
3.00	0.91
4.00	0.46
5.00	-0.19
6.00	-0.76
7.00	-1.00
8.00	-0.81
9.00	-0.28
10.00	0.37

Interpolation from data

x	Kriging, Beta = 1.5	Akima	f	Taylor3rd	CubSpline
0.00	0.0000	0.0000	0	0	0
0.10	0.0528	0.0735	0.066617	0.066617	0.066542
0.20	0.1119	0.1439	0.132939	0.132938	0.132799
0.30	0.1741	0.2115	0.198669	0.198667	0.198485
0.40	0.2380	0.2763	0.263517	0.263606	0.263317
0.50	0.3029	0.3386	0.327195	0.32716	0.327007
0.60	0.3682	0.3986	0.389418	0.389333	0.389272
0.70	0.4332	0.4563	0.449912	0.449728	0.449826
0.80	0.4973	0.5121	0.508407	0.508049	0.508383
0.90	0.5597	0.5661	0.564642	0.564	0.56466
1.00	0.6184	0.6184	0.61837	0.617284	0.61837
1.10	0.6665	0.6684	0.66935	0.667605	0.669249
1.20	0.7053	0.7052	0.717356	0.714667	0.717113
1.30	0.7539	0.7598	0.762175	0.758173	0.7618
1.40	0.7909	0.7993	0.803608	0.797827	0.803146
1.50	0.8278	0.8364	0.841471	0.833333	0.840988
1.60	0.8625	0.8703	0.875595	0.864395	0.875164
1.70	0.8948	0.9008	0.905829	0.890716	0.905609
1.80	0.9244	0.9280	0.932039	0.912	0.931862
1.90	0.9507	0.9507	0.954108	0.927951	0.95406
2.00	0.9779	0.9779	0.971938	0.938272	0.971938
2.10	0.9811	0.9811	0.98545	0.942667	0.985366
2.20	0.9852	0.9936	0.994583	0.94084	0.994342
2.30	0.9958	1.0062	0.999298	0.932494	0.998893
2.40	0.9934	1.0090	0.999574	0.917333	0.999048
2.50	0.9794	1.0047	0.995408	0.895062	0.994838
2.60	0.9707	0.9963	0.98682	0.865383	0.986289
2.70	0.9604	0.9826	0.973848	0.828	0.973432
2.80	0.9472	0.9636	0.956649	0.782617	0.956295
2.90	0.9307	0.9392	0.935	0.728938	0.934907
3.00	0.9093	0.9093	0.909297	0.666667	0.909297
3.10	0.8766	0.8756	0.879555	0.595606	0.879524
3.20	0.8393	0.8400	0.845905	0.51516	0.845766
3.30	0.7988	0.8022	0.808496	0.425333	0.808231
3.40	0.7559	0.7620	0.767496	0.325728	0.767129

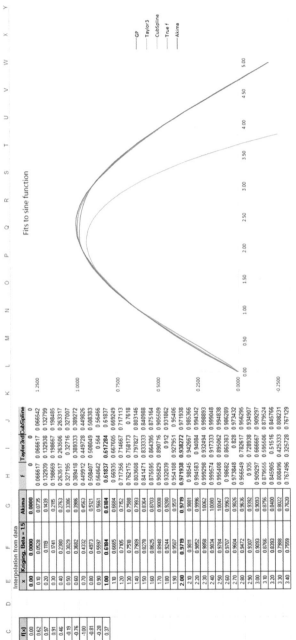

Fits to sine function

Legend: GP, Taylor 3, CubSpline, True f, Akima

Figure 2.11 Comparison different fitting methods on the sine function.

since years for problems which are even *more* difficult than most IC blocks, e.g., for geostatistical applications or earth modeling. Almost any kind of nonlinearity and number of parameters can be treated with such (almost) non-parametric methods.

So-called kernel density estimation (KDE) is a further example for such a "parameter-free" method (see Chapter 3), but actually any method comes at least with some assumptions. One tricky part KDE is the "bandwidth" setting, the setting on how smooth the fitting result should be; and for GPM there are similar issues. In low dimensions, GPMs are harder to calculate than splines, but for more complex corner setups this situation reverses. Generally runtime is not a big issue anymore; and usually the circuit simulation times are still much larger. Several benchmarks are available on different GPM algorithms, in [Guerra-Gomez2015] you can even find a circuit-related benchmark. In the examples, the rms error was in the order of 2%–10% for the better modeling algorithms. This sounds good, because the rms error at the training points would be even almost zero, so we talk about the error in the testing points. On the other hand, the local peak errors might be much larger. In GPM we can at best say that such cases are very rare, statistically. Typically the model *creation* times are in the order of seconds to minutes, whereas the model *evaluation* times (just executing Equation (2.1)) are much shorter.

In theory, you can always construct extremely difficult cases where only full-factorial would be reliable, but modern mathematical algorithms almost act with almost proven efficiency and high reliability; actually, as these methods can detect the internal errors (Figure 2.12), they will simply switch back to full-factorial in such rare cases, but keeping good speed in most real cases!

Note: The error estimate $\varepsilon(x)$ is zero at the sampling points x_i, and ε is quite large in regions with low sampling density. In addition, ε is rising quickly in *extrapolation* regions, so GPM's offer directly what designers would also assume. To some degree, the GPM error estimation is similar to the error estimation with Taylor polynomials, so it is not really magic that such error estimations are possible. If the data itself has uncertainties (right Figure 2.12), its variance V can be just add to the model tolerance ε^2.

What does this all mean to circuit design? Indeed, such adaptive "pure mathematical" algorithms are in competition to clever "heuristic" methods, like expected WC or parameter rankings to the setup and following a certain "strategy". As usual, with a clever testbench setup, more speed is possible too, e.g., if we divide our tests into fast and time-consuming ones, we could run

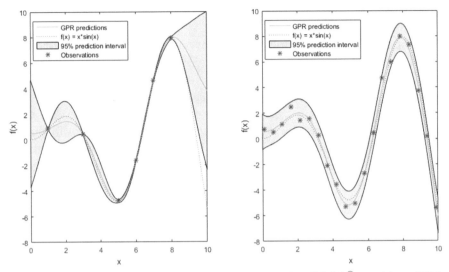

Figure 2.12 GPM applied to non-noisy and non-noisy data (Matlab® tutorial on GPM) [Matlab].

a fast method for the slow tests, but use a more intensive search method for the fast ones. However, in conclusion, the WC corner searching problem in $x_\mathrm{R} = [x_\mathrm{Rmin}, x_\mathrm{Rmax}]$ can be regarded as almost solved. Doing it manually can be a time-consuming and boring work, and unfortunately, it is needed from time to time after circuit changes or for testbench extensions. So let the computer do it for you! In [Woods2015] you can find a further benchmark on mathematical examples, and (more important) detailed descriptions for combining sampling methods with parameter screening and modeling techniques.

As mentioned, finding the worst-case among <u>statistical</u> variables x_S for given yield is important as well, but it is significantly more difficult. So let us soon inspect the statistical behavior of designs in more detail (Chapters 3 to 7). In Chapter 9, we will come back to the worst-case topic again, when we combine it with optimization techniques. Some basic local optimization techniques look similar to our different OFAT variants, whereas global optimizers often have elements of a full factorial analysis, but for efficient circuit optimization we will improve the algorithms further.

Too much praise for mathematical WCD search methods? Check out our little men versus machine showdown in subsection 2.8.2.

2.5 Monte Carlo and Mismatch

Even many corner simulations do <u>not</u> cover the full worst-case condition and also give usually <u>no</u> good yield indication. This is because there are not only variations in *global* parameters (like temperature or R_{sheet}), but also parameters could vary from *instance to instance* (like the two transistors in a differential pair, e.g., on threshold voltage). In digital designs with older technologies, typically the process variations dominate, but in analog designs and when using ultra-deep submicron technologies, also the mismatch becomes very critical. Due to this and lower supply voltages, all kinds of circuit design become more and more difficult. *Even* for a perfect layout, you have to accept a certain *mismatch*, unfortunately; actually, the statistical models even assume that you indeed strictly follow best-practice layout rules [Hastings]. Physical reasons for such statistical variations are variations in channel doping concentration, gate oxide thickness, or line roughness.

Putting all global and all instance parameters together, you often have thousands or more parameters in a single real-world block. Simulating all parameter combinations is usually simply impossible, so a 100% verification becomes extremely hard. Usually, statistical methods jump in here! The easiest and also most general form of statistical analysis is a Monte Carlo simulation. MC requires a statistical model for each variable, and based on that, random samples will be created and each parameter set will be simulated. Actually, this is similar to a corner simulation, just the parameter changes are now random in MC. MC is one method to capture the performance dispersions from statistical parameter variations. One key problem with MC is that all results depend on chance, so finding the true parameters, like the sigma σ of a normal Gaussian distribution from MC data, can be tricky. A typical situation comparing *multiple* MC runs is shown in Figure 2.13; for a *large* MC count, we can expect that most MC results will be in a very *small* tolerance region, hopefully close to the true value. However, for lower MC counts, the variations are wider, so our estimation is more uncertain. On top of that there is even no guarantee at all that the average from MC runs with a moderate count is really identical to the one taken from a huge MC runs! There could be a general bias and a bias for finite samples. A famous example (checkout Google) is that $1/(n-1)\sum(x-x_m)^2$ (with sample mean x_m) is a so-called <u>unbiased</u> estimator for the variance V, but actually $\sqrt{(1/(n-1)\sum(x-x_m)^2)}$ is <u>no</u> such "beautiful" unbiased estimator for standard deviation σ! In addition, it is seldom that your distributions look so nicely bell-shaped and symmetrical; they could be even discrete and highly asymmetric!

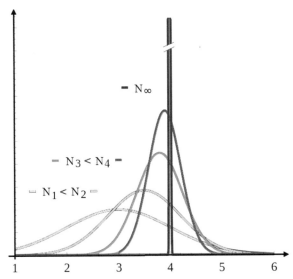

Figure 2.13 Typical behavior of MC estimates for different MC counts N.

For luck, as engineer, you can often relax and accept an error of 10% in sigma (like you do not care much whether the offset standard deviation is 5 mV or 5.5 mV); often such errors are on that level already for moderate MC counts (like n = 100). However, there are several examples where you need a much higher precision (like for verifying a yield loss on 0.01%) and/or even with 1000 samples, the bias is well above several percentages. In Chapter 3, we care more about the foundations of Monte Carlo and statistics.

Both process and mismatch parameters are basically statistical parameters x_S. And the verification has to run on these plus the usual operational range parameters x_R (like temperature and supply voltage). However, in older IC technologies, usually the technology parameters are often split into process parameters treated by corners only, and mismatch parameters treated by MC. So to limit the modeling effort and to speed-up the verification, the foundries supplied full technology corner sets like "nominal," "fast," "slow," or "slowNMOS-fastPMOS" to address the worst-case on global process parameters. For this reason, it would be (in principle) enough to run a PVT corner analysis <u>and</u> a mismatch-only MC analysis to take mismatch into account. Luckily, mismatch is often quite independent from PVT parameters, so often it is good enough if you run the MC mismatch analysis at nominal PVT conditions.

A more severe problem is that usually the foundry-provided process corners often do not fit well for analog designs. In fact, they are only composed to address the worst-case speed for CMOS logic cells! From a foundry, you will simply never get a worst-case process set for IP3 or noise figure or phase margin! Even the meaning of FF or SS for analog circuit is not fully clear, e.g., fast NMOS transistors have typically a lower threshold and thinner gate oxide (leading to larger I_{DSAT}), but the thinner oxide would also lead to a larger gate capacitance C if we use such NMOS as a filter capacitor. However, *larger* capacitance means *lower* bandwidth and thus *slow*—but the corner is named "fast"! Similar problems arise in many kinds of circuits, even in logic circuits. In current-mode logic (CML), we use NMOS differential pairs and resistive loads, and the worst-case on gain $g_m R_L$ is for slowNMOS and fastR, so a foundry-provided "worst-case" corner set with fast transistors and fast resistors would not hit the gain worst-case. So you may create your own "expected analog" critical corners to improve the situation, but whatever you do, you would need (quite many) more corners, so (much) more verification time.

In addition, you have to check to which yield such process corners are related, and it could happen that the corners are set to a 5σ limit, but in your design, you may aim for 3σ yield. In such cases, a PVT simulation would stress your design too much, and in other cases, the PVT simulation will still show less variations than the real production!

A solution is only possible if you have access to full *statistical models* for both process and mismatch, as in all modern PDKs. This would enable you to find the correct statistical worst-case sets for *your* circuit and for *your* performances of interest in your operating range!

This way statistical techniques enable full-yield verification, but one pity remains: MC results depend on chance, so in addition to model errors and numerical inaccuracies, we also have to deal with a statistical sampling error. MC has the advantage that statistical variance becomes smaller and usually acceptable if you just increase the MC sample count enough (e.g., to 1,000), and it can provide a direct yield estimate. The pity is that sometimes you may really need quite a large MC count, so a lot of time.

For luck, MC is by far not the only technique to evaluate the statistical behavior of a design. In the remaining chapters, we also talk about more advanced, more complex methods. For special cases, like pure DC analysis, advanced simulators with built-in methods to address mismatch can be used. They work similar to noise analysis or statistical hand calculations – in both

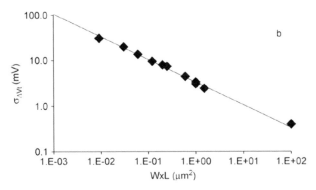

Figure 2.14   Pelgrom's scatter plot obtained from measurements in 90 nm [Tuinhout].

the individual effects add up quadratically, for noise you work with the spectral noise densities and for statistics you work with the standard deviations e.g., described by Pelgrom's area law for mismatch. Such analysis is typically much faster, but often less accurate compared to a full MC analysis.

Pelgrom's law (Figure 2.14):

$$sV_T = k/\sqrt{A}$$
$$\text{with mismatch constant } k[Vm] \tag{2.2}$$

Example: Mismatch calculation by hand
To fit an analog circuit into the supply rails, you need to keep an eye on the MOS transistor saturation voltages. To achieve a certain V_{DSAT} for a given technology and current, you may need a certain minimum W/L ratio of r $= W/L \geq 10$, so $W = 10 \cdot L$ at the limit case. The question is still how large should be W and L, and often you can obtain both according to the desired threshold voltage accuracy like $\sigma V_T \leq 10$ mV and this is related to the device area $A = W \cdot L$ and to the process and device-specific mismatch constant k (e.g., it is highly impacted by the oxide thickness). A typical value is $k = 5$ mVum (quite valid for 180 nm CMOS, modern processes are better due to lower gate oxide thickness, e.g., giving 2.5 mVum, but older processes like 1 um CMOS are worse, e.g., showing 20 mVum). So for $A = 1$ um^2, we get $\sigma V_{TO} = 5$ mV, and with $A_{min} = 0.25$ um$^2 = W \cdot L = 10 \cdot L^2$, we would be at the spec limit of 10 mV; and we can now calculate the desired $L_{opt} = \sqrt{(A_{min}/10)} = 0.158$ um and $W_{opt} = 10\, L_{opt}$.

Note: It looks that already a small MOS transistor can give a good matching, but actually multiple transistors can have an impact on the overall mismatch, and also your spec limit on offset voltage is usually not just one sigma, but for instance 5σ—all this leads quickly to quite large transistors. The consequence? The very newest technologies offer extremely small transistors; this which might be translated to big area savings, but due to the matching problems this advantage is hard to realize in analog blocks. This does not mean that no area improvements are possible anymore, but design becomes harder and "going" digital or mixed-signal is more and more a native choice.

How about finding statistical WC corners from MC? What we can do in several design environments is "picking" a certain MC sample like the most extreme point in a MC analysis (e.g., on offset or supply current) and using it as an *approximate* statistical WC corner. Next, we can even put them into the set of WC corners on range parameters x_R to form a set of approximated WC overall corners. In Chapter 3, we explain MC in much more detail, and in Chapter 7, we put several techniques together to be able to effectively find statistical worst-case corners with well-defined sigma level with good accuracy, so-called worst-case distances (WCD).

Note: If we pick, e.g., after a 100-point MC run, just the worst sample, then such sample might be equivalent to maybe 2.8σ or 3.1σ, so it would depend pretty much on chance. And the chance to get a 5σ sample is very low. To get such extreme sample, we may need millions of points, so pure MC is not efficient for this task.

Mismatch on single transistor or pairs? To measure the mismatch in the laboratory, it is good to use transistor *pairs*, because using pairs the global variations and temperature gradients will cancel out. It is best to use many such pairs for a stable statistic. So many people think mismatch simulation has also to be done on pairs or you even need to "define" pairs of transistor instances. This is not really true! Also for one *single* transistor, you can have a certain mismatch like $\sigma(V_{TO}) = 10\,\text{mV}$. Having two such transistors in a differential pair gives a total sigma that is $\sqrt{2}$ higher, so you have to look up carefully when looking for the concrete values. Almost all tools work with instance-specific mismatch constants, but in your circuit the measured mismatch for a certain output may depend on multiple devices.

2.6 Moving to a Robust Circuit Design

Once you found the overall worst-case on deterministic and statistical variables—with manual methods you typically can only approximately reach it—you can be in different situations like the design passes all specs even at all worst-cases or not. In the latter case, the design should be improved, but how can we do it? You need to minimize the sensitivities to many parameters by carefully choosing the circuit topology and by carefully sizing the circuit. Many techniques can and should be applied in general, not only if the specs are very tight.

From the circuit viewpoint, these are the major techniques for a robust design:

1. Avoid dependence on absolute parameters, e.g., by the use of current mirrors, differential pairs, and feedback.
   Choose suited topology, reduce systematic errors, increase loop gain, etc.
2. Reduce the variations caused by mismatch
   e.g., by increasing the component sizes and by applying dynamic matching techniques.
3. Apply calibration techniques
   e.g., using replica circuits in the bias part and putting VCO in a PLL loop.
4. Reduce variations
   by adding a voltage regulator, by using special bias schemes (e.g., PTAT bias gives near-constant g_m) or by the addition of cascade stages (for better PSRR or CMRR), etc.
5. Use of components with higher stability
   e.g., use certain resistor combinations to cancel TC and reduce statistical variations and use external references (like crystals or high-accuracy resistors).
6. Relax block specs, which may lead to more difficult specs in other blocks
   e.g., often a function can be easily implemented with spending more supply current.
7. To optimize the yield, you should "center" the design
   e.g., try to tweak the design so that spec violations appear with similar probability for competing specs.
8. Even in case of spec failures, the design should be still functional, and small changes in parameters (like temperature) should only cause small variations in circuit behavior (like gain).

Use less risky, well-proven circuits, and avoid "dirty" tricks (like relying on second-order effects as the TC of a certain resistor type).

9. Of course also digital designers have their tricks, like the use of dynamic voltage and frequency scaling, dynamic body bias, and using redundant circuits.

Obviously, some design techniques can be automated by using parameter optimizers (e.g., let the optimizer tweak W & L to find a combination giving a good compromise on speed and matching), whereas others (like using clever mixed-signal techniques) are much more difficult to automate. State-of-the-art EDA tools are suited for parameter optimizations, but true topology optimization is still at an academic level and limited to quite simple blocks.

Fighting Mismatch? The Pelgrom's law puts quite strong constraints on many analog designs. For instance, in a flash ADC, the input capacitance is a critical factor, because 2^{n-1} comparators are connected in parallel at the input. So getting one bit more requires $2\times$ smaller offset voltages, and this leads to $4\times$ more area. For the same speed, we end up in (roughly) $4\times$ more power! What can we do in such situations? And how else can we fight against mismatch, doing more than just spending more area? This is not a book on advanced circuits, and we mentioned already some general techniques. Especially suited for fighting against mismatch are chopping, correlated double-sampling, and dynamic matching. All this is very popular in sensor design, but also switching techniques have their limitations, e.g., due to nonlinear charge injection and kT/C noise. A less well-known method is this: Instead of using two transistors for a diff pair input stage, we could simply insert eight transistors in the layout. In a calibration phase, using switches we can try each pairwise combination (there are 28) and measure the offset. Then, just use the best one! One can easily show that this minimum pairwise offset is significantly better than a 2×4 transistor diff pair! Also the input capacitance is lower, so this technique is not bad for high speed too. The major effort is of course the implementation of switches and the calibration part. Other statistical circuit tricks? Yes, e.g., on-chip resistors may have a tolerance of $\pm 20\%$, but if you combine two resistor types which are statistically independent, then the overall tolerance reduces a bit. The worst-case would be the same, but it has now a lower probability!

2.7 An Efficient General Design Strategy

As mentioned, a simple but extremely time-consuming way to check the design for yield would be to run a large enough Monte Carlo analysis including all testbenches at <u>all</u> (environmental) corner <u>combinations.</u> The huge effort is usually only acceptable for a final sign-off verification or for simple circuits. Even looking at the huge amount of created data requires quite some time to display and to interpret the results, whereas "direct" inspection methods usually give immediately a better understanding.

Efficient <u>design</u> (i.e., search in x_D)—in opposite to pure verification (fix x_D)—requires several simplifying strategies, mainly based on divide and conquer. So let us reinvestigate our op-amp design example and refine the flow further.

At the end, we will see that we obtain a flow with high efficiency, but also with some risk and this will be quantified and minimized in the following chapters.

Let us inspect the extended flow proposal in more detail (Figure 2.15); of course in each step there is some iteration, and also overall, and there is some *overlap* among the tasks. For instance, it makes a lot of sense to include parasitics as soon as possible to your simulations (especially in RF, high-speed or low-power designs), not just when a full layout is already available. This can highly reduce the number of iterations, even if the estimated parasitics are not 100% correct. Also modern layout tools allow us to run LVS checks and DRC much earlier, or offer design rule-driven layout creation. In addition, there is also a lot of work on preparations, e.g., we assume that the design team received a process development kit (typically from the foundry) and has installed all tools, auxiliary libraries, etc. Also making a small testdesign upfront, and running through all phases makes highly sense.

Besides these preparations, also in testbench creation and modeling is truly a <u>big</u> part of the work too! The hand-sizing is usually done by formulas and tools such as Smith chart, math toolboxes, linear equivalent circuits, scattering parameters, CMOS quadratic IV law, filter catalogs, and symbolic calculations. For many standard circuits like LNAs, op-amps, bandgaps, and OTA, you will find several step-by-step instruction guides for design. Parasitics may come from package models, or derived from rough formulas like $L = 1$ nH/mm or microstrip transmission line formulas. More on this topic and related design tools is presented in Chapter 10.

One difference to the basic flow is that we do a split, i.e., we decide for a <u>design</u> phase with focus on speed (using few corners, like 3–10) and

Select suited circuit topology   experience
based on key specs re-use or compose

Do hand calculations for components   know-how
use calculator, special tools

Create testbenches   understanding
DC, AC, transient, noise, etc.
extend and automate step by step

Sweeps   debugging
verify understanding, refine circuit topology and values
check sensitivities and if you have enough margins
add package model, bond wires, pads, neighboring blocks
estimate parasitics on critical nets

MC-MM   know-how
check for, e.g., σ(Voffset) inspect histograms
use other mismatch analysis alternatively
if possible: pick extreme MC samples to create statistical corners
tune circuit to minimize variations

Corners   experience
run at expected worst case + all combinations for critical parameters   systematic
find most critical combinations for critical performances
debug and improve circuit further

Calculate yield form PVT and MC-MM analysis   experience
if possible double-check yield via MC-P and MC-All

Tweak design   understanding
at found WC corner set, best including also statistical corners
you may still need to consider some safety margins

Sign-Off   systematic
Check again the design via full PVT and full MC
Create a documentation and make a review

Layout   experience
placement and routing
LVS, DRC, extract parasitics

Final Sign-off   systematic
check postlayout behavior

Figure 2.15   Typical manual analog flow and main challenges.

understanding (doing sweeps), and we have a <u>sign-off verification</u> step with focus on coverage and reliability. This way we can reduce the simulation effort in the tweaking phase. For instance, the most critical corner in analog designs is often lowVdd+slowMOS+fastR+Tsweep, for stability, it might be fastMOS+lowT+CLsweep, and for speed, it is often slowMOS+highT+maxCL. The sweep phase is done not only on range parameters x_R, but typically also on design parameters x_D (usually a subset of the most important variables). Also model parameters might be swept, e.g., the ground inductance of a package model and the wiring capacitance on critical nets for checking the design robustness. Especially in RF designs, it is not so clear which parameters can be assumed as given and fix (like lead frame inductances) or are actually design parameters.

In the "design" loop, where we modify x_D, we have a lot of iterations. For understanding, it is best to use parameter sweeps intensively. One problem is that by tweaking the design, we can also change the critical corner combinations, so we need an update on them from time to time.

Mismatch (MM) is typically not much corner-dependent, so we can do the MC-MM analysis at typical corner (or expected WC). This analysis should be done early, because in the pure sweeps done before maybe the circuit is optimized too much for speed, so that the transistors might be too small for low offset. One advantage is that often a good interpretation of MC-MM results is possible, and usually 100 points are often enough, so quite a short MC analysis can help a lot in tweaking the design. Unfortunately, a PVT run together with such a MC-MM analysis can only give rough yield estimations. So if available, you should also do a MC analysis with "process only" for additional insights and also a MC analysis with all parameters—on the most critical VT corner. The later can also give good yield estimates, or at least an option to double-check the results.

Many modern design environments allow to save the worst MC sample as a statistical corner. Instead of (or in addition to) only making a hand calculation (like on $\pm 3\sigma$ offset), you can just put such statistical corner into the set of the usual PVT corners and size the design over it. This is a good technique and should be done for the most impacted specs and for those with significant variations from mismatch (like offset or CMRR, but usually not for phase margin or NF).

Note that some additional design margins are needed, because foundry-provided corners like SS, FS, and FF usually do not cover well analog worst-cases like phase margin PM and IP3.

The simulation effort is quite moderate, and both corner and MC analyses can run in parallel on a compute server. This is often only partially possible for more complex algorithms like gradient optimizers or worst-case distance methods (Chapter 7).

A more detailed PVT and MC analysis is usually possible for a fix design. It is usually very time-consuming on the postlayout netlist, so best include expected parasitics as soon as possible and do the postlayout simulations only at typical and few critical corners (like those on speed and stability). The technique is simple: The layout parasitics have usually no strong statistical variation, so usually you get a shift in the absolute offset voltage, but the sigma remains, and the bandwidth gets reduced maybe by some ten percent. With that <u>qualitative</u> *a priori* knowledge and your prelayout result, you could just compose the total postlayout behavior with <u>few</u> postlayout simulations only (in extreme cases just a nominal simulation). By "borrowing" information, you can get quite reliable information with much lower effort. Of course, you cannot do this for effects highly dominated by parasitics, such as cross talk.

Over-all, we need to pamper up our design, so starting the simulation part at nominal conditions is a native choice. Incrementally, add all types of variation (process, voltage, temperature, parasitics, etc.) that may matter to get an understanding. As design tweaking for optimum yield and performance requires many re-simulations, it has to be done in an efficient way with focus on the worst-case combinations.

2.7.1 Desirable Improvements

The described flow can be applied in many commercial design environments for some years, but it has also its limitations. If we do only a PVT and MC-MM analysis, we ignore that foundry-provided corners typically only cover the WC on CMOS speed, but not typical analog measures and circuits!

Options to improve:

- *Extend the foundry corner set, e.g., to include more complex cases like FF_slowRslowC. Use them for complex cases like analog filters. Another example is having mixed MOS corners for thin and thick gate (high-voltage) MOS transistors, using them is often need in pad cells or level-shifters.*

- *Consider PDK and model extensions, e.g., to be able to <u>set</u> the desired yield value (in sigma) for the PVT corners.*
- *Provide corner templates for all designers in the project to ensure a minimum coverage.*

We do a short MC analysis on mismatch only, and usually, the output performances will be normally distributed, but not always. Using ±3σ and combining the PVT and MC-MM results is essentially an <u>extrapolation</u> method with some risks. The risk is often too large for high-sigma yield targets!

Options to improve:

- *Extend the MC analysis to get also high accuracy for the yield estimation; unfortunately, this may require many more MC samples. Check out Chapter 3 for more background information and examples. An easy-to-apply MC speed-up method is low-discrepancy sampling LDS (see Chapter 6).*
- *Even no change in simulation setup is needed, if you use enhanced mathematical methods to address also non-normal cases, as described in Chapter 4.*

If we run a MC analysis and create statistical corners, we cannot know how accurate these corners will be. In addition, to get a 4σ sample, we would also need a very large MC count!

Options to improve:
Consider to create really accurate statistical corners to get rid of the extrapolation risk as shown in Chapter 7. Such statistical corners have the advantage that they can be (almost) accurate for <u>your</u> circuit and performances, so they do not lead to under- or overdesign as the standard foundry corners. Also they can include mismatch.

Such a flow is not only for design and verification; it can also provide additional design insights.

Options to improve:
Apply multivariate analysis to obtain sensitivities from MC results. Chapter 5 describes the methods and many examples, e.g., we can apply a mismatch contribution analysis with minimum overhead on simulation time.
Do correlation analysis to understand trade-off and to better estimate the total yield from the partial yields (Chapter 5).

This flow requires of course some experience and careful setting of design margins.

Options to improve:
To get an overview and concrete numbers, let the design environment do a sensitivity analysis, and this is often easier than doing sweeps manually. Many environments also support corner-dependent spec settings, so we can include design margins if we want.

As mentioned, modeling is better than margins. Modern tools offer support to add estimated parasitics to your design and can also calculate layout-dependent effects and quite accurate parasitics estimates from a partial layout.

Having a complete verification setup is also a perfect preparation for an *automatic* optimization.

Options to improve:
Define which parameters to tweak and let an optimizer do the sizing job. This is more efficient than optimization by plain sweeps (see Chapter 8).

When is a design good? When is a flow optimum? We will care for optimization in Chapters 7 and 8, but with a strong focus on circuits. A general question is indeed what we want to achieve! For instance, if we treat in a layout each individual wire length as important and want to minimize it, we can hardly reach an optimum. At best you can reach something called a Pareto optimum. Such Pareto optimum point is already reached if you have a situation where you can only improve one goal if you get worse on another goal. Obviously, in such situations, a certain Pareto optimum can be far away from your ultimate design goal where it is more important to minimize just the length of the real critical nets or maximize the amplifier bandwidth. Luckily, most optimum conditions are quite *flat* naturally, so spending too much time to exactly reach an optimum makes little sense. So often you need to be pragmatic, like "A design is good if you cannot improve it significantly anymore, if it is in spec, and if your boss likes it."

An optimum flow is also difficult to implement; it is possible only for simple problems like finding the optimum for a parabola. The worst-case might be found by running all parameter combinations and searching for the most extreme result. This is good for small problems, but often it is better to exploit your design know-how and use a slightly riskier but much faster method. In conclusion, your responsibility as designer is mostly on setting the right goals, choosing efficient methods, and improving the design step by step, e.g., starting with ideal current sources, then implementing a real MOS current mirror, adding auxiliary functions, and aligning with other blocks. At best you have always "something that works." For sure, an optimum design execution is almost only possible if you would know the result in advance! So realistically many things depend on anticipation, on experience and on your capability to exploit the information you already have—as good as possible! Plus, you should avoid redundant work or even dead ends.

2.7.2 Mr. Murphy and Mr. Beckmesser

Sometimes specific person stands for something quite specific, like Mr. Giacomo Casanova for womanizers, Robinson Crusoe for a lonesome person disconnected to the world, Mr. Edward Murphy is often made responsible if something gets wrong; and Mr. Sixtus Beckmesser is an annoying person, someone who knows everything better, without really being able to do it. However, in math and engineering, you should have some sympathy to Beckmesser, because we hope in some sentences so far some of the readers get almost a heart attack.

We devided our variables into three catagories, and we one of them is the range variable category x_R, another one the statistical parameters x_S, and last not least the design variables x_D. So far so good, but when explaining corner analysis we but the technology corners like FF or SS into the x_R category. Is this correct, just because SS is obviously neither pat of x_S nor x_D, and because the design environments let us do so?

Let us go for a simple example:
If you pick a sample from production and run the temperature *corners* according to your spec limits, then a sample needs to be in spec for all 3 temperature values to really pass the spec. If all samples pass, we have 100% yield. And if we get a fail e.g., only at T_{min}, the yield would already zero!

However, if we simulate three technology corners, like slow, typ, fast, and we have one fail e.g., at slow. What is then the yield? It is obviously not 100%, but realistically it is also not 0%! If we assume a uniform distribution we may apply the rule of proportion (leading to a kind of margin approach). Or we may simply assign each category to 33%? Or we may simply say that it is simply not possible from a corner analysis to conclude on the yield?!?

Indeed at some point we need to be a bit pragmatic: Following the classical corner approach, we would say, that one fail over x_R could already to a fail. However, realistically this is a bit too conservative, and we may overdesign a bit!

Better than nothing? But what about this: If we say a corner is standing for 5σ, but we want 7σ yield for our block. Are we then still overdesigning? Obviously not really, but what is realistic? Actually the *100% correct* way would be to model the process tolerances parameters as belonging to x_S, so they would require a *statistical distribution*, a pdf. Typically the pdf would be neither uniform, but often also not 100% normal Gaussian! This is because, if the process parameters are too bad, then the foundry would not process the chips further, so these samples would be sorted out! This typically leads to cut distributions, like a normal distribution which pdf is zero, e.g., beyond 5σ (or whatever). All these things may lead to some extra margin for you as designer, but better do not exploit this too much.

In addition, we mentioned that we can cover the statistical variables x_S by statistical models, which are known to the design environment and the simulator. Also this is not 100% true, e.g., also simulation errors e.g., due to rounding, time step limitations, etc. have a statistical nature, but they are not so easy to quantify. Uncertainty and risk quantification is a nice topic on which you should know the basics. Read the next chapters, check out further literature, have a talk with your quality and technology experts.

2.8 Design with Pictures Part One

In the past, graphical methods have been intensively used for design, like the load-line method for power amplifiers or the Smith chart for matching network design. Nowadays, there is a trend to use numerical methods, simulators, and more advanced models, but thinking in pictures often helps a lot to understand the design (e.g., its limitations, like matching network design becomes difficult if you need to start far away from the Smith chart center) and also to understand numerical algorithms (e.g., for yield analysis or for finding a DC operating point).

We will pick up the idea of designing with pictures, because "thinking in pictures" is very helpful also for understanding many statistical methods like worst-case distances (WCD). These have also a straightforward geometrical interpretation! Also normality can be easily checked graphically in a so-called normal quantile plot. For a first design example with pictures let us focus on a power amplifier (PA).

2.8.1 CMOS RF-PA Example

So, what is really important in a radio frequency PA design? Well, it is important that the PA can provide a given output power with a constant gain at a particular frequency range. The application essentially dictates these parameters, i.e., most systems have a predefined power range and frequency allocation that the transmitters should comply. For instance, a Bluetooth transmitter should be able to operate between approximately 2.40 and 2.48 GHz and should not exceed 20 dBm (100 mW) of output power. Usually also a certain minimum output power level is required, because with too small energy no reliable transmission is possible.

In fact, although only one transistor is needed to perform the required DC-to-RF conversion in a PA, the difficulties arise on doing it efficiently and, at the same time, in a linear enough way. There is a classical compromise between power efficiency and linearity, and how much linearity is really required depends also on the modulation scheme. Bluetooth has no very high demands, so usually a class AB amplifier fits. The frequency of operation should not be an issue here as a 45-nm CMOS process node will be used in our simulations. As target, we specify an operating frequency of 1.2 GHz and the 1 dB compression point at 15 dBm. Design for larger output power will often trigger a lot of problems, e.g., thermal problems, gain and stability degradation by ground and package inductances, reliability problems caused by breakdown mechanisms and electromigration, etc. In our starting example we want to avoid having one specification much more difficult to satisfy than others! Table 2.3 summarizes a small list of specifications for the design.

For the design of a class PA, a designer needs to know at least some RF basics, e.g., that an RF choke is used to provide the DC current, which will be modulated by the MOSFET. At the limit of class AB, i.e., class B, the resulting drain current should be *half sinusoidal* (transistor active during $180°$ out of the full $360°$ cycle) with some peak value I_p. So, a first step in assessing the design would be inspecting on the IV characteristics of the transistor and relating them to the output power. For the present example, we will make use

Table 2.3 Specifications for our RF PA design

Symbol	Description	Value	Unit
f_c	Operating center frequency	1.20	GHz
Δf	Bandwidth	>20.0	MHz
P_{1dB}	Compression point	>15	dBm
G	Gain	>30	dB
OIP_3	Third-order interception point	>15	dBm
$V_{DD,nom}$	power supply voltage	2.2	V
η_{pk}	drain efficiency the P_{1dB}	>15.0	%

of external inductors to simplify the design, although the pad models and bond wires need to be considered also.

For an ideal switching transistor (no saturation voltage, i.e. $V_{DSsat} = 0$) the drain-source voltage would be between 0 and $2V_{DD,nom} = 2.2$ V; and the maximum voltage is important for reliability reasons and it will often restrict our choice for a certain transistor type (like thin vs thick gate transistor). A finite V_{DSsat} helps actually a bit on voltage stress, but not much. Modern CMOS technologies have unfortunately quite low breakdown voltages, so usually the question arises if a cascode topology or a single transistor should be used. We can also think if special combinations, like a cascode with thick-gate device at the top and a thin-gate oxide device at the bottom. Such combination guarantees a high gain and can treat large output voltage swings, but has of course a saturation voltage given by the sum of both devices; and a large V_{DSsat} degrades the efficiency. So there are some trade-offs *in each* option that a designer should consider, for instance, thick-gate devices allow you to improve the voltage capability and reliability, but the f_T is reduced because, in general, the minimum length L_{min} is larger than the L_{min} of thin-oxide devices for the same process node. What about increasing L? This would typically improve on output conductance, but for an RF PA it makes little sense, because power, gain, and efficiency matter much more. To find a good design we must optimize each solution as much as possible, or we need to exclude solutions which are impractical. So the designer must know whether a certain issue (like breakdown) is in fact restrictive in the application or if it is merely a minor side effect. For instance, our frequency of operation is relatively low, so using thick-gate transistors cannot be excluded. Also some effects might be not fully covered by the target spec. For instance, isolation is not part of the spec yet, and to improve the isolation between input and output, the designer may adopt a cascode topology, maybe using thin-gate oxide devices only.

Here and in many other design aspects, the simulator can assist the designer on deciding the values of some parameters. Obviously, that needs basic understanding of what is really in play, at least at a qualitative point of view. Sometimes by visual inspection with some rough calculations, one can get a very feasible starting point, leaving the fine-tuning for a later procedure. To determine the transistor dimensions, a designer can perform a quick DC sweep and obtain the i_{DS} of a (now chosen) cascode circuit— Figure 2.16. We assume $L = L_{min}$ to achieve the highest f_T possible and start by guessing a plausible device width based on experience (at least on the order of magnitude).

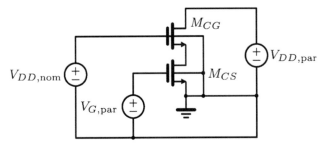

Figure 2.16 Cascode cell for DC sweeping of power supply and gate voltage at the common source.

To find the required i_{DS}, a designer can change the width in a linear way, i.e., supposing it is needed 2.5 times more current, they just increase the width of your device by the same factor. For instance, one may assume the IV of the cascode circuit in Figure 2.16 in which the transistors are equal sized— other possible solutions with unequal sizes may be explored afterward, in optimization.

Figure 2.17 shows the output of the DC simulation. If we consider the knee voltage V_K as the $V_{DS,sat}$ (obtained from the BSIM output), in this case around 600 mV, we can assume $i_{DS,max} \simeq 14$ mA for a class B PA.

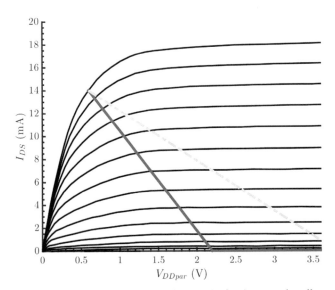

Figure 2.17 DC sweep simulation results for the cascode cell.

Under these conditions, we can get roughly the peak power $P_{\max} = V_{\mathrm{o,pk}} \times I_{\mathrm{o,pk}}/2 = 5.6$ mW(7.5 dBm), with $V_{\mathrm{o,pk}} = V_{\mathrm{DD}} - V_{\mathrm{K}} = 1.6$ V and $I_{\mathrm{o,pk}} = i_{\mathrm{DS,max}}/2 = 7$ mA. This means we can estimate a large signal load $R_{\mathrm{L}} = 1.6$ V/7 mA $\simeq 229$ Ω. Note, that this approach is quite different from small-signal design techniques, required e.g., for low-noise amplifier (LNA) design, where you focus on small-signal S-parameters and noise.

Figure 2.17 shows the draft load line (in dashed black) for the class B at 7.5 dBm output power. If we want to achieve a given peak power $P_{\max}$ that is ρ times higher than the present peak power, we need to improve the current by about ρ times, so we will need ρ MOSFETs in parallel (in both transistors of the cascode) to see whether we can achieve such peak power. The load will need to be reduced accordingly, $R_{\mathrm{L}} = 229$ Ω/ρ. For a peak power of 15 dBm, ρ should be in the order of 6, implying $R_{\mathrm{L}} \simeq 38$ Ω. However, for such a case, the peak power is already at some compression level. So, a better option is to give an extra-margin of $P_{\max}$ to $P_{1\ \mathrm{dB}}$, for instance ρ = 10. Based on these numbers, one can at least roughly predict the drain efficiency (η). Other non-ideal elements will come, but at least to have a reference, one can estimate whether it will be too far from the specifications. Taking into account the knee voltage, $\eta = (\pi/4) \times (V_{\mathrm{DD,nom}} - V_{\mathrm{K}})/V_{\mathrm{DD,nom}} \simeq 57\%$ (with $V_{\mathrm{K}} = 0$ we get the famous theoretical optimum of 78.5%).

One can quickly set up a circuit like in Figure 2.18, which is a simplified approach to validate the targeted power and load, just using an ideal input

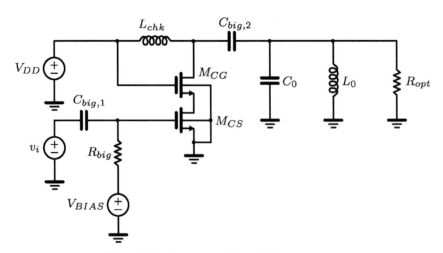

Figure 2.18　Circuit for studying basic parameters.

drive (no matching input matching network). The designer can start building a list of formulas to define some parameters—some may even be kept until the end of the design. For second order parameters we can often apply a rule of thumb. Consider for instance Table 2.4 where we listed several parameter values and respective equations. There we assumed the choke impedance as 50 times larger than the optimum load, whereas for the RF coupling capacitor, the impedance was considered 100 times smaller. Other derivations are at somehow related to specifications, e.g., the loaded quality factor (Q_L) of the output RLC network was assumed as 5. Also here the absolute value does not matter much as long $Q_L \ll Q_{spec} = f_c / \Delta f$.

It is advised to include some non-ideal aspects already from the early start. For instance, the use of finite unloaded quality factors for the inductors gives a more realistic performance (e.g., in output power and efficiency) and also helps the simulator to converge. For the present example, we will assume external inductors, so we can assume intrinsic quality factors (Q_u) in the order of 60 to 80; furthermore, it is not difficult to find inductors with a self-resonating frequency much higher than 1.2 GHz. Hence, for each inductor, one can include at least a series resistor-valued ESR $= \omega_0 L / Q_u$ (the *ESR* is not seen in schematic as a component because it is included in the inductor definition). Another important value to take into account is the bias level of the transistor. At the gate, the voltage should be above V_{T0} to achieve a class AB operation. A DC analysis will be sufficient to obtain this parameter. Note, that

Table 2.4   Parameter definition for studying the performance of the class AB PA

Parameter	Equation	Value	Unit
V_{BIAS}	$> V_{T0}$	575	mV
V_K	–	0.6	V
V_o	$V_{DD} - V_K$	1.6	V
$R_{L,opt}$	$V_o^2 / (2\,P_{max})$	38.0	Ω
Q_L	(arbitrated)	5.0	–
ω_0	$2\pi f_c$	7.540	Grad/s
L_0	$R_{L,opt} / (\omega_0\,Q_L)$	1.0	nH
C_0	$1/(\omega_0^2\,L_0)$	17.4	pF
L_{chk}	$50 \cdot R_{L,opt} / \omega_0$	252.0	nH
C_{big}	$100/(R_{L,opt}\omega_0)$	349.0	nF
R_{big}	(arbitrated)	10	kΩ
$Q_u(L_{chk})$	(arbitrated)	80	–
$Q_u(L_0)$	(arbitrated)	60	–
$ESR(L_{chk})$	$\omega_0\,L_{chk}/Q_u(L_{chk})$	23.7	Ω
$ESR(L_0)$	$\omega_0\,L_0/Q_u(L_0)$	127	mΩ

it also a good method to build up the testbench step by step, to immediately see if one change has a surprisingly large impact (e.g., in the package inductance at the source is not yet included, and it would have some impact on the PA gain).

Once all the parameters have been established, the PA circuit can be simulated using transient analysis or periodic steady-state simulations (PSS). Figure 2.19 shows the results from a single-tone continuous-wave (CW) test using a PSS simulation. It depicts two time-domain representations of the PA signals, one for the current through the channel of the transistor i_{DS} and the output voltage at the load. This time domain allows to see whether everything is as expected, e.g., whether the i_{DS} is nearly a half-sine waveform (50% conduction over the period) and the output is sinusoidal, although the spectral analysis output will provide more accurate information. In the present case, there is some influence of the triode operation, but still if one simulates the circuit with infinite and finite quality factors, one gets an output power of 14.7 and 13.4 dBm, respectively, and an efficiency of 49.5 in the ideal case against 40% with finite Q, i.e., one can identify already some performance degradation. Naturally, the circuit requires some optimization, but at least the designer can predict some rough numbers for a quick start.

Even though one can use transistor width scaling to set your peak power, such does not assure you that a different load value could in fact benefit the circuit performance, so often we require a kind of multi-parameter optimization

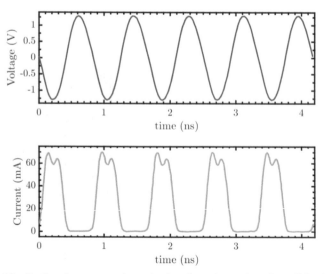

Figure 2.19   Testing the output voltage (*top*) and maximum i_{DS} of amplifier (*bottom*).

for circuits. A typical design technique for PAs is the "load pull" method, in which the complex load impedance is swept in terms of both real and imaginary part. This allows the designer to get an idea of the performance under different loads, and helps finding the optimum load. In discrete implementations, load-pull techniques (based on output tuners to vary the load) are very useful, because we would include all the parasitic elements as they are, such as lead inductance or other parasitics resultant from interconnections. However, when it comes to IC design, the simulator gets a hard job, because we need one simulation for each complex impedance! The post-processing and the plot looks the same in real world and simulation. Figure 2.20 shows the contour plots from a load-pull simulation. In Figure 2.20(a), each contour represents the load values for a given constant power level, whereas Figure 2.20(b) depicts constant PAE contours. In both cases, the lower the radius of the load locus, the higher the power (or the PAE).

As shown in Figure 2.20, to achieve the peak power and peak PAE, completely *different* load values are required, implying a noticeable performance trade-off. Also, the drive strength must be properly analyzed, so that the power gain can be optimized. Source-pull simulations can be also included to this end, but a very complex simulation setup must be implemented (4D sweeps instead of 2D sweeps). Moreover, corner analysis can also provide additional information, but the time required to perform load pull is in fact an issue. In terms of simulation, load- and source-pull techniques are excessively time expensive because the parameter sweep domain is quite vast. If we decide for a certain "compromised load" impedance, we could extend our

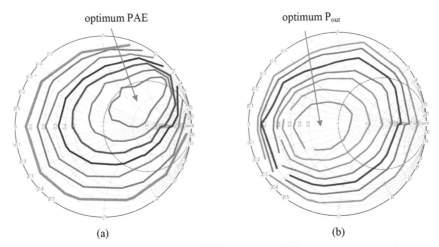

Figure 2.20   Load pull (a) constant PAE contours and (b) constant power contours.

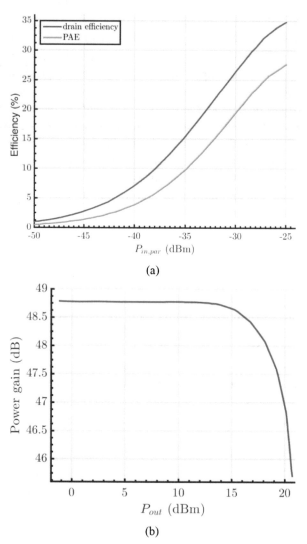

Figure 2.21   Simulations for the nominal conditions—(a) power efficiency and (b) gain.

PA design, and include a matching network, e.g., from 50 Ω standard antenna impedance to this selected impedance. Another design option would be to select a flexible enough matching network topology, and to directly tweak the *component values* in the network (together with other relevant parameters, such as transistor width), instead of tweaking the complex impedances.

After some efforts on manual tuning, a solution arises. Some nominal results such as efficiency and gain, are shown in Figure 2.16.

Nonetheless, when subject to process variations, the performance differs a lot, unfortunately! For the present example, the V_{T0} value depends on the process corner, and if we assume a constant gate bias, in some cases the operation can be class C (for instance in SS corner, because the V_{T0} is higher) or in less-deep class AB (e.g., in FF). So, the signal excursion will be different and will have e.g., some impact on P_{out} and PAE. One can choose SS to establish the starting point with some margin, or give some adequate performance margins, having in mind a yield optimization to be performed later. A sweep with different sizes of the cascode (equal widths always for the common gate and common source transistors), with the gate bias kept fixed will indicate that the number of fingers of the transistor above 20 is required. It turns out that the compression point differs in about 4.5 dB for FF and nominal.

Figure 2.17 depicts Monte Carlo results for a first manual design of the PA for some performance metrics. Although most of the samples fall in the safe side, there is still room for yield improvement. That will eventually sacrifice some performance parameters for others in order to have multiple specifications with higher yield. Such optimization procedure will be addressed in later chapters.

2.8.2 Worst-Case Search Showdown

When we introduced different algorithms, we used a CMOS inverter as example. We could be easily extend it and include further specs, like on leakage current, dynamic average current, static cross-current, area, input capacitance, output resistance, jitter, and DC gain. So even such a simple inverter can have quite many specs and many different WC combinations. And we could easily extend further, e.g., for a CMOS Schmitt trigger, we would have the same problem plus a big interest for the input switching threshold voltage and hysteresis. Also the corner set might be extended for inclusion of generator resistance and load capacitance. Or we may add a second inverter or a level shifter. In the later case, we should add the second supply as further corner variable. In an uplevel shifter, it could easily happen that not the min V_{low} + min V_{hi} case is most critical but min V_{low} + max V_{hi} case; mixed cases are often overlooked.

As we mentioned, sometimes simple circuits like an inverter can be difficult for WC finding. However, we do not want to focus so much on near-digital

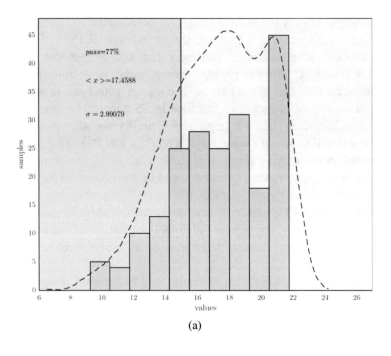

(a)

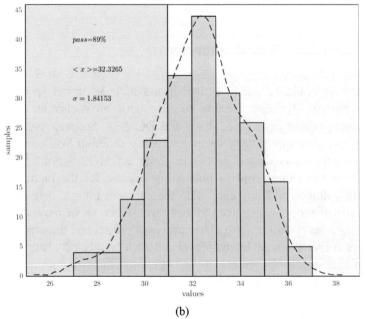

(b)

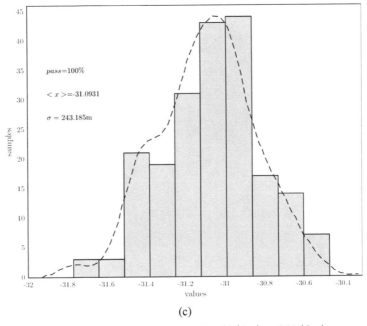

(c)

Figure 2.22 MC results for (a) IP_3, (b) $|S_{21}|$, and (c) $|S_{11}|$.

circuits, so for our showdown on worst-case search methods, we picked a second difficult example. It is a CMOS op-amp, the one which we will use later also for a statistical analysis (see Chapter 4). Actually, it is often not so clear whether "difficulty" is in the circuit complexity or because of other tricky things. For normal WC corner search, OFAT typically fails in roughly 20–60% of the specs, and one tricky performance in amplifiers is often the closed-loop gain peaking, because it is critical to several variables (like load capacitance, frequency compensation elements, etc.) and many dependencies are highly nonlinear. On stability, our dedicated example op-amp circuit itself is indeed tricky, because the frequency compensation scheme of this amplifier works with an advanced pole-zero cancellation to achieve a bandwidth as high as possible (in MC, you can also see issues, e.g., the histogram is becoming bimodal (see Figure 4.14).

To check how good the different methods work, we run the different WC corner search algorithms on this op-amp, for all specs and five corner variables. Table 2.5 shows the results of different methods applied to the closed-loop gain peaking (as very *difficult* measure). As expected, the fast OFAT method is also the least accurate one. Luckily, the user gets at least a warning:

Table 2.5 Worst-case corner search with different methods on op-amp gain peaking

WC method	Peaking	WC combination	#simulations	#simulations in total	Comments
Full-factorial	15.61 dB	C_{max}, $I_{biasmax}$, V_{DDmin}, T_{max}, SF	324	324	Simulations can run in parallel if a large compute server is available
Standard OFAT	10.32 dB (2.93 dB)	C_{max}, $I_{biasnom}$, V_{DDmin}, T_{nom}, SF	14	14	User gets a warning that OFAT solution is inaccurate
Standard OFAT around expected WC	4.8f or 13.3*	C_{max}, $I_{biasmin}$, V_{DDnom}, T_{min}, FF*	14	14	*Result depends also on user entries! User gets a warning that the solution is inaccurate
Preordered OFAT	12.74 dB	C_{max}, $I_{biasmax}$, V_{DDmax}, T_{max}, SF	13	n.a.	Error is in supply, because that was set as first variable
Preordered OFAT with one further iteration	15.61 dB	C_{max}, $I_{biasmax}$, V_{DDmin}, T_{max}, SF	15	n.a.	Refining on the first set variable

Internally, OFAT is just composing the overall WC combination from the individual worst-cases, and using a linear model, it can also estimate the performance value (2.93 dB) for this combination. In a verification step, this can be verified by just executing this composed WC combination: We get 10.32 dB which indicates that the (actually linear) extrapolation was not really accurate. One surprising result is that on this performance also OFAT around an *expected* WC does not work well, although it worked *fine* in our also quite difficult CMOS inverter example! If we e.g., start the search around V_{DDnom}, T_{min}, C_{Lnom}, $I_{biasmax}$, FF, which is a meaningful set for many op-amp designs, we get almost 0 dB as worst-case, which is completely wrong! Of course, the method could also work fine, so for another meaningful start set we get e.g., 13.33 dB, which is now indeed better than the standard OFAT method around nominal. However, although OFAT around an expected WC should work better, also for good theoretical reasons, it does not really provide high robustness.

Table 2.6 shows the corner analysis results (324 corners, nominal process corner NN was only simulated in combination with the other variables at nominal), and based on that (best by using the xls file provided at the River webpage), you can try for yourself to find the worst-cases with your own methods; you can truly double-check why certain methods do not provide the absolute worst-case.

One problem in the gain peaking behavior is that it is simply constant at zero for many corner combinations, because the op-amp is very stable and behaves like a first-order low-pass filter. Another difficulty is that peaking is quite sensitive to small parameter changes for certain parameter regions. All these special characteristics, mathematically equivalent to *high-order* functions, are the reason why OFAT (being not tool-specific) fails significantly and also older automatic search algorithms do not fully end up in the true absolute WC (although being much better than OFAT!). Fur luck, all the other op-amp performances (actually more than ten) indeed caused almost no such severe difficulties.

As a further example let us check in detail our stepwise preordered OFAT search method. The user would need to define which parameters are most critical, so the accuracy depends also on user-provided settings. Let us assume that process corners will be regarded as most important, then temperature, then load capacitance (as we know this op-amp is tricky on frequency compensation), then bias current, and last supply voltage (because analog circuits often have good PSRR). In addition, we can take an estimated WC combination into account, and for op-amp, stability this is usually FF,

Table 2.6   Op-amp corner results to showcase different WC search methods

	C_L/F	I_{ref}/A	V_{DD}/V	Process	$T/°C$	Peaking/dB	t_{rise}/s	I_{DD}/A
Nominal	1p	10u	1.5	NN	27	2.092	1.87E–08	1.43E–04
Spec	–	–	–	–	–	< 3	< 30n	< 250u
Corner-ID								
0	10f	9u	1.5	SS	–40	0	1.56E–08	1.27E–04
1	10f	9u	1.5	SS	27	0	1.77E–08	1.26E–04
2	10f	9u	1.5	SS	100	9.66E–16	1.95E–08	1.25E–04
79	1p	11u	1.7	SS	27	5.07	8.02E–09	1.57E–04
80	1p	11u	1.7	SS	100	6.193	7.48E–09	1.55E–04
81	10f	9u	1.5	FF	–40	0	1.78E–08	1.36E–04
105	10f	11u	1.7	FF	–40	0	1.42E–08	1.72E–04
106	10f	11u	1.7	FF	27	0	1.54E–08	1.69E–04
107	10f	11u	1.7	FF	100	0	1.70E–08	1.66E–04
132	100f	11u	1.7	FF	–40	0	1.42E–08	1.72E–04
133	100f	11u	1.7	FF	27	9.58E–16	1.53E–08	1.69E–04
134	100f	11u	1.7	FF	100	0	1.69E–08	1.66E–04
141	1p	9u	1.7	FF	–40	0	1.62E–08	1.44E–04
142	1p	9u	1.7	FF	27	1.43E–01	1.87E–08	1.40E–04
143	1p	9u	1.7	FF	100	1.315	2.20E–08	1.38E–04
150	1p	10u	1.7	FF	–40	2.91E–15	1.45E–08	1.58E–04
151	1p	10u	1.7	FF	27	3.41E–01	1.71E–08	1.54E–04
152	1p	10u	1.7	FF	100	1.427	2.03E–08	1.52E–04
153	1p	11u	1.5	FF	–40	0	1.56E–08	1.64E–04
154	1p	11u	1.5	FF	27	4.32E–01	1.74E–08	1.62E–04
155	1p	11u	1.5	FF	100	1.727	1.95E–08	1.60E–04
156	1p	11u	1.6	FF	–40	0	1.47E–08	1.68E–04
157	1p	11u	1.6	FF	27	4.06E–01	1.68E–08	1.65E–04
158	1p	11u	1.6	FF	100	1.599	1.94E–08	1.63E–04
159	1p	11u	1.7	FF	–40	0	1.34E–08	1.72E–04
160	1p	11u	1.7	FF	27	6.07E–01	1.54E–08	1.69E–04
161	1p	11u	1.7	FF	100	1.73	1.91E–08	1.66E–04
162	10f	9u	1.5	SF	–40	3.87E–15	1.61E–08	1.33E–04
235	1p	11u	1.5	SF	27	11.88	1.65E–08	1.59E–04
236	1p	11u	1.5	SF	100	15.61	1.80E–08	1.58E–04
237	1p	11u	1.6	SF	–40	8.786	1.43E–08	1.64E–04
238	1p	11u	1.6	SF	27	11.24	1.67E–08	1.62E–04
239	1p	11u	1.6	SF	100	13.33	8.06E–09	1.60E–04
240	1p	11u	1.7	SF	–40	8.802	1.41E–08	1.68E–04
241	1p	11u	1.7	SF	27	9.88	1.68E–08	1.65E–04
242	1p	11u	1.7	SF	100	12.74	7.37E–09	1.62E–04
243	10f	9u	1.5	FS	–40	0	1.70E–08	1.29E–04
322	1p	11u	1.7	FS	27	0	1.14E–08	1.59E–04
323	1p	11u	1.7	FS	100	0	1.19E–08	1.58E–04

maximum I_{bias}, and maximum C_L, whereas for other variables it is hard to say, so keep them at nominal (like in standard OFAT). The search would start now with a V_{DD} sweep (least important variable), with the other parameters at the expected WC. Putting the full-factorial data into Excel and using the filtering option, you can do this by hand (see Table 2.5). In a similar way you can create any WC finder you can think of, and test it! The V_{DD} sweep is picking the points #154, #157, and #160, and the WC on peaking is maximum V_{DD}. With this, we would next sweep on bias current, so running points #142, #151, and #160. The last one is redundant, so we could skip now the simulation that point. We would find now maximum I_{bias} as current WC.

Now, we go for C_L and run points #106, #133, and #160 (again available from cache). As our WCC guess on C_L was correct, still #160 remains the WC, and we continue with a *T* sweep. Here, we run #159 and #161 and get T_{max} as WC, and last, we run the process corners (#80, #242, and #323 as new points). And ultimately, we found #242 and SF as WC giving a peak of 12.74 dB. We used 12 simulations only (speed-up $27\times$), and the result is in between the standard OFAT (10.32 dB) and the true WC (15.61 dB). Looking to the corner combination, we are only wrong regarding supply voltage, whereas OFAT is wrong on two further variables! Actually, also this result is no real surprise, because the V_{DD} WCC was already in the *first* sweep, so it has the highest chance to fail; an obvious improvement would be to re-iterate: Running points #236 and #239 would end up in the correct overall WC of 15.61 dB—in only 14 simulations (speed-up $23\times$)! (Table 2.6).

A further nice experiment is to check how much the result depends on user-provided WC guess. If we assume the same variable ranking, but make no assumptions on WCC, we would get still the correct WCC! Of course, this is no proof, but shows at least to some degree the robustness of the stepwise OFAT method, especially with iteration. The reliability might be further improved at the expense of some speed reduction, by checking for multiple expected WCC or just the expected WCC and nominal settings (Table 2.7).

Overall, the op-amp example nicely shows that it is indeed extremely difficult for a designer to make a *good* educated guess for the worst-cases! Designers typically either follow the OFAT idea (which is wrong on three of five variables!) or rely on experience (which maybe does not help much for new designs and new technologies). In the stepwise preordered OFAT method, the designer can at least partially include his know-how, and we can improve accuracy and speed significantly.

Note, that also this method does not pick up all information being available from simulations: If each performance simulation would run individually,

Table 2.7   Excel screenshot for the last step in ordered OFAT plus refinement

Paramete CL/F	Iref/A	Vcc/V	gpdk09 T/°C		Peaking t
Spec					< 3 <
					fail F
3 PVT_236	1,00E-12	1,10E-05	1,5 SF	100	15,61
5 PVT_239	1,00E-12	1,10E-05	1,6 SF	100	13,33
9 PVT_242	1,00E-12	1,10E-05	1,7 SF	100	12,74

we can hardly improve it further, but usually <u>several</u> performances can be obtained from one test and one simulation, e.g., you can easily get static and dynamic supply current in one simulation. So when looking to the WC for the different specs step by step, e.g., starting with gain-peaking spec, we could <u>use</u> the previously obtained results (in this case from peaking WC search) for the *remaining* WC searches on rise time or supply current! Having this information and putting it into a performance model, we can improve speed and accuracy further. For instance, it makes sense to start the WC search with the most linear and least correlated performance. In that, the risk for finding the wrong WC combination is low, and the gathered information could be used later for the more critical specs—as it was the case for gain peaking in this example op-amp.

These are indeed the principles of work in most modern design environments! They follow an idea called "design of experiments" (DOE); circuit designers would probably call it "testbench setup". DOE aims for gaining a maximum of information for a performance model (relating the inputs, like corner variables x_R, to our outputs f) with *minimum* effort. Actually DOE covers also lab experiments, and e.g., dealing with measurement errors. In computer simulations, nowadays often "replacing" breadboarding, the errors are much smaller, or at least the repeatability is much better, so there is even a new scientific field for such methods, so-called DACE, design and analysis of computer experiments! Here the focus is e.g., more on complexity and space filling, not so much on making stable estimations in the presence of measurement errors.

Several reliable and efficient standard DOE methods exist, and most require only a minimum user input. For instance, note that the stepwise OFAT relies mainly on ranking only, not on true quantitative modeling; so at some point, clever mathematical methods can indeed exploit the obtained simulation results <u>better</u> than humans, and advanced algorithms will usually outperform manual approaches; especially if the problems get really complex (large number of variables, high degree of nonlinearity, strong correlations). Actually, this situation for worst-case finding is a nice similarity to circuit optimization (see Chapter 7)! Also DOE has become a wide field in math, including many more methods than corner analysis and Monte-Carlo. For instance, for linear model fitting we need to run two points, if we *know* the relationship is linear, but to check linearity we need at least three points; and these should be placed *not* too close, if there are influences like numerical noise or nonlinearities. This means better go *early* to the extremes; and this is what designers do since years in corner analyses! Another classical result is that for a polynomial fit over a fix interval the optimum point placement is often related to Chebyshev polynoms. So to some degree DOE and math can also help designers in testbench design; and some further DOE results will be presented in the chapter on Monte-Carlo sampling methods, where we go *beyond* pure random MC. In Chapter 6 we will see that DOE is also useful for statistics, but note, there is no free lunch: If we optimize a set of points for a certain task, like modeling of a second order system, then we may have to make some trade-offs regarding other goals, like yield analysis.

Note that unfortunately any adaptive or stepwise ordered approach will take overall, for many specs, more points than standard OFAT. Also the OFAT speed-advantages will reduce if we have many performances, and if want to hit their worst-cases. This is just because for different performances we will usually find different WC combinations. This is obvious, but it does happen quite slowly, e.g., in realistic corner setup we can easily have hundreds of corner combinations, so even if we have 30 specs and maybe 20 different WC combinations this "fill-in" effect will be quite moderate.

In OFAT you actually make a (linear) model around a center point, and some worst-cases might be far away from that. This leads to difficulties because for the modeling of *large* deviations (starting from the center) you really need high-order models, and all kinds of models for this task have a larger "extrapolation" risk than a model which starts already a point close to the worst-case. Also here, we will find similarities in the following chapters, like when comparing response surface models and the WCD methodology (Chapter 7). The most advanced methods are adaptive, so the previously obtained simulation results are used to make decisions for the next simulations. This gives many more options for better speed and accuracy, like you have more design options when using feedback techniques or other clever circuit design concepts!

2.9 Questions and Answers

1. Compare your typical design approach with the one shown in Figures 2.1, 2.4 and 2.9. In which parts do you spend most of the time? Where is reuse possible?

   The answer to these questions depend probably highly on block type and technology. Sometimes 50% is in testbenches and finding a topology and sometimes 90% in design tweaking an almost fix topology, and you need a lot of experience to quantify this upfront. Often there is a trade-off, like you can set up complex testbenches more efficiently by first running them with Verilog-A models, but the model creation also takes some time.

2. We mentioned that verification is only a subproblem of design: Which design is better suited for operating from $0°C$ to $85°C$: one that works till $105°C$ and then fails completely for strange reasons, or one that works fine till $95°C$ and then slightly leaving spec, but still being acceptable till $175°C$?

   Obviously, pure verification at corners is not enough, e.g., if you run your simulations from $0°C$ to $85°C$, you would never ever have found a problem!

3. Discuss the typical analog design trade-offs on different examples like ADC, op-amp, or just a single common-source stage!

   Look at subchapter on biasing and transistor sizing.

4. If you see a certain mismatch in a MC analysis, can you improve it with a better layout?

   Usually not, the mismatch constants are already obtained from an almost perfect layout, using best practices like dummies and common-centroid placement! So you can hardly improve it further.

5. Can you "beat" Pelgrom's law on mismatch?

   Yes, with a DAC and a digital calibration unit, you can compensate any offset. Or you can apply switched capacitor or dynamic matching techniques! A nice new method is "combinatorial matching," e.g., split your diff pair into 2x4 subtransistors, using all 2x4 devices gives you the usual $2\times$ offset improvement, but using the best 3-tuple out of four gives an even lower offset (like 20% beyond Pelgrom)! Extending this to 8 out of 12 can give you an even larger improvement rate!

6. How much speed-up can you expect by using modern adaptive worst-case corner finding methods?

The more complex the problem, the more room for improvements, so speed-up depends mainly on the number of variables and the number of points for each variable, but also on the variable and performance behavior and on correlations. OFAT is the fastest and riskiest strategy, and it is often 100× faster than full-factorial; so adaptive algorithms give typically a speed-up against full-factorial of 2× to 20×. High nonlinearity can slow down advanced methods, so avoid. binary outputs (they are also bad regarding optimization (see Chapter 7).

7. Discuss design trade-off in different circuits like op-amps, active filters, or ADCs. Which trade-offs are very hard, for which you may find a workaround typically?

   Usually things related to power, noise, and speed have hard physical limits, but technologies with higher breakdown voltages and mobilities would still give an improvement. Search in IEEE papers for figure of merits!

8. Discuss the transistor sizing approach for our RF PA design. Which sizing criteria make sense, which not?

   If your technology is slow, and you have to go to the limits, then probably inspecting f_T is the best starting point. Make the transistor large enough a achieve an on-resistor low enough for good efficiency, but making the width larger, will often lower the f_T, so you need to find a good compromise. As w_T is g_m/C_{in} this type of sizing is also highly connected to capacitance and g_m based biasing; and, as mentioned, the starting point is clearly R_{on} based sizing. Flicker noise or mismatch can be often ignored, but mismatch has an impact on the design of the bias network, at least if no power control loop is available.

9. Try to visualize and understand the different corner setting methods. Checkout literature on design of experiments (DOE). Are there (beside OFAT) designs which show no exponential rise in number of points regarding the corner variables?

   If you need to model a polynomial behavior, then the number of corner points should rise with the number of coefficients, so quadratic for 2nd order functions. One design doing so is the Box-Benken design (Figure 2.23). Here we create block-wise variable combinations and have indeed a quadratic effort.

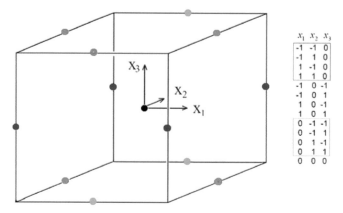

	x_1	x_2	x_3
	-1	-1	0
	-1	1	0
	1	-1	0
	1	1	0
	-1	0	-1
	-1	0	1
	1	0	-1
	1	0	1
	0	-1	-1
	0	-1	1
	0	1	-1
	0	1	1
	0	0	0

Figure 2.23　Box-Benken point set for three variables.

2.10 Rules for Corner Analysis

We described typical manual semi-automated design flow approaches. Later, we will pick up many of these ideas, and in the next chapters, we dig into the details, because also the simple-looking individual tasks like MC analysis have different faces and are far from trivial.

Rules for Corner Analysis:

- Remember the limitations that a corner analysis cannot really give a yield estimate. It cannot treat mismatch, and it cannot replace statistical analysis like Monte Carlo.
- Try to get a full overview of circuit performances and influencing parameters ASAP. Solve problems step by step, best starting with the most critical ones.
- Use OFAT sweeps for circuit understanding and debugging, but also consider sweeps with the remaining other parameters <u>not</u> at their nominal values, but at the real critical ones.
- Consider using sweep features directly offered by simulator analysis; this reduces the netlisting overhead and sometimes it also leads to better convergence.
- Do not forget variables to sweep, and use enough values, especially for difficult variables like temperature. Note that still such one-dimensional sweeps do not cover correlations! For this, a more detailed corner analysis, beyond OFAT, is required.

- Also sweep circuit parameters to check how much you can influence the circuit performances. Also check for variations in external components, like SMD elements. Sometimes (like for AC or DC performances) you can use also the simulator sensitivity analysis for this task.
- Make the ranges extreme enough to really let the circuit <u>fail</u>. Try to understand <u>why</u> the circuit stops to work, like "transistor N8 goes out of saturation."
- Make sure that all specs are fully understood and that all designers in the team use the same (minimum) ranges on temperature, supply voltage, etc.
- Include known important worst-case combinations ASAP, like $lowV_{DD}$+SlowMOS+fastR which are often most critical for saturation.
- In case of convergence problems, consider a transient analysis and sweep the parameter over time (like temperature or supply voltage). Sometimes you need special features to do so, the so-called dynamic parameter or Verilog-A models.
- Make your testbenches as realistic as possible, but step by step, e.g., include neighboring blocks, add a package model, and insert estimated wiring capacitances.

2.11 Summary on Worst-Case Corner Search

In subsection 2.8.2 we described methods which combine designer's knowledge with standard techniques, and we get some improvements on speed and reliability, e.g., demonstrate on a CMOS inverter. However, also these methods can fail, e.g., in difficult cases, like our op-amp gain peaking example. In this example, we have also nicely seen that too *greedy* methods (like OFAT or OFAT around an expected WC) can fail even quite dramatically, whereas modern adaptive methods work (at least) almost satisfactorily. So for the topic of WC finding, tools can outperform designer's anticipation capabilities; that is just why we have them.

On the other hand we have seen that clever tool setup *can* provide big improvements on time and accuracy. So if the circuit simulation runtimes are long, such methods "inspired by manual techniques" make still sense, because in some cases methods *without* any *a-priori* knowledge cannot really compete on speed, even if they are adaptive. Related to simulation effort, we have a linear grows with number of parameters for OFAT methods, whereas the reliable full-factorial method has an exponential effort. For moderately difficult problems, we can expect that advanced, adaptive methods have *roughly* a quadratic behavior; so the advantages over full-factorial become larger the *higher* the *complexity*. This is again a strong argument of just using such methods.

Of course, one can think of performing a much bigger, more representative benchmark [Guerra-Gomez2015], but the result would not be so much different compared to our few examples; and also the criteria weightings might be different. For instance, if your company has a huge compute server, the speed in terms of *number* of simulations of a WC search algorithm would not matter so much. Here, the ability to run the simulations *in parallel* matters more, and in this case, fix (non-adaptive) algorithms like full-factorial have clear advantages. When the problem is *extended* to include also the statistical worst-case (Chapter 7), also big servers will be pushed to their limits—even more with the inclusion of circuit optimization. Only fix strategy algorithms have the advantage of *full* parallelization capability for simulations, so with a huge compute server full-factorial would be even faster than the fanciest adaptive algorithm. However, usually practical reasons prevent using the full-factorial method, e.g., you typically do not want to bother your colleagues too much by taking the exhaustive approach and occupying the full server for a "stupid" block verification.

In Chapter 11 we will give an outlook on further techniques, not yet available in commercial design environments, but the question is usually: Aren't the universal, pragmatic solutions we have already good enough? Or is the problem so difficult to design and to simulate (like finite element simulations) that more specific methods are worth thinking?

For circuit design the already available WC corner methods are indeed fine for all pragmatic engineers. Therefore probably more research and development efforts will go into other directions, for dealing with more complex problems!

PART II

Basic Statistical Design Techniques

3

Classical Monte-Carlo and Data Analysis for Yield

Normalized Yield Loss of MC Analysis

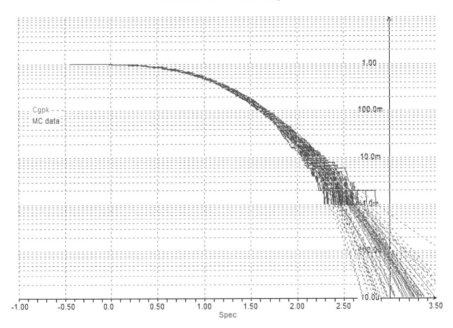

A good statistical method can give a speed-up against running all combinations of parameters. Monte-Carlo is so such a technique and practically the most general one. For this reason, most designers use it since many years, but we also want to give clarifications on which uncertainties are usually involved with the different methods. First we focus on MC estimation methods for single real values, like the partial yield or the mean or standard deviation of a certain performance.

Statistics is often regarded as a boring topic. The statistical theory seems to come with a wild bunch of concepts and special terms. Without computers many people would probably agree on this. However, having a computer enables you to do virtual experiments, and running those many times, till you got the feeling "now the results are indeed quite stable." This way there is no need to do any special calculation: Just setup your problem in a kind of virtual testbench, run it, and collect the data. This way you can build up a very good feeling for statistical data and the uncertainty coming with it.

Unfortunately, too often in real design work it is the other way round: You set up your Monte-Carlo analysis, wait for some hours, inspect the results briefly, and you are done—without much reflection and without detailed and critical result interpretations. Figure 3.1 shows how different even a simple uniform (rectangular) distribution can look like. It also shows nicely how random MC can be. In Chapter 6 we will also check for randomness in higher dimension, with further surprising results. Of course, also in Gaussian distributions there is such randomness *intrinsic* to the *sampling* process!

On the other hand, in spite of runtime problems and special statistical terms, there are good reasons to do problem solving in a statistical way, and surprisingly sometimes MC can be quite efficient! It can be much faster than some people claim and also some speed-up techniques can be applied further— some with very low risks, some with more, some with big speed-up, some with speed-up only in certain cases. We will tell you how to get as much as possible from MC step by step. In Chapters 3–5, we use classical MC techniques available in all circuit simulators, no need for expensive tools!

> Our goal is: *You should get a feeling for statistics, as you have a feeling for circuits!*

This is important, because we have to go beyond pure descriptive statistics, and statements like "The proportion of fail samples is 2% in our current MC run"! "Speed-up" sounds good, but is sometimes risky! Sometimes speed-up methods work straight forward, but in other cases there is no one-to-one comparison possible, because two algorithms come with different prerequisites. All-in-all, we as designers have many options and for sure, in reality you have to deal with uncertainties, probability, and statistics, so in the first chapters we focus on the consequences of this for design verification, e.g., yield analysis. Later we extend this to multi-variate analysis, addressing the correlations between performances and statistical variables. This is a bit more complex but can also lead to deeper design understanding! It is also more difficult but doing it efficiently has triggered the invention of several advanced techniques beyond random Monte-Carlo. Later we also extend the

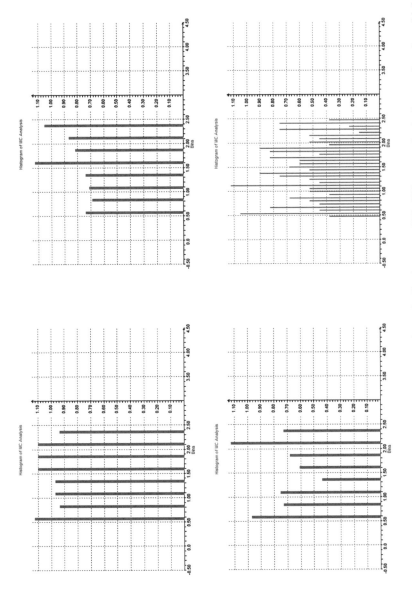

Figure 3.1 Four histograms generated with MC program (uniform distribution, $n = 256$, different seeds, the last one has also a higher bin count—giving less smoothing).

Table 3.1   When to use what regarding basic statistical techniques

Class	Analysis	Limitations	Applications
MC univariate analysis	Classical diagrams (histogram, cumulated histogram, quantile plots)	Your eye has to decide	First inspection if the design works meaningful, debugging
	Sample yield	Need a lot of points for stable statistic	Yield verification
	C_{PK}	Data has to be normal distributed	Yield verification. Use the generalized C_{PK} for non-normal data.
MC multi-variate analysis	Classical diagrams (scatter plots)	Your eye has to decide	Check if parameters are correlated or not
	Correlations, contributions, performance model coefficients	Usually many points needed for stable results	Understand relationships between statistical variables and performances

techniques to support not only yield analysis, but also design optimization for yield improvements.

Some math should not be skipped, so we want to build up intuition about probability density functions (pdf), Monte-Carlo, yield estimation, confidence intervals, etc., but also some terms and measures—less well known in the circuit design community—are also very useful and basically simple, like percentiles, sampling methods, estimates, cumulative distribution function (cdf), worst-case distances, normal quantile plots, etc.

What do we want from numerical algorithms in general and statistical methods in specific?

- Of course speed matters, but usually the time spent is highly dominated by circuit simulation times (including netlisting and extraction of performances) and <u>not</u> by internal runtimes of the statistical parts. This means we need trustable results with a moderate count of simulations. Otherwise we cannot use the methods during the design tweaking phase or in an optimization loop.
- Accuracy matters as well, the results from a MC analysis depend on chance and vary statistically. Usually there is a trade-off between speed and accuracy, but algorithms have also some numerical and systematic errors. Such errors should not increase much for nonlinear problems, at high yields or non-normal distributions.
- For application to complex real-world designs, we also need scalability. Algorithms with strong increase in simulation effort for complex designs

featuring many variables (like $>1{,}000$) are usually quite limited in application.

- We also need robustness, because circuits often work in a highly nonlinear way. Random data can contain outliers, and also simulations can fail due to nonconvergence. To be scalable, efficient, and robust we often need adaptive algorithms and models. For instance, ranking methods are usually much more robust than classical least-square methods, but this comes with the price of lower efficiency.
- The results should be easy to understand and come with an accuracy indication. For instance, trusting results based on strong assumptions (like data is normal) or extrapolation is riskier than results based on mild assumptions (like pdf has finite variance) and interpolation.
- The setup should be easy, and the results should not depend too much on user settings. Bad settings should be reported including suggestions for an improved setup.

Note [Keynes]: We will deal with many formulas and definitions. From school you may remember that probability itself has been often introduced a bit arbitrarily by the axioms of Kolmogorov, similar to the geometry axioms from Euclid! There are meaningful *other* interpretations on what probability or geometry "is," but luckily very often for engineers such philosophical details do not matter much! As we can use in our computer near-ideal random number generators we have almost no limitations, whereas in reality often the concept of probability as a kind of limit "frequency of occurrence" is not so easily applicable, because some unknown parameters change the probabilities over time.

For Further Reading:
Around Monte-Carlo there is a lot of literature. As a starting point, best pick the references related to circuit design, but actually it is very interesting to see also MC working in other fields of science and engineering. *Note*: If you search for "yield estimation" you will often find pure electrical engineering papers, in general or for math literature it is better to search for "percentiles."

- R. E. Walpole, R. H. Myers, S. L. Myers, K. Ye, *Probability & Statistics for Engineers & Scientists*, 9th Edition, Prentice Hall, 2012.
- D. M. Lane, *Online Statistics Education: An Interactive Multimedia Course of Study*, (http://onlinestatbook.com/2/estimation/confidence.html).
- H. Schmid and A. Huber, *Measuring a Small Number of Samples, and the 3σ Fallacy: Shedding Light on Confidence and Error Intervals*, IEEE Solid-State Circuits Magazine, vol. 6, no. 2, pp. 52–58, 2014.

- S. Kotz and N. Johnson, Process Capability Indices, Taylor & Francis, 1993.

3.1 Corners vs. Monte-Carlo

In a corner analysis the design is stressed at well-defined critical parameter combinations, and this is quite a straightforward scheme. However, what is really Monte-Carlo? How general is it? Besides that Monte-Carlo is a city in the south of Europe, we found no single best definition.

This is the nicest statement about Monte-Carlo I ever heard:

One single Monte-Carlo point can tell you more about the circuit, then hundred nominal simulations.

In MC we <u>mimic</u> our real-world system in a computer and use statistical models to include production variations. Even if models are not accurate, MC is useful because we can check our design on robustness. So even one MC sample result is much closer to real world production samples than a nominal simulation, which is actually (partly) an artificial best case (e.g., regarding mismatch)—a kind of Potemkin village. So do not fool yourself and skip doing a MC analysis! A nominal simulation is actually simply not so much "nominal" as you may think! It is more a concept for testing ideas, and to put the design in an almost ideal state. If designers create a real prototype, e.g., on a PCB, they will not create something close to a nominal simulation, they will create <u>one</u> Monte-Carlo sample!

One important measure for robustness is the production yield, but also others (like mean and standard deviation sigma of our output performances) can be of designer's and quality engineer's interest. Therefore, MC has found a huge number of applications, like in weather forecasting, chart analysis, biology, etc.

However, mathematicians have another view on MC; here you find things like "<u>Monte-Carlo integration</u> has this and that characteristics," so basically it works "amazingly well," e.g., completely <u>independent</u> on shape and dimensions and correlations! So MC is integration? The good thing with math is you can really <u>prove</u> something under certain well-defined prerequisites. And indeed the yield can be related to an integral, and we can prove accurate convergence of the sample yield to the true yield. In a wider sense MC is any technique making use of random numbers for solving problems! The problem itself might have no real relation to random numbers, e.g., integration is a perfect example (see Figure 3.2), like also possible with many other methods (like Simson's rule, etc.).

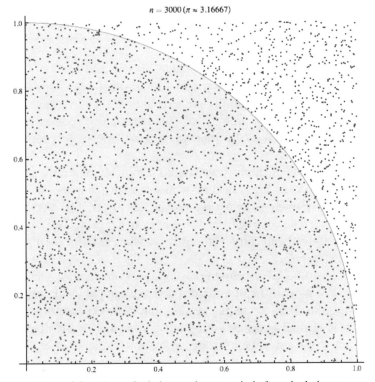

$n = 3000 \, (\pi \approx 3.16667)$

Figure 3.2 Monte-Carlo integration on a circle for calculating π.

Note: The mathematicians seem to "love" integrals and the yield, because there we have proven convergence! You <u>cannot</u> prove that the sample mean, standard deviation, median, etc. will converge in general! In addition: The simplest way of doing MC would be just to run it and look to the performances plots graphically, just to get a feeling how large the performance spreads are, e.g., by placing markers. However, to get histograms you also need an automated performance evaluation (e.g., to get the 3dB-bandwidth BW). For yield analysis you also need specifications (like BW > 20 MHz). These additional entries are almost a prerequisite for all automated methods, so we will not always mention them (Figure 3.2).

One can really prove under very mild assumptions (namely that the samples are identically and independently distributed—i.i.d.) that the MC convergence speed is $1/\sqrt{n}$ for the yield integral! This sounds that MC speed is quite moderate compared to algorithms like Newton-Raphson having quadratic convergence. For example integration by Riemann's sum (giving $1/n$ speed) or even by Simpson rule is significantly faster, but amazingly <u>not</u> if you do that in <u>higher</u> dimensions (each random variable gives one new dimension) as we have in real circuit problems!

Note that the good behavior of Monte-Carlo also in cases of high complexity is a huge advantage over practically all other algorithms! A corner analysis becomes more difficult if you need to treat many variables, but MC yield estimation not! This problem of "dimensionality" comes back to us and to anybody if we want to improve Monte-Carlo or just replace it by something faster!

For its general applicability, we should see in MC not only integration, more something like an art of dealing with statistics in a numerical way. In a computer we have many more options and access to variables than we would have in statistical data coming from a vote or from a fab! And we can also take ideas from analytical and combinatorial approaches to tailor the algorithm to our problem structure.

It looks like MC is inaccurate due to slow convergence, but it can be even worse. For more difficult estimates than the sample yield, we may need many more quite fuzzy prerequisites and maybe we cannot prove that the distribution is normal but only asymptotically normal or we simply cannot easily tell the $1/\sqrt{n}$ convergence starts with a low n like 20 or with a large n like 200,000. In some cases, already simple estimates like the sample mean will <u>not</u> even converge at all (inconsistent estimates)! On the other hand, advanced MC schemes may give a much higher convergence rate, but only under restrictions and it may happen that also the convergence stops at some point, like beyond 20,000!

Luckily MC can often be done this way that some self-checking for accuracy is possible (beside just to make a "golden" run with 1000× more points). A simple way is splitting the data in two equal parts, evaluate them separately, and then compare the results. This kind of cross-correlation analysis is the simplest, not the most efficient one, and many other ways exist. A very crafty method is "bootstrap," which we will discuss in Chapter 5.

As mentioned, behind the scenery MC uses statistical models (see Figure 3.3), so each statistical parameter is described by its probability density function (pdf); in many of the cases as normal Gaussian distribution given by a certain mean μ and standard deviation σ.

However, the designer's real interest is usually in the <u>performance</u> variations, and in between both there is a long often highly nonlinear circuit simulation. So there is usually a kind of curtain between the user getting just the MC results and the true population defined by the statistical models and the often very complex circuit behavior! In MC (but not in a production)

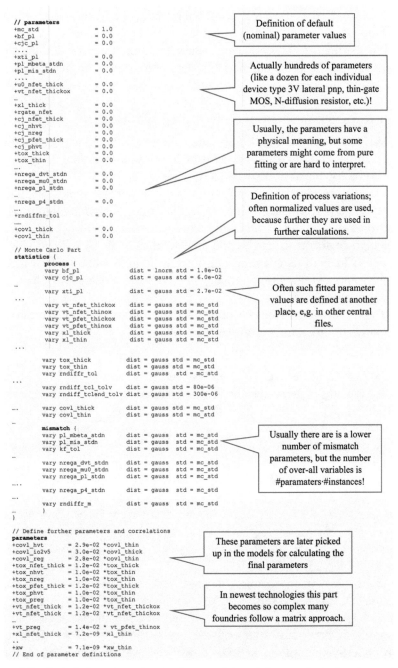

Figure 3.3 Statistical section of a simulator model card for a typical ultra-deep sub-um process.

we can inspect the statistical models and can obtain the <u>exact</u> value of mean and sigma for all statistical parameters, but we cannot do that for the circuit performances. It is not even clear what kind of distribution the circuit performances follow! As circuits often act nonlinear, the originally Gaussian distributions usually appear "distorted," so becoming non-Gaussian at the output!

To give a first summary of corner vs. MC and statistics:

- Verification using foundry-provided corner combined with environmental corners is only leading to a trustable verification if your design is pure CMOS logic and if mismatch has almost no impact!
- Foundry-corner-based verification is inaccurate for typical analog circuits and performances, so it can lead to under-design. It might also lead to over-design, e.g., if the foundry corners are related to 6σ, but you design a high-performance circuit and you are already happy with 2σ yield.
- For yield analysis you need statistical techniques, but by only inspecting the sample yield you need many MC samples, especially for high-yield verification ⊗.
- The corner method may become time-consuming too if you have to cover many variables and performances.

Is "Worst-Case" a precise term? Indeed if something is bad, you can probably make it still even worse, but of course if our design spec is for a certain temperature range like –40–125°C, it makes not much sense add too much further margins. Only some margin can be usually justified due to model inaccuracies. When we talk about WC it is something like a "realistic" WC, i.e., we search for the WC parameter combination <u>within</u> the specified environmental ranges and with a certain (minimum) <u>yield</u>. Pure "digital" WC corners sets like FF, SS, FS, and SF will only represent the speed WC for CMOS logic (for a certain yield like 5σ). To extend the corner idea for analog many PDK model set-ups come with further corners, like slowR, fastR, slowC, and fastC. So in principle including also these and all combinations in a corner verification gives you a further insurance. However, the effort increases a lot and you can still not treat well mismatch and correlations. Also on "sigma" you will typically over-design, when combining the 5σ slowR with 5σ-FF and 5σ-slowC corner. Quantifying the overall sigma of such combination is not easy and would rely on further assumptions. Better directly use statistical methods and use corners more to test design improvements and ideas, from time to time, and as small double-check.

Table 3.2 Overview on corner analysis vs. Monte-Carlo

	Corners Process	MC
Simulation effort for pure CMOS	Low	Medium
Simulation effort for large # of device types	High	Medium
Check timing for full-custom digital designs	Yes	Not efficient
Correct device correlation	No	Yes
Check operating conditions of analog designs	Yes	Yes, but harder to analyze
Check analog performance variability	Inaccurate	Yes
Estimate production yield	No	Yes
Estimate for worst-case performance	Inaccurate	Yield-related
Obtaining process parameter sensitivities	Too inaccurate	Yes[1]
Obtaining parameter & performance correlations	No	Yes[1]

[1] See Chapter 5.

With MC methods or gathering and analyzing measured data, you can almost never <u>prove</u> anything, at best only disprove. For instance, your analysis based on assuming a normal Gaussian distribution might be completely meaningful, but this does not mean that the data is really coming from a normal Gaussian pdf, it might be probably also from a Gaussian mix or from a Gaussian distribution with cut at $\pm 9\sigma$, or from a Student-100, etc. Only if you would fully disprove <u>all</u> such alternatives, you might be able to convince other people that in this case assuming a normal Gaussian is really the only correct method. Typically at some point you have to <u>take</u> the risk of relying on statistical methods, as you also take some risk in relying on models, etc.

Note: We named this subsection "Corners vs. Monte-Carlo," but later (in Chapters 7 and 9) we will see that both methods can be combined, which means we can make the—native and quite fast—corner verification methodology more accurate by fully adapting it to our analog problems and to include mismatch. This way we get better understanding and can also solve difficult problems efficiently like the verification of high yield targets (like 6σ or 1 ppb).

3.2 Questions and Answers: Test Yourself

There are some limitations to MC and also other questions come up, especially as many designers have at least some basic knowledge about statistics which they may remember:

1. What would happen if all our element parameter distributions would be uniform instead of Gaussian? How would that change the histogram of PSRR or I_{DD}?

   It would usually change not much, so the histograms may still look quite Gaussian! This is due to "central limit theorem" CLT. The uniform pdf has a clear cut (roughly at 1.5σ, whereas the normal pdf has no limit), these cuts would almost disappear. For instance, a differential pair could give 3σ maximum offset roughly, because one transistor could be at 1.5σ in the worst-case and the other too, giving 3σ in total. And the <u>more</u> variables are involved the <u>less</u> the cuts have an impact! Already summing e.g., ten uniform variables will give a distribution which is very similar to a true (uncut) normal Gaussian distribution!

2. The "central limit theorem CLT" tells us that if we add the samples from many different statistical variables we will anyway approach the <u>normal</u> distribution to a high degree—even independently on the distribution shape of the <u>original</u> variates! So also a MC analysis on a circuit design should give normal histograms?

   No, because the CLT comes with further restrictions like need for finite sigmas of the individual distributions. Also in circuit design we often do not add up enough statistical variables to obtain really a good approximation; and circuits do not always simply add variables, also multiplication and division appear!

3. Assume the requirements for the CLT are fulfilled, how fast will we approach the normal distribution?

   Also this depends on several factors: If we add samples from a uniform distribution, then a good approximation is often already reached if adding just 10 samples. However, this approximation is usually only good in the distribution center, like μ ±2σ, but not at 5σ!

4. MC is an almost universal method if you are interested in the yield—that is mathematically proven. And another assumption is usually that if we <u>extend</u> the number of MC points, we can <u>always</u> improve accuracy on estimates like the mean.

   Even this assumption is not correct, it is not true for the mean on a Pareto distribution or for the standard deviation of a Student-2! In both cases we would observe that the sample variations grow instead of becoming smaller.

5. Is MC working correctly if we have an <u>infinite</u> number of statistical variables in our design? Can we allow a random number of random variables?
   Good news, random MC will still converge, e.g., the yield estimation would be not impacted at all, but unfortunately many MC extensions run into problems (e.g., LHS and LDS, see Chapter 6).

6. If the distribution of a certain output performance is not Gaussian, can we still <u>make estimations</u> (e.g., on yield) from this MC data?
   Also here MC is flexible and reliable! For instance you may assume another specific distribution (like lognormal or Weibull) or use more general theorems like Chebyshev theorem (which makes no pdf shape assumptions, only the variance has to exist)!

   Chebyshev's theorem states that the proportion of observations falling <u>within</u> k standard deviations σ of the mean μ is at least $1 - (1/k^2)$ (for $k \geq 1$), so the ±3σ area covers at least only 90% (much less than 99.7% for a normal distribution!).

3.3 Important Definitions and Concepts

To prepare a MC analysis, the design environment or the simulator needs access to statistical models (see Figure 3.3). And typically the technology parameters follow a *continuous* probability density function pdf (x), e.g., they show a normal or lognormal behavior but there are also many other well-known distributions and all have their meaning and application. pdf(x)dx gives us the probability P that the random variable is within the interval $(x, x + dx)$, so the pdf is related to the *frequency* of occurrence.

When taking random samples $(X_1, X_2, X_3, \ldots X_n)$ from the pdf we can put the data into a histogram and get a staircase approximation to the probability density function pdf. The cumulated histogram, showing the yield, is giving an approximation to the cdf—the cumulated distribution function (sometimes also called integrated distribution function). This approximation is called the empirical cdf and is a staircase-shape monotonous function, like the cdf starting from y (first sample) $= 0$ to y (last sample) $= 1$.

Figure 3.4 shows the Cauchy distribution, not the normal distribution! This gives an example that it is quite easy to choose the wrong distribution. Actually *many* distributions have a center and tail regions (featuring the rare events causing pain); and look bell-shaped.

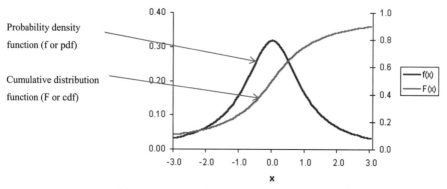

Figure 3.4   Pdf and cdf of a typical continuous distribution.

Also discrete pdfs exist, e.g., the sample yield or dice can only take discrete values. Mathematically the differences often do not matter, e.g., both kinds can be displayed in a histogram, and we use the same formulas for yield estimation, mean, standard deviation, etc. Only the concept of density is more difficult in the discrete case, because mathematically we require Dirac pulse functions; the cdf is a staircase function for discrete distributions (like the empirical cdf, Figure 3.5d.).

Manipulating Histograms? To visualize MC data the histogram is a good starting point, but one big question is how to choose the number of bins. If you want high resolution you need to select a large bin count (like 30 for 200 MC points). This gives quite a noisy histogram with many small peaks, so if you want to demonstrate that your MC data depends highly on chance this is a good method! If you want to demonstrate that you can trust the MC data a lot better use a very small bin count. Actually the optimum number of bins depends on MC count and on distribution type. For normal data and not too small counts, Sturges formula might be used bin = log2(n) + 1, but it smoothes quite much, so you will probably not see if your distribution has two or more modes! Many programs use bin = $\sqrt{n}$ or $2n^{0.33}$ (Rice's rule). Note, that the cumulated histogram has several advantages: You can directly readout the yield and the binning does not matter so much, as even with bin count = n you would still get a monotonic plot, i.e., actually there is no need for binning! A general problem with histograms is that the tail region that dominates the yield is hard to check in the cumulated plot just the deviation from 1.0 counts, and 0.1% is almost impossible to read out.

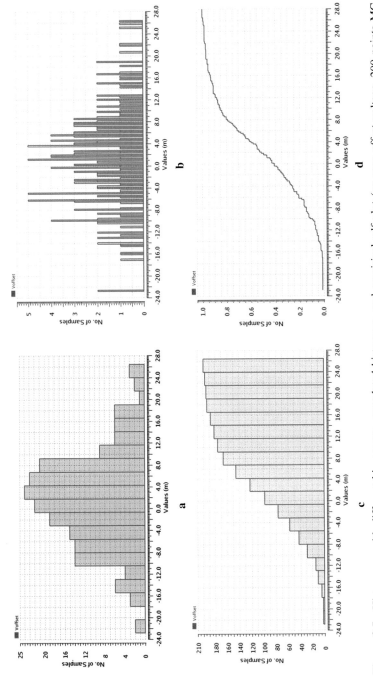

Figure 3.5 Histograms with different bin counts, cumulated histogram, and empirical cdf plot (op-amp offset voltage, 200 points MC run)—tail region marked yellow.

The cdf behaves like the yield, so it starts at $y = 0$, ends at $y = 1$. Often the x-range is from $-\infty$ to ∞, but sometimes it is limited (as for the uniform), to positive values or a certain range.

$$\text{cdf}\,(s) = \int_{-\infty}^{s} \text{pdf}\,\mathrm{d}x \qquad (3.1)$$

The cdf and pdf are programmed into the random generators of the simulator, and the parameters (like mean and standard deviation for a normal distribution) are read from model files. Often the reverse task is required, like you want to hit a certain yield $Y = 90\%$ so the cdf is 0.9, and now you want to search which spec-setting is required to hit this point. This requires the inverse function to the cdf; the cdf^{-1} is usually just called the percentile function. For the uniform distribution, the pdf is constant and the cdf (or cdf^{-1}) is a linear ramp. For a normal Gaussian distribution the cdf or cdf^{-1} is nonlinear, but if we have uniform random variables between 0 and 1 we can generate any other kind of distribution by using the cdf^{-1} as transformation. This is not necessarily the easiest way to practically generate random numbers for a certain wanted distribution (like normal, Cauchy, exponential, lognormal, etc.), but for the-oretical analysis this can be helpful, so later in the chapter about advanced Monte-Carlo sampling methods we focus often on uniform distributions.

Prometheus – Johann Wolfgang von Goethe:

Bedecke deinen Himmel, Zeuss,   *Cover thy spacious heavens,*
Mit Wolkendunst.   *Zeus, With clouds of mist.*

It looks that Zeus followed Prometheus "advise", and covered not only the sky but many other things too. Statistics can be interpreted in different ways, like talking about "frequencies of occurrence" or use it in where we have a "lack of information".

Note, the parameters defining a certain distribution are fix numbers—sometimes known (if you inspect the model setup files), sometimes unknown (if the sampling involves a nonlinear circuit simulation)! We need "tricky" inference techniques to obtain such true fix parameters, if we only have the random samples available, and the accuracy of such *inference* can be quite limited.

Many things rely on modeling—not only MC models—so designers often need to add some extra-margin for this, like make wires wider than needed acc. to electro-migration or IR drop requirements or let circuit work to 20% higher f_{clk} than needed or add 0.25 dB because of some test equipment limitations.

Also, MC analysis requires margins due to confidence interval widths. Fortunately, even if the models are not perfect, using them is the best you can do and they can help to make a design robust against changes (like in T, in V_{DD} or from mismatch). Truly at some points designers have to trust or make a decision how much they trust (like defining a confidence level) or use another algorithm, run more MC points, etc.

Many variables follow a normal Gaussian distribution, and the pdf of the standard normal distribution is given by:

$$N(x, 0, 1) = \frac{1}{\sqrt{2\pi}} e^{-x^2} \tag{3.2}$$

"Standard" means its mean is zero and the standard deviation is unity. The cdf of a normal distribution is related to the error function, which is unfortunately not easy to express in terms of simpler functions; it is just a new function like Bessel functions, etc.

If we have a sample Y from $N(0, 1)$, we can get normal distribution with mean μ and standard deviation σ $N(\mu, \sigma)$ by this linear transformation:

$$Y' = \mu + \sigma Y \tag{3.3}$$

μ is a measure of location and σ a measure of scale.

Moreover, note that this transformation can be applied also for many non-Gaussian distributions, so the concept of location and scale can be used quite generally. Not only the mean is a measure of location, and the standard deviation is not the only measure of scale. For each class of distribution fitting to the concept of location-scale there is an optimum estimator, for instance for the Cauchy distribution the median is stable, but the sample mean would not even converge!

Linear circuits like amplifiers perform the same linear transformation (often unfortunately we often do not know the parameters). So if we look to linear measures in an amplifier (like voltage and current, but not power or level in dB), also the output measures will be normally distributed, just scaled, and shifted. A diode characteristic is often an almost exponential function that would lead to lognormal data. Leakage currents often follow this kind of distribution.

Normal or Gaussian? Should we call the discussed famous distribution the "normal" distribution or "Gaussian" distribution? Normal fits a bit better because also several generalizations of the "normal Gaussian" distribution are called Gaussian as well! In opposite to the normal Gaussian distributions, these feature more parameters, so they are more "flexible," e.g., can also be tweaked to fit for asymmetric or long tail data. The simplest and most important "generalized" Gaussian is the Student's t distribution (so it's not called according to Carl Friedrich Gauss, but another famous statistician who has published a work on it under the pseudonym "Student").

3.4 Expected Values

One major outcome from a MC analysis is getting <u>sample</u> values—usually displayed in a histogram—for the different circuit performances. These samples, either a single MC result or the whole collection, depend on chance, but what the user wants to know are the real distribution parameters behind them. Besides the distribution parameters itself (like mean μ and sigma σ for a normal Gaussian distribution), also other characteristics are very essential and can be accurately calculated if we know the pdf analytically. This is often not the case unfortunately, so we aim for a statistical estimate, which is an estimate of a property of a distribution, calculated from given samples from the distribution. It is quite important to realize that the parameter μ is not something dependent on chance, but a sample estimate like the sample mean m depends on chance, as the whole data set depends on chance.

Let us start with an example: If we roll dices, we are often interested in the <u>expected</u> value E (or mean value or average value), when betting on dices. It can be easily calculated; the pdf is discrete and we assume a fair dice with $\text{pdf}(i) = 1/6$.

$$E\left(X\right) = 1 \cdot 1/6 + 2 \cdot 1/6 + 3 \cdot 1/6 + 4 \cdot 1/6 + 5 \cdot 1/6 + 6 \cdot 1/6 = 3.5 \quad (3.4)$$

We can easily generalize this example to make it applicable for other cases:

Expected value $E\left(X\right) = x_1 P_1 + x_2 P_2 + \dots \text{ or } \int x \cdot p \cdot dx$
Remember: $\int p \cdot dx = 1$
$$\quad (3.5)$$

$$\text{Mean } \mu = \int_{-\infty}^{\infty} xp \, dx = E\left(X\right) \quad (3.6)$$

Lookup: The mean can be calculated for most random distributions in general. The same name μ is often also used for the location parameter for a normal distribution, but there is a function parameter. Also for lognormal distributions we often use a parameter named μ, but in this case it is <u>not</u> identical to the expected value!

Another measure of location is the median or the mid-point, and a measure for the width of a distribution is e.g., the spread or the standard variation σ.

In the general case, the expected value is an integral, and we can express other important definitions by using integrals or expected values:

$$\text{Variance } V = E([X - \mu]^2) =: \int_{-\infty}^{\infty} (X - \mu)^2 \cdot p \cdot \mathrm{d}x \qquad (3.7)$$

$$\text{Standard deviation } \sigma = \sqrt{V} \qquad (3.8)$$

$$\text{Yield } Y = E(\delta\,(x)) = \int \delta\,(x)\,p\mathrm{d}x \quad \text{with} \quad \delta\,(x) = 1 \text{ if circuit pass else } 0 \qquad (3.9)$$

As mentioned, mathematically the overall yield is given as the full parameter space volume integral over the product of the indicator function δ and the joint pdf (note: in almost all real cases we have more than one statistical variable, so the pdf is a function of multiple variables x). For independent random variables the joint pdf is given as the product of the individual pdfs. However, taking correlations into account is actually also easy, because you can usually *decompose* the overall distribution into <u>independent</u> "principal" variables (using so-called principal component analysis PCA), and in fact this is often done in the model files anyway.

The indicator function gives a 1 in the pass (or acceptability) region (the region where all performances are in-spec) and 0 in the fail regions. As the indicator function is 0 in the fail regions, we can alternatively also calculate the yield as volume integral over the joint pdf only over the pass region.

All these measures like mean, standard variation, yield, etc. rely on the true distribution pdf (and their integrals), but as circuit designers we usually do not know the pdf of our outputs and usually we cannot accurately integrate (only finite sums)! Actually, all the formulas look similar, and we can indeed use the same methods for integration, but not all methods converge equally well and a method may work fast and accurate on the mean and variance, but not on the yield. The reason is simply that the pdf is often a smooth and easy to integrate function; and this is often also true for many circuit performances like offset voltage. However, the indicator function needed for yield analysis is nonsmooth, so we can expect more difficulties. So especially for yield and high-yield analysis many special techniques have been developed, whereas for getting the mean or variance just Monte-Carlo integration is often good enough (although not perfect, regarding speed).

3.5 Estimates, Bias Error, and Confidence Intervals

Remember: Usually the real measures of the circuit performances (pdf, mean, variance, etc.) are not available analytically, we can only estimate them from our actual MC result. This means that any estimate (e.g., the mean of the MC data) depends on chance! The circuit is actually doing a mapping from the (element) parameter space (often containing thousands of variables) to the performance space (often a dozen).

Some important estimators for the measures we discussed in the previous chapter are:

$$\text{Sample yield} = n_{\text{pass}}/n \tag{3.10}$$

$$\text{Sample mean } m = 1/n \sum x_i \tag{3.11}$$

$$\text{Sample variance V} = 1/(n-1) \sum (x_i - m)^2 \tag{3.12}$$

$$\text{Sample standard deviation } \sigma = \sqrt{V} \tag{3.13}$$

$$\text{Sample median (50\% point): } \text{cdf}_{\text{emp}}(p_{50}) = 0.5 \tag{3.14}$$

Estimates are not the same as the true distribution values or expected values, actually even different names should be used, like mean μ vs. sample mean μ, but often this is not done due to laziness, unfortunately. Often the laziness comes with small risks only, because s and σ might differ by only 5%, but in other cases, like yield analysis, the differences can be much larger.

The big question is: If mean and sample mean are not the same, how much can we trust such so-called statistical estimates? Actually, we can even use different estimators for the inference on the mean μ, e.g., for a Gaussian distribution the mean and the median are identical, so should we use the sample mean or the sample median for inference on parameter μ?

As an engineer you know it is often not good enough to have only a point estimate, you also need an error estimation. Usually there are two kinds of errors:

1. Uncertainty due to statistical variance

   - Reduce it as much as you want by increasing the number of MC samples

2. Systematic errors (bias)

   - $1/n \sum (x_i - \mu)^2$ is also converging to variance but has finite-sample bias
   - Outliers impact the mean much more than the median

Systematical errors often cannot be reduced so easily by just increasing the MC count. However, if you can indeed run a huge MC analysis you can estimate the error in yield estimation quite well (because the sample yield has no bias error). We can also do many MC analyses and look to the variations in the estimates from one MC analysis to the other. By running many thousands of MC analyses, we can "easily" find out in which interval 90% of the estimates are in, but unfortunately this is very time-consuming. Indeed such *statistical* variations can be treated by so-called confidence intervals; in many cases you can calculate confidence intervals giving a lower and upper confidence bound CI = [LCB, UCB] also <u>without</u> doing such huge repeated MC analysis. However, the user must be aware of the fact that also confidence intervals are derived from the available MC data depending on chance, which means that also confidence intervals depend on chance and are nothing else than estimates [Hoekstra]! In addition, you almost never get 100% confidence that the true measure (yield, mean, standard deviation, etc.) is in a certain range. Such statistical uncertainties lead to the need of a kind of statistical design margin, e.g., even if your sample yield and your yield target is 99%, you still should not fully trust it! What you can trust (more) is the lower confidence bound LCB, which might be only 97%. So actually you <u>need</u> some amount of *over-design*, because only if your design is a bit better giving 99.7% then the lower confidence bound (LCB) might be indeed equal or above your 99% target. So in this case we actually work with 0.7% over-design; how large this statistical over-design margin is depends on the used estimator (like sample yield, C_{PK}, etc.) and the number of MC points. Later, in Chapter 5 when discussing worst-case distances WCD, we will also learn about statistical methods *without* such sampling error—so in theory without need for statistical design margins and so potentially less or even zero over-design.

There are also many other aspects in our inference, like: Can we guarantee that for an almost infinite number of MC points the error will really approach zero, or will there be a remaining bias error? How much will the calculation be impacted by outliers? In many cases the mean is more efficient than the median, but the median is far more robust against outliers!

Truly these aspects are important for EDA software implementations and need careful tweaking. There is simply no best estimator regarding efficiency, bias, robustness, and calculation effort. Only for a certain class of distributions some algorithms may outperform others, but usually you can always provide counter example cases, so only quite complex algorithms are flexible enough to deal with difficult real-world problems. In effect the progress in EDA tools is often in such details, not directly observable by the user.

A MC key problem is that the variance in the estimates is often quite large. More advanced techniques, beyond MC, are typically much better in this aspect, but they may come with more assumptions and if these assumptions are not valid such advanced methods often introduce a significant bias error! For this reason designers using such advanced methods should be clearly aware of <u>which</u> underlying assumptions have been taken and if that is compatible to the design under investigation and the wishes on accuracy.

How to measure "speed-up" and "design efficiency"? EDA vendors are often asked how large efficiency improvements are in a new software version or by using a new feature. In math this is interestingly by far <u>not</u> so easy to tell compared to the use of a faster compute server. "Speed-up" sounds always good, but is sometimes risky! And the risk is sometimes hard or impossible to quantify. In the statistical sense speed-up often means "variance reduction". Lowering the standard deviation by $2\times$ usually translates to the option to use $4\times$ less simulation points, but only if the variance reduction can be achieved *without* adding a systematic (bias) error. In addition, one assumption often used is that we have the "normal" $1/\sqrt{n}$ relationship, which is also <u>not</u> always the case.

Sometimes variance reduction methods work straight forward: If you measure something in lab, you hope that your single measurement value x_1 is close to the real value. Of course doing the measurement again can lead to another value, and e.g., always taking the last value is not a too bad approach, but if noise is present we can expect quite significant variations still. To get a more stable estimate we could take the *average* value; and another approach would be to ignore all extreme values, so taking the *median* value. However, not all cases are simple like this, and not always it is so easy to see (or even calculate) the gain in accuracy, to check for prerequisites, and to clarify the advantages and disadvantages.

In circuit design you can do even more than a clever data *analysis*, e.g., you can inspect the statistical model parameters, you can create clever testbenches and do hand calculations on offsets; or we can run not only random sets for the setting of statistical variables, but set them in a systematic way. Some methods are based on sampling and with confidence intervals, we can tell about accuracy, but we <u>cannot</u> easily quantify if an assumption like "data is Gaussian" is valid! Also we sometimes need to compare statistical and non-statistical methods, and hard error limits like $\epsilon < \epsilon_{max}$ should be treated differently than statements about variance, and the choice of test cases has a big impact on statements about speed-up too.

If one estimate has $1/\sqrt{n}$ convergence and second one has $[\log(n)]^s/n$, then the speed-up of the second one might be impressive, but actually it depends on which settings of n (e.g., representing number of simulations) and s (e.g., number of statistical parameters) is regarded as meaningful. Of course, we hope that we pick realistic values, like designers can effort running n = 500 points, but seldom 500M ones. Also the type of circuits can vary a lot (having s ranging e.g., from 1 to 30 or more). In addition, during the design tweaking phase we may take more risks than for sign-off.

Last not least, sometimes the speed-up is a bit theoretical, e.g., maybe you simply do not really need to know the sigma of an offset voltage with 0.5% accuracy which may require indeed 10,000 simulations using an old standard method; or a new method needs 10 times less simulation points, but it cannot run all these in parallel like the other old-fashioned "slow" method.

3.6 Basic Data Analysis for Normal Gaussian Data

If the data is normally distributed you can make quite easily a detailed data analysis, using many school book techniques, but how to check for normality? The easiest way is an eye inspection of the histogram, which provides a picture approximating the probability density pdf of the performance data. Here the problem arises on how many bins you should set: More bins leads to more noise, but too few bins can make an inspection also difficult. Most difficult is to decide whether the data follows a normal distribution also regarding tail shape, because here you have often only few data points and also it is not easy to decide if a certain curvature is normal Gaussian (i.e., according to e^{-x^2}) or not. So looking to many histogram examples is a good training, but we can do even better.

Indeed, the so-called normal quantile plot solves the problem of curvature inspection, because it shows a kind of transformed cdf plot (actually the x-axis is the inverse normal cdf, also named z-score, and the y-axis is the sorted data).

For normal distributions the data should fit to a straight line in the (normal) quantile plot. Also an interpretation for nonstraight lines is quite easy. For instance if there is a clear lower limit the quantile plot will start horizontally (see Figure 3.6).

If data is long-tailed, the quantile plot gets shaped like arctan, for short tails it would look like hyperbolic tangent (like $I_C(V_{in})$ of a differential pair),

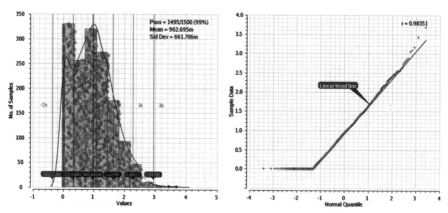

Figure 3.6   (a) Histogram and (b) normal quantile plot from an op-amp – tail region marked in yellow.

for asymmetric (skewed) data we get a kind of J or reversed J shape. For more details, look at Figure 3.7; it also shows the relation to the kurtosis k (normalized central 4th order moment of the distribution, more in Chapter 4 when we discuss non-normal data).

The "trick" is usually how to know if a deviation in the quantile plot is just a random effect or really a systematic non-normality. For this reason

	All but a few points fall on a line	Outliers in the data
	Left end of the pattern is below the line while the right end of the pattern is above the line	Symmetric, long tails at both ends Typically k>3
	Left end of the pattern is above the line while the right end of the pattern is below the line	Symmetric, short tails at both ends Typically k<3
	Curved pattern with slope increasing from left to right	Skewed to right Typically skew>0
	Curved pattern with slope decreasing from left to right	Skewed to left Typically skew<0

Figure 3.7   How to interpret normal quantile plots.

some tools (like RealTime MC) add (sample yield) confidence intervals to the quantile plot, but if the non-normality is only mild, then a decision is usually still difficult. In particular, in the tail regions, a decision is often tough to make because there we have usually only a few samples, so *any* statistic will not be very stable. In Chapter 4 we will discuss non-normal distributions in more detail, and we also demonstrate *numerical* tests for normality (Figure 3.8).

If we are <u>sure</u> that the data follows a normal distribution, <u>then</u> we can calculate also confidence intervals (CI) quite easily, because the normal cdf and pdf is analytically tractable. Due to this, like we can calculate the sample mean directly from the data, we can also calculate CI from the data. Note that we calculate the CI for the sample estimate like sample mean m, not for the fix (but often unknown) distribution parameter μ! Like the sample estimate also the CI depend on chance! We can only expect that in average it will correct, according to its confidence level. If our assumption on normality is violated, then also the CI calculation will get wrong. In such cases confidence intervals from a normal approximation can be at best approximated CI and usually better methods exist (like bootstrap, Chapter 6).

Another method to get a confidence interval would be just doing our MC analysis again and again. For instance to calculate the CI on sample mean m, we can do a MC analysis very often, with different seeds but with the same count n. And we will observe that also the sample mean m will be <u>approximately</u> normally distributed.

In this case the standard error (SE) as the standard deviation of the sampling distribution of a statistic (most commonly of the mean) is given by $SE = s/\sqrt{n}$ (with standard deviation s). And the confidence interval for a confidence level

Load data into first column A, start at column 2. *Sort* data in ascending order.

Label the second column B as *Rank*. Enter ranks, starting with 1 up to n

3rd column C is *Rank Proportion*: C2 = (B2 – 0.5) /COUNT(B$2:B$n) + copy to remaining rows.

4th is column *z-score*. Use normsinv function: D2 = NORMSINV(C2)

Copy 1st column to the 5th column to make chart creation easier

Select the 4th and 5th column. Select chart wizard, then scatter plot.

Figure 3.8 Excel steps to show normal quantile plot.

of 95% will be around the mean with $\pm1.96\text{SE} = \pm1.96s/\sqrt{n}$; so $\pm2s/\sqrt{n}$ is a good rule of thumb for the 95% confidence interval (for the sample mean of a normal distribution)—compare this to Figure 1.11.

Note ☺: Often we are only interested in single-sided specs or single-sided CIs, so only the upper "outliers" degrade the yield, so the risk for being out of desired range like $\geq 2\sigma$ is actually only app. 2.5% not 5%. If you are already happy with lower confidence then the CI becomes narrower, but you take more risk, so usual confidence levels range from 75% to 99%. The higher the confidence level, the lower the risk making a wrong decision, but actually already small deviation in the model can lead to severe errors also in the confidence intervals!

A detailed analysis has been done by William Sealy Gosset on normal distributions. He derived that the confidence interval on the sample mean m is related to the Student's t distribution, and the factor 2 in our formula is slightly a function of the sample count n. In older days engineers used lookup tables, now almost all statistical tools and EDA software directly provide the user the confidence intervals based on Student's t distribution; "Student" was the pseudonym which Mr. Gosset used for his article.

Also the CI for the standard deviation can be calculated analytically, and again we could double-check it by repeated MC (like in Figure 3.9).

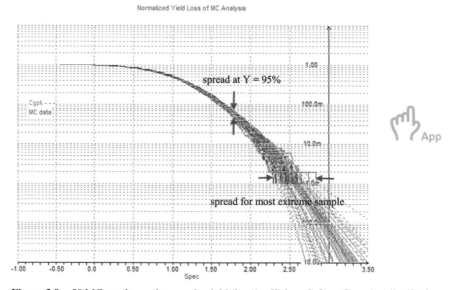

Figure 3.9 50 MC results on the sample yield ($\log(1 - Y)$ in red) for a Gaussian distribution.

The CI for s is related to the chi^2 distribution. As a rule of thumb, for $n = 200$ the 95% CI on s is app. $\pm 10\%$. Actually the distribution is slightly asymmetric, so the real value is $+11/-9\%$. This is no surprise, because upper "outliers" are more likely than lower ones (there will be no negative values).

Note: You can find an incredible number of papers on confidence intervals CI, but most of them are related to the mean. However, circuit designers are usually most interested in the standard deviation, like for offset voltage, or in the yield. CIs on these are a bit more difficult, but usually also available in design environments. In particular, the tail region of a distribution is of high interest and here different CI methods can really give different results. In tail regions like beyond 99% any statistic will rely on quite few samples, so CI become wide, maybe too wide to make design decisions! This is a major motivation for the advanced methods in Chapters 6 and 7.

Functions and Distributions. Most people think of the probability *density* function pdf when talking about distributions like the normal or uniform ones. This is just because the pdf looks like the most important graph, the histogram, giving the *frequency* of occurrence. The histogram is also good for identifying distributions by eye inspection. However, we have seen that for check on normality, for yield estimation or confidence intervals also other functions matter. Maybe go through this chapter again, and look up what function is for what! Often indeed the *cumulated* distribution function cdf (the integral of the pdf) is even more helpful, because it is directly related to yield and to the probability that a variable X falls into a certain interval [a, b]. The reason why the cdf is not so famous is just that it often *looks* not so characteristic, because the integration e.g., smooth out edges, and the point of highest density is much harder to see. Also the *inverse* cdf (the percentile function) matters. For instance, the factor "2" used in our approximation for the 95% confidence interval is coming from the Student-t distribution, but not its pdf but from the inverse cdf. Actually knowing this (and not much more) is already extremely helpful when doing statistics.

3.6.1 The Yield Estimation Problem

We discussed basic estimates and confidence intervals. And one outcome was that for simple estimates like sample mean or standard deviation we usually do *not* need many MC points—often 200 points are enough to decide if offset voltage is small enough, but how many points are needed to verify the *yield* with a certain significance?

Using the <u>sample yield</u> the accuracy (e.g., Y confidence interval) depends on Y itself and on the sample count n. For $\pm 1\%$ absolute accuracy ΔY and a yield of 50% app. 11000 points are needed (confidence level at 95%—quite a typical value), and at 98% yield we need roughly 800 points. Unfortunately also 800 simulations is not that low, and 1% absolute error in yield is not that accurate, because it matters much if your loss is 1% or 3%, or even 0.1% vs. 2.1%! Actually looking to the yield loss or to the error in terms of sigma yield estimation is generally more stable in the distribution center than for tail regions.

Focusing on the *relative* yield loss error we would need to look at $\log(1 - Y)$, and Figure 3.9 is showing this for a repeated MC run. Looking to the spread in y-direction gives quite a native feeling on how "instable" MC results can be (a constant Δy in this plot is related to a certain fix relative error, due to the log y-axis).

Note: In this plot the spec is set to 3.0 giving a true yield of 4σ. Therefore, we almost never get a fail within n = 1024 MC points, so at some point we reach $Y = 1.0$ and $\log(1 - Y)$ becomes infinite (the red curve plot stops there). The green curve is an extrapolation, and in the extrapolation region the variations are even larger.

If your circuit is well designed, then often a short MC run shows <u>no</u> fails, so the sample yield becomes 100%, but of course this is typically too optimistic. To be on the safe side you may ask again for a confidence interval. The yield confidence interval problem has been solved by Clopper-Pearson, resulting in a quite complex formula using the beta function, but <u>if</u> you have <u>no</u> fails, then already the "Rule of Three" is a very good approximation for the 95% CI:

$$\text{CI}(Y) = [1 - 3/n, 1.0] \qquad (3.15)$$

This way we can also easily calculate how many points the MC run should include till we can decide with 95% confidence if the design fulfills a certain desired yield:

$$n \geq 3/(1 - Y_{\text{target}}) \qquad (3.16)$$

Already this simple formula demonstrates very well our verification problems, because it will lead to the fact that high yield verification needs a lot of MC points, usually much more than for obtaining stable values for sample mean and standard deviation of the performance values! Look at Table 3.3 for more

Table 3.3 Verification with sample yield according to the rule of three

Yield in Sigma	True C_{PK}	Single-sided Yield Loss	Number of MC Points (if No Fails)	Comment
0 sigma	0	50%	4	is of low interest
1 sigma	0.33	15.9%	17	is of low interest
2 sigma	0.67	2.3%	130	the minimum realistic yield target
3 sigma	1	0.14%	2200	often used as target
4 sigma	1.33	0.003%	95K	often the limit for pure MC sample yield
5 sigma	1.67	290 ppb	10M	typical for blocks in high-volume chips
6 sigma	2	1 ppb	3G	typical for memory

details, it includes also the process capability index C_{PK} which has a strong connection to the "yield in sigma" (see next Section 3.6.2). Also note that there is no simple reciprocal relation between yield loss (failure rate) and sigma; it is a special nonlinear relation you have to be aware of. For instance from 2σ to 3σ we have roughly to divide by 20 to treat the loss, but from 5σ to 6σ it is already 290 (look also at Table 1.3).

A distribution with a constant failure rate is the exponential distribution, playing a key role in radioactivity. One distribution with such tail behavior but a "Gaussian center" is the so-called logistic distribution; if you want a power law tail instead, you would end up in the Student's *t* distribution. Actually many distributions exists, and all have their meaning and application.

Already these numbers will typically lead to long simulation runs, but if you have indeed failed samples the lower yield confidence limit will be even lower and we need even more MC samples (see Figure 3.10, some more details and further screenshots can be found at [Iastate]), and the convergence rate will be $1/\sqrt{n}$ – not $1/n$ as the simple rule of 3 may suggest!

Note ☹: The are many confidence interval approximations in the mathematical literature. Look up that the interval from a normal approximation can be very bad [Schmid], because the sample yield distribution can be highly non-normal! Do not use it, better use Clopper-Pearson or Agresti-Coull (both are slightly on the pessimistic side compared to the "Rule of Three").

If the design is perfect (like $>6\sigma$, giving almost no fails) and we want to ensure just 3σ only we will need typically app. 2000 points, but if the design is truly only 3.15σ we need app. 16,000 points to make an accurate enough decision based on counting failed samples. If we use the C_{PK} (next chapter) we would need only approximately 1100 points. However, if we allow no

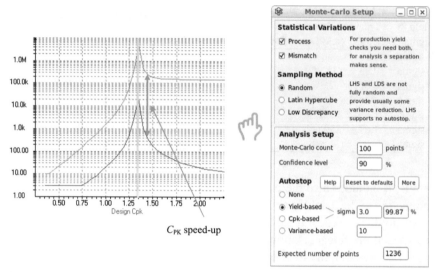

Figure 3.10   Required random MC count to verify a 4σ design based on sample yield vs. C_{PK} [Iastate].

design margin both sample yield and C_{PK} would require an infinite number of points. So we have an over-design vs. speed trade-off.

The Binomial Distribution. Looking to yield means dealing with pass and fail only, i.e., doing investigations on a binomial distribution. Its "accurate" confidence interval has been first calculated by C.J. Clopper and E.S. Pearson, in 1934. Although many laws exists which give the normal Gaussian distribution a strong preference, especially for a large number of samples, it is by far not the only important distribution. If we use the normal approximation to the binomial distribution for the confidence interval, we get the simpler so-called Wald interval. It is easier to calculate and often used, but unfortunately it can be far too optimistic. For instance, having a C_{PK} of 1.5 and a huge set of 100,000 MC points the CI from the normal approximation would be still 3× too optimistic on yield loss, for IC design better forget the Wald interval (even the "Rule of 3" is better in this case).

The reason for the large number of points at high yields is that such yield analysis based on outer samples, the tail samples—and these are rare; any statistics based on them will have quite large variations and will be quite unstable. When looking just to 3σ verification, one may still accept to run 2,000 to 16,000 samples at least for fast-running testbenches, but during the design phase you typically need many such runs, and remember already for a failure rate of 1000 ppb (0.0001% or 4.75σ) the numbers get huge: we need 1,000,000 samples as a real minimum giving us hope just only to observe a fail (no confidence interval included!); having no fails (so the design should be even much better than 4.75σ) the 95% confidence limit would dictate us 3,500,000 samples, and for a design margin like 0.33σ (roughly 5× less loss) we would typically need 6,000,000 MC points, and for less over-design even more!

There are many known attempts to solve this general yield verification problem. One is to go back to the original yield definition and solving the yield integral by other means than MC. For instance, a numerical integration by Riemann's sum would have 1/n convergence rate if we would only have one statistical variable, and Simpsons's rule would be even faster! However, both methods will slow down in higher dimensions – and may become even slower than MC. On the other hand, we will also demonstrate methods which require theoretically even no such "design margin," so coming in theory with almost no need for over-design.

3.6.2 Sample Yield vs. C_{PK}

If we can assume a normal distribution, we can solve the problem of verifying the partial yield also in another way (Figure 3.11). We can calculate a much less quantized estimation by creating a Gaussian fit to the data and we can calculate a yield estimate from this fit by using the cdf of the normal distribution, which is related to the error function. The process capability index method is exactly doing this, and the big advantage of the C_{PK} is that we can even get a realistic yield estimate (so below 100%) if there are no fail samples! This way we can obtain a yield estimate with tighter confidence interval, but the user should be aware of potential systematic errors.

Note: To get the total yield from the different C_{PK}s for each spec we need multi-variate techniques and correlations. This will be discussed in Chapter 5.

The C_{PK} is given as:

$$C_{PK} = |USL - \mu|/3\sigma \quad \text{(for single-sided upper spec limit)}$$
$$C_{PK} = |\mu - LSL|/3\sigma \quad \text{(for single-sided lower spec limit)} \quad (3.17)$$
$$C_{PK} = \min(|\mu - LSL|/3\sigma, |USL - \mu|/3\sigma) \quad \text{(for double-sided specs)}$$

Check if data is close enough to a normal distribution

Approximate data with a normal Gaussian distribution
For this estimate the parameters via sample mean and sample standard deviation

Calculate the C_{PK} value according to spec type and value

Calculate $Y(C_{PK}) = \frac{1}{2} \cdot (1 + \mathrm{erf}(3 C_{PK} / \sqrt{2}))$

Figure 3.11 C_{PK} flow (prerequisite is of course a stable process).

Note: From the mathematical view point the use of the min function is a horrible approach, because there would be no sensitivity to the nondominating spec-side till we reach the balance. A much better approach—also in conjunction with optimization—is to calculate two C_{PK}s for both spec-sides and then calculating back to yield for both. To combine them into one we just have to add the yield loss and can again transform back from yield to C_{PK} via inverse error function! Later we will have a similar problem for the worst-case distance (WCD) approach, and we can solve it similarly.

The C_{PK} formula is actually a kind of normalized spec margin or performance margin approach. The normalization is done in terms of sigma of the output distribution, this sigma and the "yield in sigma" are only the same if we really have a normal Gaussian output (performance) distribution, which often cannot be assured so easily. The whole approach is a continuous one, whereas the sample yield is more a yes/no or 1-bit ADC approach. From circuit design you know 1-bit ADCs have higher quantization noise but no nonlinearity, whereas multi-bit ADC have much better SNR, so the C_{PK} method is more similar to analog or multi-bit ADC style! Also for optimization and debugging avoid binary specs; it is simple waste of information! Sometimes it might be indeed "native" to use a spec like "power-up circuit OK", but in such case better look to start-up time directly and create an "analog" spec!

The C_{PK} measures the relative performance variation (e.g., due to mismatch and process variations); actually by putting the distance of spec limit vs. mean in relation to the standard deviation, for double-sided specs it also takes the spec-centering into account. One practical advantage in using the C_{PK} instead of yield values is that the C_{PK} values are more manageable, e.g., $C_{PK} = 1$ means 0.135% loss and is equivalent to a spec distance of 3σ. A C_{PK} of 2 means 1ppb loss, which is already a very small value. Usually you require at least a C_{PK} beyond unity.

As the C_{PK} depends on μ and σ, we can also easily calculate a C_{PK} variance $V = σ^2$ and the C_{PK} confidence interval (±2σ for 95% confidence).

$$σ \, C_{PK}^2 1/9n + C_{PK}^2/2n \tag{3.18}$$

$$\text{e.g.,} \, σ \, C_{PK} \cong 5\% \text{ at } n = 250 \, \& \, C_{PK} = 1$$

Even for $C_{PK} = 2$ the variance of the C_{PK} (σC_{PK} in Equation (3.18)) is quite small for already moderate MC counts. This means that using the C_{PK} we only need MC counts in the order of hundreds, even for high-yield verification! Whereas yield estimation by sample yield $Y = n_{pass}/n$ would require often billions of points (Figure 3.12).

One may wonder whether the C_{PK} method—equivalent to a Gaussian fit—is the "best" method. In fact, it really is, in some way, but only if it is really sure that the data comes from a normal Gaussian distribution! In this case also the method for determining the two parameters, mean and sigma,

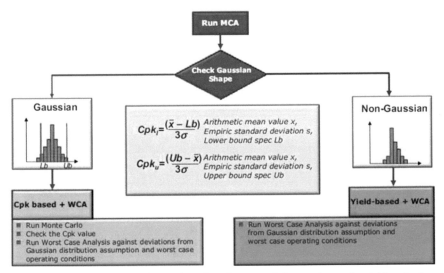

Figure 3.12 Classical split flow using C_{PK} for normal data and sample yield as backup (in addition we can use most advanced methods described in following chapters) (courtesy: MunEDA).

by the well-known estimates sample mean and sample standard deviation is optimum, e.g., more efficient than the use of the median instead of the mean. The underlying theory is fundamentally given by the concept of maximum likelihood (ML). In ML estimation (MLE) we determine the parameters so that the probability to get the given data is maximized. At this point we recommend looking to dedicated mathematical literature, and we also want to mention that although the concept of maximum likelihood sounds so general, there are still some cases in which it should be complemented with further techniques. For instance, ML parameter estimation is often quite sensitive to outliers and often also not the easiest method (e.g., it is hard to apply for a tri-angular or a Cauchy distribution).

The Moment Method. To categorize distributions we can e.g., look to the pdf or to the quantile plot, but we can also do it based on the so-called distribution moments. Look at the equations for mean and variance; these are the first and second moment of the distribution. We can simply extend the idea by using higher exponents. The 3rd moment is called skew s and measures the asymmetry, and the 4th order moment is the kurtosis k, measuring the tail behavior (to some degree). Usually the moment are taken around the mean (central moments), also the higher moments are usually normalized to the standard deviation. This avoids getting too extreme numbers, and makes the "shape" measurement independent from the distribution scale. For instance a Gaussian distribution has a kurtosis of 3.0 and zero skew. In the past, moment fitting was the most commonly chosen method for data fitting, but MLE is even more general (e.g., it can be applied even if the higher moments are infinite, like for the Cauchy distribution).

Table 3.4 lists also other methods for yield estimation, as a non-normal distribution can be non-normal in many different ways there is almost no single best method. Also the result interpretation is usually more difficult than for the normal case: For the C_{PK} we have to deal with the two distribution parameters for location and scale plus the spec limit; for non-normal cases things become more complicated.

3.6.3 Confidence Interval-Based Autostop for MC

The user is interested in MC results having a certain minimum accuracy, which is related to the width of the confidence intervals. As confidence interval

Table 3.4 MC yield estimation techniques

Name	Method	Limitations	Comment
Sample yield	Nonparametric yield estimation	Large variance, so wide CI	Standard method in design environments, no bias error
C_{PK}	Gaussian fit	Only accurate for normal distributions	Standard in QA
Kernel density estimation (KDE)	Kernel density fit (almost nonparametric)	Limited extrapolation capabilities, so hardly suited for high-yield estimation	Difficult setting for smoothing bandwidth, available in math packages
Multi-parameter fit	e.g., [Lange] using generalized lambda distributions	Bad fit e.g., for multimodal distributions or for long tails with cuts	Available in advanced design environments
Nonparametric fit plus tail modeling	e.g., KDE and Pareto fit to tail	Limited accuracy for Gaussian distributions or for long tails with cuts	No easy interpretation of parameters,
Generalized C_{PK}	See Chapter 4 and [Weber]	Limited accuracy e.g., for long tails with cuts	Available in RealTime MC

calculation for the sample yield is quite simple, this has been exploited in many design environments to implement a kind of MC autostop feature. This could reduce the setup effort for the designer, e.g., he/she only has to set a certain minimum and maximum number of points, a certain yield target to be verified, and the simulator "decides" when to stop.

Figure 3.13 show the MC run of a very good design with no spec fails, so the yield and the lower CI limit always increases. This way we will cross the desired yield level at some count *n*.

A spec-fail would push down *Y* and CI limits (see Figure 3.14).

The position for spec fails of course depends on the random MC walk, so on MC seed value. In this case the design is too bad to achieve desired yield target; so even the upper yield confidence bound UCB is worse than the yield target! If your design is really bad, like giving a fail already in the first five MC points, the autostop could come very early. Here the 95% CI is approximately [0.08,0.90], so if your target yield is 90% or higher, the MC autostop would be triggered. Statistically this is correct and you would save a lot of time. On the other hand, it might be inconvenient, maybe the yield is low due to quite an uninteresting spec that is not fully confirmed or you are also interested

Yield *lower* confidence bound vs n   Yield estimate $Y = n_{pass}/n$

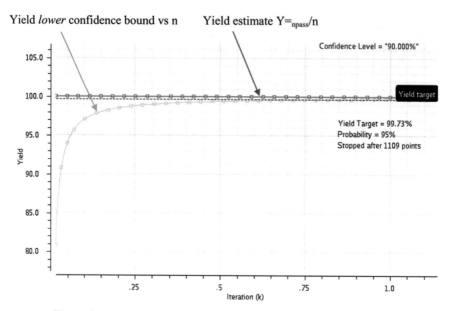

Figure 3.13   Sample yield confidence limits for a MC run with no fails.

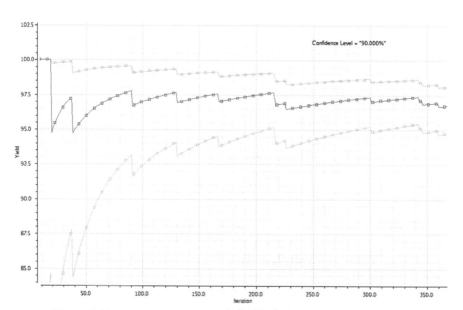

Figure 3.14   Sample yield confidence limits for a MC run containing fails.

in looking to histograms, for which you should have many more points. It could also happen that the autostop comes very late, just because your target yield and your actual design yield are close together. In such cases, autostop does not help on speed but of course on accuracy. For these reasons, even when using autostop, the specification of a minimum and maximum count (or runtime) makes sense.

An autostop feature might be also implemented based on other confidence intervals, like on sample standard deviation s (Figure 3.15) or when reaching a certain accuracy level for a contribution analysis (Chapter 5). Usually also plots for checking how stable the mean estimation is are available, but often the MC mean m is of lower interest, because it is for Gaussian distributions close to the nominal simulation (Figure 3.16). As you know how stable m is depends on σ.

CI width follows the chi^2 distribution and approximately a $1/\sqrt{n}$ law. Also you can see that it is quite symmetric if it is tight enough (like for large n). With larger confidence level (like 99%) the CI would be larger. A meaningful autostop criteria could be e.g., obtained from the relative error on the standard deviation, like $|UCB(s) - LCB(s)|/s < 0.05$. Note: Often CI calculations for mean and sigma are based on normal approximations, so you may add some safety margin to be prepared for non-normal distributions (see next chapter).

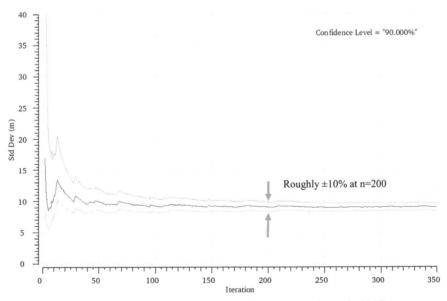

Figure 3.15   Sample standard deviation confidence limits for typical MC run.

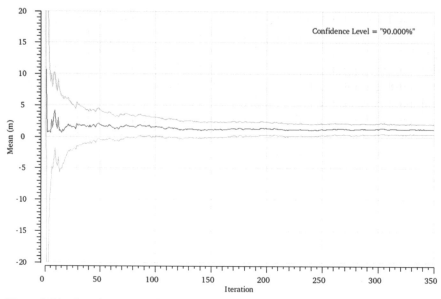

Figure 3.16   Sample mean confidence limits for typical MC run (same circuit as in Figure 3.15).

Essentially all most advanced statistical analyses typically feature such autostop (although they may execute much more steps than a pure MC analysis). This way the user does not need to know in advance how many simulations are needed; instead he/she sets a certain accuracy level for a certain estimate, like sample yield or standard deviation, or maybe more advanced estimates like C_{PK} or generalized C_{PK} (see Chapter 4) or correlations (see Chapter 5).

Testing in Quality Assurance. In MC with autostop, we do basically one test and decide whether we should continue our analysis or not. In quality assurance this is a standard technique too, but often reality creates more problems. Imagine that e.g., the testing of chip is costly or even destructive. How should a company check if a delivery e.g., of 1,000,000 parts, is fine or not? For cost reasons, it makes sense to test only a small subset, and if e.g., all chosen samples are fine, we could extend the testing (more samples, more costly tests). This is called sequential testing, and in principle designers or EDA tools can do the same, e.g., going beyond simple MC autostop. For instance, it may make sense to exploit that some tests in the MC setup run quickly (e.g., simple DC or AC simulations), but others take much more simulation time (transient noise analysis, load-pull analysis, etc.).

3.7 Questions and Answers

1. In PDK's normal distributions often have cuts for the process parameters, e.g., at 5σ (so you will never get samples for these variables beyond $\pm 5\sigma$). Can you still use the C_{PK} and sample yield?
   Yes, the sample yield works for all kinds of distributions anyway! The C_{PK} is slightly too pessimistic if your distribution is a normal distribution with cuts.

2. Can MC handle correlations correctly?
   Yes, there is no problem at all on this. For some advanced non-MC statistical analyses correlations might cause problems, but not in random MC.

3. How many points do you need to get a stable sample standard deviation and C_{PK}, like $\pm 10\%$?
   There is no hard limit; it depends on confidence level. For near-normal data, you typically need only about 200 samples to make the CI $\pm 10\%$ with 95% confidence. For very long-tail data the CI might be much wider!

4. How many points do you need roughly to get in your MC result a sample that is as extreme as $+3\sigma$ or -3σ?
   There is again no hard limit for a uniform distribution you would <u>never</u> get a sample 3σ off from the mean! For a normal distribution you may need 100 to 300 samples typically, it depends (pretty much) on chance. Unfortunately you need many more points for 6σ!

5. Imagine you get 1,000,000 result samples like the height of a good, so you can calculate mean and sigma, maybe to 1.70m and 0.1m, respectively. So we can also calculate the approximated 95% confidence interval to $1.70\text{m} \pm 2 \cdot 0.1/\sqrt{1,000,000} = 1.70\text{m} \pm 0.0002\text{m}$! This means the CI is very tight! Imagine you ask 1,000,000 people about the height of the emperor of China, and you would get these numbers, do you believe you can get the height this way to an accuracy of $0.0002\text{m} = 0.2\text{mm}$?
   Think a bit before you check out Google!

6. How can I calculate the spec setting for a certain yield? *Via C_{PK} you only have to solve the C_{PK} formula for the spec limit, so it is easy, but only correct if you really have a normal distribution. Via sample yield you can do so too, but the result is much more quantized and only acceptable for low yields, like 90% for a MC run with 100 points.*

7. Imagine you are in a certain city and you know all taxis are numbered from 1 to n with no gaps. You take a trip and from time to time you see a taxi. How can you estimate the total number of taxis from your observations? *We can expect that all numbers appear with the same probability, so we do estimations on a discrete uniform distribution. One method is to calculate the mean value and multiply it by 2, but this is <u>not</u> the best way to do it. Do you have ideas for faster convergence? What about taking the maximum?*

8. Figure 3.10 shows an almost <u>flat</u> curve for the sample yield at high yield levels. Please look at it and explain why this makes sense! *For high design yields we will get fails only with a very low probability, and it matters not much anymore if the design is 5σ or 6σ if you want to verify only 3σ. For the C_{PK} this is not the case, because it really exploits the spec margin.*

9. We use the term spec or performance margin; and we use differences; why not ratios? For which type of distributions it would be better using ratios? *With ratios you run into problems with performances which can have both signs. For lognormal distributions using ratios would be optimums!*

10. Imagine you have an outlier in your almost Gaussian data and you are using the C_{PK} method for yield estimation? What happens if you have an upper spec only, but an outlier at the negative (non-spec) side? *Indeed the mean and even more the sigma would be impacted, so a large sigma could decrease the C_{PK}, although the outlier would be a pass sample! Also look at cpk.xls from the River webpage for experiments.*

4

Monte Carlo and Non-Normal Data

We extend the basic methods to address also non-normal data, because using the normal approximation will often lead to severe over- or underdesign for circuits. Distribution-free estimations are also possible, but usually lead to much wider confidence intervals. One example of an advanced non-normal yield analysis is the application of the new generalized process capability index C_{GPK}.

If we have no normal distribution, what else can we assume? And how accurate can our estimates, e.g., on yield be with a limited number of samples?

Indeed, having a good guess on what type the distribution is (such as lognormal, uniform, and Gaussian mix) always helps to improve estimation accuracy. If we assume "nothing", then we can use distribution-free estimates like the sample yield and have to live with its wide confidence interval and there is a need for large n [Schmid]!

If we assume no specific shape but have a good estimate for the standard deviation, we can use the Chebyshev theorem, so actually even for non-normal distributions, there is a clear theoretical foundation of the spec distance method

175

for yield estimation. However, unfortunately, the Chebyshev method leads to wide confidence limits (see Figure 4.6), whereas the Gaussian fit may lead to severe systematic error. So you can try to fit the data to another model (instead of a normal Gaussian one).

For Further Reading:
Older and basic statistical literature focuses (too) often on normal distributions, but nowadays also non-normal analysis has found a huge interest. Also, the topic of which model to choose is a hot one, and new techniques like model selection or model averaging (or fusion) have been created.

- Lange, C. Sohrmann, R. Jancke, J. Haase, B. Cheng, A. Asenov, U. Schlichtmann, *Multivariate Modeling of Variability Supporting Non-Gaussian and Correlated Parameters*, IEEE Transactions on Computer-Aided Design of Integrated Circuits and Systems, Vol. 35, No. 2, pp. 197–210, Feb. 2016.
- Yield Prediction with a New Generalized Process Capability Index Applicable to Non-Normal Data, Weber, S.; Ressurreicao, T.; Duarte, C., IEEE Transactions on Computer-Aided Design of Integrated Circuits and Systems, Vol. 35, No. 6, June 2016, p. 931ff.

4.1 Examples of Non-Normal Distributions

A simple non-normal case is the uniform distribution (e.g., fitting often well for many discrete components), and if this is the case, we can get even $1/n$ convergence instead of $1/\sqrt{n}$. If we add samples from multiple uniform distributions, the result is *not* again a uniform distribution, but actually a good approximation of the normal Gaussian distribution. This is due to the central limit theorem, so sometimes nature helps us to apply well-known Gaussian approximations! However, circuits not only do summations or differences! The sum of normal variates is again normal, but actually the sum of two uniform variables is giving a triangular distribution, and the sum of two lognormal variables is not lognormal!

Of course, MC results can look more difficult, e.g., having two modes ("peaks")—here we may better assume a mix of two normal distributions (having already five parameters in total). In such cases, we may need 10× more samples compared to the fully normal case and in extreme cases may be even 1000× more (depending on yield, mix ratio, etc.). Gaussian mixes can be highly non-normal (in opposite to summing normal variates, which still gives normal Gaussian distributions), and in circuit design, they can occur if our circuit has different modes (or states) of operation, like a multiplexer giving Gaussian outputs in both cases, but overall providing a mix of both due to duty cycle. In subsection 4.8.2 we give some more examples.

Non-normal methods are needed because circuits really create such non-normal distributions more or less all the time! And *whatever* we assume is almost never the true distribution type! With statistics only, you can select a type which gives a good fit and good *prediction*. The only thing we can have "confidence" in is that at least the data <u>might</u> come from our model, and based on that, we do estimations with a certain confidence level. Unfortunately, even if the data pass a normality test, it might be still non-normal or a non-normal model might fit even better. In a MC analysis, many of the technology parameter distributions might be indeed normal or lognormal or uniform, but not your <u>output</u>—for many reasons:

- You have a measurement in dB
- You look for filter passband ripple or settling time
- You have a circuit with 2 modes, often giving a <u>mix</u> of 2 distributions
- DNL of a flash-ADC is defined by max (V_{offset})
- Delay of a CMOS gate $\sim 1/(V_{\text{GS}} - V_{\text{TO}})$
- Looking to leakage current
- Using |x| in a specification (e.g., $|V_o|$ is half-normal)

This list shows quite nicely that if something "special" happens in your design or just only in your test bench setup, easily non-normal performance data will be generated! However, unfortunately, it is not always easy to understand <u>which</u> of the causes have actually taken place. Non-normal data are a frequent case, and only sometimes strange distributions indeed clearly indicate design weaknesses.

Experience shows that roughly 35% of all analog measures are so non-normal (look at Figures 4.12 to 4.14 for examples), and even for a moderate yield (like 99.8%), estimations based on the normal assumption may become significantly biased. Only sometimes, the user can easily avoid non-normal data (e.g., by not using a spec in dB). Often, it is the circuit *itself* creating a certain nonlinearity leading to non-normal data. Of course, even circuits regarded as linear (like passive filters) can create non-normal MC data, if you inspect performances such as overshoot, phase margin, and poles and zeroes. It does <u>not</u> mean that it will always happen and you can never trust specific methods and you would always need to stick to pure random MC and sample yield: often indeed, the Gaussian approximation is not so bad, and usually, there is a fix part and the statistical variation is something small on top, like $1 + \Delta$, with $\Delta \ll 1$. If we would apply a nonlinear operation like $f = 1/x$, we would end up in $f = 1 - \Delta + \Delta^2 + \dots$ and still in a very mild form

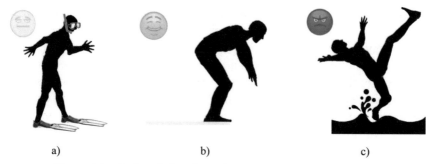

a)   b)   c)

Figure 4.1   (a) Overdesign, (b) optimum design, and (c) design fail.

of nonlinearity due to $\Delta \gg \Delta^2$! Unfortunately, there is no guarantee for this smooth behavior, because there might be stronger nonlinearities, many variables, many correlations, etc.—or you have to go more in the direction of not so small Δ due to circuit specification and technology limitations.

Actually, there is <u>no</u> clear limit, protecting you from making a non-robust design, to be out-of-spec and to have no need for non-normal techniques—there is a smooth transition, a slippery gray area (Figure 4.1)!

What is an outlier? This is a sample that would <u>destroy</u> your fit to the model you "assume", so actually you have to *decide*! It is best to inspect simulation results of the related MC point manually! The usual rules, like remove points "beyond 6σ", are only useful if you can assume near-normal data. The sample yield is quite insensitive to outliers, and they just lower the overall yield a bit, whereas the C_{PK} could heavily impact even by one outlier. For the generalized C_{PK}, you can use the method recommended in [Weber]. The decision if a certain sample is an outlier is not always easy to made, and it could happen that just this "outlier" is truly an indicator that the currently used model is too simple, actually even wrong! In physics, a new model can be a real revolution, and some examples are the quantum Hall effect found in 1980 and the prediction of outer planets.

So if you feel your MC run contains an outlier, then <u>debug</u> it and try to understand and to "repair" the circuit! However, it could be that it is too much effort. If you keep all data including outliers you may overdesign this way. If you completely ignore an outlier you may underdesign, so another method called winsorization is sometimes also usefull: If e.g., the outlier is too extreme (like 1E100) just cut it back at least to the 2nd most extreme sample. This way you can also make sure that e.g., the mean calculation becomes more robust against outliers.

In Chapter 7, we will address the problem of extreme samples to some degree again and in an accurate way by introducing worst-case distances— generation of corner samples with dedicated yield level!

4.2 Identification of Non-Normal Distributions

With the normal quantile plot, we can identify a normal distribution, in less critical cases even with a histogram. Also, numerical tests are available, but any test based on sampled data has its uncertainty. Already small deviations to normal behavior can lead to significant yield errors, so you should get a feeling when the normal assumption might be still applied with acceptable errors and when not.

Check for Normality? In addition to the normal quantile plot, also pure numerical tests for normality are available, e.g., the Jarque–Bera test (JB based on skew s and kurtosis k).

$$JB = n/6 \cdot (s^2 + (k - 3)^2/4)$$

JB is quite a powerful test, and it combines two measures which can also be easily interpreted by themselves: skew s is a measure of asymmetry, and kurtosis k is a measure of the relationship between inner and outer samples. Symmetric distributions have a skew close to zero, and the normal Gaussian distribution has a kurtosis of 3. If JB is large (like beyond 7), we can usually assume that the data are significantly non-normal. In such cases, we do not use the C_{PK}. Please also inspect the spreadsheet example Figure 4.10 for JB calculation.

Identification also for other distributions can be useful, because often the circuit performances do not follow a normal distribution. Leakage current follows often an exponential law, so assuming here a lognormal behavior is much more meaningful than assuming normal data. So a good analysis for this special case is making a fit to a lognormal distribution and using it for yield estimation. Such approach will typically lead to more accurate estimations in terms of variance and systematic errors. However, in the general case one problem is that it is often hard to say which law we should assume, e.g., multiple effects can have an impact, or already the transistor models use very complex functions. Another interesting problem is how can we treat such mixed cases, e.g., a weighted sum of normal and lognormal variates that can vary smoothly from a perfect normal distribution to a full lognormal behavior.

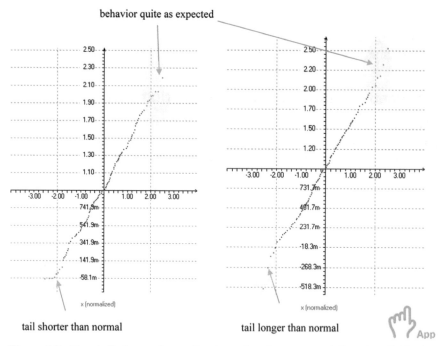

Figure 4.2   Two typical normal quantile plots taken from a normal Gaussian distribution ($n = 256$).

For statistical problems with one variable, we always have a 1-to-1 relation from x_S to f, but unfortunately that would not be the case for multiple variables $x_S = (x_{S1}, x_{S2}, \dots)^T$. The problem of treating multiple statistical variables will be covered in this Chapter 4.

Figure 4.2 shows how difficult it can be to identify a normal distribution via quantile plot; with 256 points, it can be still hard to decide whether the behavior at $\pm 2.5\sigma$ is Gaussian or not. Figure 4.3(b) shows the quantile plot for a Student-4 distribution (look also to subsection 4.8.1); the yield error in sigma (indicated by the blue arrow) is already roughly 0.5σ, but the quantile plot is just starting to become distinct.

Let us now investigate how we can improve our estimations when dealing directly with an *arbitrary* output distribution.

4.3 Non-Normal Data Analysis via Generalized C_{PK}

We have already inspected two different yield estimation methods, but only the sample yield has no systematic error for non-normal data. On the other

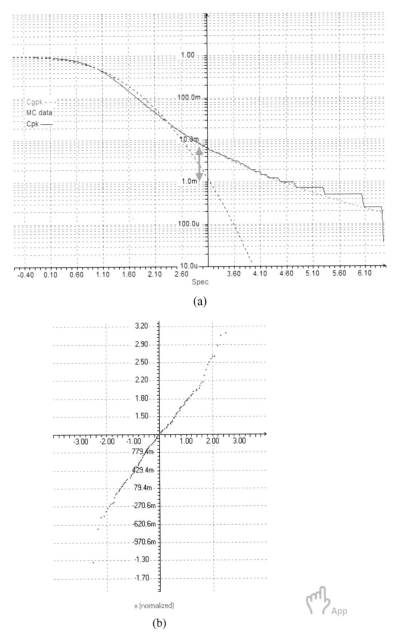

Figure 4.3   (a) Log $(1 - Y)$ for Student-4 and normal distribution fit (averaged MC run with n = 4 K, C_{PK} = 1, but true C_{PK} = 0.82) and (b) quantile plot for Student-4 (n = 256, not averaged).

hand, the C_{PK} allows us to interpret also small MC data sets efficiently, and a lower MC count gives the designer a speed-up in making <u>design decisions</u> (Figure 4.4).

The question is: can we obtain a similar speed-up also in the general non-normal case, e.g., by making a <u>more detailed result evaluation</u>?

The old state of the art on process capability index is to make a Gaussian fit, thus extracting the <u>two</u> distribution parameters μ and s, and the normalized spec distance $(USL-\mu)/\sigma$ gives us a yield estimation. This is a <u>distance</u> method, and the good thing is that intuitively the yield is indeed well correlated with the spec distance—although it is not the only measure!

To address this problem, a new generalized C_{PK} has been developed [Weber], which features one more parameter "t" to describe also the tail behavior. This way, the generalized C_{PK} is quite accurate also for <u>non-normal</u> data, whereas the C_{PK} can be <u>easily</u> systematically wrong by 50%, in cases where the bias of new is only 5%.

Instead of fitting a normal Gaussian pdf, we fit a "generalized Gaussian" pdf. Also, the fit is not done on the whole data, but only to the spec-sided part—starting at the distribution mode (instead of the mean).

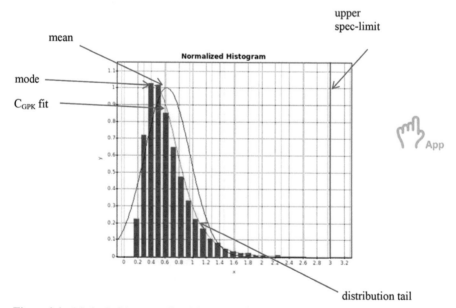

Figure 4.4 Method of the generalized C_{PK} (Gaussian fit is shown as blue curve; it is obviously bad).

Some distributions have multiple modes; here, the C_{GPK} would start at the spec-sided mode. Doing the parametric fit on spec-sided samples has several advantages; for example, non-spec-sided outliers will have no impact, and our fitting function can be formulated much easier, so that indeed, one more parameter (namely t) compared to the "old" C_{PK} gives a dramatic improvement (Figures 4.5 and 4.6).

With the tail parameter *t*, we can model a much wider range of shapes (Table 4.1).

Note: Some of these distributions are exactly included in the model, and others are only approximated. The parameter *t* is actually a parameter of the model cdf (see [Weber]); and there is no simple formula as e.g., for the standard deviation, but we can apply MLE or moment fitting. However, Table 4.1 shows that *t* can be still easily interpreted, just as a normalized tail parameter; complementing the location and scale parameters.

Of course, there is no free lunch: as the C_{GPK} has one more parameter to fit, the statistical variance becomes larger in near-normal cases compared to

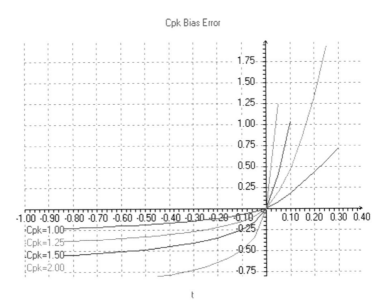

Figure 4.5   Comparison of C_{PK} versus $C_{\text{GPK}} - C_{PK}$ bias error for symmetrical cases (C_{GPK} error is zero): as the C_{GPK} is also a distance method, we can also calculate the spec limit for a given target yield.

C_{PK} assumes <u>linear</u> relation spec versus C_{PK}.
Stable but too simple!
Sample yield is correct, but reaches
Y=100% (C_{PK}=inf) because a 1024-MC run
gives not enough extreme examples. And its
CI is very large!
Chebyshev limit is also not tight enough for
good estimations and works only if σ exists.
C_{GPK} makes a very good yield prediction!

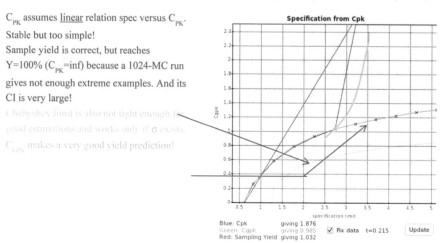

Figure 4.6   Calculating back to spec with different methods (lognormal data used as example).

Table 4.1   Distributions and tail parameter

Distribution	Tail Parameter t	Comment
Uniform	−1	short-tail, low kurtosis
Triangular	$-1 \leq t < 0$	
Parabolic		
Typical bimodal distributions like staggered Gaussian		
Gaussian	0	e^{-x^2} tail, $k = 3$
Peaky distributions like stacked Gaussian	$1 \geq t > 0$	
Student-t, logistic, etc.		
Cauchy	+1	$1/x^2$ tail, infinite k

the C_{PK}. This can be nicely seen if we look to the correlation between yield, C_{PK} and C_{GPK} (look at the scatter plot, Figure 4.7).

As expected, there is a strong correlation with the yield, but the C_{PK} is more stable than the C_{GPK}. So the C_{PK} is still preferable for clearly normal distributions, but for already small deviations, the C_{PK} advantage of lower variance is compensated by its much larger bias error. This bias–variance trade-off is very typical and almost impossible to avoid. Actually, the C_{GPK} is a clever mix of parametric and nonparametric methods, and in opposite to a pure nonparametric modeling (e.g., using the empirical cdf),

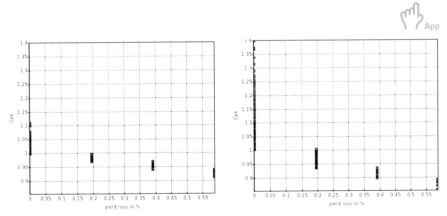

Figure 4.7   Correlation between yield, C_{PK} and C_{GPK} (normal data, $n = 512$, $C_{PK} = 1.0$, 256 MC runs).

nonparametric and tail modeling [MacDonald], or just using a more complex model [Lange], we can efficiently model many difficult distributions like bi- or multimodal distributions and we still fully include the normal Gaussian distribution.

Figure 4.8 compares the sample yield, C_{PK} and C_{GPK} sample count for verification as a function of the true guaranteed C_{PK} (by 95% confidence

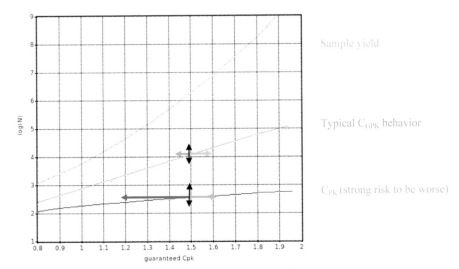

Figure 4.8   Random MC yield verification count for different estimators (design margin 0.375σ).

interval). It shows that the sample yield is highly inefficient especially for high yields, but as mentioned, using it requires no extra-margin for any bias errors due to model limitations (look at the horizontal arrows in Figure 4.8). It can easily happen that the C_{PK} is wrong by more than 1σ, whereas the C_{GPK} bias is usually 5–10× lower. Note that the green curve for the C_{GPK} is only an average, as it depends also slightly on the distribution type (vertical arrows). In the succeeding subchapters, we will give some concrete examples (for measured production data and for certain mathematical distributions).

Alternative Distribution Fitting Methods. There are indeed many ways to fit data to a certain distribution (we mentioned MLE and moment fitting). Instead of using the C_{GPK} concept we could also do it a bit differently [Weber]. We could also try to fit over the entire data (like the C_{PK} does), or we could also only model the tail (Table 3.3). For instance, we could assume a triple Gaussian mix to model distributions with up to three modes, but obviously a high flexibility comes with the price of many parameters. The advantage of modeling only the tail is some more flexibility and potentially high accuracy in this region of interest! But one big question is where to "start" the fit and where the tail begins? Having enough data, it is indeed possible to answer that question, but to some degree, it results in a model which depends on the fit for quite few data points, just the tail points. So the price for low-bias errors is typically having quite large confidence intervals. For the tail modeling, typically the generalized Pareto or generalized extreme value distribution is assumed. Also when using the C_{GPK} concept, we could plug-in different distributions, or we may extend the concept with one more parameter to be able to model the shape of the mode and the one for tail independently. Also almost completely nonparametric fits are possible, e.g., based on KDE, but typically they are not well suited for high-yield estimation.

4.4 Analyzing Real Production Data

Of course many statistical methods are not only applicable to MC results, but also applicable to real data measured in production. The data in this example (Figure 4.9) have been taken from [Shinde] and come from a USB2 squelch circuit: trip point has been measured on $n = 3,999$ silicon samples.

Let us do two analyses: according to [Shinde], let us inspect what we can estimate if we do not use the full measured data, but only a subset of 25 samples. This is often regarded as minimum requirement for a normal

USB_SQUELCH_SEARCH

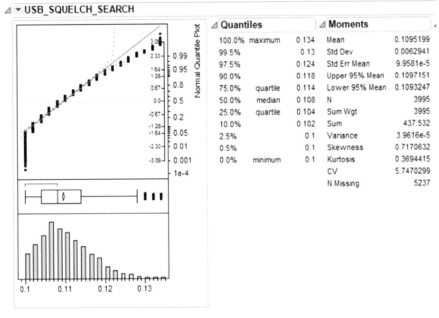

Quantiles		
100.0%	maximum	0.134
99.5%		0.13
97.5%		0.124
90.0%		0.118
75.0%	quartile	0.114
50.0%	median	0.108
25.0%	quartile	0.104
10.0%		0.102
2.5%		0.1
0.5%		0.1
0.0%	minimum	0.1

Moments	
Mean	0.1095199
Std Dev	0.0062941
Std Err Mean	9.9581e-5
Upper 95% Mean	0.1097151
Lower 95% Mean	0.1093247
N	3995
Sum Wgt	3995
Sum	437.532
Variance	3.9616e-5
Skewness	0.7170632
Kurtosis	0.3694415
CV	5.7470299
N Missing	5237

Figure 4.9   Data from a fabricated USB interface [Shinde].

data analysis. In many such cases, designers can expect near-normal data for good reasons, because usually the trip point of a comparator is dominated by device mismatch—and mismatch can usually modeled well with normal distributions. Later, let us check this analysis against an analysis taking the full statistical data into account (Figure 4.10).

The original Intel conference paper exemplified already a yield estimation on a subset of n = 25 samples. The authors obtain a sample C_{PK} of 2.04 (6.12σ). By visual inspection of the histogram, the authors regarded the data as <u>normally</u> distributed and they predict 4σ as lower yield 95% confidence limit. This means although the sample C_{PK} is 2.04—indicating a very good design—a statistical analysis based on the assumption of normality can only guarantee a C_{PK} of 4/3 = 1.333. The difference between 4σ lower CI limit and 6.12σ sample C_{PK} would go to zero for $n \to \infty$.

This difference looks like a good "safety margin", but this simple analysis does not take some important aspects into account: sample skew is s = 0.71, and the critical specification limit is at the long-tail side. This leads to a <u>too optimistic</u> C_{PK} yield prediction! If we apply the Jarque–Bera normality

```
C107:   = KURT(C2:C102)        – a measure of tail behavior
C108:   = SKEW(C2:C102)        – a measure of unsymmetry
C109:   = 100/6*(C108*C108 + C107*C107/4)
```

C109	▾	⋮	×	✓	*fx*	=100/6*(C108*C108+C107*C107/4)

◢	A	B	C	D	E	F
97		0,3951461	-0,265931241	0		
98		0,9646656	1,807599664	0		
99		0,9106683	1,344881804	0		
100		0,2933183	-0,543716415	0		
101		0,0116437	-2,268687314	0		
102						
103	Avg	0,504315	0,02401982			
104	Stddev	0,2847205	0,96873295			
105	Yield in %			99,0000		
106	Cpk			-0,851965164		
107	Kurtosis		-0,123452811			
108	Skew		0,130741339			
109	JB		0,348390781			

Figure 4.10   Spreadsheet example for the calculation of the Jarque–Bera normality test.

test for the whole data set, we can obtain a value for JB beyond 200, which clearly indicates non-normal data, but for n = 25 JB is only 2.2 (indicating only very mild non-normality). A large JB value indicates that we should not apply a Gaussian fit, but better apply the new generalized C_{PK}. This predicts a true C_{PK} of 1.30—instead of 2.04! And of course also the CI for the generalized C_{PK} would be even lower, being at 1.20. Overall, the designer's conclusion should be that the design definitely *needs* significant improvements. And it is unfortunately not enough just to measure more samples to tighten the confidence interval!

Note: In this example, the sample yield is still 100%, because there is no fail even in 3999 points! The sample yield confidence limit is approximately 99.88% and is equivalent to a true C_{PK} of 1.02, which is worse than the CI of the generalized C_{PK} (which is at 1.20). The Clopper–Pearson yield limit (also used in most EDA tools for yield confidence intervals) for 1 fail in 3999 samples would be 98.6% only (equivalent to a true C_{PK} of 0.73!). Also look up: Jarque–Bera is one of many normality tests, neither the best, nor the worst. It is actually quite powerful which means that it needs usually not that many points to make a decision. On the other hand, there is simply no best normality test, because deviations to normal data can be of many kinds.

4.5 Yield Estimation for Non-Normal MC Data via $C_{\mathbf{GPK}}$

In the previous example, we inspected measured production data, and let us now inspect MC data from an operational amplifier. Such amplifier is basically a linear circuit, where most designers would expect to find quite normal histograms. However, we will see that also such classical analog circuit can easily create non-normal data (Figure 4.11).

A complete CMOS op-amp with tricky feed-forward frequency compensation has been designed (but not fully optimized) and verified for almost all common specifications in a big MC run for mismatch only ($n = 1500$). The PDK does not offer global variations, but these would of course lead to wider variations and potentially even more non-normal data.

Many histograms are indeed near-normal (like the one for current consumption and offset voltage), but here are also some interesting non-normal histograms, where we really need the C_{GPK}. In the Figures 4.12 to 4.14 you will find some examples taken from [Weber2016], e.g., Figure 4.12 shows HD_3 data, where the C_{PK} is too pessimistic.

Figure 4.13 shows a second example, looking to the peaking of the closed loop gain; here, the C_{PK} is too optimistic.

A third example is shown in Figure 4.14, the 3-dB-BW (the graph of amplification versus frequency has two peaks due to the feedforward scheme—and the circuit is really functional, no bug).

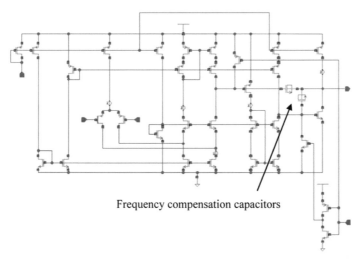

Frequency compensation capacitors

Bias part   Input stage   2nd stage   3rd stage   output stage and enable logic

Figure 4.11   Inspected op-amp circuit.

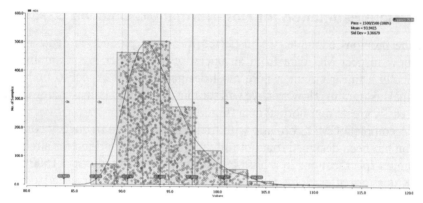

Figure 4.12   Third-order distortion in dB (C_{PK} too pessimistic due to spec at short tail) [Weber2016].

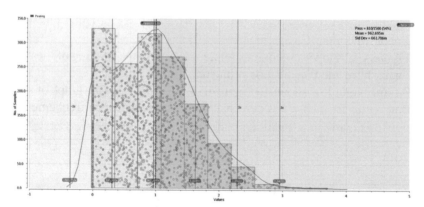

Figure 4.13   Gain peaking in dB (C_{PK} too optimistic, spec on long tail) [Weber2016].

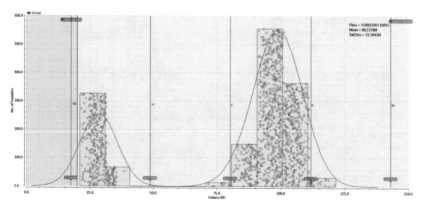

Figure 4.14   3 dB bandwidth (C_{PK} too pessimistic) [Weber2016].

4.6 Questions and Answers

1. How can I check my MC data if it is following a normal Gaussian distribution?
   Inspect the normal quantile plot or apply a numerical test like the one according to Jarque–Bera. JB = n/6(s² +(k − 3)²/4).

2. What is the distribution for the output voltage of a two-resistor voltage divider made of discrete resistors?
   For discrete elements, it is not realistic to assume a Gaussian distribution, and a uniform distribution is usually more realistic. The sum of two uniform variables gives a triangular distribution! Actually, the output voltage is a nonlinear function of the two resistors, but the nonlinearity is not that large, so indeed we can expect a near-triangular distribution.

3. Is the generalized C_{PK} having a systematic error?
   The C_{GPK} includes a much wider class of distributions without such bias error, but if the data are not part of the model pdf, we get some bias. Usually, it is only 10% of the normal C_{PK} bias, like for a lognormal distribution. In such cases, and also on a uniform distribution, the C_{GPK} bias makes the yield estimation a bit too pessimistic—so you are on the safe side, but overdesign a bit.

4. Is the C_{GPK} method an interpolation or extrapolation method?
   It is both, depending on the yield level! In opposite to many other extrapolation schemes, the C_{GPK} makes a very meaningful extrapolation. Also the C_{GPK} method might be combined with WCD methods to get rid of the extrapolation risk.

5. $\pm 3\sigma$ around mean is equivalent to capturing approximately 99.73% of the distribution if the data are normal, but how much is it for a Student-5 (which looks very similar to a Gaussian distribution, look at subsection 4.8.1)?
   Although the kurtosis k is still moderate for the Stundent5 distribution, and the normal quantile plot indicates no strong non-normality, the yield is pretty much less, approximately 98.8%—so only approximately $\pm 2.5\sigma$. So the error in terms of sigma is approximately 20%, thus often much larger than the confidence interval reports! Remember, confidence intervals do not quantify such systematic errors.

6. Can we extend the C_{PK} and C_{GPK} concept also to multiple performance?
   Yes, this is possible and for multivariate pure normal distributions, several solutions exist. These are usually good enough for process monitoring, but in circuit design you have very often to deal with

non-normal distributions. In Chapter 5, we present a good approximated solution.

7. Can it happen that running one more point in MC, which gives a *pass*, leads still to a <u>decrease</u> C_{PK} for a spec like CMRR >40dB?
   Yes, it is possible, and assume we have a short 50-point MC run with sample standard deviation of 2 dB and mean 60 dB, so C_{PK} was 10. Imagine the next MC point is at 100 dB—it would shift the mean down, but the standard deviation would increase even more, so overall the C_{PK} could become worse! In pure Gaussian outputs, such event would be extremely rare, but non-normal data can give such surprises.

8. Imagine you have a Gaussian output y for a certain performance in a MC run. Now you take this in dB, which kind of distribution will you get?
   It will be a new distribution, and it is <u>not</u> the lognormal distribution!

9. If we add many samples from independent uniform distributions, we end up by central limit theorem with a normal distribution. Would this also be the case for other distributions?
   For instance, even when adding exponential distributions, we would lose the asymmetry; and the left side short would become longer, whereas the longer right tail would become shorter (e^{-x^2} instead of e^{-x})! However, e.g. the Pareto distribution is asymmetric too, but has infinite variance, so here the CLT would not work.

10. Discuss which kind of problems can be solved with pure random MC and sample yield?
    Check runtimes, accuracy, design improvements, inclusion of corners, which additional analysis should be done, etc.

11. Imagine you have a Gaussian distribution with mean = 0 V, so we get in a nominal simulation usually also 0 V. Now we apply the exponential function to this output, leading to exp(0) = 1 at the new output. However, what happens in MC? Will the mean be also at 1?
    No! The median will be there; and it is now different from the mean (average) value! If our nonlinear function is non-monotomic even the median will usually not be preserved.

4.7 Rules You Have to Know for Monte Carlo

When setting up a MC analysis, one big general question is how to decide on required number of points for certain target accuracy? Actually this problem is not only related to MC but to taking statistical samples in general, like

Table 4.2 Overview on basic normal and non-normal MC techniques

Analysis	#points	Comment
MC for checking mean and standard deviation of performances	e.g., 100	Some mild non-normality allowed. For extreme distributions, bad or no convergence
MC for checking sample yield	app. 2K for 3σ	See Table 3.2
MC for yield via C_{PK}	app. 200	Data should be highly normal, especially for high-yield targets
MC for yield via generalized C_{PK}	app. 500	The number of points depends slightly on yield level and distribution type
MC for checking distribution type	>50	Depend on how accurate you want to model modes and tails; to differentiate between similar distributions, you may need >1K points
MC for correlation analysis	$100 - {>}1000$	Dependent on number of variables involved

for production data inspections. It essentially depends also on <u>what</u> kind of estimate you are interested, e.g., 1% yield accuracy is good if the design has a yield of 50%, but it is not good enough if the target is 99.7%! Many measures also depend on <u>distribution shape</u>, and this can never be <u>fully</u> known <u>upfront</u> (Table 4.2).

So best <u>know</u> some basic rules and their prerequisites and make a <u>MC test run</u>, check histograms, and look to confidence intervals. If the CI is $2\times$ too wide, then increase the number of points by approximately $4\times$.

Rules for <u>any</u> kind of data:

1. You can trust the sample yield $Y = n_{pass}/n$ (because it is a distribution-free estimate).
2. But CI of Y is large. If there are no fails, then CI limit is approx. given as $3/n$.
3. In random MC, there is no dependency on number of statistical variables for estimates like Y, μ, or σ, but of course it might be the case for correlations (Chapter 5).

Basic rules for <u>near-normal</u> case:

1. Most frequently used: 95% confidence interval
   (so 5% risk of false decision)
   + assuming a normal Gaussian distribution
   + assuming $n \gg 1$ (like 50)

2. Then, e.g., the 95%-CI of the <u>mean</u> μ becomes roughly $\pm 2\sigma/\sqrt{n}$. This is two times the standard error $SE = \sigma/\sqrt{n}$.
3. So to know variance on mean μ, you need to know σ.
4. Also σ has a variance: $\sim 1/\sqrt{2n}$, e.g.,
   $n = 50$ gives 10%, so if $\sigma V_{offset} = 10\,\text{mV}$, it is typically within $8 \ldots 12\,\text{mV}$ with 95% confidence => not so bad
   $n = 200$ gives 5% => often good enough
5. Other measures (such as correlations or the mode) may need more points (like 1000).

Rules for significant <u>non-normality</u>:

1. Apply tests, e.g., via Jarque–Bera and normal quantile plot.
2. σ might not converge for long-tail distributions!
3. σ variance is usually (roughly) proportional to $\sqrt{\text{kurtosis}}$ (4th moment).
4. Do not trust the C_{PK}! Use C_{GPK}, sample yield or the methods from Chapter 7.
5. If data is not very non-normal, then CI width and sigma follow still often follow the $1/\sqrt{n}$ law. But bias errors can follow any law, and even for infinite n they might be an error (e.g., using the CPK for yield estimation on non-normal data).

4.8 Design with Pictures Part Two

The non-normal distribution examples in this Chapter 4 were quite obvious; that is, by careful visual inspection, it was quite clear that the data are not normal and that a data analysis based on normality would lead to bad results. However, sometimes it is not so easy to decide whether data are normal or not. And in high-yield cases, even small deviations can lead to significant errors, because a Gaussian fit and the C_{PK} based on that are kinds of extrapolation method. An interesting question is how large is the risk that a <u>deviation</u> to the normal distribution can be seen "late"?

In our USB fab data example, the Jarque–Bera JB value was huge for $n = 3,999$, but for the same skew s and kurtosis k with lower MC count, JB will drop and there is a "gray" area (like JB = 1..4) where it is quite likely that the data is indeed normal, but the confidence is still quite low, or vice versa the data is non-normal, but too close to normal, so that most people would apply methods based on the normal assumptions, mistakenly.

One further problem with JB and many other normality tests is that they do <u>not</u> take the yield level into account! So the risk of "seeing" non-normality late and the error in yield prediction depend on how big the non-normality is but also on <u>how much</u> you "extrapolate". For instance, to "see" the difference between a Student-100 and a normal distribution, you may need 100,000 samples, but between Gaussian and uniform maybe only 50–100 samples. There is even no such thing like a confidence level for such investigation, but at least we can provide guidelines and train ourselves with examples. In statistics and yield verification just multiplying the sigma is risky, actually even when using normality tests.

Note: With the RealTime MC program that complements the book, you can do such investigations in a very short time, because a much faster simulator is built-in, than just plain SPICE.

4.8.1 Normal versus Student-t versus IH Distribution

The normal distribution has a kurtosis of 3 (note Excel gives 0, because the 3 is subtracted internally), whereas the other two distributions which we have chosen as example, the Student-t and the Irwin–Hall IH distribution, feature a parameter to adjust the kurtosis. In our RealTime app, we can tweak this parameter to obtain a kurtosis of 3.3 for the Student-t and 2.7 for the Irwin–Hall distribution.

Actually, the Student-t is a very important distribution, e.g., for confidence interval calculations, but also the IH has a relationship to the normal distribution. It is composed of the sum of uniform distributions, and if we would add up an infinite number, we would end up in the normal distribution (Figure 4.15). Note that many distributions (even discontinuous ones) can be connected to the Gaussian distribution this way, just due to central limit theorem (CLT)! You can also extend the normal distribution, by adding parameters to adjust the shape and to introduce an asymmetry. Actually, there is no single "best" way to do this, and many such generalized Gaussian distributions exist. There are also distribution families with no tight connection to the normal distribution, but in spite of that, they can be still very similar (e.g., the logistic distribution).

Looking to the (smoothed) histograms of all our three examples, you can hardly see any difference (Figure 4.16), just because even for n = 1024 the tails are very hard to inspect visually. Even in the normal quantile plots (Figure 4.17) you really need a huge number of points to identify the distributions.

The JB value for checking normality is approx. 4.5 (pretty close to the gray zone), and if we set the spec for a C_{PK} at 1.0, we get the standard deviation of only 2.5%, i.e., n is large enough to really get a stable C_{PK}.

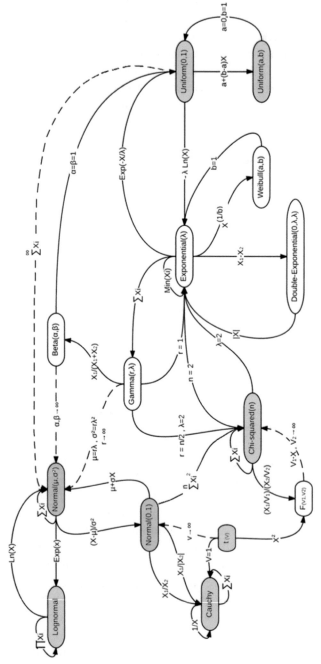

Figure 4.15   Some important distributions and their relationships.

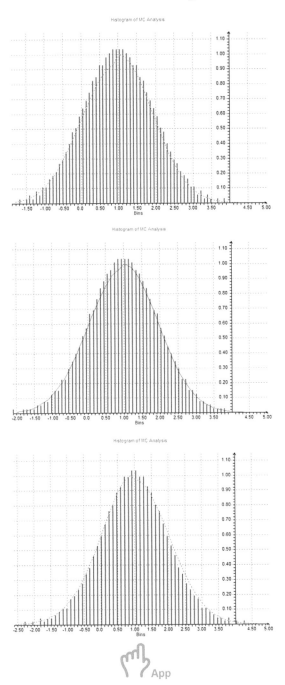

Figure 4.16   Histograms for the three inspected distributions (averaged).

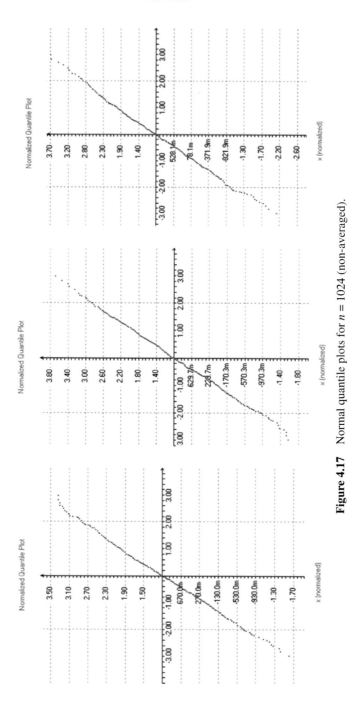

Figure 4.17 Normal quantile plots for *n* = 1024 (non-averaged).

The question is now how large is the systematical C_{PK} error, especially at higher yields. This can e.g., be checked by running a very long MC analysis using 64 K points and using the generalized C_{PK}, which has a highly reduced bias error. We set the spec limit to obtain a C_{PK} of 1.5, and for the Student-t distribution (Figure 4.18), the C_{PK} yield loss prediction is too optimistic by 2.5 orders of magnitude (whereas the C_{GPK} has no bias error in this case)! For the IH distribution, it is vice versa and the C_{PK} estimation is too pessimistic by approx. 30× (Figure 4.19).

Notes: At some point, the red curve (showing the sample yield) drops to infinity, because it is still 100% even for $N = 64$ K. The green curve is the result of yield estimation by the generalized C_{PK}, which makes a very meaningful extrapolation.

In conclusion, an MC run with 1024 giving normal data according to standard tests and applying the C_{PK} can give <u>still</u> give big yield errors for $C_{PK} \geq 1.5$. Having no fails, the Clopper–Pearson lower confidence bound LCB would be only at 99.64% (equivalent to $C_{PK} = 0.896$). The C_{GPK} LCB for the Gaussian case is approximately 1.2—and this is (without stronger assumptions) what you can "guarantee" at best. The C_{PK} LCB is 1.43 for normal data, but even this (and not only the point estimate of 1.5) is still too optimistic compared to the true yield and true C_{PK} being at 1.235.

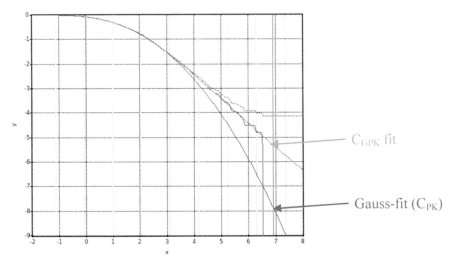

Figure 4.18   Plot of $\log(1 - Y) = f$ (spec) for Student-t distribution.

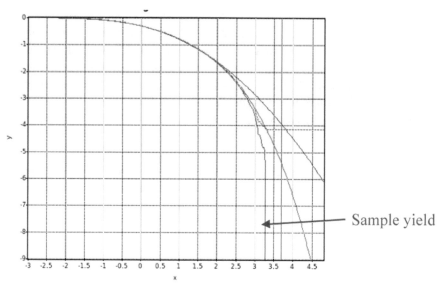

Figure 4.19   Plot of $\log(1 - Y) = f$ (spec) for Irwin–Hall distribution.

4.8.2 Calculations with Random Numbers

Can we calculate with random numbers, like we do with real or complex numbers? Yes, you can, but indeed some rules will *change*, and only few will remain the same. For instance, taking a random variable X with normal distribution and multiplying it by 2 gives a normal distribution with doubled σ. However, adding two *independent* normal random variables (of same σ) gives only $\sqrt{2}\cdot\sigma$, so X+X is not always equal to 2X. Also taking the difference is special, because X−X gives us the same distribution as X+X for independent *standard normal* variables! Also note that also the rule X+X = $\sqrt{2}$X would only work for Gaussian variables, not for uniform or lognormal ones (here we would get a *change* in the distribution type); (only) for Cauchy variable we would indeed observe X+X = 2X.

Taking $\exp(X)$ gives us the lognormal distribution, but adding two independent lognormal variables gives <u>no</u> lognormal distribution again! However, interestingly adding two Cauchy distributions gives us a Cauchy distribution, so to some degree the normal and the Cauchy distribution are special. Taking the absolute value of a standard normal distribution gives us a *half-normal* distribution, but taking the difference from these would give us again a normal distribution.

What about more difficult operations such as multiplication and division? For instance, dividing two normal independent normal variables gives another distribution, which is the Cauchy distribution! Actually the division spreads the distribution a lot, so the result (the Cauchy distribution) has much stronger tails than the normal distribution. The Cauchy tails are so strong that even mean and sigma does not exists; it looks a bit like a normal distribution with many outliers.

Adding two independent *uniform* variables gives us a triangle distribution; and adding really many independent uniform variables gives us a very good approximation to the *normal* distribution; and this is true for even <u>any</u> infinite sum of random variables; just *finite variances* are required. So even adding e.g., lognormal variables (being quite asymmetric) would end up more and more

Table 4.3 Overview on calculations with independent random variables

X_1	X_2	Operation	Result	Comment
Std-Normal	–	–X	Std-Normal	$\mu = 0$, $\sigma = 1$
Normal	–	exp(X)	Lognormal	Log(X) is not lognormal
Std-Normal	–	$|X|$	Half-normal	$\mu = 0$, $\sigma = 1$
Normal	–	$|X|$	Folded-normal	Appear for performances like $|V_{offset}|$
Std-Normal	–	X^2	$\chi_1{}^2$	Chi-square, important for confidence intervals
Normal	Normal	$X_1 - X_2$	Normal	Mean substracts, variance adds up still
Uniform	Uniform	$X_1 + X_2$	Triangle	Mean and variance added
Triangle	Uniform	$X_1 + X_2$	Quadratic	Mean and variance added
Cauchy	Cauchy	$X_1 + X_2$	Cauchy	Location and scale add up
Std-Normal	Std-Normal	X_1/X_2	Std-Cauchy	Very wide tails
Half-normal	Half-normal	$X_1 - X_2$	Normal	
Lognormal	Lognormal	$X_1 + X_2$	Not lognormal	New distribution, looking slightly more normal
Std-Normal	Chi	X_1/X_2	Student-T	important for confidence intervals
Std-Uniform	Std-Normal	X_1/X_2	Slash	Similar to Cauchy

in a normal distribution (which is symmetric). Again the Cauchy distribution is special, because it has no finite variance. The central limit theorem CLT will tell us even more, because if we know the mean values and variance of the original distributions, we can calculate the over-all mean and variance just as the sum of the "input" distributions. And the normal distribution with that parameters will often give an excellent fit to the sum distribution. However, this fit is usually only good near the distribution center, not in the tail regions.

Actually on all these things there is quite nice material available in the internet! By creating little testbenches and running MC analysis can find such relationships directly from circuit simulations. For instance, simulating a multiplexer with two inputs driven by normal distributions, you can obtain a Gaussian *mix*. Such mixes are often multimodal, so not normal Gaussian at all. With Verilog-A you can perform almost anything you want, because it also supports random number generation, even for "very" special distributions, and of course it also supports many math functions. Although Verilog-A does not support so many distributions (e.g., no Cauchy distribution), you can often easily create whatever you want with moderate effort by calculations with random variables (see Table 4.3).

PART III

Advanced Statistical Design Techniques

5

Multivariate Statistical Analysis for Design Insights

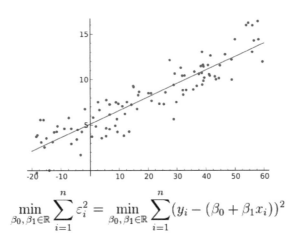

$$\min_{\beta_0, \beta_1 \in \mathbb{R}} \sum_{i=1}^{n} \varepsilon_i^2 = \min_{\beta_0, \beta_1 \in \mathbb{R}} \sum_{i=1}^{n} (y_i - (\beta_0 + \beta_1 x_i))^2$$

In this chapter we discuss how we can treat and analyze multiple variables and outputs in statistical analyses. Important applications are the calculation of sensitivities and correlations from MC results. Such techniques are also in use for more advanced yield verification methods, which we will discuss in Chapters 6 and 7. In design modeling is often a key task, and it could fill books on its own, so we will focus on an overview and several specific methods related to variation-aware design. Univariate performance analysis is a kind of blackbox modelling; it might be very good, but a multivariate analysis can turn blackbox to whitebox modelling, giving more insights, more speed and accuracy.

If we would just add (or subtract) the sample values from two or more simple normal Gaussian distributions, we would <u>still</u> end up in a normal Gaussian distribution. So we simply cannot tell from MC *output* data that the histogram was created by one, two, or whatever many variables! Also if our design would act as divider on normal variates giving Cauchy variates (see Figure 3.4), we cannot tell whether our circuit is a univariate Cauchy random generator or whether it acts as divider on two normal variables. So what could be the benefit of a multivariate analysis? Two good examples are sensitivity analysis and finding performance correlations to understand the design trade-offs. A third application is performance modeling, in this we try model e.g., a certain output performance (like power-supply rejection ratio PSRR) as a function of the (statistical and/or non-statistical) input variables.

Note: Up to now all estimates and the convergence speed and variance are not impacted by the number of statistical variables and design complexity, so random MC sample yield verification for 90% yield and a certain CI is as fast for a simple diff-pair as for huge blocks—in terms of simulation count (not in time of course)! This independence on complexity will not be the case for the more advanced analysis!

Designers often ask for the sensitivities (actually a linear performance model) of the different performances to the different "design" variables, like transistor width of M1 or to sheet resistance or temperature, etc. Sometimes you are lucky because the simulator itself might have a built-in sensitivity analysis, which runs typically quite fast, but unfortunately this is too often not the case. As a workaround, you can do a short sweep Δx (e.g., 1% or 0.1σ) of the individual parameter of interest and look to the change in performance Δy. This one-factor-at-time parameter varying technique (OFAT) is not only accurate, but also quite slow if you have to inspect many parameters.

In school you have been taught:

> *"If you have n parameters, you need n equations."*

One obvious problem is that if you want to analyze the sensitivities of your circuit to all statistical parameters, the OFAT method becomes slow, because real designs often have thousands of statistical parameters describing mismatch. If you have 50 transistors in your design and each has two mismatch parameter (like for modeling of threshold voltage and mobility), you would end up in 100 mismatch parameters in total, plus "some" more for process variations like further hundred. In modern process nodes (below 40 nm) you find typically even many more variables, like more than thousand for process and e.g., six for mismatch for each transistor. Couldn't we use MC techniques to obtain sensitivities in a "smart" way? As MC is applying little changes Δx and we collect the circuit response, then a correlation analysis could give us the desired sensitivities! One obvious problem is that if you have thousands of statistical parameters, but only 200 MC samples, how we can obtain thousands of sensitivities? But you can determine at least quite accurately statistical estimates for the maybe top-5 sensitivities, and that is often 99% of what the designer really wants to know (Figure 5.1 and Figure 5.10)!

The simplest multivariate analysis is a linear regression, so the result allows a linear approximation of the circuit behavior; in good cases such a model can explain 99% of the true circuit behavior and only 1% remains as model error e. Such basic regression is possible if more or equal data points are available than variables, then just a best *fit* is the result. Best fit means lowest error e, and different criteria could be used like minimum worst-case error or minimum average quadratic error (rms). Actually the minimization could be done by any optimization algorithm (see Chapter 8), but exploiting the special structure of the error function could give a significant speed-up compared to universal optimizers.

The general modeling flow is shown in Figure 5.2. In modern design environments, advanced options exist e.g., to extend the model to become quadratic. Such techniques are usually called response-surface modeling (RSM). In addition, sparse approximation methods exist, which can be applied if too few MC samples are available. There are also algorithms that focus directly on the relative importance of parameters [Groemping]; these are quite robust even for highly nonlinear problems, if many parameters exist and only a moderate sample count is available. Such response (output variable $f(x)$) model is usually not based on physics or structures, like a SPICE transistor model, it is usually a pure empirical model.

If no good model can be found, you can at least apply *non-parametric* multivariate methods, e.g., based on ranking, and you will be still able to make a correlation analysis. Also, the well-known median ("50% point") is a rank estimate, and the idea of sorting and ranking can be applied quite generally. Advanced research [W. Zhang] and commercial implementations typically use mixed methods and include clever error control mechanisms.

Another clever technique is to collect the impact (contribution) of the different statistical variables, e.g., all the ones coming from one transistor or a certain sub-block (like bandgap or LNA), then providing a combined output for this group, like for transistor N4 or LNA1! Truly the number of variables *decreases* this way, and we can still make a reasonable, accurate contribution and correlation analysis for complex circuits [Li].

What would happen if the design is so nonlinear that no good model can be found? This can happen in many such correlation analysis or model

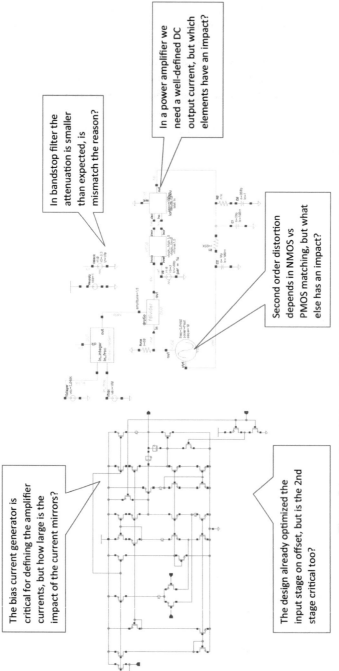

In bandstop filter the attenuation is smaller than expected, is mismatch the reason?

In a power amplifier we need a well-defined DC output current, but which elements have an impact?

Second order distortion depends in NMOS vs PMOS matching, but what else has an impact?

The bias current generator is critical for defining the amplifier currents, but how large is the impact of the current mirrors?

The design already optimized the input stage on offset, but is the 2nd stage critical too?

Figure 5.1   Topics which can be addressed by a mismatch contribution analysis.

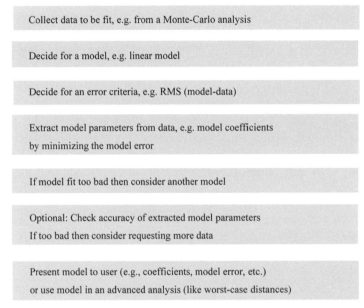

Figure 5.2 Typical modeling flow.

building whereas pure random MC yield verification even works in the case of <u>infinite</u> number of variables, or when the <u>number</u> of random variables is also <u>random</u>! In such truly special cases, any multivariate analysis is indeed very difficult, and we may better switch to the more basic techniques. A further difficulty occurs if the simulator and statistical algorithms simply have no access to many statistical variables, which might be the case in a transient noise analysis or in the presence of strong numerical noise. Here we would have performance variations from statistical variables in the circuit (like those describing mismatch, e.g., causing excessive DNL) <u>and</u> further variables acting as random noise generators in resistors and active elements or in the simulation algorithm itself.

Another example for a multivariate analysis is to inspect the relations between different <u>outputs</u>, like leakage current and speed. The total yield depends on the correlation between the partial yields, so if we know the partial yield (e.g., from C_{GPK} or another dedicated method like WCD, see Chapter 7) and the correlation, we can make a better *total* yield estimation. Or we can better understand our design, like if there is room for improvement or if we already have a good balance between competing performances and specs.

In conclusion, several kinds of multivariate analysis are possible and are of high interest, but it could also happen that such analysis can become quite time-consuming, e.g., an MC analysis itself has a linear rising effort according to the number of samples n—and you can even run the circuit simulations in parallel. The effort for obtaining model parameters for a multi-dimensional nonlinear model relating m statistical parameters and k outputs based on given MC data with n points is often very big, and it often rises approximately quadratic with m. Clever methods are required to limit the effort by focusing on the real important correlations, like looking to those performances and parameters with really significant changes and critical specs.

For Further Reading:

- Xin Li and Hongzhou Liu, "Statistical regression for efficient high-dimensional modeling of analog and mixed-signal performance variations," *45th ACM/IEEE Design Automation Conference (DAC'2008)*, Anaheim, CA, 2008, pp. 38–43.
- Andre Lange, Christoph Sohrmann, Roland Jancke, Joachim Haase, Binjie Cheng, Asen Asenov, Ulf Schlichtmann, "Multivariate Modeling of Variability Supporting Non-Gaussian and Correlated Parameters," in *IEEE Transactions on Computer-Aided Design of Integrated Circuits and Systems*, vol. 35, no. 2, pp. 197–210, Feb. 2016.
- W. Zhang, T. H. Chen, M. Y. Ting and X. Li, "Toward efficient large-scale performance modeling of integrated circuits via multi-mode/multi-corner sparse regression," *47th ACM/IEEE Design Automation Conference (DAC'2010)*, Anaheim, CA, 2010, pp. 897–902.

5.1 Multivariate Probability Density Functions

To lean about multivariate statistics, we should start on how to describe their probability density function. As for single variable statistics, let us start with a normal (Gaussian) distribution, just with two variables. One variable could be the differential offset voltage of an amplifier and the second variable could be the offset in common-mode voltage. Both are often normally distributed and in a circuit we may have the case that some transistors causing the differential offset are the same as those causing the common-mode offset, so we can expect some correlation ($|c| > 0$). In other circuits, the effects might be almost independent, so we can expect little or no correlation ($|c| \ll 1$). How can we mathematically capture this effect? Instead of generating a set of single random numbers, we generate two variables x_1 and x_2, and the distribution is now according to a 2-variable density function. Looking only to the samples of X_1, we would find a certain mean and standard deviation, and accordingly for X_2. As for the univariate case, we can start from a standard-normal distribution (having zero mean and a standard deviation of unity), and interpret mean and standard deviation as linear transformation (m makes a shift and s makes a scaling operation); for the two variables, we can just take again such linear transformation, but now it is a linear *matrix* transformation. This we can apply on samples coming from a multi-dimensional standard-normal distribution, just to get any multi-dimensional normal distribution—of arbitrary mean, standard deviation, and correlation. In such matrix trans-formation, the shift by the mean (now a vector μ) is easy to interpret, so most analysis is around the matrix part **B** doing a scaling and also usually a rotation.

$$Z \sim N(0, I)$$
$$X = ZB + \mu$$

(5.1)

where B is a n $\times$ n matrix, and one can show that the multivariate pdf can be written according to Equation (5.2).

$$\mathrm{pdf}\left(x, m, \sum\right) = (2\pi)^{-n/2}(\mathrm{BB^T})^{-0.5} \exp\left(-\frac{1}{2} \cdot (x - \mu)^T \mathrm{BB^T}(x - \mu)\right)$$

$$\mathrm{with} \sum = \mathrm{BB^T}$$

(5.2)

Σ is called the covariance matrix, and it is a symmetric positive semidefinite n $\times$ n matrix (one row and column for each parameter). Therefore, we can interpret such multivariate normal pdf always as a kind of bell-shaped distribution, just in n dimensions. More picture we present in chapter "Design with Pictures Three" (and in part six, together with optimization – there the Hessian matrix **H** has a similar importance). If Σ is a pure diagonal matrix, we have no correlations and the bell shape is axially symmetrical. If we have nonzero nondiagonal entries, we have correlations and the bell shape is rotated to some degree.

Like the variance or standard deviation, we can also estimate the covari-ance matrix from statistical data. The problem is usually that the matrix can be very big, so many coefficients have to be calculated, and that requires usually many points, more points than for a univariate analysis!

Note that the matrix entries of Σ do not directly represent the correlations ρ, but there is a close relation. For the 2-dimensional case we can write:

$$\Sigma = \begin{pmatrix} V_1 & V_{12} \\ V_{12} & V_{12} \end{pmatrix} = \begin{pmatrix} \sigma_1^2 & \sigma_{12}^2 \\ \sigma_{12}^2 & \sigma_{12}^2 \end{pmatrix} \tag{5.3}$$

$$\rho^2 = \frac{\sigma_{12}^2}{\sigma_1 \sigma_2} \text{ with } |\rho| \le 1 \text{ (Pearson correlation coefficient)} \tag{5.4}$$

Note: In our book we use ρ for the *Pearson* correlation coefficient (many programs use p to avoid display problems), whereas c is used for correlation in general (e.g., also non-parametric measures).

The formula to estimate the covariance $\text{COV} = V_{12}$ from data is very similar to the one for the variance V:

$$\text{COV} = 1/(n-1) \sum (x_{i1} - \mu_1)(x_{i2} - \mu_2) \tag{5.5}$$

We can also estimate ρ directly (instead of using Equation (5.4)):

$$\rho = 1/\sigma_1 \sigma_2 \cdot \sum (x_1 - \mu_1) \cdot (x_2 - \mu_2) \tag{5.6}$$

With some additional math, one can prove that contours of constant probabilities lie on concentric ellipses. For $\rho = 0$ the ellipsoid axes will be in parallel with the coordinate system axis. The Pearson correlation coefficient r is ranging (like correlations c in general) between -1 and $+1$, and linear transformations will not change its value. For $r = |\rho|^2$ we have full correlation among the two variables, so knowing one of them gives us also the other accurately. If the offset voltage V_{off} and the threshold voltage of transistor N1 have $\rho = 0.5$, then $r^2 = 0.25 = 25\%$ of the variance in V_{off} can be explained by $V_{\text{TO}}(N_1)$! So a correlation analysis is very useful to find out how much each statistical *contributes* to the total variation (contribution analysis).

From basic statistics, it is known that the sum of two normal variables X and Y is still normally distributed. This is also the case for nonzero correlation. Without correlation the variances just add up, whereas with correlation we get

$$V(X + Y) = V(X) + V(Y) + 2\rho\sqrt{V(X)V(Y)} \tag{5.7}$$

So, whenever $\rho < 0$, then the variance is <u>less</u> than the sum of the variances of X and Y. For full correlation $\rho = +1$ we can just add the standard deviations (like we add up for non-statistical tolerances in the worst-case).

Such multivariate analysis could be extended to non-normal distributions, but usually anyway most of our statistical variables in the SPICE models are normally distributed. So, very often calculations are done with normal distributions, and if needed a transformation can be applied. In the statistical variable space, often such transformation can be found easily (in opposite to the performance domain), because the underlying physics is usually known.

5.2 Correlation

From public opinion polls or from inspections like "Should I become a pipe smoker because they live longer than nonsmokers?" you know that correlation does *not* directly mean that there is a true *causal* relationship! If we have <u>no</u> correlation among two random variables, then knowing one variable <u>cannot</u> help to make more accurate estimations on the other one. However, in case of correlation we could indeed improve our estimations! Also in older design environments getting correlation tables was available; and the problem on how to interpret correlations tends to be less difficult for circuit design than for medicine, because designers can setup their testbenches distinctly (and anyway everything relies on simulation models).

There are actually many different ways to measure and quantify correlations, e.g., we may inspect the yield on leakage current and speed in a logic circuit. Usually the parts having high speed unfortunately also have high leakage, so if the yield loss on both specs is 10%, the total loss might be indeed close to the worst-case of 20%, because we have a strong negative correlation on yield. One obvious problem here is that if you have relaxed specs and/or a low MC count, then maybe one or both partial yield losses are zero, so we cannot obtain the "yield correlation", and usually we want that the correlation is a measure on the design, on the relationship of variables, and not really on the spec limits. What we could do is relating the two performance variables leakage and speed in a <u>model</u>, e.g., a linear model, and this way correlation becomes just a coefficient in this (multivariable) model! This leads to the co-variance matrix for normal Gaussian distributions, and these kinds of methods are called parametric correlation.

Unfortunately, in circuit design linear models might be very inaccurate, and if you apply a linear model to a nonlinear system you may find <u>no</u> significant correlation although strong correlations are present! You may try in such cases more complex models like quadratic models or you can apply transformations, but this is time-consuming (which transformation to take?)

and already a quadratic fit takes more time and is usually less stable. For these reasons, also nonparametric correlation measures have been created. These are not based on a certain model, but usually on ranking, so we do not apply the correlation analysis directly to the performances, but first rank one variable and inspect how similar the data points are ranked in the other variable. This way all nonlinearities have no impact at least as long as the relationships are monotonic. So this is a quite robust method for parameter classifications, and often used as starting point for further investigations, like on parametric correlation, model selection, etc. Figure 5.3 gives an example, note that with lower correlation |c| the point cloud would look much less like a "string", more like a wide cloud.

The effort in model creation depends highly on the number of variables (but not all may really matter) *and* nonlinearities. The accuracy requirements may differ a lot, e.g., often for just getting a ranking it does not matter if a certain variable contribution is 30% or 33%, but e.g., for an optimization on the sensitivity you typically need a higher accuracy. The accuracy is often high in areas where many simulation results exist (like in the distribution center), but it is often much lower for the distribution *tails*, which is very critical for yield and failure rate estimations! So for simple cases 50 simulation points are often regarded as a kind of minimum, for more complex cases we can expect a linear rise according to the number of statistical variables, but for very complex circuits (typically with more than 50 transistors) usually some effort reduction is possible, because simply not all variables are important, but

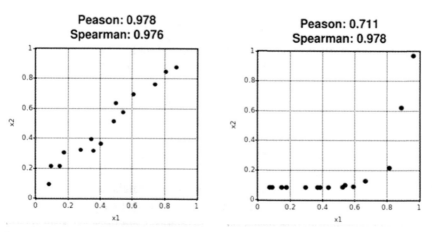

Figure 5.3 Pearson and Spearman (ranking-based) correlation coefficient for different 2D data.

maybe only 10%. Of course, for real circuits we usually need many models, one for each performance.

The higher the accuracy requirements to achieve not only a ranking but e.g., also a high yield estimation, the larger the model generation effort, but still there is usually some extrapolation risk that you have to be aware of. A concept reducing such extrapolation risks for yield estimation is worst-case distances WCD (see Chapter 7): Here the model fitting is around the fail region (and not the distribution center), and the model pass-fail border is linear. So if the true circuit pass-fail border is also linear, there would be *no* (local) model error regarding yield estimation!

In modern design environments, also classical response surface models (RSM) become better and better, due to more experiences, more computation power, and careful algorithm tweaking. For instance, high-order polynomial models can be used very efficiently. In approximations around a fix point, Taylor polynomials are best suited; for approximations in a fix interval, Chebyshev polynomials are best for given maximum error limit; in conjunction with statistics and normal distributions, so-called Hermite polynomials are best suited and used intensively.

We have seen that correlations can cause difficulties in finding the overall variance or the total yield, but powerful algorithms exist to *decompose* the problem into one without correlations: Principal components analysis (PCA) is a technique that starts with correlated variables and ends with uncorrelated variables, the principal components, which are in descending order of importance and preserve the total variability. PCA is of high interest for semiconductor foundries and their modeling teams, because from PCA results you can find out which and how many variables are required to create accurate statistical models for circuit design.

Correlation vs. Dependency. If we have one random variable x_1 and set $x_2 = f(x_1)$, then also x_2 becomes a random variable. However, x_2 is fully dependent on x_1. For linear f we would also get a Pearson correlation factor of either $+1$ or -1. However, for nonlinear functions, like a quadratic one, the linear correlation factor might be much smaller, although we would still have fully dependent variables. So correlation is mainly a measure of linear dependence. In conclusion: Zero correlation is no guarantee for independency, but if there are no dependencies the correlation must be zero!

Now imagine, you get some data and calculate the Pearson correlation factor from it, like $\rho = 0.15$. Does this mean we have indeed such correlation and also some dependency? Actually <u>not</u> 100%, because for small sample counts we would have also significant random variations, so also two completely independent random variables may have some random correlations. This is similar too many other situations, e.g., a normal Gaussian distribution might show some skew, just by chance.

Unfortunately, the situation could become even more difficult for more than two variables. In principle, we could look also for correlation among matrices, but even this would not cover all kind of dependencies. It is a bit like when moving from real numbers x to complex numbers z. We can do a lot more, but we lose also something because we cannot even say if the imaginary unit i is larger or smaller than zero.

5.3 Regression and Multivariate Modeling

A so-called linear regression can be regarded as the origin of many modeling methods. The easiest example is fitting a straight line, so a linear model with one variable x, to data pairs (x_i, y_i). Having two points we could just directly calculate the two parameters a_1 (slope) and a_0 of our model $f(x) = a_1 x + a_0$. However, in a MC simulation or from measurements we have usually many points to fit, so our set of equations becomes over-determined. Unfortunately it is usually impossible to reduce the model error ε_i to zero for each point, so starting with an attempt like $f(x_i) = a_1 x_i + a_0 + \varepsilon_i$ is a more realistic approach.

The simplest way would be to ignore many points and to select the two "best" ones, but this would often lead to quite arbitrary results; and in the presence of measurement errors, like noise, we would also throw away valuable information, which would lead to reduced accuracy. So one way to go is to define an error <u>critera</u>, and to <u>minimize</u> the error ε. One popular criteria is the average quadratic error, or the square-root of it, the rms error. This way bigger deviations $\varepsilon_i = f(x_i) - y_i$ would have *more* impact on the fit than small errors. So outliers could have quite a significant impact. Even more critical would be using the maximum error as criteria, and a more robust method would be to minimize the mean *absolute* deviation. So over-all the rms criteria is a often a good compromise; and it also leads to a simpler parameter estimation!

This was the situation till the end of the 19th century, so people used it because it was the easiest method. However, indeed there are further good arguments to minimize the quadratic error, and e.g., not the maximum error! One key reason is that if we assume a normal distribution for the error ε, then indeed the minimization of the rms error leads to the *best-possible* result! And in this case the model coefficients are almost directly related to the mean and variance of the data, which appear almost all the time in conjunction with Gaussian distributions. If no outliers are present, then also the assumptions of a normal Gaussian error distribution is quite native, e.g., because due to the central limit theorem ε would always approach a normal distribution if the error arises from summing many small errors (of finite variance). If we assume a normal error distribution then also the very general method of maximum likelihood (ML) would give us the same formulas, and the solution would be the most likely one, the one with *maximum probability*.

The obtain the solution for the model parameters (so-called regression coefficients) a_1 and a_0, calculate first the two sample means of x and y, and then use these equations:

$$a_1 = COV_{xy}/V_x = \Sigma(x_i - x_m)(y_i - y_m)/\Sigma(x_i - x_m)^2 \qquad (5.8a)$$

$$a_0 = y_m - a_1 \cdot x_m \qquad (5.8b)$$

These equations can be found by calculating ε^2 and minimizing it by setting its partial derivatives $\delta\varepsilon^2/\delta\varepsilon_1$ and $\delta\varepsilon^2/\delta\varepsilon_0$ to zero. Indeed the formula for the slope parameter look similar to the one for correlation, and you can double-check them for yourself by inspecting special cases.

Using these equations is straightforward, so let us inspect one typical and one example with some difficulties.

In Figure 5.4a we fit a linear model to well-behaving MC data, whereas in Figure 5.4b there are some difficulties. In both cases the data might be e.g., the output voltage of an amplifier versus the input voltage, so we model a performance $y = V_{out}$ as a function of a variable $x = V_{in}$, actually it does not matter if the variable is a design variable or e.g., a statistical variable (like V_{TO}). One obviously special behavior in Figure 5.4b is that the noise *increases* with x, so its variance is by far *not* constant. Note, that our approach was "having an error ε on top of y", but we assume that x is *known*, having *no* error. And we assumed that the error is an *additive* term. Maybe here another approach could fit better, e.g., a multiplicative error behavior? A second critical point is the data in the lower left corner, which seems to be slightly clipped! Maybe the measurements get wrong because the equipment was not suited

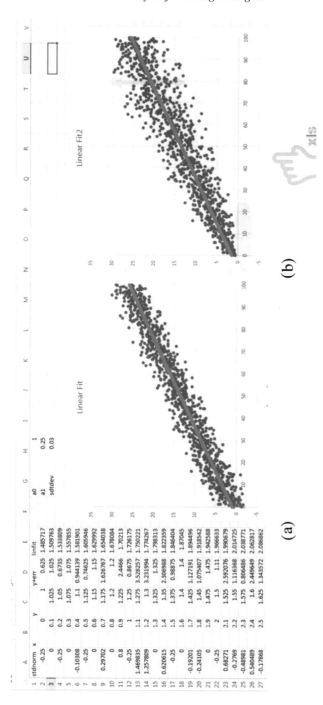

Figure 5.4   Two different measurement results with linear fit.

for negative values? Also in non-technical real world statistical examples we have sometimes to deal with so-called *censored* data. For instance, we want to calculate the distribution of travel expenses, but the report systems reports the exact numbers only for expenses if they are above 100$, for all lower ones we only know the number of these. Luckily such problems are seldom in MC, but not impossible, e.g., if you measure a delay time, and your simulator stop time is too short. Interestingly even for such difficult situations clever well-founded statistical methods are available.

What about having the error in y versus in x? Indeed assuming an additive error in y fits much better to real-world, simulations and MC problems. For instance, in MC we use random samples e.g., for V_{TO} or R_{sheet}, but still these random samples values are <u>fully known</u> to the simulator!

On top of all these issues, we may also ask why we only look to the error in $y = f(x)$, not e.g., regarding the first derivative? Also for such problems solutions are available, but again this usually less relevant for circuit design, because unfortunately from most simulations we do not get the derivatives will little effort. And if the errors are significant, then the derivatives are often even less accurate.

An interesting question is what could be the origin of the remaining the model error ε. In our amplifier, it could e.g., be pure electrical noise, or it could be that we forget to measure and include further *influencing* parameters, like supply voltage or load changes. In the latter case we can *improve* the fit with a *multi*-dimensional performance (response) model. Such generalization to n-dimensional linear functions $f(x, \mathbf{A})$ is possible by using matrix techniques, and also polynomials could be used as model in a similar way. In MC we are even in a better situation than a designer in a lab, because here measurements errors are much smaller (no error from measurement equipment, but only much smaller numerical errors), and also the influencing parameters are well-known.

In our case of an amplifier using a linear model was meaningful, but if we would also like to model the nonlinear behavior, like compression, an extended model is required. Indeed Figure 5.4 seems to indicate that there is some gain reduction at high input values, so e.g., a quadratic model could capture this effect. Arbitrary nonlinear models require iterative methods ("nonlinear least-squares"), just optimizers to minimize the error (see Chapter 8).

One interesting thing in models is that they allow better design under-standing, e.g., knowing about the sensitivities (our coefficient a_1 in the linear model) can help us to improve the design or at least to understand its limitations. If our simulations are accurate, if we include all important parameters and if we can find an accurate enough form of the model, then the model would give us almost the same performances as the simulations done for calibrating the model. And in addition, we could use the model for interpolations, extrapolations, and searches, e.g., for worst-cases or optimum behavior! A main advantage is often that e.g., such model-based optimization or search could run much faster because model equations are typically much faster to evaluate than performing true circuit simulations, this is even true for highly complex models, which may include dozens of parameters and many nonlinear terms. Later, we will pick up modeling techniques several times for advanced tasks, but now let us also discuss some further modeling tricks and direct applications.

Why we need "tricks"? We mentioned that "in MC … the influencing parameters are well-known", but this "well-known" only means that we know all the model input parameter names, just *all* parameters in the netlist (and in the simulator models) can influence the simulated performance of our system. In addition, the all values are available in principle too, but of course a model based on all parameters will be too complex by far, and for extraction (model calibration) we would need to run a huge number of time-consuming circuit simulations.

5.3.1 Variable Screening and Model Choice

A more complex model can typically give a better fit to the data, but the more the variables, the harder the estimation. Often MC data is so "noisy" that it just looks like only a high-order model can give a good fit. However, using too many model parameters can end up in a fit following the randomness too much. This is called *over-fitting*; and it should be avoided, because the model would become unsuitable for *predictions*, unsuitable sometimes even for interpolations, and even more for extrapolations! If you look to real-world data, like from fabrication or from looking to nature, it is often uncertain which kind of model should be used; e.g., in medicine, we simply do not know how many parameters are present and what the dependencies are! A good guideline is following the law of parsimony (sparingness) and only introduce as many parameters as really needed as a minimum—physical laws are typically simple!

This is a key problem for any researcher! However, doing a MC analysis we often have the opposite problem: We know the simulation setup and have to deal with many parameters and highly nonlinear models and circuits! However, which model should we create to understand, to simplify and speed-up the design process? In a first step, we can run a linear regression on all parameters, but then the problem is to make the algorithms adaptive enough to give a good balance between model bias (systematic error) and (statistical) variance. This can be done by adding a parameter screening step before we continue with modeling.

In this second step of modeling, we need to decide to apply only a simple linear model or e.g., a quadratic model. With a pure linear model, we have the risk that the model does not fit well, because maybe the circuit has also strong quadratic characteristics, and if we just generally apply quadratic modeling, we have more parameters to determine and need more MC points for the same accuracy. So often we should create an adaptive model that is linear in some variables, quadratic in few important variables, and even ignoring many less important parameters [Shan2011, Moon]. Actually the goal is to be accurate enough for making performance predictions for saving simulation time and to allow the designer the interpretation of the circuit behavior, not to create a model that is as accurate as possible—for this better stick to the netlist! As also the netlist-based simulation gives us a model, and we would create a new simplified performance model, the whole process is often called meta-modeling.

The problem of model selection occurs also in yield estimation, e.g., whether we should use the sample yield, the C_{PK} or the generalized C_{PK}. For instance, we could either choose the most promising approach (model *selection*) or we may even try to combine the different results (model *averaging*). Both have their application, and for both some mathematical foundations exist, but unfortunately the problem is harder than simple confidence intervals. Check out e.g., papers or books on decision theory and Bayesian statistics [Jaynes1995]. In most design environments the user get some feedback of the model accuracy, the goodness of fit (GOF). One numerical criterion for the GOF is the *coefficient of determination r*. It measures the fraction of variation in the data which is captured by the response model.

$$r^2 = 1 - 1/((n-1)V) \cdot \Sigma(f(x_i) - y_i)^2 \qquad (5.9)$$

Note: r is *constructed like* a correlation coefficient, but a model fit will never give negative values, and also things like covariance make little sense here.

r^2 is zero if the model has an average quadratic error ε^2 identical to the variance V of the data. Remembering that the variance is average quadratic error regarding the mean y_m, we can say that in this case we would just assume a model of *zero order* (just $f(x_i) = y_m$). For simple performances like offset voltage we can expect that already going for a linear (first order) model would improve r^2 e.g., from 0 to roughly 0.9 or higher. This means the model "explains" 90% of the variation in the data. With no model or with a zero order model, we would need to capture all the variance with pure statistical methods, and we would e.g., need to accept a certain uncertainty represented by confidence intervals. With a good model we can reduce this uncertainty (and often quite dramatically), which means that modeling is to some degree a kind of noise reduction method! In Chapter 6 we will use MC and modeling together for a more stable yield estimation. This is one method to enables Monte-Carlo analysis for regions of higher yields (like $4-6\sigma$, depending on complexity) with acceptable simulation effort.

A simple single criteria like r^2 could be sometimes misleading, so it is usually best to inspect the fit directly in a scatter plot. Even if r^2 is low, there could be still an option for model generation, just obviously the currently applied model fit is not good. However, inspecting other model parameters or another form could lead to improvements.

Another disadvantage of r^2 is that it does not take the model *complexity* into account. This means you cannot use r^2 directly as a criterion for model selection or parameter filtering, because in most cases r^2 would become higher and higher by making the model more complex, till you may end up in fitting to "noise" (over-fitting). A better model selection with the purpose of performance predictions is possible with other numerical criteria like the AIC (Akaike information criterion) or BIC (Bayesian information criterion, e.g., [Jaynes1995]). Both combine the goodness of fit with the number of parameters into one value. This allows a quite solid comparison of models with different complexity, and e.g., when to stop modeling refinements.

5.3.2 Variance Contribution Analysis

An immediate application of a multivariate analysis is finding the sensitivities and the (relative) contributions of a design. For statistical variables, we are usually interested in the different contributions of each statistical variable (or a variable group) to circuit behavior, like "mismatch in V_{TO} of M2 creating 30% of the total offset". This is the primary application, and it is also a perfect example on how Monte-Carlo can give additional insight to the design, by identifying the critical devices (like transistor N3 or resistor R5) with *largest* impact on performances. For this analysis, no specs are required, so it can be done at a very early stage of the design (in opposite to a full yield analysis).

The first step for such contribution analysis is to assume a certain model—being a function of the statistical parameter—and compare it to the data. Of course we aim for finding the optimum coefficients in our model that minimize the model prediction errors (often the rms error is used). Once we found the model, we can check how much impact each parameter has and provide a sorted list as analysis output (Figure 5.5).

Of course a model, created via linear regression, would usually not have zero error, so some uncertainty is still present, and the sensitivity results can be trusted only if the goodness of fit is high enough. If e.g., the coefficient of determination r^2 is too small, then the model might be not good enough (see Table 5.1 for a case where a linear model is applied mistakenly) or the results are too noisy. However, look up: r^2 is often still close to unity (like 0.91, which means that 9% of variation is <u>not</u> explained by the model) *even if* the fit in the *tail* region is *bad*. This can happen because r is usually dominated by the mass of samples in the center region. For pure contribution analysis, this is only a small problem, but later we use such multivariate models also for other purposes like yield estimation, and here the tail region is very important because it contains the rare events causing problems.

To get more stable results (especially for mismatch variables), often some tricks are applied, like we can put instances in parallel together, and also for full subcircuits we can obtain the contributions, and they are of course also more stable than the contributions on individual instances.

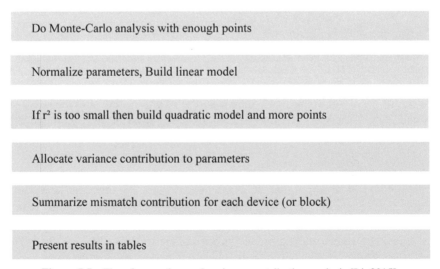

Do Monte-Carlo analysis with enough points

Normalize parameters, Build linear model

If r² is too small then build quadratic model and more points

Allocate variance contribution to parameters

Summarize mismatch contribution for each device (or block)

Present results in tables

Figure 5.5   Flow for an advanced variance contribution analysis [Liu2015].

Table 5.1 Variance analysis results on a nonlinear testcase using different models[1] and MC count [Liu2015]

Golden Result (n = 2000)		Proposed Method (n = 500)		Linear Model (n = 500)	
Fit	$r^2 = 0.900$		$r^2 = 0.901$		$r^2 = 0.247$
/I0/M2A: deltoxn	43%	/I0/M2B: deltoxn	40%	/I0/M2B: deltoxn	0%
/I0/M4A: deltoxp	38%	/I0/M2A: deltoxn	39%	/I0/M2A: deltoxn	3%
/I0/M4B: deltoxp	5%	/I0/M4B: deltoxp	6%	/I0/M4B: deltoxp	0%
/I0/M1A: deltoxp	5%	/I0/M4A: deltoxp	5%	/I0/M4A: deltoxp	5%
/I0/M1B: deltoxp	1%	/I0/I1/M6: delvthn	1%	/I0/I1/M6: delvthn	0%
/I0/M7A: deltoxn	1%	/I0/I1/M7: delvthn	1%	/I0/I1/M7: delvthn	0%
/I0/M 7B: deltoxn	1%	/I0/I4/M1: delvthp	1%	/I0/I4/M1: delvthp	1%
/I0/I1/M1: deltoxp	0%	/I0/I4/M2: deltoxp	1%	/I0/14/M2: deltoxp	3%
/I0/I1/M1: deloxp	0%	/I0/M7A: deltoxn	1%	/I0/M7A: deltoxn	0%
/I0/I1/M1: delvthp	0%	/I0/M7B: deltoxn	1%	/I0/M7B: deltoxn	3%

[1]The data was achieved with setting non documented internal environment variables. A normal user would use the automatic mode and would never see the bad results from pure linear fit as output.

The classical application of such contribution analysis is inspecting the "bad guys" in a circuit, e.g., those responsible for most of the offset voltage. In a non-optimized design, often a few transistors dominate the offset, so if you make their area 4× larger you can halve their offset, and if these transistors are really the dominating ones, then also the overall circuit offset would be reduced to almost 50%. To double check this, you can run MC again and also the contribution analysis.

Note that there is no restriction to DC performances, and you may also look to the contributions of a filter characteristic or any transient behavior! In principle no full re-run of the MC is needed: If you keep the same seed value of the pseudo-random number generator and only change parameter values (but not circuit topology), then it would be theoretically enough to re-simulate just a few extreme corner samples, because also these should show the offset reduction. This way we approach the idea of statistical corners and worst-case distances (see Chapter 7).

In modern processes and complex circuits, it can easily happen that it becomes hard to make the number of MC points larger than the number of statistical variables and the model parameters, so classical (linear or nonlinear) regression cannot be applied. Besides nonlinearity and the problem of which kind of model should be used, this becomes a further problem. One approach to maintain efficiency is using so-called orthogonal matching pursuit OMP [Tropp2007, Liu2015] and advanced ranking methods [Moon] (Figure 5.6).

The idea is to focus on the most important relationships, so that we can find with a moderate number of simulations (like 100) still at least the top-5 contributions. This way it is indeed possible to provide useful design insights in a very efficient way: Imagine you have a big series connection of resistors of two types, e.g., 400 pieces of 1 Ω and 4 pieces of 100 Ω, all having 1% mismatch tolerance. The 100 Ω resistors dominate the overall tolerance behavior, but with simple OFAT simulations you would need at least 405 simulations to verify this! With OMP and a 40-point MC run, you would also identify the four big contributors correctly. The price we have to pay is that the accuracy would be limited, e.g., actually all four 100 Ω resistors need to have the same contributions, but due to the randomness of MC it would be not the case, and the errors might be e.g., 30%. However, this would be still enough to identify the top contributor correctly, and you can now run OFAT only on *these*, so overall you need with this trick only $40 + 5 = 45$ simulations to actually get an accurate result. Of course, for resistor circuits any designer could easily apply some hand calculations, so we will later demonstrate this technique in much less trivial nonlinear circuits.

In [Moon] further (quite nonlinear) engineering examples of moderate complexity could be found, and also the problem of false discoveries (unimportant variables not filtered out unfortunately) and false non-discoveries (strong variables will be screened out accidently) is addressed. To optimize the screening the algorithms need to be carefully tweaked to find a good compromise between screening accuracy and speed. As rule of thumb the total number of sampling point has been set (in a quite conservative way) to $n = n_1 + n_2 = 5 \cdot s + 2 \cdot p$ (s = total variable count, p = active, dominating variables). The worst-case on screening errors happens usually if there are many parameters with small effects "in the same direction". Then there is a strong risk that almost all such parameters will be regarded as inactive, but this way also there significant over-all effect gets ignored. This is quite a severe problem, because also in circuit designs such cases are not rare. Think e.g., of a voltage divider built with many small resistors in series,

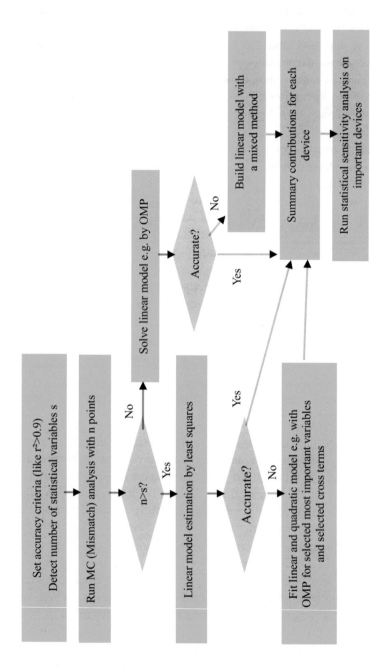

Figure 5.6   Typical flow chart of an advanced contribution analysis [Liu2015].

or of a flash ADC where each offset can cause DNL errors, but the offset may come from quite many parameters. Exploiting structural or hierarchical information we could improve further (even with automated recognition techniques, see Section 9.4.5). In addition, we can add another iteration after the second last step in Figure 5.5, or by using adaptive sampling methods. More on advanced sampling methods beyond grids and random sampling in general can be found in Chapter 6.

5.4 Adaptive Sampling and High-Dimensional Models

In 5.4 we end up with a two-step approach of MC plus OFAT for an accurate mismatch contribution analysis. This is actually suggesting the next step of automation in future tools; and this will be "adaptive" sampling. This would also enable to extract more complex models with an affordable number of simulations. Such complex models are often called high-dimensional model representations HDMR, and they have the form (5.6).

$$f(\mathbf{X}) = f(x_1, x_2, \ldots, x_n) = f_o + \Sigma f_i(x_i) + \Sigma f_{ij}(x_i, x_j) + \ldots$$
$$+ f_{1..n}(x_1, \ldots x_n) \tag{5.10}$$

Notes:

- The 1st and last entry are unique, so no need for the sum symbol here.
- Usually we expect the model is valid for a certain finite parameter space, e.g., $0 \le x_i < 1$ (box space).
- In the modeling task we fit *first* the low-order terms, so that the higher-order terms are only used to model the remaining low-order model *errors* ε_i. So we start the fit for f_0, the move on with 1st order terms, 2nd order terms, etc. (often using a recursion).
- The functions can take any form, no need to restrict e.g., to polynomials
- This representation is also often used to check different sampling algorithms [Kocsis1997]. If high-order terms are significant, then the variance reduction e.g., via low-discrepancy sampling is limited (Chapter 6).

Example: Using 2nd order polynomials

$$f(\mathbf{X}) = f_o + \Sigma a_i \cdot x_i + \Sigma\Sigma b_{ij} \cdot x_i \cdot x_j + \cdots + \Sigma c_i \cdot x_i^2 \tag{5.11}$$

> **Sensitivity-driven design and other methods sensitivity analysis.**
> Variation-awareness and sensitivity-driven design have a strong connec-
> tion. We mentioned two so-called perturbation-based techniques to obtain
> sensitivities; one is OFAT the other is a Monte-Carlo-based. Both have
> their limitations, so one may wonder for more options. Indeed, a simulator
> performing a sensitivity analysis is often doing something different, using
> the so-called *adjoint network* method. This allows a faster analysis by
> reusing the results of a DC or AC analysis. This method is also very
> accurate, but the major problem is that it works not (well) for transient
> analysis.

Notes:

- This form is often used, also in the matrix form (see Chapter 8 on optimization)
- Only f0 can model a constant term, so we can estimate $f_o = \int f(\mathbf{X}) d\mathbf{X}$, then we can continue the estimation in a similar way for the other terms.

This approach is used successfully in many areas of engineering, because on
the one hand it is very general but also in many real applications the influences
of high-order terms is quite small, so we can keep the numerical effort still quite
moderate. In addition, following (5.6) mathematicians have proven some nice
statements about sampling and modeling. The optimum "simulation points"
to extract the model depends on the model characteristics, but often these are
not fully known, e.g., we often do not know exactly if a linear model would
fit, or which variables give a quadratic contribution. So *adaptive* sampling is
becoming a very native solution. If, on the other hand, we know that we have
to apply a e.g., linear model, we could ask now indeed for the best simulation
points. Actually different definitions of "best" exists, but looking for minimum
variance is the most popular choice, and that leads to point sets with "large
coverage". For example, we could sample x_1 at 0, 0.5, 1 and ask which point
should we *skip* to still have a variance as low as possible? The mathematical
solution is that you should skip the middle point(s) to span a covered "volume"
as large as possible; and this is in synch to what designers also do anyway in
corner simulations.

Actually, very often math gives us a good backup for our manual design
methods, but also the limitations become often more clear. In this case, we
could also take another model, like a pure second order model, and indeed
the optimum point set could change, so indeed the approach of only covering

the extreme points would not always be the best one. It is "practically" the best one, because in reality most designs behave quite linear, or at least more linear than pure quadratic. This is also the reason for the success of advanced (but still fix) sampling methods, which we describe in Chapter 6. Also note: The best point set for extraction depends on the model, so if we do not know the model we cannot choose the best points. This looks like a "chicken and egg" problem, but it is not that bad, because also starting from a non-optimum set of simulated points gives us useful information about how the model should look like, so *iteratively* we can solve also such difficult problems mathematically, step by step, like designers do. In [Woods2015] you can find some detailed algorithm descriptions combining advanced sampling, screening, and modeling techniques; and it includes a small benchmark.

5.5 Multivariate C_{PK}s

The classical process capability index C_{PK} and the generalized C_{GPK} treat <u>one</u> specification and <u>one</u> output data set at a time. Usually a datasheet contains many specifications, but often only a few of them cause the biggest headache for designers. With the new C_{GPK} they can treat the specifications efficiently, but sometimes there are cases in which the yields of the individual performances are quite easy to fulfill, but the overall yield might still be surprisingly quite low, because the performance characteristics compete and are difficult to fulfill on the same time. The so-called multivariate C_{PK} can be calculated and will take correlations into account.

In the univariate case, there is a one-to-one relation between the yield and the specification limit (acc. to inverse cdf, i.e., the percentile function), whereas for the general case infinite specification combinations exist to obtain a certain total yield! For one dimension a spec is always a simple interval like [LSL, USL], whereas in multiple dimensions not only both boxes (mathematically spoken hypercubes) but also ellipsoids could make sense.

These issues and several others (like treating non-normality) make the application and even the definition of multivariate C_{PK}'s more difficult, especially as in the design phase not all specifications might be fully clear. Therefore, there is yet no standard on multivariate C_{PK}s.

Note: If you search for "yield estimation" you will often find pure EE papers, in general or for math literature it is better to search for "percentiles".

5.5.1 Total C_{PK} Estimation via Correlations

Having no standard method does not mean that you cannot do anything. As mentioned, if we have the C_{PK}s or C_{GPK}s for all our performances, we can approximate the total yield and so the overall C_{PK} by just taking the minimum (worst-case) among C_{PKi}, i.e., we would simply ignore all correlations. This calculation would be identical to just assuming positive correlation c $= +1$, so *non*-fighting specs. An alternative method would be to calculate the related yield loss for each C_{PK} and to add all losses to be on the safe side, so assuming *fighting* specs. A compromise would be assuming *no* correlation, just to multiply the individual yields, but also this tends to be (a bit) too pessimistic.

As we now know well how to deal with correlations, we can even do better, and without additional time-consuming calculations or simulations, even for moderately high yield targets.

Example: If you have a single upper spec limit and $C_{PK} = 1.0(3\sigma)$, this is equivalent to a yield loss of 0.135%. If a second C_{PK} on another performance is $C_{PK2} = 2.0(6\sigma)$, then the total yield is highly dominated by the smaller C_{PK}, so also the total yield is very close to 3σ or the total loss is 0.135%, so taking min (C_{PKi}) is a good approximation for the total yield. However, if the 2nd C_{PK} is also 1.0, then the total loss depends on correlation, and the total loss may range from 0.135% to 0.27% (2x uncertainty)! In terms of sigma (or C_{PK}) the error is equivalent to approximately 10%.

Note: Such calculation could be done quite easily also for correlations different from -1, 0 or $+1$, but also the measurement of correlation would have an impact, because the correlation might be nonlinear and different for tail and center regions of the distribution.

Of course, for more performances and specs this kind of error could grow significantly. What helps is that often few C_{PK}s dominate anyway, so usually the systematic yield loss error is well below the maximum of $2\times$, like only 10–20% (e.g., on the 0.135%). On the other hand, it could also happen that especially in a well-optimized design the spec margins are quite well balanced, so that not just one spec dominates. A good rule of thumb (look at Table 5.2) for $C_{PK} > 1$ (equivalent to $>3\sigma$ or $<0.17\%$ loss) and only two performances is this: If the second worst C_{PK} is $+0.15$ larger than min(C_{PK}), then the correlation effect is below 20% in terms of yield loss and smaller than 0.06σ.

Table 5.2 Total yield for two C_{PKS} or WCDs and different correlation factors

Y_1	C_{PK1}	Y_2	C_{PK2}	Correlation	Total Yield	Comment
3σ	1	3σ	1	−1	2.781σ	Worst-case
3σ	1	3σ	1	−0.5	2.781σ	Very close to worst-case
3σ	1	3σ	1	0	2.784σ	Close to worst-case
3σ	1	3σ	1	0.5	2.793σ	Close to best case
3σ	1	3σ	1	1	3σ	Best case, identical to min(WCD) but usually too optimistic
3σ	1	4σ	1.33	−1	2.994s	Worst-case, not far from best-case
3σ	1	4σ	1.33	0	2.994s	Close to worst-case
3σ	1	4σ	1.33	1	3σ	Best case

Note, all these errors are also present in many advanced methods like worst-case distances (WCD $= 3C_{PK}$ for Gaussian distributions), because they also only address the *partial* yield problem, not the total yield.

Several methods are known to address the problem, but many of them are rather complex and usually unnecessarily compute-intensive such as principal component analysis (PCA). A simpler general solution could look as follows: The total yield is a function of the two partial yields Y_1 and Y_2 and the correlation factor c, being between −1 and +1. The cases $c = -1, 0, 1$ are easy to solve:

$$Y(c = 0) = Y_{\text{nocor}} = Y_1 \cdot Y_2 \qquad (5.12)$$
$$Y(c = 1) = Y_{\text{bc}} = \min(Y_1, Y_2)$$
$$Y(c = -1) = Y_{\text{wc}} = \max(0, Y_1 + Y_2 - 1)$$

Note: The min or max function would work for Y in sigma or e.g., in percent, but in these equations better use the Y in absolute terms like $Y = 0.997$ (look at Table 3.2 for the relation between sigma and percentage loss).

For arbitrary c, we can simply interpolate using a power function fit through these three known combinations. The correlation c might be calculated as Kendall's tau (non-parametric correlation measure, see Wikipedia), or we could also exploit the fact that our power fit allows an inversion (look at Figure 5.7), so if we tighten the specs artificially till each partial yield is 90%, we could get all partial and the total sample yields *directly* by counting, and this way we could also calculate *back* to the correlation c. Note: If one method gives $c = -0.9$ and another −0.95 this has usually not much effect on total Y (see Table 5.1); and the results could be used for further internal error calculations.

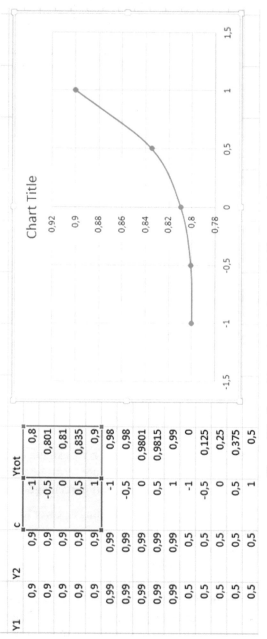

Y1	Y2	c	Ytot
0,9	0,9	-1	0,8
0,9	0,9	-0,5	0,801
0,9	0,9	0	0,81
0,9	0,9	0,5	0,835
0,9	0,9	1	0,9
0,99	0,99	-1	0,98
0,99	0,99	-0,5	0,98
0,99	0,99	0	0,9801
0,99	0,99	0,5	0,9815
0,99	0,99	1	0,99
0,5	0,5	-1	0
0,5	0,5	-0,5	0,125
0,5	0,5	0	0,25
0,5	0,5	0,5	0,375
0,5	0,5	1	0,5

Figure 5.7 Spreadsheet showing the total yield vs. correlation.

Example:
$Y_1 = Y_2 = 90\% \Rightarrow Y(1) = 90\%$,
$Y(0) = 81\%, Y(-1) = 80\%$,
e.g., $Y(c = 0.5)$ gives 83.85% yield
$Y_1 = Y_2 = 99\% \Rightarrow Y(1) = 99\%, Y(0) = 98.01\%, Y(-1) = 98\%$,
e.g., $c = 0.5$ gives 98.15% yield

An extension to an arbitrary number of partial yields is possible too, but as mentioned often one or two C_{PK}'s dominate anyway—and only among these the correlation has a certain impact (Table 5.3). If the difference in partial yield is $> 1\sigma$, the error from correlation is typically below 0.006σ, which means that even if we had this error multiple times, the overall impact would be still small, compared to the sampling error (or other errors, see end of Chapter 7). In addition, we could repeat the correlation calculation multiple times till the overall best case and worst-case would be close enough together. Table 5.3 gives some examples for higher dimensions and for 3σ, note that for higher yields the impact of correlation becomes smaller (in terms of sigma), which is a side effect of the $\int e^{-x^2}$ law between C_{PK} and yield.

Note: It looks that by neglecting some high partial yield results we would always be (a bit) too optimistic on the total yield, but actually there are also effects leading to some pessimism. For instance, taking the minimum C_{PK} (or C_{GPK} or WCD) creates some negative bias, if we deal with sample estimates instead of true values. This is because the minimum function is nonlinear, having a negative curvature.

So once we have calculated the total yield and "overall sigma," we could easily <u>add</u> the required correlation-dependent margins in sigma or terms of (generalized) C_{PK} to the <u>partial</u> yields. This way we can optimize

Table 5.3 Total yield for multiple C_{PK}'s of 1 and different correlation factors (case $c = -1$ is close to 0)

Dimension	Correlation	Total C_{PK}	Loss Error	Comment
2	−1	0.927	2×	Minimum C_{PK} to get 1.0 overall is 1.068
2	1	1.000	0	
3	−1	0.883	3×	Minimum C_{PK} to get 1.0 overall is 1.107
3	1	1.000	0	
4	−1	0.850	4×	Minimum C_{PK} to get 1.0 overall is 1.133
4	1	1.000	0	

meaningfully not only on partial yield(s), but also on the total yield, without the requirement for huge MC counts, and even for cases with strong correlations and non-normal distributions.

5.5.2 Total C_{PK} Estimation via Blocking Min

Simply using $\min(C_{PK})$ as estimate for the total yield can be quite inaccurate because it ignores correlations. But instead of using the performance correlations, we can also combine the MC data for different performances in another way, by using the so-called "blocking min" approach. Remember, for the total sample yield we simply check each MC sample if it fails at <u>any</u> of our specs, and one fail is enough, i.e., the worst-case counts. On the other hand, we know that using only pass-fail information leads to wide confidence intervals as we simply remove information! So what about making a kind of "overall spec-margin" approach, a kind of "analog" overall pass-fail in the style of the C_{PK}? Following the approach from [McConaghy] we can first <u>normalize</u> all performances like zero means performance hits exactly the spec, <0 means spec violation and >0 means pass, and the larger the spec margin the larger the <u>relative</u> spec margin m. For instance, for a sample x and an upper spec limit we could use $m(x) = (USL - x)/\sigma$. We can do this for one MC sample for <u>each</u> spec, e.g., BW may give $m = 0.5$, PM gives $m = 0.2$, etc. Now we can apply the min-function (giving the worst-case) in the same way as we do for total yield for each MC sample, but as we have now a <u>continuous</u> measure we can expect tighter confidence intervals, and we can better treat cases with no or few fails! This way we nicely include all correlations in a very native way. In [McConaghy] the authors claim that this method is reliable and trustable, but actually there are also weaknesses: To treat <u>all</u> the different specs correctly, we really need a correct normalization, but unfortunately using the standard deviation sigma σ is only acceptable for near-normal data. Having long-tailed data (like for a Student's, Pareto or Cauchy distribution) it could happen that the sample standard deviation becomes inconsistent and would never converge to a finite value. This way one difficult performance may corrupt the whole yield calculation.

Another problem is this: we get a new distribution from $\min(m)$ and this distribution is quite non-normal <u>even</u> if the original MC data is normal, so using the C_{PK} for it would create further inaccuracies. [McConaghy] proposes the use of kernel densities; Figure 5.8 gives an example of moderate difficulty: One performance gives normal data, one is lognormal, and both have the same partial yield (for specs set to 2.5 in Figure 5.8).

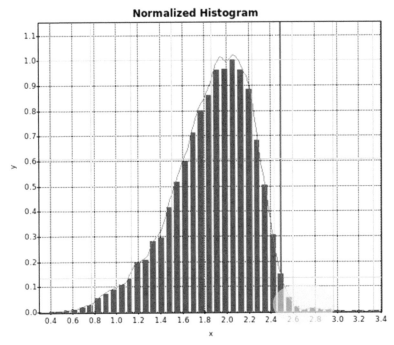

Figure 5.8 Blocking-min histogram and KDE fit for one normal and one log-normal distribution (using the standard deviation for normalization) leading to an overall difficult to handle distribution.

The long tail of the lognormal distribution leads to a "kink", but thanks to the use of 16K points and well-set KDE bandwidth the fit looks meaningful. However, in reality you are seldom in such a comfortable situation, because having less points leads to much more noisy data and increasing the KDE smoothing interval leads to a significant bias error at the spec value, where just the "knee" is! In general, KDE is also not good at all for extrapolations, and the error from potentially bad normalizations is not treated at all by KDE. Please also look at Figure 5.18 for further KDE examples, and to Figure 7.13 in chapter on k-sigma corners. The later shows the KDE application for yield estimation in comparison to other methods like C_{PK}, sample yield, etc.

A better idea would be to use the C_{GPK} (Chapter 4) for the total yield estimation based on $\min(m)$, plus an improved normalization among the different specs. A more robust, less bias-generating scale measure would be using percentile differences like $p_{97}-p_{50}$ (p_{97} gives the point on the x-axis giving 97% yield or cdf $= 0.97$) instead of σ or to perform the scaling based on pdf estimates. This way we can normalize at least almost correctly, so that the "kink" would disappear almost, even for the sketched critical case of long-tailed or asymmetric distributions, and based on that, we can indeed obtain similar good results as by using the correlation method explained in the previous subsection, even with lower effort.

A small advantage for using correlations is that they usually also help on understanding the design and the trade-offs, but a correlation analysis takes (slightly) more time (usually still much less than circuit simulations).

We could also combine the two methods, using the blocking min method to get an estimate, then using the correlation method and include only as many correlations till we get a certain yield accuracy. This way the designer could also immediately see how important the correlations are and which of them are relevant.

5.6 Design with Pictures Part Three

Let us now inspect a more difficult circuit example. A contribution analysis is often helpful for design insights, especially doing it for mismatch effects, because as a designer you can influence mismatch to some degree (more than e.g., global process variations). Sometimes the results are more or less trivial, e.g., the offset voltage is usually dominated by the input transistors, but actually this was true only in older technologies. For instance, in bipolar or BiCMOS you have transistors with high gain like 40 dB, so indeed the 2nd stage has only a very small impact, but in modern CMOS you are in a more difficult situation, and you need to keep an eye on many more transistors!

Of course, still some circuits are only critical in a few aspects, like for a bandgap reference usually DC accuracy matters most, so it is easy to overdesign for other characteristics (like by spending a lot of chip area). Also for many special blocks a solid theory is available, like for sensitivities of LC filters.

An interesting block is a comparator, because it is often equally critical in several aspects like area (if you need many, like in an ADC), speed, accuracy, kick-back, noise, metastability, etc. In particular, a latched comparator is also very critical on layout parasitics too, so let us take such a block as DUT for our advanced techniques like correlations, (mismatch) contribution and (later) worst-case distances WCD and optimization.

5.6.1 Latched Comparator Sensitivity Analysis

Some things are hard to demonstrate, e.g., in digital circuits mismatch often has only a minor impact. And some circuits (like CMOS NAND) almost always work—almost independent on sizing and technology. However, this is not true for typical analog circuits; only a few very basic circuits like maybe a 10-transistor OTA are uncritical, so let us also choose a more difficult comparator topology, being not so easy to design by hand (Figure 5.9).

The components usually come by default already with MC models, but as the comparator is critical to layout capacitances too (see [Geiger] for a detailed offset analysis on a similar comparator), we have added rough parasitic estimates with a certain sigma (like 10%) to the schematic. Now we can run a random MC analysis, and a contribution analysis for sensitivities (on roughly 100 parameters) from the MC data, after it [Weber2015]. It is also possible to run a sensitivity analysis by simple OFAT techniques (e.g., we can shift each parameter by $\Delta x = 0.1\sigma$ to get $S = \Delta y / \Delta x$), so we can compare both methods. Doing so in Figure 5.10 we can see some inaccuracy in the mismatch contribution results; this is typical. Besides increasing the MC count, we will present in Chapter 6 methods to improve the accuracy without more runtime (e.g., using so-called low-discrepancy sampling LDS).

We can sort on sensitivities, and of course small variations in a performance usually indicate low sensitivity and a stable design, whereas big variations may indicate potential problems or point to instances which are just natively critical (like input transistors are usually critical on offset).

To check for nonlinearities, we should look to V_{offset} and to $|V_{\text{offset}}|$, for the latter we get non-normal data and we expect that the contribution analysis becomes more difficult, because here at least a quadratic model needs to be fit. We can also expect more difficulties for performances which are impacted by many transistors or for those where the simulation accuracy is not perfect (like effective input capacitance in our specific example).

Running 200 samples is usually acceptable for such small blocks; using only 50 points would lead to a really inaccurate contribution result for $|V_{\text{offset}}|$ (indicated from the tool by r^2 not available). For $n = 200$ we get a kind of just acceptable accuracy for difficult measures (e. g. we get $r^2 = 0.9$) and for well-behaving outputs even quite an accurate fit ($r^2 > 0.99$). In the latter, we see that instances which should have identical contributions (e.g., input diffpair on offset) have at least very similar contributions (like 20% vs. 18%). The results for hysteresis are similar (40% vs. 42%). Figure 5.10 shows a typical instance-related output.

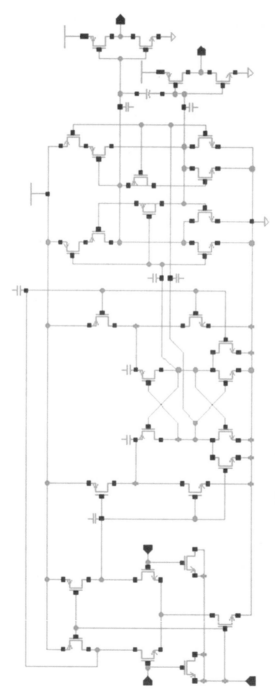

Figure 5.9 CMOS multi-stage latched comparator schematic.

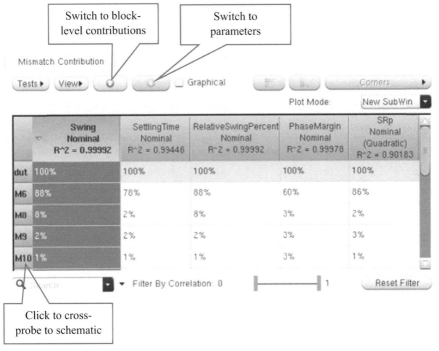

Figure 5.10 Typical contribution table with instance-related output [Deepchip2014].

Once you inspected the instances (like N1, P2, R3, etc.) on their contributions, you can often also switch the hierarchy level in the environment to e.g. the contributions on the (transistor) parameters itself. This allows to check whether V_{TO} or mobility or area w · l is more critical. In many cases designers have a good feeling about this, but in current sources it might be not so clear if V_{TO} or mobility is more critical. Knowing such relations can help to improve the circuit, by thinking instead of trial and error.

In Figure 5.11 an according typical table regarding the statistical parameters is shown. In our circuit, we can e.g. observe that for hysteresis the V_{TO} matters much less than wl-parameter.

To get a good overview also in complex designs often several spreadsheet-like filtering and sorting options are available.

Relating the contribution results to real circuit design problems, the most interesting results are here probably on offset (because this is a key factor on most comparator designs) and on hysteresis. The latter can be often made very small, due to reset transistors. However, actually the hysteresis in our circuit

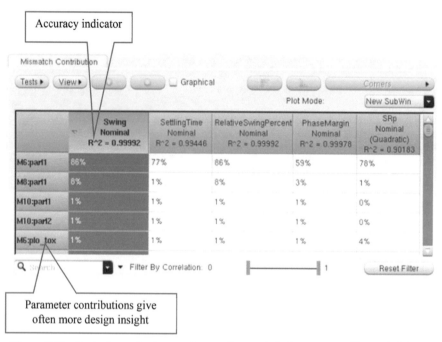

Figure 5.11   Typical contribution result regarding statistical parameters [Deepchip2014].

was *not* that small, even under nominal conditions. To really find the root cause we needed a mix of trial and error, thinking, manual sweep techniques and contribution analysis: Interestingly the output NOR latch is introducing a significant amount of hysteresis and that is the third amplifier stage! Already this is a nice result, because typically offset and hysteresis are regarded as a problem for the design of the first stages only! So the circuit has been extended by two "dummy" MOS transistors around the output latch (Figure 5.9). This gives us some control on adjusting the hysteresis.

Another interesting part is often the sensitivity to the *parasitic* capacitances at the different *nets*, because this can serve as layout guidance. Here the results were as expected, the first stage output net is most sensitive, and based on that we may setup later a layout constraint (look at Chapter 10 for more about constraints) for the net capacitance, to allow only a certain amount of mismatch like 1% (Figure 5.12). Actually this result is no big surprise, because high-impedance near-input nets are usually critical, but having numbers on *how* much offset comes from transistor mismatch and *how* much from layout is again very helpful.

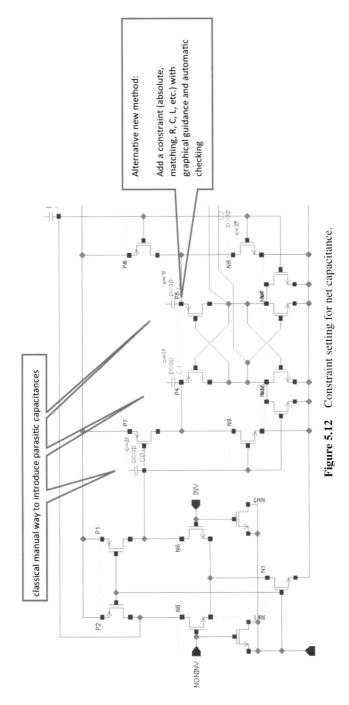

Figure 5.12   Constraint setting for net capacitance.

Note, that in other designs, like those without pre-amplifier in front of the latch, might be much more sensitive to layout asymmetries than our a bit more "tricky", more complex comparator. One example of such more basic dynamic comparator would be a pure so-called "strong-ARM" latch [Kobayashi].

If you want to get rid of contribution inaccuracies due to MC sampling, we could also run an OFAT-based contribution analysis. Here the contributions from the differential pair transistors are indeed really identical (Figure 5.13). However, such OFAT analysis needs as many simulations as statistical parameters; so if the circuit is more complex, the MC method is more efficient.

A correlation analysis among the outputs is interesting too. We could check if measures like offset and or delay (PSRR, hysteresis, etc.) and are correlated or not. A good starting point is looking to scatter plots. Figure 5.14 shows that hysteresis and offset are not much correlated, so we can hope that we are able to optimize both quite independently.

The nominal performance is usually close to the MC histogram center or close to the sample mean, but there are also exceptions, e.g., a stable op-amp design has $90°$ phase-margin and but even best designs would hardly exceed $95°$, but of course "outliers" might be well below $60°$, so the scatter plots

Parameter Name	V_{offset}	I_{DDavg}	t_{delay}	V_{hys}	C_{ineff}
PM9:pltw	0.06%	0.89%	0.13%	41.22%	0.00%
PM8:pltw	0.06%	1.45%	0.02%	41.22%	0.00%
NM4:pltw	0.15%	0.96%	10.09%	7.25%	0.00%
N10:pltw	0.15%	0.19%	0.00%	7.25%	0.00%
N1:pltw	0.00%	0.27%	39.80%	1.22%	1.94%
P12:pltw	0.06%	1.23%	0.46%	0.18%	2.18%
P11:pltw	0.06%	1.07%	0.14%	0.18%	0.00%
N1:pvt	0.00%	0.26%	7.60%	0.18%	2.54%
NM7:pltw	13.51%	1.17%	0.91%	0.13%	2.35%
NM6:pltw	13.51%	1.80%	0.21%	0.13%	2.22%
NM7:pvt	3.64%	1.88%	0.97%	0.08%	2.18%
NM4:pvt	0.00%	1.55%	3.11%	0.08%	2.18%
N10:pvt	0.00%	0.77%	0.40%	0.08%	2.10%
PM9:pvt	0.00%	0.00%	0.00%	0.08%	0.00%
PM8:pvt	0.00%	0.00%	0.00%	0.08%	0.00%

truly identical contribution according to circuit symmetry

Figure 5.13   OFAT-sweep based contribution result (sorted for V_{hys}).

become quite unsymmetrical, and a histogram would be highly non-normal! In our comparator a similar effect comes from defining the offset spec in a nonlinear way as $|V_{\text{offset}}| <3$ mV, giving also highly skewed statistical data (Figure 5.14, left histogram). Sometimes this non-normality is introduced a bit artificially, but sometimes this even helps: For instance we can expect no strong *linear* correlation of PSRR to linear offset, e.g., even for strong physical correlation we can usually expect that a sample with –10 mV is as bad on PSRR as one with +10 mV. So if we use $|V_{\text{offset}}|$ as spec, instead of V_{offset}, we will find this type of correlations easier. However, there is no free lunch because the mismatch contribution is more accurate on V_{offset} (get r^2 in the order of 0.99) than on $|V_{\text{offset}}|$ (delivers only about 0.9).

There are many more examples for nonlinear correlation, e.g., between phase margin and overshoot (in a second order system you can even calculate one from the other) or between rise time and bandwidth (the relation $BW_{3\,\text{dB}} = 0.35/t_r$ is quite accurate for any filter order and type; it does not matter much if you have a Butterworth or Chebyshev filter or if the filter is RC, LC or transmission line based!) (Figure 5.15).

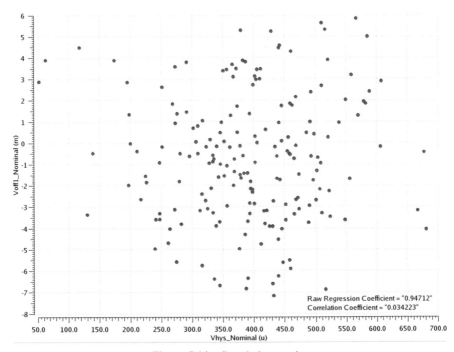

Figure 5.14 Correlation results.

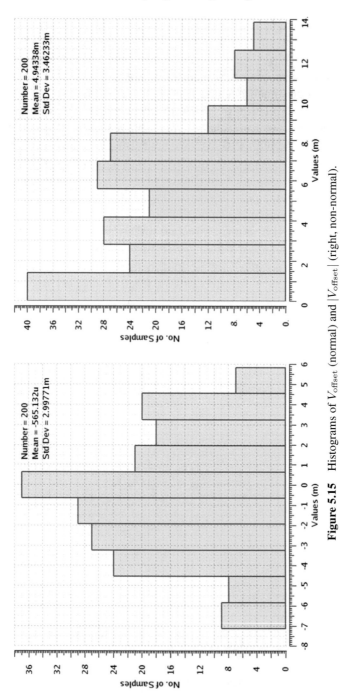

Figure 5.15 Histograms of V_{offset} (normal) and $|V_{\text{offset}}|$ (right, non-normal).

5.6.2 More on Covariance & Co.

This chapter started with many formulas, even with matrices. So let use present here some more pictures to get a better understanding for correlations. As mentioned, correlation is a difficult topic, so let us focus on the bi-variate normal case. If you are not sure if such an analysis would fit, then <u>always</u> plot your data, inspect scatter plots before doing blindly some calculations.

In the bi-variate case we get *pairs* of samples, like X_1, X_2, X_3, ..., X_n with $X = (x, y)$. x might be the total amplifier offset voltage and y e.g., the offset in the 1st stage only, so it is meaningful to assume Gaussian distributions. From the data we can directly calculate the estimates for the variances, V_x and V_y, and for the covariance $COV = V_{xy}$. Maybe we find a negative correlation, because one of our amplifier stages is inverting. So we can put the calculate estimates into Σ, our covariance matrix. Note, V would be here not in Volt or mV, but in V^2 or mV^2. For numerical investigations it usually a good idea first to divide the data e.g., by 1 mV to get numbers easier to handle, e.g., $V_x = 4$, $V_y = 2$, $V_{xy} = -2$. This would be equivalent to standard deviations of 2 mV (total offset) and 1.414 mV (1st stage).

$$H_{red} = \begin{pmatrix} 4 & -2 \\ -2 & 2 \end{pmatrix} \tag{5.13}$$

$$\rho = \frac{\sigma_{12}}{\sigma_1 \sigma_2} = -2/\sqrt{8} = 0.707$$

Let us now really put example data into a scatter plot, and inspect how everything is related. For this purpose you can download an Excel file corr.xls at the River webpage. There we created uniform random variables, converted them to Gaussian, and made a scatter plot. To let the example follow our theoretical investigations close enough, the sample count is chosen to a quite large value of n = 1000 (see Figure 5.16).

What we can see immediately is that there is no 100% correlation, just because not only the first stage (y-axis) has an impact on the total offset (x-axis). What we get is a point cloud, fitting to a rotated ellipse. Actually the *contours* for constant pdf(x, y) are ellipses.

In the sheet we just programmed the 1st stage as normal distribution with standard deviation $\sigma = \sqrt{2}$ (J2). For generating the total voltage we take –J2 and add a 2nd normal distribution with the same sigma, so actually also the

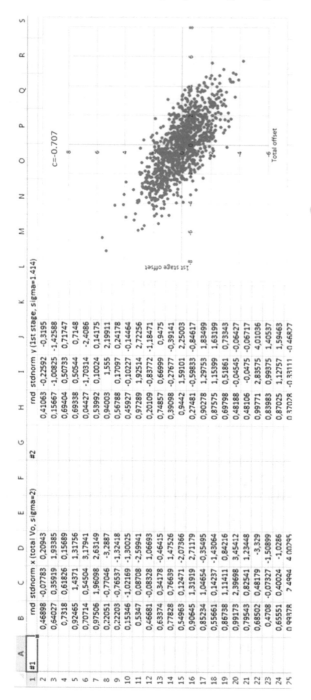

	A	B	C	D	E	F	G	H	I	J
1	#1	rnd	stdnorm	x (total Vo, sigma=2)			#2	rnd	stdnorm	y (1st stage, sigma=1.414)
2		0,46898	-0,07783	0,20943				0,41063	-0,22592	-0,3195
3		0,64027	0,35919	1,93385				0,15667	-1,00825	-1,42588
4		0,7318	0,61826	0,15689				0,69404	0,50733	0,71747
5		0,92465	1,4371	1,31756				0,69338	0,50544	0,7148
6		0,70714	0,54504	3,17941				0,04427	-1,70314	-2,4086
7		0,97506	1,96098	2,63149				0,53992	0,10024	0,14175
8		0,22051	-0,77046	-3,2887				0,94003	1,555	2,19911
9		0,22203	-0,76537	-1,32418				0,56788	0,17097	0,24178
10		0,15346	-1,02169	-1,30025				0,45927	-0,10227	-0,14464
11		0,5347	0,08708	-2,59941				0,97289	1,92514	2,72256
12		0,46681	-0,08328	1,06693				0,20109	-0,83772	-1,18471
13		0,63374	0,34178	-0,46415				0,74857	0,66999	0,9475
14		0,77828	0,76639	1,47526				0,39098	-0,27677	-0,39141
15		0,54963	0,12471	-2,07366				0,9442	1,59101	2,25003
16		0,90645	1,31919	2,71179				0,27481	-0,59833	-0,84617
17		0,85234	1,04654	-0,35495				0,90278	1,29753	1,83499
18		0,55661	0,14237	-1,32064				0,87575	1,15399	1,63199
19		0,86738	1,11411	0,84216				0,69798	0,51861	0,73343
20		0,99173	2,39698	3,45412				0,48188	-0,04545	-0,06427
21		0,79543	0,82541	1,23448				0,48106	-0,0475	-0,06717
22		0,68502	0,48179	-3,329				0,99771	2,83575	4,01036
23		0,4708	-0,07327	-1,50899				0,83983	0,99375	1,40537
24		0,65551	0,40024	-1,0286				0,87025	1,12757	1,59463
25		0,99378	2,4994	4,00295				0,37028	-0,33111	-0,46827

Figure 5.16   Scatter plot for an amplifier example.

remaining stages have the same standard deviation as the 1st stage. As result, both parts have the same contribution, which is also indicated by $r^2 = 0.5$. Note that editing the spreadsheet the random values will be regenerated, and with n = 1000, there is still significant uncertainty e.g., in the sample correlation factor.

The angle of rotation of the ellipse depends on the scales, but e.g., with an eigenvalue analysis we can calculate actually all the properties of the ellipse. So what are the eigenvectors and eigenvalues? You may get them using a math package or even from a Web page, the eigenvalues are all real due to the symmetry of Σ. Generally, the eigenvalues of a matrix A are defined by $Ax = \lambda x$. This equation leads for a 2×2 matrix to a quadratic equation for λ which can be solved analytically.

$$\lambda^2 - \lambda(a_{11} + a_{22}) + (a_{11} \cdot a_{22} - a_{12} \cdot a_{21}) = 0$$

$$l = 3 \pm \sqrt{(9 - (8 - 4))} = 3 \pm \sqrt{5}$$

The value in calculating such ellipses is that these are directly related to probabilities. They are a kind of equi-potential lines for the (two-dimensional) pdf, and when looking to the area inside we get a direct relation to the yield. So e.g., having the ellipses for different yield levels (in percent or sigma) gives us direct feedback on how wide-spread the distribution is. The eigenvalues and eigenvectors of Σ are:

$$\lambda_1 = 0.7639320225: \quad (0.618033988749895; 1)^T$$
$$\lambda_2 = 5.2360679775: \quad (-1.618033988749895; 1)^T$$

Note: In the appendix you can find a web-based tools for this task.

The rotation angle of the ellipse principal axis against the coordinate system is $\alpha = \arctan(1.618/1.0) = 58°$, which can be checked in Figure 5.17, where we applied the same axis scales. The square-root of ratio of the eigenvalues is 0.382, which also gives the ratio of the ellipse axis lengths, so for the least and most sensitive direction.

The good thing is that even without these calculations, correlations can be understood quite well. However, knowing these basic things quite well can also help for other tasks. For instance, we can continue our analysis and calculate the covariance error ellipses according to a certain confidence level. Note, in our case the ellipses are not related to design yield, because we have no specs in our example! Our calculation is similar to the confidence interval analysis on variance or standard deviation, and this is related to the chi^2 distribution (see Section 3.5 on confidence intervals).

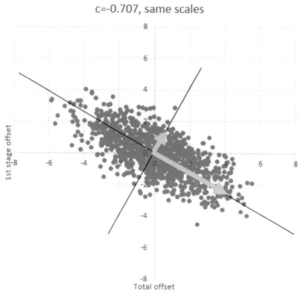

Figure 5.17 Scatter plot with the principal axis overlaid.

5.7 Questions and Answers

1. Taking an extreme sample from a MC run, like the one with maximum offset. Can we expect a strong correlation to the joint pdf?
   Not really! Usually few variables dominate the circuit behavior for a certain performance, and the others often have little impact. So the many nonimportant variables (e.g., mismatch on power-down transistors) will have an almost random setting in such extreme MC sample!

2. How many points do we need to get a stable mismatch contribution analysis?
   This depends on circuit nonlinearity and number of statistical variables, so it is hard to give concrete numbers. This is because you are typically most interested in finding e.g., the top-5 components with major impact on performances. So accuracy depends also on how much few variable <u>dominate</u> above others, and also nonlinearity has an impact. For typical analog circuits you should use 500 points and LDS (see Chapter 6). In many EDA implementations you could just let the algorithm itself select the count automatically (e.g., based on r^2).

3. Designers expect for good reasons that often there is a correlation between offset and common-mode rejection. However, it comes out that your tool does not report a significant correlation. What should you do?

   Best inspect the scatter plot directly. Tools usually report the linear correlation factor only, but it is not unrealistic that we have low linear correlation, but strong non-linear correlation. If the scatter plot has no really shape or pattern, we probably have indeed no kind of correlation. Look to Figure 5.3 for examples.

4. Could it happen that 3 statistical variables like a, b, c have <u>no</u> pair-wise (mutual) correlation at all, but among all 3 there is 100% correlation?

   Yes, statistically this is possible, and an indication how complex the topic of correlation can be – and how much intuition can fail. One consequence of this fact is that the correlation tables you get presented are often better than analysis which ignore correlation, but still such 2D tables can be far from presenting a whole picture. Here is an example: Have two standard uniform variables U_1 and U_2. Create a 3rd variable $U_3 = (U_1 + U_2)$ mod 1. U_3 is also a standard uniform variable, and of course it is fully dependent on the other two, but actually U_1 vs. U_2 and also U_1 vs. U_3 and U_2 vs. U_3 are not correlated at all. All mutual scatter plots look like a 2D uniform box. In verilogA you can easily run this example in many circuit simulators; or check out corr.xls at the River webpage.

5. What would happen if we ran a contribution analysis on a transient noise testbench?

   The contribution analysis can only take into account the statistical variables belonging to model and instance parameters, but not the current and voltage noise sources. So it can only calculate and rank a part of the variances! Mathematically you can see this from the quite low r^2 values in such analysis, if 50% of the output voltage variation is from pure electrical noise, the r^2 is limited to 0.5 (it might be lower due to other limitations like nonlinearities which are not 100% correctly modeled.

6. Kernel density estimation (KDE) is often used to perform a "non-parametric" fit to a histogram, but actually there is some freedom in selecting the kernel functions and the bandwidth. Only if we know that

the data is e.g., Gaussian, we can really achieve good fits with fix rules. *Check out further literature and look at Figure 5.18. For distributions with edges or multiple modes you cannot apply much smoothing. For too less smoothing it can happen that the fit becomes bimodal although the true distribution is unimodal!*

7. We use often the offset voltage of an amplifier as example for a contribution analysis, but of course there are actually many more applications. The good basic feature is that we can reuse our MC results. So if you can simulate something, you can get the contributions. Give some more complex examples and alternatives.

   Looking to op-amps, e.g. also the behavior of class-AB stages can be significantly impacted by mismatch, also the operation of rail-to-rail input stages. For OTA's the g_m could be impacted, which would e.g. have an impact on filter time-constants and Q-factors. A more complex example could be the current mismatch of a charge-pump, which often impacts the spurious response of a PLL. Another classical is a bandgap reference; here designers need to know where improvements are needed, e.g. in the current mirrors, for the resistors or e.g. in the loop amplifier. Some simulator have built-in mismatch analysis, so if your performance is e.g. directly a DC or AC output current or voltage, such analysis can be an alternative.

8. Imagine you transfer a circuit from an older technology to a new one, like a FinFET process. What can we expect regarding the contribution of each transistor?

   New processes come typically with many more statistical variables for each instance, so the sensitivity of each is harder to estimate, and we require more simulation points. In a contribution analysis these individual contributions are collected for each device instance, so the results become more stable again. However, over-all definitely more runtime is required. Also some accuracy loss is realistic, e.g. because modeling and filtering becomes more difficult.

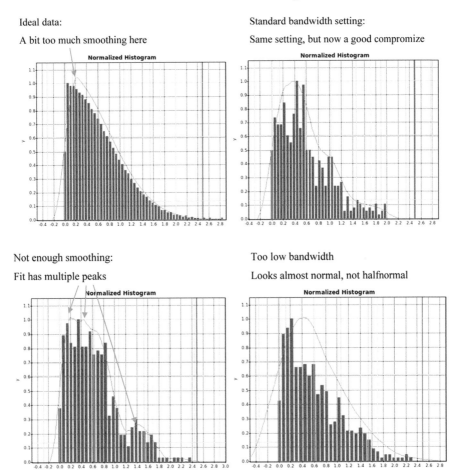

Figure 5.18   Ideal data and MC data with n = 512 samples, and KDE fits with different bandwidth settings.

6

Advanced Sampling Methods

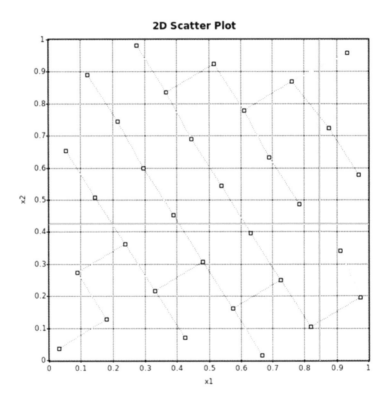

In this chapter, we discuss how we can speed-up yield verification, and MC in general, by using advanced *sampling* techniques. For the user, the setup effort for such techniques is very small. We will also discuss two further ideas: one is using a multivariate model to estimate the circuit performance *before* actually running the circuit simulations, and the other is complete "synthetic" Monte-Carlo, so-called *bootstrap*.

Interestingly in school children learn about statistics without Monte-Carlo and apply basic analytical calculus instead and use combinatorial approaches.

Direct calculations are obviously much faster, so can we apply similar techniques also to circuit design? Unfortunately a combinatorial approach would lead to huge efforts, like requiring a corner analysis with thousands of variables. We mentioned that one key point for Monte-Carlo success is that it simply mimics the production variations, but actually in the <u>simulation</u> we have even more access to the variables and also more *control* on *setting* them, so we should exploit this! First we do this in a "static" way, i.e., we decide *upfront* on which points to simulate.

A further extension is running a short "pilot" MC analysis to gather information; and next we can construct a model to *predict* circuit performance (so without further time-consuming circuit simulations). Now you may simply take this response-model-based estimate for your design decisions or (much better) you could double-check e.g., the pass-fail yield estimation by simply running the circuit simulation with the model-selected points. For instance, we could skip the simulations for samples far away from the spec limits! With the executed simulations we can obtain an error estimate; and we can improve our performance model further. If the model assumptions are valid, we can expect a big speed-up, whereas if the assumptions are invalid such flow may fail or it would go back from "intelligent" Monte-Carlo to simple Monte-Carlo! To some degree such advanced statistical methods give us the best of both worlds: MC robustness and flexibility, plus speed-up by applying iterative techniques and avoiding unnecessary simulations. Actually such "sorted" MC can be regarded already as a big step in the direction of dedicated *high-yield* methods, which will be discussed in Chapter 7, but before this we explain so-called bootstrap techniques, because bootstrap is often used in many such advanced statistical algorithms.

For Further Reading:

Monte-Carlo is a very general approach, but also advanced methods become more and more popular, so a lot of materials are available, especially in relation to circuit design, but also search e.g., for design of experiments, discrepancy, sampling, statistical blockade, bootstrap.

- Digital Nets and Sequences: Discrepancy Theory and Quasi–Monte Carlo, J. Dick, F. Pillichshammer, 2014, Cambridge University Press.
- J. Jaffari and M. Anis, "On Efficient LHS-Based Yield Analysis of Analog Circuits," in *IEEE Transactions on Computer-Aided Design of Integrated Circuits and Systems*, vol. 30, no. 1, pp. 159–163, Jan. 2011.

- M. Sobol', D. Asotsky, A. Kreinin, S. Kucherenko, "Construction and Comparison of High-Dimensional Sobol' Generators," in *Wilmott*, John Wiley & Sons, Ltd., no. 56, pp. 64–79, Nov 2011.
- A. Singhee and R. A. Rutenbar, "Why Quasi-Monte Carlo is Better Than Monte Carlo or Latin Hypercube Sampling for Statistical Circuit Analysis," in *IEEE Transactions on Computer-Aided Design of Integrated Circuits and Systems*, vol. 29, no. 11, pp. 1763–1776, Nov. 2010.
- A. Singhee and R. A. Rutenbar, "Statistical Blockade: A Novel Method for Very Fast Monte Carlo Simulation of Rare Circuit Events, and its Application," 11th DATE Conference, March 10–14, 2007.
- Efficient Trimmed-sample Monte Carlo Methodology and Yield-aware Design Flow for Analog Circuits, Chin-Cheng Kuo et al., DAC 2012, June 3–7, 2012, San Francisco, California, USA.

6.1 When to Use What?

Table 6.1 gives an overview on the described statistical methods in both Chapter 6 and 7, including hints when to use them. Note, if we state something like "...limited to approximately 3σ" it does not mean that the algorithm will completely fail above 3σ, but it is likely that there are e.g., more efficient

Table 6.1 Overview on basic and advanced statistical techniques for circuit design

Analysis/Methods	Outcome	Limitations	Applications
MC & picking worst sample	Very rough worst sample + all usual MC outputs	Not really accurate & limited to app. 3σ	Quick tweaks, but double-check with further MC analysis
MC and sample yield (optionally with LDS or similar methods)	Yield estimate and confidence interval + all usual MC outputs	Time-consuming for high yield targets	Verify low yields, creating a golden reference if able to run long MC analysis
Solve yield integral with numerical integration	Accurate yield, spec pass-fail hyper planes	Extremely time-consuming for high dimensions (>20)	Creating a golden reference, no commercial tool available
MC and C_{GPK} (or other CDF fitting methods like tail modeling)	Yield estimate and confidence interval + all usual MC outputs	Some extrapolation error (e.g., too pessimistic for Gaussian data with cuts)	Verify yields up to app. 5–6σ, creating a silver reference with moderate MC analysis

(Continued)

Table 6.1   Continued

Analysis/Methods	Outcome	Limitations	Applications
Bootstrap	Generation of synthetic MC results, relying on resampling with replacement	Initial data must be large enough. Error increases for estimates which rely on few tail samples (like high yields, kurtosis)	Confidence interval generation is the major application, also often used internally to check accuracy of other algorithms
(Mismatch) Contribution analysis	Get contribution of each instance (transistor, resistor, block, etc.) to output performance variations	With moderate MC count only accurate for the major variables	Get insights to circuit behavior, do it to identify the most important parameter to be optimized, most useful for mismatch, because this can be easily tweaked by designers (in opposite to process behavior)
Sensitivity analysis based on OFAT	Get sensitivities directly by sweeping each parameter individually, e.g., by 1% or 1σ.	Correlations need more complex sweeps. Large simulation effort if many parameters need to be inspected.	Get accurate insights to circuit behavior, do it to identify parameter to be optimized.
(Model-based) Sorted MC for yield verification	Yield	Accurate enough for design phase & limited to app. 3σ–5σ	Efficient yield verification, with relaxed accuracy or higher effort also for higher sigma
(Model-based) Sorted MC for corner generation	approximated WC distance	Accurate enough for design phase & limited to app. 3σ–5σ	Approximate statistical corner generation, with relaxed accuracy or higher effort also for higher sigma)
Worst-case distance WCD	Yield and accurate WC distance	Very accurate (no sampling error) up to high σ values like 6–9, speed limited for low σ	Sign-off verification for high yield, statistical corner generation for optimization
Sigma-scaled sampling SSS	Yield and approximated WC distance	Accurate enough for design phase & limited to app. 4–6σ, recommended for complex designs	Verification for high yield, with relaxed accuracy or higher effort also for higher sigma

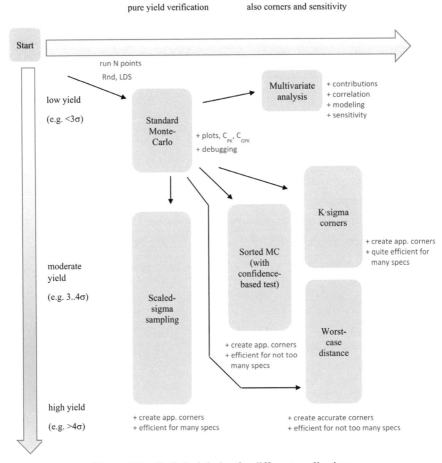

Figure 6.1 Statistical design for different applications.

methods, so using it for 4σ makes probably only sense for small blocks where the simulation times are short. In addition, real EDA tools often use a mix of methods to get overall a higher speed-up which could enable the verification of higher yields (Figures 6.1 and 6.2).

6.2 Advanced Monte-Carlo Sampling Schemes

Actually, this chapter is not really about classical Monte-Carlo! To find out how a design behaves under parameter (statistical and/or deterministic) changes, we can follow several different strategies of taking samples and

Random sampling

Slow 1/√n convergence, mimic real-world

behavior

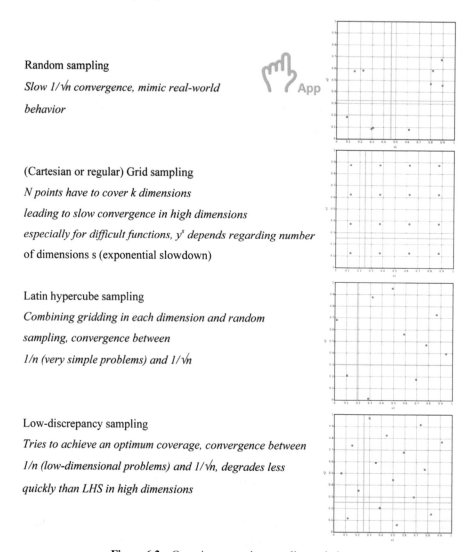

App

(Cartesian or regular) Grid sampling

N points have to cover k dimensions

leading to slow convergence in high dimensions

especially for difficult functions, y^s depends regarding number

of dimensions s (exponential slowdown)

Latin hypercube sampling

Combining gridding in each dimension and random

sampling, convergence between

1/n (very simple problems) and 1/√n

Low-discrepancy sampling

Tries to achieve an optimum coverage, convergence between

1/n (low-dimensional problems) and 1/√n, degrades less

quickly than LHS in high dimensions

Figure 6.2   Overview on major sampling techniques.

run simulations on them, like one-factor-at-time (OFAT), full-factorial or just random (classical MC), and each has its advantages and limitations. Initially we wanted to keep that chapter much shorter, because in circuit design environments the enablement of a certain sampling method is often just not more than a mouse click in the usual Monte-Carlo setup window! However, for good reasons, there is a mass of reports and resources available on this topic, and most of them promote big accuracy benefits over classical random Monte-Carlo.

The topic of sampling creates also a nice bridge between MC and corner techniques, because both (and many other advanced techniques) are adressed by so-called *design of experiments* methods (DoE). On the other hand, we have to mention clearly, the advantages of pure static sampling methods regarding variance reduction in real complex circuit designs are often quite *limited*!

So <u>what</u> could be the reasons for that mismatch in research papers and practice? And can we expect some further improvements in the future?

Up to now we assume that we run a MC analysis and look to the result data, trying to do a "clever" <u>analysis</u>. If we run MC many times and average, we can reduce statistical errors dramatically, so what about averaging "up-front," directly on the statistical parameters itself, instead on the MC results?

When doing a MC analysis, we typically assume that we work like nature does, e.g., that the samples are truly random. So internally pseudo-random generators are used to generate almost ideal random sample points, and such sets of random points are used in the MC simulations. In very old MC environments, no good random number generators have been used, so that also random MC was sometimes not as good as expected. However, those days are gone, and modern pseudo-random number generators are almost perfect and have huge period lengths like $2^{19937} - 1$ or higher. The only key difference of pseudo-random versus ideal random sampling is that for pseudo-random, we can define a *seed* value which makes the sequence *reproducible* (e.g., for debugging) without need to store all points.

However, indeed we can often improve the convergence speed (variance reduction) and the accuracy if we do <u>not</u> use purely random or pseudo-random variables—switching from random-MC to quasi-MC!

Yield is an integral, and we mentioned already that integration by Riemann's sum has usually 1/n convergence (Simpson's rule is often even much faster, but is quite dependent on the smoothness of the integrand), instead of $1/\sqrt{n}$. The reason is that in random (or pseudo-random) methods often some (random) *crowding* effects appear, so it can happen that the statistical space is <u>not</u> as well covered as one might "expect"! The idea of quasi-MC sampling methods is to improve the coverage by a kind of "optimum spreading" technique. Such point sets are a topic in design of experiments (DOE); and in this context often the point sets are called a "design". So math can help in testbench design as well! To avoid confusions we prefer the term point *set* over "design".

Nonrandom sampling schemes can lead to more stable estimates, e.g., for sample mean and standard deviation of the performances or for the correlations in a multivariate analysis! And vice-versa, for the same level of accuracy you need less sampling points, less simulations, so you are able to make design decisions in shorter time! Of course, there is no free lunch and we will also discuss in which cases the speed-up from such advanced "quasi-random," more "well-spread" sampling schemes is limited. For getting literature, search e.g., for *low-discrepancy sampling*, phrases like "optimum spreading" are inventions by marketing people, but it indicates the idea quite nice.

Note: An alternative to just sampling the s-dimensional statistical space would be a <u>direct search</u>. For smooth problems, the effort grows often only according to s^2 (and not exponentially).

Monte-Carlo Tricks like "Averaging Upfront"? If we ask someone to do a sketch for a uniform distribution, we will probably get a histogram which looks like a rectangular box. However, what you *get* by sampling from a uniform distribution is *different*, really different (look at Figure 3.1, which compares different samples)! So why using such random samples? Why not "ideal" numbers like equally-spaced samples. This idea, also arising from solving integration problems, will lead us later to so-called low-discrepancy sampling LDS!

However, also other improvements are possible: If we take such "bad" truly random samples, we could still improve with another little trick: For instance, an ideal amplifier should have zero offset, so if the sample points have zero mean, we *should* get this in simulation too. However, unfortunately i.i.d. random samples do not have zero mean, usually. So we could simply shift our random numbers by this amount to correct the error on mean! Cool? We could also scale our point set to get the correct standard deviation! These ideas are very similar to so-called latin hypercube sampling, which is even a bit more tricky.

However, with *all* such tricks you will *not* solve all problems, and the improvements will become smaller and smaller the *more* dimensions we have, and the more nonlinear our circuit performances become! This one face of the so-called "curse of dimensionality". For instance, LDS almost works perfect in one dimension and also two dimensions (See Figure 6.13), but at some point (depending on algorithm details, nonlinearity, number of points and dimensions) the problems start again. Users have to be aware of this! Luckily, in many cases such "tricks" work also in circuit design, but not always? and sometimes there are further risk, not present in slow random methods.

6.2.1 Cartesian Grid Sampling

A random sampler gives only a $1/\sqrt{n}$ integration convergence. A clearly better set of points for one-dimensional integration is an equidistant set, as for integration with rectangular or trapezoidal rule giving $1/n$ convergence. So far so easy, but interestingly a simple rectangular grid—similar to a typical full-factorial corner analysis—is not optimum at all for two dimensions already! Figure 6.3 is showing why: A simple rectangular grid aligned with the axis variables (Cartesian or regular grid or lattice) is good for simplifying algorithms like reducing memory consumption, applying fast Fourier transform, etc., but it is not "covering well" difficult functions, and only for slowly varying functions (maybe like V_{DD} or T over a short range) such grid approach is fine. Amazingly a random grid can be clearly better, the only pity is that a random approach cannot give a guarantee, and indeed with so-called quasi-random numbers we can further improve.

Note that this "special" advantage of random sampling is quite substantial, and if we think a bit more it will be not so much a surprise anymore: In the grid scheme we treat each variable in the same way, like in a chessboard of 8×8 points. However, in real designs often few variables dominate, and having only 8 distinct values instead of 64 for such important variables is a clear disadvantage of the grid against random sampling! Actually the $1/n$ speed in 1D went down to $1/\sqrt{n}$ due to the square-root-law chessboard relationship, which means in 2D the rectangular rule gives no guarantee on being faster than random MC anymore! In 3D or higher dimensions s the slowdown will even go on further according to $1/n^{1/s}$. This is called "the curse of dimensionality", which appears in many facets. The situation is not only critical for simple grids, but also if we e.g., inspect multivariate Gaussian distributions. With non-Cartesian grids we *can* improve, sometimes significantly, sometimes only a bit, but general limitations remain for high dimensions!

In special test cases a simple grid approach can even fail almost completely (Figure 6.4).

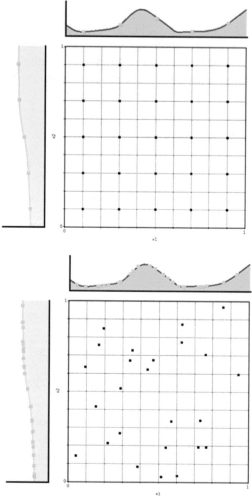

Figure 6.3   2D sampling by using a rectangular grid vs. random numbers. x_1 is an important parameter, and x_2 is a noncritical one.

We already noticed that the gaps in 2D grids become much larger than in one dimension; we have only $\sqrt{n}$ points to cover each variable, instead of n (like 8 versus 64). And in higher dimensions everything would become worse, e.g., in 8 dimensions and with 10,000 points we would only get roughly 3 points in each dimension! This problem of dimensionality can be reduced by using other sampling schemes, but it cannot be 100% solved for nonrandom

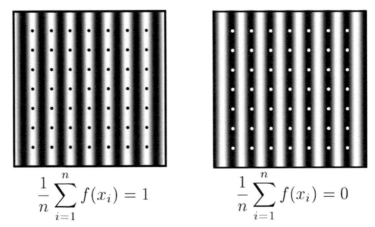

$$\frac{1}{n}\sum_{i=1}^{n} f(x_i) = 1 \qquad \frac{1}{n}\sum_{i=1}^{n} f(x_i) = 0$$

Figure 6.4 2D integration on special functions can fail completely while using a rectangular grid.

schemes. So high-dimensional problems remain difficult, whatever fancy tricks or mature math algorithms you apply!

Figure 6.5 shows the convergence of random MC sampling vs. grid sampling in one dimension; as expected random is much slower. However, if we use the grid approach on our MC introduction example on the unit circle (Figure 3.2), we see that already in two dimensions the grid method loses its speed advantage compared to random (Figure 6.6)!

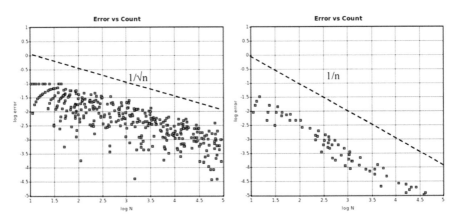

Figure 6.5 Integration speed for pure random MC (any dimension) and for grid sampling ($d = 1$ only).

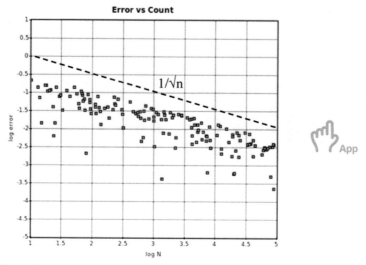

Figure 6.6   Integration speed slowdown for grid sampling in two dimensions.

A nice method extension, which could relax some problems at least, is so-called *jittered* sampling (JS). In this we use still the grid of hyperboxes, but <u>inside</u> each one we put the samples randomly (Figure 6.7). Something similar is also done usually in the even more popular latin hypercube sampling (Figure 6.8).

6.2.2 Latin Hypercube Sampling LHS

A simple orthogonal grid approach is limited, and a pure random approach has slow general convergence. What about a mix? One well-known quasi-MC sampling scheme doing so is latin hypercube sampling (LHS), sometimes also

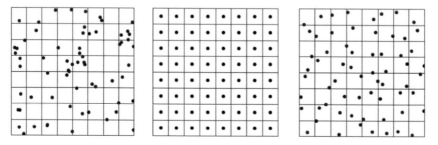

Figure 6.7   Random sampling, grid sampling, and jittered sampling (grid-based stratification).

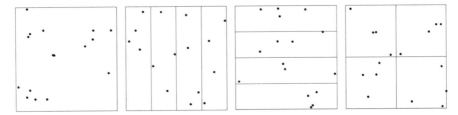

Figure 6.8 Random sampling and three different kinds of stratified sampling.

called n-rooks random sampling (which is actually the better name!). The idea is to do random sampling but with making sure that the points give somewhat less gaps and crowdings as pure random sampling. In LHS this is achieved by placing the points similar to n rooks on a chessboard which do not threaten each other, so in each row and column we have one rook, but only one! Such LHS configuration can be directly used instead of a random set (centered LHS) or we can apply also a random jitter inside each cell (randomized or jittered LHS).

Note: Here in the book we usually present formulas or pictures from non-jittered LHS, but in simulators often jittering is applied. The differences are usually minor. At the River webpage you can download a spreadsheet example lhs.xls with examples of jittered and non-jittered LHS data and plots.

If we would take samples from a <u>pure</u> random uniform statistical variable, and if we split its range into equal histogram bins, it could happen that in some bins just *more* (or less) samples fall than into others. This is causing gaps and crowdings, and this causes the slow $1/\sqrt{n}$ convergence. At least to some degree, LHS is avoiding this kind of bad coverage by using a grid of n intervals for each variable. This is the trick: For 2D and e.g., n = 64 we get only 8 boxes in a Cartesian grid, but in LHS we divide just each variable in n boxes—avoiding gaps <u>in each</u> variable pretty well.

As mentioned, MC is basically integration, but integration by rectangular rule is much faster, and with LHS we just make MC a bit closer to a grid and faster on convergence. LHS guarantees indeed quite a good distribution <u>within</u> each dimension individually, and e.g., simple statistics like the sample mean often converge much faster than in random MC. However, it comes out that such LHS or n-rooks scheme is unfortunately far too simple to speed up also on difficult test cases, with many dimensions, strong nonlinearities and

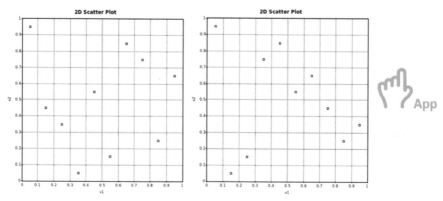

Figure 6.9 Two 10-points-LHS sets (non-jittered) in 2D with quite different 2D coverage (no randomization inside boxes for easier comparison).

correlations. Figure 6.9 is showing why: two different two-dimensional LHS point sets are shown, but LHS set b is *not* covering well the 2D space; only for each *individual* 1D dimensions, both sets are good *by construction*.

For these reasons, LHS may give a good speed-up for simple cases, like if in a filter the cutoff frequency is mainly impacted by R_{sheet}, but not in many other more complex circuit design cases.

A second LHS problem is this: From the algorithm, LHS is based on a kind of prebinned random number generator, and only a full set of latin hypercube n points gives really the beneficial, somewhat better coverage. If we would stop our simulations earlier (See subsection 3.6.3 on MC auto-stop), then the equal binning structure would get lost, and we would have no speed-up against full random MC anymore. Also if we would decide that we need more MC points, LHS runs into problems, because an extension of the grid with re-use of the existing MC results becomes difficult.

Figure 6.10 shows this aspect of LHS behavior: the right plot shows that after executing all 512 points the LHS error is indeed significantly smaller (small y-variations in multiple runs), but for a stop after half of the points the advantage is much smaller. The left plot shows the random sampling behavior: Due to the multiplication by $\sqrt{n}$ the error envelope is almost constant, according to the usual $1/\sqrt{n}$ convergence, as expected. Comparing Figure 6.10a and b, it looks that if stopping at $n = 512$ LHS is indeed *much* better (like >90% less variations), this is mainly due to the very simple test case in which just one statistical variable is dominating (so it is basically a 1D problem). LHS is a near-random scheme which provides well-distributed

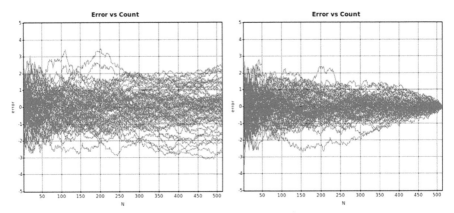

Figure 6.10 Error of the mean (multiplied by $\sqrt{n}$) for random and LHS.

samples for <u>each</u> individual statistical variable, but LHS is <u>not</u> good enough in multi-dimensional problems.

Note: If we apply jitter sampling on top of the LH set, the speed-up show in Figure 6.10 would slightly decrease, but we get also one advantage: Without jitter, the most extreme point in the interval $(0,1)$ is just defined by $1/(n + 1)$, so we never get a 6σ sample in a 1D 100-point LH set. However, with jitter this problem disappears like in random MC.

In a bad LHS implementation, we may even get biased LHS results if stopping early. Modern LHS algorithms feature internal randomizations to avoid such bias, but some limitations exist and in many benchmarks LHS has been outperformed by so-called *low-discrepancy* sampling (LDS) (e.g., [Kuche2011]). At least in principle LDS solves indeed some problems which still exist in LHS: First, LHS results have still a statistical nature, so at best we only achieve a variance reduction but we cannot tell anything about *hard* error bounds. Second, actually LHS <u>is</u> a one-dimensional LDS method, but not more; with LDS we have an option translating the benefits of LHS also to (somewhat) higher-dimensional problems.

Simple LHS is often indeed inferior to LDS, but also optimized LHS algorithms exist [Affair2011, Joseph2008], partially picking up LDS ideas, sometimes going even well beyond that. Also note that LDS is a very general term (see next subsections), and speeific LDS algorithms can differ a lot, in how they work (e.g., optimization vs construction-based), and in which level of "quality" they achieve [Lemieux2008]. The simplest approach to improve LHS, LDS, or any sampling scheme would be to generate many point sets and to filter-out all the bad sets, like those having significant (linear and quadratic) correlations or showing patterns or gaps. However, in many dimensions this would be an extremely wasteful approach.

One aspect in multi-dimensional sets is the correlation between the different variables. As we know, correlation means reduced variance, so two heavily correlated variables behave almost like only one statistical variable, i.e., the effective dimension and the coverage of the statistical space gets reduced. For ideal random variables or LHS we should have no correlation, because all dimensions are independent, but for too low n and large s we can have quite large correlations just by chance, in our point current set.

One further negative side-effect of correlation is this: If we add two independent standard normal variantes, we get again a normal distribution, just with sigma of $\sqrt{2}$. However, in case of significant correlations we get a wrong value for the standard variation, so overall we could get MC results with significant errors in variance and regarding C_{PK}.

On the other hand, correlations are only just *one* factor that matters; knowing that the sample standard deviation is correct does *not* make sure that yield integration is correct or that the statistical space is covered well. For this, the so-called *discrepancy* D matters.

6.2.3 Discrepancy of Point Sets

We have seen that for difficult, highly varying functions the integration error becomes large, but the function and its variation V is related to our circuit analysis, and often we cannot change it. The second factor regarding (absolute) integration error ΔI is related to the sampling itself. This second key factor is the so-called discrepancy D; the theoretical basis for this is the Koksma-Hlawka (HK) inequality [Dick2014], giving an integration error bound.

$$\Delta I = \left| \int f dx - 1/n \cdot \sum f \right| \leq D(x_1, x_2 \ldots x_n) \cdot V_{HK}(f) \qquad (6.1)$$

Note: The concept of total variation *V* over a certain interval is easy to capture in one dimension by integrating $|df/dx|$, but for higher dimensions many integration paths are possible, and multiple *V* definitions exist. V_{HK} stands for variation "in the sense of Hardy and Krause." One problem is that for some important functions V_{HK} is not bounded, so the error limit provided by Equation (6.1) would become useless. Luckily this does not automatically mean that the error would become indeed infinite, but it is one factor which could reduce the advantage of advanced MC sampling schemes. Actually, other variation definitions exists (e.g., by Vitali, Frechet, or Arzela), but they are often harder to connect with integration error bounds and discrepancy, or they may give less tight bounds (similar to Gaussian approximation vs. Chebycev limit). Also note that all sample values have the same weight in the summation, like in the sample yield evaluation or for simple integration by rectangular rule. So in theory by introducing some weighting we may improve further—an idea similar to techniques we discuss later (e.g., importance sampling or sigma-scaled sampling).

It is quite intuitive that redundant simulations, so crowdings in the statistical space, should be avoided; also gaps, because they would lead to bad verification coverage. So how can we describe this mathematically? As we can generate any distribution (like a normal Gaussian) from a standard uniform U_{01} via transformation (using the inverse cdf, so-called percentile function), people defined also the discrepancy D for simplicity on uniform numbers between 0 and 1. Other cases can be derived from that easily.

In such unit box of dimension s (we use s to avoid confusions with discrepancy D), the volume is just unity, so the ideal point density (points per "volume" in 3D) would be $1/n$. If we would check the point density of a pure random set generated by a an ideal standard uniform random number generator, we would find that only for the whole $(0,1)^s$-box we trivially and surely get the desired density, but doing the density check for *subregions* we would find random deviations! And the worst-case deviation is commonly defined as discrepancy D. So for an "optimum spread" point set we want such homogeneity for <u>all</u> kinds of sub-boxes $B = (a_i, b_i)^s$.

$$D = \max |(\text{points in B})/n - \text{Volume}(B)|$$
$$= 1/n \cdot \max |\text{points in B} - n \cdot \text{Volume}(B)| \qquad (6.2)$$

Note: Like there are different measures for variation *V*, there are also different ones for discrepancy *D*. One other popular definition is to restrict the boxes to $(0, a_i)^s$, so that we always start at the origin (so-called star discrepancy), or instead of rectangles (boxes) we may also use circles (spheres). Also the boundary treatment can be different, e.g., 0.999 and 0.001 are highly separated if we only have the uniform distribution itself in mind, but if we apply a variable transformation this may change. Also for applying a transformation according to the inverse normal cdf it makes a difference if we inspect D e.g., in (0,1) or [0,1]. Star discrepancy is often used because it is just easier to handle, but the general discrepancy theory even deals with many more topics related to the problem of discreteness (like rounding errors, sets for color coding, etc.) (Figure 6.11).

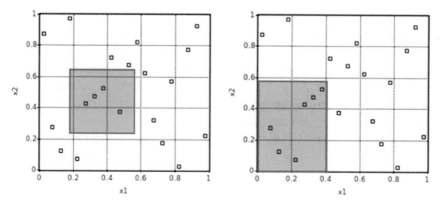

Figure 6.11   Sketch for box-based discrepancy evaluations on 2D point sets.

In one dimension $s = 1$ we can quite easily calculate D: Taking a set with $n = 4$, like 0.2, 0.4, 0.6, 0.8, we could set our testing "box" B just in a "gap" with "volume" of almost 0.2, so the number of points in B would be still zero and we get $D = |0–4 \cdot 0.2|/4 = 0.2$. Is this the worst-case? We could also check the box from 0.2 to 0.8, and we get $|4–4 \cdot 0.6|/4 = 0.4$. Indeed the best 4-element set would be 0.125, 0.375, 0.625, 0.875 with $D = 1/4 = 0.25$. So such an equidistant set can have a discrepancy in the order of $1/n$, which is an excellent low value; and for large n we can reduce it to a value *as small as we want*. This also means (acc. to Equation (6.1)) that we can make the worst-case integration error as small as we want. Real random numbers are much worse on D (e.g., 6× larger in one dimension). Unfortunately, checking D in high dimensions is <u>much</u> more difficult than in one dimension, because we have to inspect *all* point sub-boxes with to find the one with highest discrepancy. Already in 2D this is a difficult task for typical n like 500. For large n only iterative algorithms are practical, and so their runtime depends on accuracy too [Thiemard2001].

Also the construction of sets with low discrepancy is difficult in high dimensions (but possible). In principle, for each value of n in the s-dimensional hypercube, there exists a point set with the lowest discrepancy of all point sets. That discrepancy must be larger than the so-called Roth bound. This means that LDS is often better than random, but especially for large s the accuracy benefit and the resulting speed-up is limited.

$$\text{Roth bound: } n \cdot D_8 \geq C_s \cdot (\log n)^{(s-1)/2} \tag{6.3}$$

In conclusion, using sets with low discrepancy D would be helpful for improvements on integration by Monte-Carlo. On the other hand, notice that also D is only <u>one</u> criterion with impact on accuracy, and it is a worst-case criterion. The positive aspect is that Equation (6.1) gives a *stronger* limit than the pure statistical limits for random sampling, but note two sets might be similar on the *worst-case* discrepancy and one might be still better in average (e.g., if applied to a bigger benchmark set) on integration accuracy. Figure 6.12 shows two sets of similar 1D discrepancy: a random set on the right (with its typical crowdings) and a set created by so-called Poisson disk sampling on the left. The worst 1D gaps are marked with yellow lines, and they are not really pronounced, but of course the more homogenous Poisson set gives typically more accurate estimations. Note that of course also the 2D discrepancy is different, but that is very hard to check due to exponentially rising calculation effort. Poisson sampling makes sure that a minimum distance to the nearest neighbor is guaranteed (so we observe no crowdings!), but the creation is based on a random process still (e.g., based on "dart throwing"). One problem in Poisson set creation is that you can easily manage to guarantee a minimum neighboring distance (e.g., by ignoring points which violate this

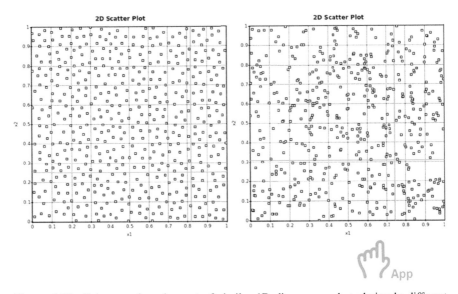

Figure 6.12 Poisson and random set of similar 1D discrepancy, but obviously different homogeneity and 2D discrepancy.

condition), but you cannot guarantee good coverage of the whole $(0,1)^s$ box for a <u>given</u> number of points n, so the sampling has to stop once the coverage is reached, and you just have to use *that* number of points you have at that point in time.

6.2.4 Low-Discrepancy Sampling LDS

LHS is still quite random; with LDS we go further away from randomness, and also instead of MC we should talk about quasi-MC. The core idea is to minimize the discrepancy of point sets even further, by construction! Within some limitations, LDS is very successful, and one reason for the many literature available on quasi-random numbers is that they are not only used for Monte-Carlo. Other applications include sampling of 2D or 3D pictures and scenes, design of experiments, and optimization. Also nature "invented" kinds off "optimum well-spread" sampling: The retina of an eye is not a rectangular grid of light receptors (like in a CCD chip of a digital camera), and it is also not uniform random. It is something in between; and this helps to get <u>both</u> a good resolution and almost no artifacts (like Moiré patterns). The retina pattern is similar to so-called Poisson sets. Also when looking to throwing darts (or making a snapshot of raindrops falling down to earth) the assumption of a true uniform random distribution is not the best one, because the finite diameter of the darts prevents too tight neighboring positions (or at least lower the probability for this) and the events are not independent if we do not remove previously thrown darts. Figure 6.13 compare a uniform random point set with LHS and LDS. Note, that LHS is not much different from random, already in 2D examples! In the scatter plots we marked the coverage "gaps" in yellow, good LDS sequences have almost no gaps and crowdings.

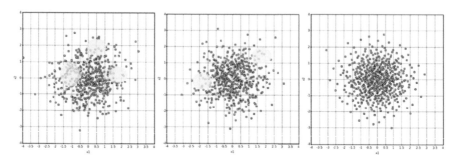

Figure 6.13   Gaussian 2D scatter plots from different sampling schemes (n = 512: random, LHS, LDS).

In one dimension we know the optimum sequence, so what about two dimensions? In two dimensions some near-optimum discrepancy sets are indeed already known (look at Figures 6.14 and 6.15), and they look like (slightly shifted) lattices or crystals. Note that in higher dimensions we can define "distances" or "volumes" in different ways, so also the definition of discrepancy becomes more complex, i.e., different papers may treat discrepancy in different ways and also what an "optimum" discrepancy is becomes a difficult task. In 1Δ the gap is just $\Delta x = x_2 - x_1$, but in higher dimensions we can measure the length of Δx using different norms; the

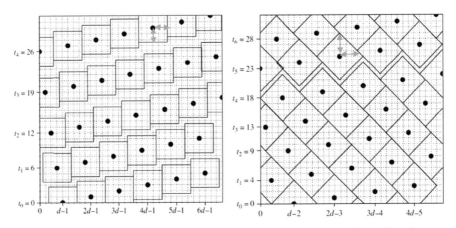

Figure 6.14 Optimum 2D grids for L_∞ and L_1 discrepancy, $n = 33$ [van Dam].

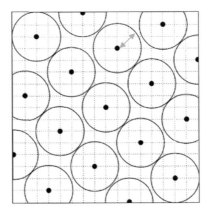

Figure 6.15 L_2-near-optimum grid for $n = 17$ [van Dam].

Eucledian norm $l^2 = \Delta x^2 + \Delta y^2$ is only *one* option among others. Adding the magnitudes, so setting the exponent equal to one, gives the so-called Manhattan norm (adding each co-coordinate), and using a large exponent would give the largest entry the biggest contribution, leading ultimately to $l = \max(|\Delta x|, |\Delta y|)$.

One measure that is helpful too is *entropy*; like in nature optimizing on entropy leads to a state in which the points have a certain minimum distance, thus keeping the potential energy low.

Do you feel discrepancy is something difficult? Yes, it is. What we have marked in Figure 6.12 are the widest gaps in the projections to the axis. So for a uniform distribution, to get good coverage we should place the samples in a way that makes the largest *uncovered* space as small as possible. This means we minimize the so-called dispersion δ, which is somewhat different to discrepancy δ. Only low discrepancy guarantees low integration errors, low d alone does not! However, for luck, low discrepancy always implies low dispersion (but not vice versa). Note that the HK error bound gives us directly the sampling error of the empirical cdf to the true integral.

Another option to look how "similar" distributions are, and also at "disorder" in general, is using the Kullback-Leibler divergence and the (relative) entropy! Optimizing on entropy (follows $\Sigma\, p \log p$) gives also lattice-like point sets, and entropy has also a strong link to other statistical concepts such as maximum likelihood estimation (MLE). Interestingly entropy is also strongly related to the so-called minimum spanning tree MST, which provides the shortest connection of all points. If the MST is long and has small variations, we reach a kind of equivalent to an energy or entropy optimum! The MST is luckily easier to calculate then the discrepancy, so it is very practical. At the chapter intro we presented a scatter plot from a near optimum 2D point set, generated by the use of the golden ratio, and giving a near-optimum MST!

For graphical applications also other criteria are popular based on spectral characteristics. Based on that we would e.g., find that a so-called Poisson spectrum contains more high-frequency parts (so-called blue noise) than pure random sets.

Indeed Fourier transform is not only good for complex number calculations and for filter design; it is also about sampling, patterns and reconstruction of signals, so there is a strong connection to statistics too

[Pilleboue]! For instance a translation of Fourier series (using sine and cosine) to discrete series is the so-called Walsh series which can help to construct space-filling point sets. In circuit design and Monte-Carlo the link to Fourier series is not so obvious, but LDS is also used in ray tracing, picture data construction, rendering and compression. In normal ADC design we use ideal (non-random) uniform sampling and linear filters for reconstruction. Clever statisticians extend this to advanced sampling schemes with less problems on aliasing, ringing, etc.

Using Fourier analysis results we can also explain MC convergence rates nicely, like the $1/\sqrt{N}$ behavior for random MC or the improvements with LDS. For instance, accurate yield integration is often very difficult because the indicator function (giving 1 for a spec pass and 0 for a fail) is non-smooth and therefore difficult to reconstruct with use of few sample points. This is especially true if the circuit performance includes many difficult functions (equivalent to high bandwidth signals), because sampling all of them accurately is a very hard problem in general.

A native extension beyond random and quasi-random sampling would be <u>adaptive</u> sampling by putting more samples in the critical regions (where the integrand changes mostly), and indeed we do so in the chapter on high-yield estimation. Also other techniques are possible and quite similar: For instance, in MC integration we give each pass/fail result the same "weight" 1/N, but advanced integration methods like Gauss integration can obtain faster convergence on smooth integrands by using non-constant weights. So there is no stop in aiming for a faster, more intelligent MC!

The discrepancy of pure random sets (or sequence, see next subsection)) is quite large (Equation (6.4)), and if a set follows Equation (6.5) it is called a low-discrepancy set (or sequence).

$$\text{For random: } D_\infty = \mathrm{O}(\sqrt{[1/n \cdot \log(\log n)]}) \qquad (6.4)$$

$$\text{For LDS: } D_\infty = \mathrm{O}(1/n \cdot [\log n]^s) \qquad (6.5)$$

Some things should be noted:

- Defining LDS by Equation (6.5) does not tell anything about whether it is random or deterministic; indeed simple LDS generators are not random anymore at all, but others indeed include some randomization.

- Note that for random sampling the dimension s has *no* impact, but for LDS (and LHS) it <u>does</u> have one (although not as strong as for a simple grid).
- Unfortunately, Equation (6.5) is only about the limit behavior for large n, it makes no statement about accuracy for low or moderate counts.
- For the simple Cartesian grid D depends also on s, for s = 1 we can achieve a discrepancy in order of 1/n, but for s = 2 we are at the same level as random MC (order $1/\sqrt{n}$). Generally D follows $1/n^{(0.5+0.5/s)}$.

If you know up-front, that your quasi-MC analysis requires 100 points, then we know already a method for optimum 1D discrepancy: Just use an equidistant set of points (which is also a 1D non-jittered LHS)! However, also using also for the second statistical variable the <u>same</u> 1D-optimum set is very <u>bad</u> (Figure 6.16 showing a "collapsing" 2D point set, composed from two identical equidistant set, plus transformation to a normal distribution); it is even worse than the simple Cartesian grid; we would get no correct statistics for the

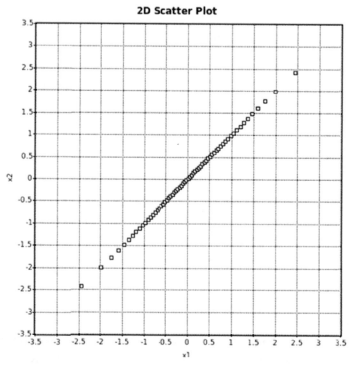

Figure 6.16   64-points Gaussian 1D set and an attempt to use it in 2D.

difference $X_2 - X_1$, so it would not work for designs in which correlations are significant!

Note: Figure 6.16 could be a random LHS as well, similar to Figure 6.9, but for luck such cases are extremely seldom for realistic run counts like n ≥ 50. However, in principle this would the (almost the only) disadvantage of simple non-jittered random LHS; we could really have full correlation c = 1, which would be less likely in pure random sampling.

Already in the 1950s these problems have been discussed, and one nice LDS solution comes with the use of *different prime* numbers in each dimension (Halton set). One drawback is that this does not fully avoid "patterns" in higher dimensions, and a full coverage also requires a certain *minimum* number of points, just because in high dimensions also the prime numbers become larger and larger. So-called nets or lattices are an alternative approach, but another problem, and a more general one is e.g., this: If we want to cover the whole s-dimensional space, we should have at least in each dimension a point below and above the mean (so-called binning optimality), and this also in combination with all the other $(s - 1)$ variables. However, as the number of combinations is 2^s, we actually need this as "minimum" number of points $n = n_{\min} = 2^s$ at least if we want to be "prepared" for truly full s-dimensional behavior (our section with questions and answer give a short example). Again, in this context even pure random MC with its typical $1/\sqrt{n}$ accuracy dependence looks not so bad anymore. In Figure 6.17 we observe good space filling for the first dimensions using the Halton sequence, but at higher dimensions we see a strong degradation in 2D space filling for this

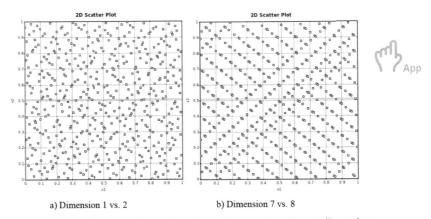

a) Dimension 1 vs. 2   b) Dimension 7 vs. 8

Figure 6.17   Halton set based on prime numbers in two dimensions.

quite old method. Modern methods behave better, but actually the minimum number of points to achieve a certain discrepancy level increases roughly linearly with the number of dimensions.

What helps is that usually not so many variables impact each different performance, just by intentional circuit construction, where dedicated elements have dedicated functions (like R and C set the filter bandwidth, but hopefully <u>not</u> many other elements). So it could make much more sense to optimize not so much the full s-dimensional discrepancy, but the discrepancy in low-order subspaces. This could often work better in many realistic test cases, but of course the HK error limit would not be fully applicable anymore and also finding a construction method becomes harder.

Hint: At the River webpage you can download a spreadsheet example quasi_random.xls with examples of Halton, Faure and other sampling methods. Inspecting these sets you can out quickly for yourself how much you can trust the different schemes.

Modern LDS generators use several different advanced techniques to avoid patterns and to maintain a speed-up over random-MC [Lemieux2008]. However, as mentioned theoretical limits exist and <u>also</u> LDS suffers on the problem of dimensionality, so it could happen that in some design cases LDS offers a clear speed-up like 4×, whereas in others (maybe just <u>another</u> performance of the <u>same</u> circuit even) any benefit is hard to measure.

An example for an LDS failure? We mentioned that a Cartesian grid can be a bad approach, even worse than random MC, but isn't the almost perfect LDS set of Figure 6.15 much different? If we *rotate* the set (being a kind of lattice) a bit we would run into the *very similar* problems as for a Cartesian grid! The only advantage is that our design would be quite special, e.g., being sensitive to a certain linear combination of random variables. You can see that in the *worst-case* also in LDS almost nothing is <u>guaranteed</u>. Actually minimum angle dependency would favor more randomized schemes like Poisson disk sampling, etc., instead of pure optimization on discrepancy (Figure 6.18). In [MacCalman2012] similar angle investigations has been made regarding LHS, confirming Figure 6.18b also for (linear) model parameter estimations.

In very rare cases you might be even worse than with pure random MC. In graphical rendering applications, there are several examples. Here also so-called aliasing effects are critical; e.g., if rendering a diagonal chessboard. If

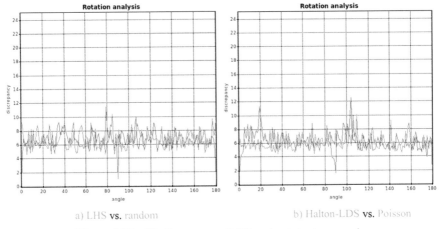

a) LHS vs. random   b) Halton-LDS vs. Poisson

Figure 6.18 1D-discrepancy of different point sets vs. angle.

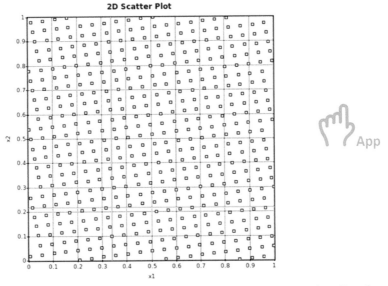

Figure 6.19 Modern, but unscrambled LDS generator in higher dimensions (7 vs. 8 and 512 points).

optimizing too much on D, we end up in visible artifacts, and to avoid them other criteria need to be included for the sample generation.

Fortunately, not many circuit problems have so strong variations as such chessboard examples, but even for realistic circuit design cases some general

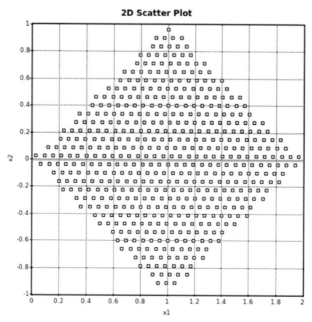

Figure 6.20 Sum vs. difference of uniform LDS samples may show bad discrepancy in x_2-direction.

LDS problems exist: in Figure 6.20a and b we built the sum and the difference of two dimensions (Figure 6.19) and get triangular distributed data, and this has a clearly non-optimum discrepancy! Actually the gaps increase similarly to Cartesian grid sampling, like we sample not with a density according to $1/n$ (as in one dimension or with LHS in each dimension) but only as $1/\sqrt{n}$.

In a real circuit we have usually normal distributions, so if we convert the uniform LDS points to normal variables, and build the difference—as a simple differential pair would do—then this difference *should* be also a normal variable. Therefore, we could simply transform it back to uniform and check again the 1D discrepancy! We would typically find that this discrepancy is again *worse* than the 1D discrepancy of the original, individual quasi-random numbers for each dimension itself! So we usually can only "hope" that this degradation is moderate, and that we are better than a pure random sampler. Modern LDS generators have indeed clever algorithms to avoid such problems, at least as much as possible. This is a complex topic because several criteria count, e.g., low correlation correlates with low discrepancy, but only

LDS in each class, but
not overall

LDS overall, but random
in each class

Low discrepancy overall,
and in each class

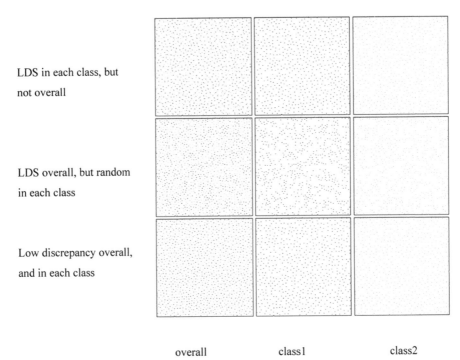

overall   class1   class2

Figure 6.21   Multi-class LDS set, suitable for error calculations or extending the MC run.

weakly, i.e., low correlation does not guarantee low discrepancy and vice versa
[Joseph]. Many reports only investigate on uniform sets, but some problems
(like tail characteristics) are even much more critical when looking to Gaussian
data than to the raw uniform data.

In addition, some things will not work well with quasi-MC in general,
e.g., accuracy *checks* are more difficult, like splitting the MC results into
two parts and doing a cross-correlation analysis for variance investigations.
This is because quasi-random numbers are <u>not</u> random (also not even pseudo-
random) and they would *not* pass some tests for randomness—they are just
"too good." To be able to do splitting and cross-correlation you would need
multi-class or sliced LDS sets, having low discrepancy inside each class (or
slice), and aligning all together in a big LDS set. In other application domains
(like imaging or LHS generation with the inclusion of discrete variables) such
techniques have been already introduced, but again it is not a trivial task for
high dimensions.

6.2.5 Sequences versus Sets

In random MC the samples should be *independent,* so we can stop it at any time, but in a 1D-optimum LDS set with equidistant samples or in LHS this would be not the case.

For instance, the set 1/3, 2/3 is optimum in one dimension and for $n = 2$ (in [0,1]), but if we want to extend it to $n = 3$ we would need to shift all points, so we need to re-run all simulations! So an *extendable* scheme, a true *sequence,* is preferable if you are not fully sure about the required total number of points! Mathematicians get rid of that problem by inventing not only optimum LDS sets, but LDS sequences which have low discrepancy even if we stop earlier, or if we need to extend them. A simple 1D example is this sequence:

$$n \ = \ 1 : 0.5$$
$$n \ = \ 3 : 0.5, 0.25, 0.75$$
$$n \ = \ 7 : 0.5, 0.25, 0.75, 0.125, 0.625, 0.375, 0.875$$

This sequence (so-called van der Corput sequence of base 2) has the advantage, over the set generation methods for known n, that we can fully re-use existing simulation points! The only pity is that for $n = 4$ or 5, etc., we need to accept some coverage compromises, and also the mean is not exactly 0.5 (as it should, and as it would be for LHS).

The sequence idea can be extended to higher dimensions, and because such LDS sequences allow an autostop capability, such enhanced quasi-random sequence generators are now standard in modern design environments and simulators. The problem is of course that any sequence generator can never be as good as a full set generator. So in principle it can even happen that the "older" LHS method could outperform LDS, like for estimation of the mean of a certain performance, on a design with moderate complexity. Interestingly, one can also show that a sequence with dimension s behaves (almost) like a set with dimension s – 1, so the benefit on discrepancy D is a dimension reduction by one. Using the Roth bound or the bounds for discrepancy for known LDS schemes (like Halton), we can also quantify in which order we would improve with the set; it is actually a factor of roughly log(n), that is the "free lunch" for throwing away an anytime autostop feature.

In [Matousek98] you can find a benchmark for different LDS schemes, different dimensions s and total point count n. One result is that the higher the dimension s, the higher the minimum point count to see an advantage of LDS against random; actually already for moderate s like 10 we need at least roughly 1000 points to see a speed-up in the order of 2. This result is fully in sync with the discrepancy theory, but the latter typically only gives asymptotical results, which can (in absolute terms) differ to the concrete low-count behavior of certain LDS methods. The degradation for large s is also in sync with the mentioned patterns appearing in classical LDS schemes (Halton, Faure, Sobol, etc.), and these effects can only be *reduced* in more advanced algorithms (e.g., by smearing out the patterns by randomization), but not fully avoided.

6.2.6 Summary and Comparison of Sampling Methods

Teaching electrical engineers, we received many questions on latin hypercube and low discrepancy sampling, like "is LHS *always* better than random?", or "what happens if we stop a MC run manually, can we still trust the results?" We hope these things become much more clear now; if not check out our list of questions and answers.

Since some years, LHS and improved LDS algorithms have been developed and implemented into EDA tools and simulators [Cools]. Using them is usually very easy; and designers can benefit from reaching the same level of accuracy already with a lower number of samples, so with shorter overall simulation times. That is the promise, but we have seen there are several difficulties and many different criteria (See Tables 6.2 and 6.3), also inspect the excellent summary on the state-of-the-art in [Pronzato2012]. Notice that to some degree in the future the differences between LDS and LHS will gray out, and actually e.g., orthogonal LHS picks up many LDS ideas [Rainville 2012]. So classical LHS is no true low-discrepancy method at all, but enhanced LHS often feature this property and could outperform many LDS schemes. For instance, in [Viana2006] an algorithm is presented which starts with an LHS set, but becoming extended picking up the "originally" LDS idea of lattice construction. However, most modern algorithms are a combination of construction and optimization (e.g., [Ebert2015]—presenting also detailed benchmark results).

As mentioned, many LDS problems also appear in LHS. LHS starts with a grid and uniform sampling, so to use them in a normal design environment they will be usually converted to normal Gaussian distributions. If we inspect each statistical variable individually, we would find more stable values compared to normal Gaussian random sampling for simple statistics such as mean, to a smaller amount also for higher moments like variance, skew, and kurtosis; this is desired. However, when we built differences (or sums or products, etc.) this "stabilization" of estimates will get almost lost. In opposite to basic LD schemes (like Halton or Faure) this is not so easy to observe visually

Table 6.2 1D projections of sampled 4-dimensional Gaussian data (n = 10, LHS is non-jittered)

1D Scatter Plot for the Four Statistical Variables Only	1D Scatter Plot for all Variables and all 2nd Order Mixed Terms	Comment
1a	1b	All "strings" look equally random. The mixed terms behave like the original variables!
Random		
2a	2b	Perfect plots only for the variables *itself*, because here we actually do sampling without replacement, giving the *same* set overall in each dimension! Random behavior for mixed terms
LHS		

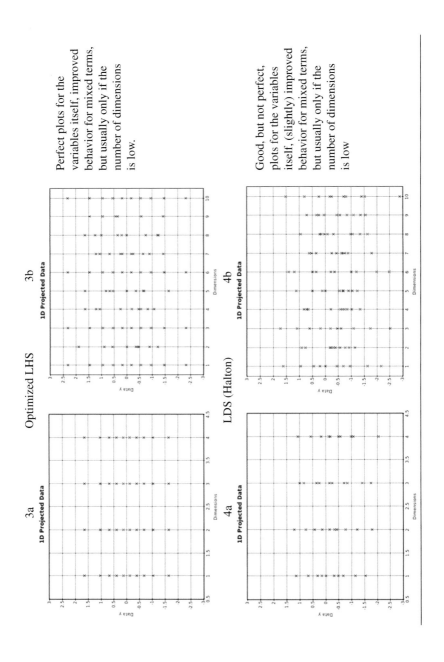

Optimized LHS

3b

Perfect plots for the variables itself, improved behavior for mixed terms, but usually only if the number of dimensions is low.

3a

LDS (Halton)

4b

Good, but not perfect, plots for the variables itself, (slightly) improved behavior for mixed terms, but usually only if the number of dimensions is low

4a

Table 6.3 Comparison of different sampling techniques

Criteria	Random	LHS	Optimized LHS	LDS Set	LDS Sequence	Comment
Convergence rate	$1/\sqrt{n}$	$a_1/\sqrt{n} + b_1/n$	$a_2/\sqrt{n} + b_2/n$	$(\log n)^{s-1}/n$	$(\log n)^s/n$	a, b depend on the nonlinearity of the function (coefficients in HDMR Equation 5.10), s is the number of dimensions
Able to treat high-dimension problems accurately	Unlimited	Highly limited, switching back to random	Limited, switching back to random	Limited, often patterns are visible	Limited, often patterns are visible (slightly earlier than for sets)	For most designs LDS is applicable, but do not use simple schemes such as Halton
Support for auto-stop	Unlimited	Limited	Possible, but not always implemented	Limited	Yes	LHS switches usually back to random if we stop earlier
Error control	By confidence intervals	Difficult				Bootstrap can be a further option

Speed-up on mean	None	Often high	Often high	Often high	Usually moderate	
Speed-up on standard deviation	None	Low	Moderate-high	Moderate-high	Moderate	Often also for correlations
Speed-up on sample yield	None	Practically only in low-yield region and simple cases	Practically only in low-yield region	Practically only in low-yield region	Practically only in low-yield region	
Sample creation time	Very small	Small	Large	Small	Small	
Sample creation memory	Very small	Moderate	Large	Small	Small	
Gaussian number generation	Many options, no restrictions	Transformation of uniform variables to normal distribution by inverse cdf (small limitations on accuracy and speed)				
Availability	Standard MC feature	Often a standard MC feature	Usually not available in EDA tools	Usually not available in EDA tools	Often standard MC feature	In design environments and simulators

aspatterns, but just in a statistical way. Looking to each dimension individually would get maybe in few percent of the sample sets, e.g., moment deviations (against an ideal normal distribution) beyond $\pm 1\sigma$, whereas for variable differences (still theoretically Gaussian distributions!) we would get e.g., 50% of such deviations! This is close to pure random sampling, so in designs where such variable differences matter (random) LHS *loses* the speed advantages. For good LDS schemes this degradation is usually smaller.

We started the chapter with 2D scatter plot pictures to give the reader a feeling for what is what, but how can we see this in circuit designs? Here we often look to histograms, and unfortunately the binning count has a significant impact on how "good" a histogram looks, on top of all potential other imperfections. So let us inspect Table 6.2 showing us the data itself, in a kind of "string", without binning. In picture 2a) LHS data is shown for each Gaussian variable, and picture 2b) shows all *differences* of Gaussian variables (scaled by $1/\sqrt{2}$). The number of dimensions is set to d = 4, so that we get 6 combinations, plus the four dimensions itself.

Note, that n is set to 10 to simplify the visual inspection, but in combination with d = 4 we are in a similar situation as in circuit design, i.e., the advanced methods work, but not perfectly. For larger n bigger improvements are possible, but often the available runtime is too limited to enter that region. And for higher complexity (large number of dimensions) we would have more problems; and actually not only second order terms would matter, also high-order mixed terms (which are hard to visualize).

With a better "smoother," more equidistant distribution of the sampling points—a set with lower discrepancy—we can improve the coverage; and in many benchmarks LDS has already demonstrated a speed-up of roughly $2\times$ to $8\times$ in real circuit designs (whereas standard LHS usually degrades earlier, especially in cases with higher complexity or stronger correlations) [Singhee2010]. We also showed that integration via Cartesian grid sampling slows down a lot already on simple 2D cases like our "π example." In this, LHS and LDS would behave significantly better, but already for $45°$ line as spec border at least LHS would already show some further slowdown.

Look-up: Enabling a certain method like LDS or LHS just with a click in the design environment also means that the user usually has no real further options to *influence* the performance; this is quite in contrast to other more advanced methods, which we inspect later!

Overall often LDS is a cheap lunch, but be aware of these limitations [Sobol]:

- For a certain sample size n there is <u>no guarantee</u> that low-discrepancy sequences will outperform random MC, so there is also no guarantee on shorter confidence intervals. The HK error bound is only valid under some very hard to check restrictions. If the prerequisites are not fulfilled, then typically only some (beneficial) *variance reduction* is provided.
- Error estimation is difficult with LDS in general. And even if someone can mathematically prove faster convergence (like $\log(n)/n$ instead of $1/\sqrt{n}$) under certain prerequisites it does not guarantee a speed-up also for low and more realistic MC counts.
- LDS speed-up is limited for the sample yield, because sample yield as estimator with discrete character generates a kind of additional "quantization noise." Another view is that the integration error depends on the function variation V and the discrepancy D, but the pass-fail indication function has a large variation.
- Competing requirements are forcing compromises, e.g., a <u>sequence</u> with option for auto-stop is less optimum regarding discrepancy than a <u>set</u> for given known number of points.
- Simple LDS generators create significant artifacts in higher dimensions s, and real circuits may have hundreds of statistical variables! This will reduce the LDS speed-up and can even cause systematic errors (at least in rare cases). A workaround to get back the theoretical advantages would be *sort* the variables [Singhee2010], i.e., to treat the most important variables with low-order sequences with no such artifacts! However, still the LDS speed-up is limited if n is not much larger than the effective dimension s_{eff} (and this is a fuzzy term—not well known and depending on circuit, model complexity and specification). As mentioned, complex circuits can have over-all still many important variables, especially when looking to many performances!

Remember, low discrepancy is only *one* criterion, and the cure of high dimensionality comes with many faces. In opposite to random MC the required LDS sample count n increases with the number of dimensions *s*, i.e., a too low count may come with artifacts like patterns, correlations, bad coverage in sub-spaces, and potentially also with bias errors—at best you switch back to random sampling. If you want too much from a method, you take a strong risk to fail.

Often you can expect that in larger sample sets you can find more "extreme" statistics, but for correlations and some other measures it is the other way round: Generating a 32-dimensional random (or LHS) set of 64 points, you will find that the worst-case correlation factor is quite large (roughly 0.4

in average), but for $n = 512$ c_{max} went down to about 0.12 only! In a Halton LDS set the increase in worst-case correlation is much stronger, e.g., for the same n and worst-case correlations the problem has the same severity already for $s = 13$ (instead of 32)! As mentioned, there is actually a trade-off between criteria like discrepancy D and correlation c (or good spectral performance, effort in sample generation, etc.), and even for discrepancy or correlation there are several measures which you could optimize.

In addition, most mathematical proofs on LDS are related to *integration*, which was also our starting point, but users do much more than pure integration with MC data. Parameter estimation by maximum likelihood is one important example, so for such tasks we leave clear mathematical grounds to some degree. Also note high-yield estimation (like $>6\sigma$) gets not immediately faster with LHS or LDS, just because both aim only for more stable statistics and more accurate integration, <u>not</u> for getting more fail samples with less MC points. Interestingly, beside mean estimation this is also one of the cases where LHS sometimes indeed outperforms LDS: For a random MC run with 500 points we can expect in average roughly a sample *spread* of 6.1σ, but of course we can also get runs which much less variation, like 4.7σ. In LHS this gets highly improved for the statistical variables itself, like roughly to $6.2\sigma \pm 0.2\sigma$, whereas in many LDS generators less improvements are available. So <u>if</u> the problem is so simple that such simple statistics matters, LHS would win.

Figure 6.22 is giving an example of a practical LDS generator: we get some small but at least visible numerical artifacts in dimensions higher than 100. Simple LDS generators show such problems already for dimensions like ten!

At the River Publisher website we uploaded an Excel file which allows to make some basic LDS experiments using a simple type of sequence generator (Halton sequence). There, the problem of unwanted patterns occurs much earlier.

Besides all these findings and restrictions, all methods have one key advantage: We can simply *combine* the discussed sampling schemes with further "speed-up" methods, like using them inside a sorted Monte-Carlo analysis (next subsection), or in conjunction with the C_{GPK} (Section 4). In addition, we can expect further improvements in the near future regarding "built-in" error calculation, stronger suppression of artifacts in high dimensions, exploiting circuit hierarchy and parameter rankings, etc. Figures 6.23 and 6.24 show the state-of-the-art regarding LDS for some realistic testcases.

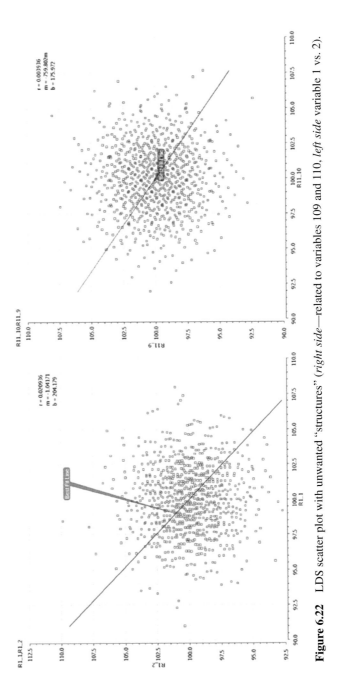

Figure 6.22 LDS scatter plot with unwanted "structures" (*right side*—related to variables 109 and 110, *left side* variable 1 vs. 2).

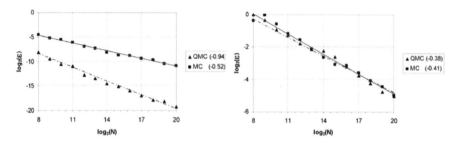

Figure 6.23 LDS integration speed-up for a simple case and a difficult one.

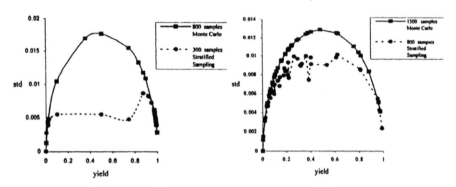

Figure 6.24 LDS variance reduction versus yield [Hassan].

Note, that LHS and LDS are designed for improving the accuracy for the *integration* of functions. In Chapter 2 we mentioned other sampling methods for *corner* analysis; for this the focus is finding the worst-case, design understanding, sensitivity investigations, actually performance modeling. In Table 6.4 you can find a comparison of several methods. There are also many other methods, like so-called D-optimal designs (created to minimize the variance in modeling), but those are beyond the scope of this book, and often a mix of these basic methods.

6.3 Design with Pictures Part Four

Using our 2D example for getting π by Monte-Carlo integration, we would observe that both LDS and LHS are much superior to random. Figure 6.25 shows that for an error of 10^{-3} we would obtain a speed-up in the order of $100\times$ for large but still realistic MC counts, for larger errors like 10^{-2} and lower counts still $10\times$ is clearly realistic.

Table 6.4 Different sampling methods and their applications

Method	Application	Comment
Random sampling	Mimic nature, work with statistical confidence intervals, etc.	Most universal method, slow $1/\sqrt{n}$ convergence
LHS	Integration	Accuracy improvements only if the variation of one variable dominates
Optimized LHS or LDS	Integration, Space-filling	Accuracy improvements if effective dimension is not too large in relation to sample count n
OFAT	Linear models, design understanding	Linear effort, not suitable for addressing mixed terms like $x_1 \cdot x_2$
Full-factorial	Space-filling, worst-case analysis, modeling	Exponential rise in effort regarding number of variables (dimension s)

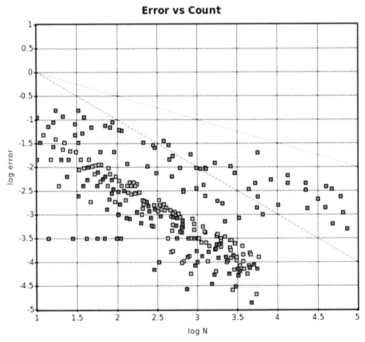

Figure 6.25 Random (gray), LHS (red) and LDS (yellow) behavior on simple 2D integration problem.

6.3.1 Experiments on Small Testcases

For more complex examples, we can indeed see at some point the $\log(n)^s$ dependence regarding dimension s which is limiting the speed-up. So if we e.g., use a high-Q 8-element LC bandpass as DUT and watch for a difficult performance like passband ripple, we would observe at best only a moderate reduction of the standard deviation by LHS and LDS, like 1.52×. Table 6.5 gives some examples looking to C_{PK} and C_{GPK} (on which LHS and LDS work better than on the sample yield). For a simple one-variable case (does not matter if Gaussian or not) and $n = 512$ we can achieve an LHS speed-up beyond 100×. In using an optimized LH set, we can achieve roughly 5× for a 3-variable Gaussian mix, but almost no speed-up for the difficult LC filter.

As expected, the speed-up usually larger for simple statistics (like C_{PK} or standard deviation) compared to the more complex C_{GPK}, but also this is to some degree an example how much the devil can be in the details: In the table we report the speed-up based on pure variance reduction, and for the filter and the C_{GPK} this speed-up is 0.5, so LHS was worse regarding standard deviation. However, the C_{GPK} distribution is asymmetric (like also for sample yield), so if looking to the lower confidence interval limit, the situation would be significantly better [Weber2016], like giving a speed-up of 1.1×.

The optimized LH set had lower correlations than standard pure n-rook LHS, and good LD sets would behave similar. For realistic, even more complex examples (like a $g_m C$ version of the bandpass or an ADC) the speed-up against random unfortunately disappear for $n = 512$, and only be present for very large n (only for these $(\log n)^s / n$ would become smaller than $1/surdn$).

Table 6.5   Speed-up on C_{PK} and C_{GPK} by using near-orthogonal LHS for different test cases

	Sampling	C_{PK}	C_{GPK}
	Gaussian mix (3 variables)		
mean	Random	0.86	1.01
sigma		3.19%	7.05%
mean	Optimized LHS	0.86	1.02
sigma		0.88%	4.23%
σ-Speed-up LHS vs. rnd		13.2	2.8
	LC bandpass (8 variables)		
mean	Random	1.41	0.99
sigma		4.74%	6.79%
mean	Optimized LHS	1.44	1.04
sigma		2.00%	9.59%
σ-Speed-up LHS vs. rnd		5.6	0.5

6.3.2 LHS and LDS for Contribution Analysis

Besides the mentioned LDS and LHS limitations, there are also several cases in circuit design where e.g., LDS is <u>indeed</u> useful, giving a significant speed-up. Looking just to histograms you can often not immediately see the advantages against random sampling, but one positive example is often the contribution analysis (See Chapter 5): with random MC the contribution result for each variable or circuit component will be quite noisy, e.g., with 800 MC points you may get the top-5 contributions accurately enough, but with LDS you may achieve the same accuracy often with already 200–400 points. This speed-up— coming here with almost no negative side-effects—is really helpful because it could mean a time saving of one hour or more.

We mentioned it already, for the sample <u>mean</u> the LHS or LDS speed-up is often even larger, but in most situations designers are much more interested in the standard deviation and extreme samples than in the mean (the mean would be anyway shifted already in a corner analysis, and the MC mean is often close to the nominal simulation).

Concrete numbers from a typical example run are given in Table 6.6 for our 32-transistor CMOS latched comparator.

We see that contributions which should be identical (e.g., input diff-pair) have at least very similar contributions and indeed LDS and LHS give for the same count somewhat more stable results. So here the LHS/LDS trick of hoping for a simple low-order problem structure works quite fine.

6.4 Synthetic Monte-Carlo: Bootstrap

In statistics, bootstrap (BS) is something different but in some way also similar to what it is in circuit design. The idea is this: A statistic from MC gets usually better if <u>more</u> data points are available. Could this be exploited

Table 6.6 Check for speed-up by LDS and LHS against random MC for a contribution analysis on latched comparator ($n = 200$, asymmetry in symmetrical contributions in percent)

	Random Sampling	LDS	LHS
Linear offset contribution from:			
most sensitive parasitic cap	3%	<1%	1%
V_{TO} of input transistors	4.4%	0.4%	1%
Hysteresis contribution from:			
Dummy transistor wl parameter	2.8%	0.8%	0.4%
Latch NMOS wl parameter	0.9%	1.2%	0.4%

without doing new time-consuming simulations? Can we do more with the data than calculating simple estimates like sample mean, sample median, sample standard deviation, sample yield, etc.? Can we generate artificial "new" data in a clever way to improve our estimations magically or get *further results*? Yes, we can! At least we can try, in fact the majority of such nowadays quite popular techniques falls in the class of "bootstrap," being a kind of virtual or synthetic Monte-Carlo: generating new synthetic data from existing data (e.g., from MC simulation or from experiments).

For sure one thing would not work as expected: if we have data samples like $x_1, x_2, x_3 \ldots . x_9, x_{10}$ and we want now to generate "new" data by putting them in an urn and picking them out randomly, step by step; this would only lead to a new sequence but of the same data! So sample mean standard deviation would be completely unchanged! The assumption of having a certain pdf leads to samples that are identical, and independently distributed. However, if we pick out samples from an urn step by step, this would be *not* the case. What

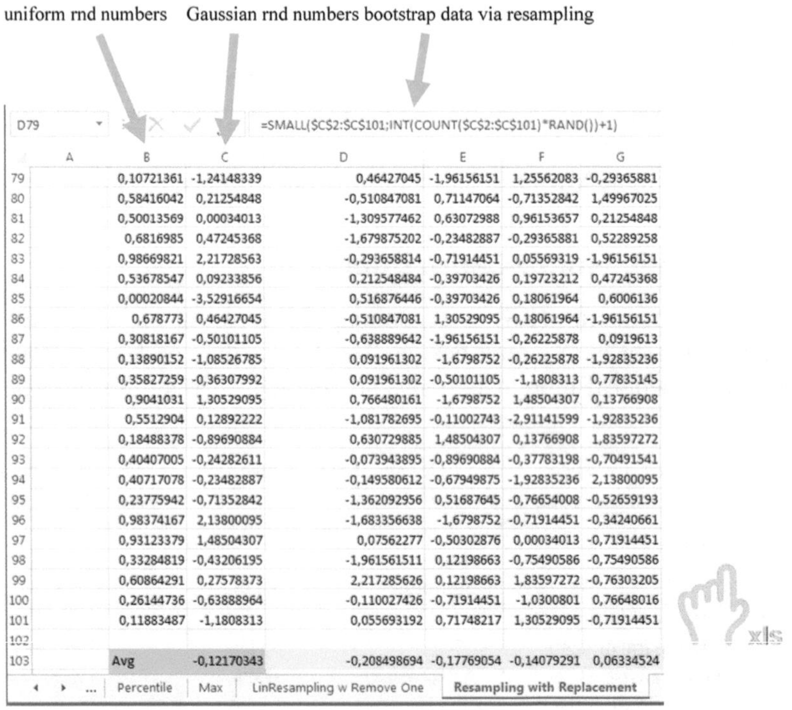

	A	B	C	D	E	F	G
79		0,10721361	-1,24148339	0,46427045	-1,96156151	1,25562083	-0,29365881
80		0,58416042	0,21254848	-0,510847081	0,71147064	-0,71352842	1,49967025
81		0,50013569	0,00034013	-1,309577462	0,63072988	0,96153657	0,21254848
82		0,6816985	0,47245368	-1,679875202	-0,23482887	-0,29365881	0,52289258
83		0,98669821	2,21728563	-0,293658814	-0,71914451	0,05569319	-1,96156151
84		0,53678547	0,09233856	0,212548484	-0,39703426	0,19723212	0,47245368
85		0,00020844	-3,52916654	0,516876446	-0,39703426	0,18061964	0,6006136
86		0,678773	0,46427045	-0,510847081	1,30529095	0,18061964	-1,96156151
87		0,30818167	-0,50101105	-0,638889642	-1,96156151	-0,26225878	0,0919613
88		0,13890152	-1,08526785	0,091961302	-1,6798752	-0,26225878	-1,92835236
89		0,35827259	-0,36307992	0,091961302	-0,50101105	-1,1808313	0,77835145
90		0,9041031	1,30529095	0,766480161	-1,6798752	1,48504307	0,13766908
91		0,5512904	0,12892222	-1,081782695	-0,11002743	-2,91141599	-1,92835236
92		0,18488378	-0,89690884	0,630729885	1,48504307	0,13766908	1,83597272
93		0,40407005	-0,24282611	-0,073943895	-0,89690884	-0,37783198	-0,70491541
94		0,40717078	-0,23482887	-0,149580612	-0,67949875	-1,92835236	2,13800095
95		0,23775942	-0,71352842	-1,362092956	0,51687645	-0,76654008	-0,52659193
96		0,98374167	2,13800095	-1,683356638	-1,6798752	-0,71914451	-0,34240661
97		0,93123379	1,48504307	0,07562277	-0,50302876	0,00034013	-0,71914451
98		0,33284819	-0,43206195	-1,961561511	0,12198663	-0,75490586	-0,75490586
99		0,60864291	0,27578373	2,217285626	0,12198663	1,83597272	-0,76303205
100		0,26144736	-0,63888964	-0,110027426	-0,71914451	-1,0300801	0,76648016
101		0,11883487	-1,1808313	0,055693192	0,71748217	1,30529095	-0,71914451
102							
103		Avg	-0,12170343	-0,208498694	-0,17769054	-0,14079291	0,06334524

Figure 6.26 Bootstrap in Excel® (columns D to G, look at boot.xls at the River webpage).

we would need to do is picking the sample out (randomly!), make a notice on result and then <u>putting it back</u>! This is called "resampling *with* replacement," and it is the core of generating bootstrap samples (actually sets)! Note that this kind of bootstrap does not explicitly assumes a certain "model," so it is called nonparametric bootstrap; and it works quite general. In our book we focus on nonparametric univariate bootstrap, but even much more complex schemes like multivariate parametric BS are possible.

Bootstrap via resampling with replacement:
    Original MC data: e.g., x_1, x_2, x_3, x_4, x_5
    Resampled data set (bootstrap data): e.g., x_2, x_5, x_3, x_2, x_4
    Another bootstrap run: e.g., x_4, x_1, x_3, x_2, x_4

Note: We can have multiple identical values and some original samples may get lost. Also note that we can do the bootstrap as often we want with little effort! Figure 6.26 shows that even a bootstrap implementation in a spreadsheet program is possible. The only limitation is that it is not so fast anymore if you want to apply bootstrap e.g., 100 times on thousands of data points. Another method to generate "new" data sets is the leave-out-one method. Also this can be used to check how much variation exists in the data or how stable our estimations are, but unfortunately leave-out-one, or a variant like leave-out-10%, works *not* as good as BS, re-sampling *with* replacement!

    Bootstrap looks tricky, but as bootstrapping in circuits, it has its clear benefits, its pro and cons. Actually, bootstrap only sounds tricky, but is very simple to do, it is very crafty, and it has a solid fundament:

- Bootstrap is <u>not</u> limited to purely normal Gaussian distributions!
- In a bootstrap sample the sample mean will be often slightly different compared to the original data, but that difference is indeed a good indication for the <u>real variation</u> from one MC run to another one! For these reasons bootstrap is a good method to estimate <u>confidence intervals</u>!
- Bootstrap works well if the data are large enough, so that the statistic you are looking for (mean, standard deviation, etc.) does <u>not</u> depend much on <u>few</u> values. If this is not the case, and you look for a statistic like the maximum (which often depends on a single sample) then bootstrap could <u>fail</u>.

For these reasons—and as it is easy to run them on a computer—bootstrap techniques are nowadays incorporated in many kind of algorithms, as internal error prediction element, helping to decide internally if we can stop an iterative algorithm or not. Of course also bootstrap techniques get improvements over the years, to get internal error predictions or to apply it to more than just confidence intervals. For instance, resampling by replacement can be made directly on the samples, so actually nonparametric, but we could also make a parametric fit and generate artificial samples from the model, instead by sample resampling. This is called parametric bootstrap. Also it comes out the simple bootstrap typically *underestimates* the variations a bit, so by adding additional "noise" we can reduce that kind of bias error. Mathematically, bootstrap is the same as sampling from the empirical cdf (which has a staircase form), so the smoothing actually makes the empirical cdf closer to the usually more realistic smooth real cdf (which is unfortunately unknown).

How much time does bootstrap take? Originally, i.e., in the 1970s, bootstrap was called a "computation-intensive" method. Indeed using bootstrap for confidence intervals needs more computation power than using classical formulas, like those based on Student's t or chi^2 distribution. However, nowadays this effort is not really large anymore. For instance, bootstrap may take a second instead of a ms, but this does not matter compared to the time to run circuit simulations being often in the order of minutes.

6.4.1 Bootstrap Application Examples

The classical bootstrap (BS) application is confidence interval estimation. For normal data and simple estimates like mean and sigma well-known formulas are available, but as mentioned confidence intervals are harder to derive for non-normal data, especially if you even do not know of which type your data is (e.g., your MC data is coming from a difficult circuit) or if your estimate is quite difficult (like the generalized C_{PK}).

So for simple estimates on normal distributions we can easily compare bootstrap CIs against the classical ones (e.g., based on Student's t or chi^2). Actually many such benchmarks exist, and BS is quite accurate if the number of samples is large enough (like >50). Also many methods exist to make the BS more accurate (we mentioned the addition of noise). With such extensions BS become really useful and are often applied internally as sub-algorithm to other methods like those for high-yield estimation. So BS acts often in the background, and you can do it for yourself in Excel®. So maybe it is most interesting just to know well when it works *not* so perfect. Table 6.7 shows the

Table 6.7 Accumulated errors of simple bootstrap on normal distribution ($b = 512$, true $C_{PK} = 1.5$)

			Error in		
n	Mean	Sigma	Kurtosis	C_{PK}	C_{GPK}
256	$<\sigma/50$	−12%	−0.661	+0.21	>1.75
512	$<\sigma/50$	−9%	−0.505	+0.14	+1.26

results of an experiment: We run a big MC analysis on a certain distribution and extract estimates for mean, sigma, and kurtosis. Now we generate BS samples many times and report the same estimates, but now from the synthetic BS data. Actually you will see that BS is quite accurate on mean and sigma, but on kurtosis the error is significant. If we do now the BS again, but not from the original MC data, but from the BS data, the BS errors would *accumulate*—like the noise or distortions from an analog tape copy taken from a copy. We do this BS experiment 10 times, so that the errors become nicely visible. Besides kurtosis, the C_{GPK} bootstrap error is also significant, so the simple BS should be enhanced to make it work well for the generalized C_{PK} [Weber].

6.5 Fast Monte-Carlo by Sample Sorting

Beside contribution analysis, a second nice application based on a multivariate MC analysis is to use the created multivariate model to predict *how* extreme a certain combination of the statistical variables is regarding circuit performance. In older software packages such response surface model (RSM) has been *directly* used to predict the yield. However, such approach is risky. Even if the goodness of fit is high (like $r^2 > 0.98$), still the yield estimation could be misleading, e.g., because the pdf tail behavior might be difficult. So these tools have not reached popularity in circuit design. A better way is to use the model and to *double check* the pass-fail estimation from the model with further real circuit simulations. By re-ordering the MC samples we can actually create a kind of "high-sigma MC" by "blocking" simulations of low interest! You may also call it sorted MC or model-based statistical blockade. If the model indicates an offset voltage close to the typical behavior, then the simulation will probably give not much new insight compared to a nominal simulation, which is usually done already many times by the designer! However, if the model predicts an offset voltage of e.g., $> 3\sigma$, then this can be a sample close to the spec limit or beyond, and it can be very interesting for debugging purposes and for yield verification. Actually running simulations with parameter combinations ending up in the distribution center is quite some waste of time, because it is almost redundant. Removing this "redundancy" can speed-up the verification significantly (like 10× for 3σ and blocks of moderate size, more at higher sigma levels), by doing a kind of 2-step sorted MC analysis (Figure 6.27).

Following this idea, we do not apply pure random MC anymore, i.e., not only a random number generator is driving the simulation. This MC speed-up technique is something *beyond* pure result evaluation, like using the C_{PK}, and we can also combine the sorted MC or C_{GPK} with other clever techniques. One limitation remains: Also when using sorted MC, we still have to deal with the confidence intervals related to the sampling process, but the speed-up from skipping non-spec-critical simulations usually allows using a (quite) large sample count n and this way we can achieve (much) tighter confidence intervals CI = [LCB,UCB] with moderate simulation effort. Like for LDS or LHS, one problem is to define if and how much you really improve on CI width and if we create a certain, undesired bias error.

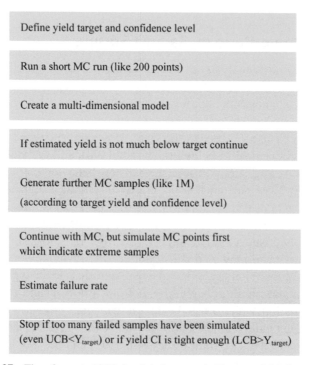

Define yield target and confidence level

Run a short MC run (like 200 points)

Create a multi-dimensional model

If estimated yield is not much below target continue

Generate further MC samples (like 1M)
(according to target yield and confidence level)

Continue with MC, but simulate MC points first
which indicate extreme samples

Estimate failure rate

Stop if too many failed samples have been simulated
(even UCB<Y_{target}) or if yield CI is tight enough (LCB>Y_{target})

Figure 6.27 Flow for sorted MC, it might be extended by a model-refinement step.

For very high yield targets, like 6σ or more, the sorted MC speed-up would increase further (like $>100\times$) but more advanced methods (like worst-case distances WCD) may offer even more. And for low yield targets like $<99\%$ the sorting speed-up is very limited, because with few MC samples like 100 you can hardly create an accurate multi-dimensional model, and there would be not many remaining MC samples to be sorted!

Different stopping criteria could be defined; e.g., instead of inspecting the sample yield confidence interval, we could also stop on reaching a certain "sigma corner level." So we could sort and simulate so many samples till we hit the 3σ or 5σ limit of all performances with a certain accuracy like ±5%. These statistical corner samples can be used by the designer for debugging, circuit improvements, etc.

In the next chapter, we will discuss other options for moving further away from MC and getting a speed-up also by other means. Also note that sorted MC is not a dedicated method for finding true statistical worst-case corners, in opposite to worst-case distances (see next subsections).

On the other hand the sample reordering idea is bright and several commercial and non-commercial solutions already exist and have demonstrated its usefulness very successfully.

6.5.1 Advanced Features and Example Run

A detailed description and many results of MC with reordering can be found in [Singhee2007], [Kuo] or [McConaghy]; the sorted MC speed-up depends not only on yield level but of course also on how good the generated multi-dimensional model is. [Kuo] proposes the use of a linear model only, and using only the important variables. This reduces the internal computation times and improves the efficiency already significantly, but to some degree at the cost of reliability. However, an interesting feature is that the sorted MC method is quite robust, at least to some kind of errors [McConaghy]: If the model predicts no fail samples, then we could just increase the MC sample count till we get fail samples also in the model prediction. If the model is too pessimistic, then mainly the speed-up is reduced, but the results are still quite trustable due to the double-check mechanism. On the other hand, there is also some risk that the model does not predict "difficult" fail areas, so some systematic yield over-estimation might happen still (but seldom). In [Singhee2007]—representing an older but already advanced implementation—you will find an example that even a model from 30,000 initial MC points might be <u>not</u> accurate enough! As the sorting is related to *each* performance and spec, the risk for such errors

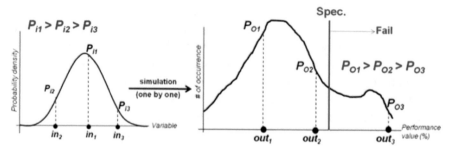

Figure 6.28   Typical probability relationships in a MC analysis [Kuo].

might be not that small, especially for real-world designs with significant complexity, many partially complicated specs, high nonlinearity, etc. Also note that unfortunately such model inaccuracies always lead to too optimistic yield estimations, so you are on the risky side.

In modern commercial tools you find several enhancements [Kuo, McConaghy] e.g., the algorithm might be modified to use already for the model <u>creation</u> more <u>extreme</u> MC samples (problem: the joint pdf is not really a good indicator for extreme performance values) or by doing a grid-based search (e.g., in Cartesian or polar coordinates; problem: quite inefficient for complex problems, points cannot be really used for other purposes like histograms, confidence intervals, etc.). Such improvements give some but limited accuracy improvements, and we need extra calculation time, so we may reduce the overall speed. On the latter, we may improve further by checking for *redundant* samples in the critical near-spec regions (Figure 6.29). Unfortunately, such cloud analysis for sample grouping and cutting out redundant samples will never speed-up much, because such samples are simply seldom for high yield targets (Figure 6.30).

As mentioned, the model extrapolation risk *increases* for higher sigma levels—a nice method to avoid this effect is introduced in a later chapter by using *upscaled* sigmas for the statistical element variables. Certainly also the modeling techniques itself are becoming better and better regarding speed and accuracy. On the other hand, the EDA vendors just <u>have to</u> improve on this, and on reliability, to really keep pace with increasing design complexity present in circuits, technology, and systems!

"Keep pace with increasing design complexity"—too much marketing speak? Actually not! Indeed some methods are quite old and working well in older technologies since some years, but on comparing e.g., mismatch

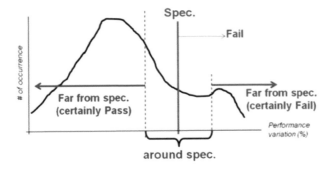

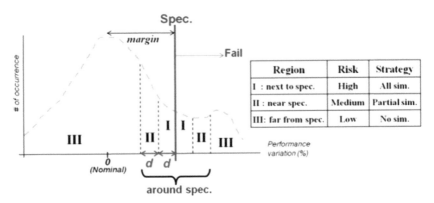

Figure 6.29   Categories in output distribution and their treatment [Kuo].

contribution results in older whereas newer program versions will often show good progress. Also other well-established features like WCD (see next chapter) have been continuously improved, by using better optimization algorithms or by providing automated setting for the initial MC count. Often users get big improvements, but in the user interface it is just one more checkbox.

On the other hand, sometimes the math behind the design problems is very difficult. For instance in sorted MC the model creation is challenging; having an *exponential* rise with the number of statistical variables. So often we need to restrict the search space to maybe few dominant factors, e.g., by setting thresholds which unfortunately introduce some model inaccuracies. And also the set of model base functions must be tractable, but our real circuit design might behave more special. And even if this model creation part works fine, we still have to create all the MC sampling points, also, for using the sample yield, with *exponentially* rising effort on the yield target in sigma! All these samples have to be applied to the performance models, so even if one model evaluation takes only a second (instead of minutes or hours for true circuit simulations), we could still have runtime problems (remember 6σ is roughly 1 fail in 10^9, and 10^9 seconds are roughly 300 years). So we usually need several further speed-up methods like parallel computing and e.g., pdf estimations based of further assumptions. This is one example and just one solution of fight of mathematicians and engineers against the "cure of dimensionality". In the next chapter we will also inspect further intelligent methods, e.g., sigma-scaled sampling. This avoids the need for multi-dimensional model creation, and it also breaks down to exponential MC count law to only a linear relationship.

The easiest way to reduce the extrapolation risk is surely to update the models from to time, just to take also the simulated points from sorting into account. We mentioned that for each output such performance model is required, so often these models have *different* accuracy, e.g., it might be easy to create a model for offset voltage but difficult for overshoot. A good method is to use what you have and first use the most accurate model (e.g., indicated by r^2), so we could sort on this and run these MC points for which the model predicts the worst performance (lowest yield). This way we get more data to allow also improving the fit for the difficult outputs, step by step. So overall the designer gets with sorted MC an almost fool-proof method, almost as easy to use as pure random MC!

Figure 6.31 shows the setup and Figure 6.32 a typical histogram from MC analysis with sample reordering (done on a LC bandpass filter). The yield target is 99.99%, so a normal MC run may require more than approximately 30 K points. In sorted MC 50 points will be run, then a model will be created, and then only roughly further 20 sorted points will be simulated. This way the histogram looks bimodal (having two "peaks"), which is a bit artificial because the true histogram is unimodal in our test case (Figure 6.32b, from a 1200-points random MC analysis). Actually this "strange" histogram helps to understand how MC with sample reordering works.

In this example the speed-up looks really big, like 70 points vs. 30 K, so $428\times$. The total runtime was not that much faster because the model generation and the sorting takes some time (like few minutes), but the speed-up is still

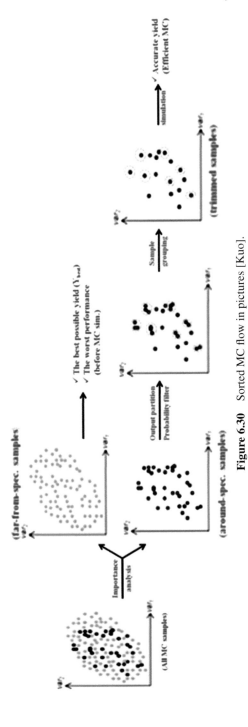

Figure 6.30   Sorted MC flow in pictures [Kuo].

Figure 6.31   MC with sample reordering, setup in an artificial UI offering further speed-up options.

very significant. However, look-up: In this example case we had 7 fail samples so that the sorted MC run actually proved that even the upper CI limit is <u>below</u> the yield target. For 99.87% we would approximately hit the edge, and the real speed-up would be now $2K/70 \approx 28\times$. For classical analog blocks like OTAs, op-amps or bandgaps you would often find similar speed-ups although these are often more complex. You may wonder why the speed-up with modified yield target was lower? Look into our chapter on questions and answers; the result is within expectations, and speed-up still amazing.

The C_{PK} speed-up would be in *similar* regions (See Chapter 3), but the C_{PK} has a large bias error because our MC data is non-normal, so we would require the generalized C_{PK}, giving a somewhat lower speed-up of maybe $5\times$. Of course the internal calculation time is much lower for both C_{PK} and C_{GPK}.

For higher yield targets (like $\geq 5\sigma$) the advantage of sorted MC and C_{GPK} would be even larger, whereas for low yield targets (like 2σ) the speed-up would almost disappear, and actually the artificial bimodality in Figure 6.31

a) typical result (some randomness present, especially in sorted run)

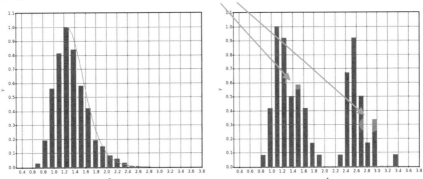

b) ideal result (no randomness, no model errors, smaller required margin)

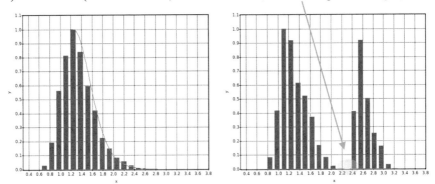

Figure 6.32 Normalized MC histograms for bandpass ripple (full and sorted run).

would also disappear, because the sorted samples would overlap a lot with the initial MC run.

Besides looking to the histogram, we can also inspect the output performance vs. MC sample (Figure 6.33): Sorted MC starts as normal MC, so we see the typical variations, approximately 68% of the points are within $\mu \pm 1\sigma$, so we observe a moderate number of both good and bad design samples. However, "unfortunately" we see only few or no extreme samples (because here the first part of the flow according to Figure 6.26 stops after 50 points already). Then the model will be created and only the sorted samples will be really simulated, starting indeed correctly with the *true* worst-case sample *detected* by the model! Then the 2nd worst, 3rd worst, etc. point will be simulated; and from this data the yield confidence interval via Clopper-Pearson can be

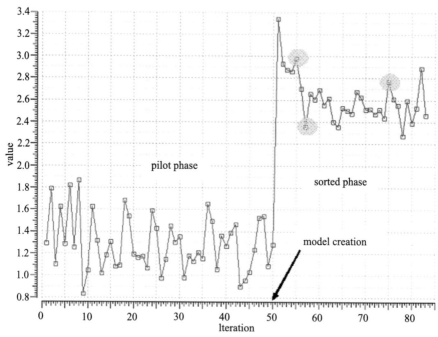

Figure 6.33   Performance plot vs. MC iteration for a typical sorted MC run (spec at 2.75 dB).

calculated as usual; and we get a yield verification as usual, just faster. Again looking to the details is interesting: point #55 is going up a bit, and #57 is not extreme as the model predicts! So actually the model is not 100% correct (see also #82). Fortunately, by simulating just some more points than actually needed we could *still* get quite sure regarding yield verification. And the algorithm can internally check its own accuracy quite nicely and use the new simulation point results to update the model if needed. So this type of error can be reduced to an acceptable level. Many details can be also found by inspecting the log file (Figure 6.34).

If something works in a *random* fashion, we can usually expect that doing it in a *systematic* way would be even faster! Such non-random way has not been found in all kind of problems, but luckily this is not the case for yield analysis!

Actually the sorted MC analysis still pushes random (or e.g., LDS quasi-random) variables to the simulator, so we could <u>further</u> speed-up if we not only sort the samples, but even *set* them. This is one key idea behind many truly high-yield algorithms—learn more about them in the next chapter.

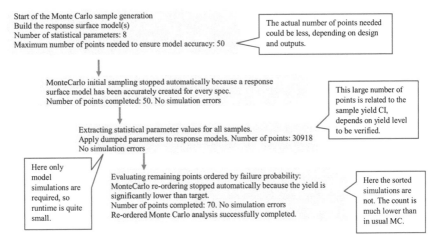

Figure 6.34 Example run of sorted Monte-Carlo executing the flow in 6.30.

6.6 Questions and Answers on Sampling and Sorted Monte-Carlo

As mentioned, the first parts of this chapter are related to advanced MC methods, still based on sampling. Often you can combine advanced sampling methods with general analysis (like contribution analysis) or further speed-up techniques (e.g., for high-yield estimation by sigma-scaled sampling or worst-case distances—see next chapter).

1. What is the typical speed-up against random MC by LDS? On what does it depend?
   The speed-up can be large, like >10×, but for realistic circuit designs it is usually much lower, it can degrade highly for problems with larger number of statistical variables. Also if the measure itself is "noisy"— like the sample yield—the LDS speed-up is only moderate. Often the LHS variance reduction is lower than that of LDS.

2. Check out Figures 6.5 to 6.7 showing the integration error of random vs. grid sampling. How would the error plot look like for LDS or LHS set of a fix length like 1024?
   For an LDS <u>sequence</u> it could look quite similar, with hopefully lower variation in general. For a <u>single</u> latin hypercube <u>set</u> it would 1st slowly improve, then went down quickly when reaching 1024.

Check Figure 6.13 on LHS vs. random for a further understanding. This plot is similar but just slightly differently "scaled."

3. What do you think of this method? Why using advanced statistical methods in case I need 0.01% yield resolution or 99.99% yield I could just run MC with 10 K points for verification!
   10 K is usually only possible for smaller circuits, also the yield lower confidence interval limit is much worse than 99.99%. Advanced methods (C_{GPK}, worst-case distances WCD, etc.) can achieve the verification task in shorter time and WCD can also give <u>accurate</u> corners suited for an optimization.

4. Explain the differences between LHS and LDS. What is the general problem in higher dimensions?
   LHS is still close to random, only regarding the one-dimensional projections it is improved. LD sets are typically quasi-random, looking like lattices or crystals. In addition most LD methods are extendable, so-called sequences. Also check out Google! You will be surprised to see how many nice articles you will find, often to completely different topics! Also our help inside the RealTime app is a good starting point.

5. In random MC you can combine two results into one to get more stable estimates, e.g. for yield. However, can you do so also for LHS or LDS?
   Actually this is a bit risky. In random MC you need two different seed settings to make sure that the data is independent and really different. For randomized LHS the method should also work, but not e.g., for using the simple Halton LDS generator and randomizing only across the dimensions (at least in some cases like for sample yield, low MC count and a high yield level).

6. LHS is always at least as fast as pure random, at least in average. However, is this true for LDS as well?
   Usually yes, but although LDS often outperforms basic LHS, in circuit design cases there is no real guarantee. One problem is that (by definition!) LDS wants just to guarantee a certain convergence speed for larger MC count n. So for low n we may have no speed-up, and the minimum n depends on the "effective" dimension of your testbench and n_{min} can be large, like 10.000 for difficult cases. Also almost no LDS generator is really perfect, often patterns occur, and some circuits might be sensitive to such artifacts.

7. We mentioned that for full s-dimensional coverage the number of samples n would rise exponentially with s, like n $> 2^s$. Can you give an example function which is really so difficult?
   To sample simple functions like f(X) = x_1 + x_2/10 + x_3/100 we need only good coverage for the dominating term, so LHS would work almost perfect. For more balanced weighting factors, LHS would run into problems, but LDS might still work if s is moderate. For real difficult functions like ɸ max(0,x_i) also LDS would not work well for realistic sample counts. The worst-case of such functions is hard to hit, because at least all individual random variables need to be positive, the chance for this and for s = 20 variables is p = $1/2^{20}$ ≈ 1ppm.

8. LH sets can be created with little effort, you can get them by construction. However, is it also possible to create a latin hypercube set from an <u>existing</u> set?

   Yes, this is possible and called latinization. So you may generate a e.g., low-discrepancy set and transfer it to a LHS. Unfortunately, the originally low discrepancy would be increased, but usually only by an acceptable amount. So if you look at Figure 6.18 for projected 1D discrepancy vs angle, then you can always improve D for 0° and 90°, but it is unfortunately more difficult to improve in regions of large discrepancy.

9. LHS (or LDS) is not as general as random sampling; there are simply more random combinations than LH sets. Isn't LHS this way missing some effects?
   Not really, also the number of LH sets rises almost exponentially with n and number of dimensions s; actually there are $(n!)^{s-1}/(s-1)$ different LH designs. This makes unfortunately the optimization of LHS quite time-consuming.

10. Create a testbench e.g., with five equal resistors having Gaussian distributions. Setup outputs for the individual resistor values, and e.g., for R_1 + R_2 and for the sum of all resistors. Now run MC on mismatch only with LHS. What do you expect regarding the histograms?
    The histograms for the individual resistors should look closer to an ideal Gaussian distribution. This is because the stratification by LHS should work best for this case, especially if in your models only one statistical variable is used for resistor mismatch. For outputs impacted by many variables, the LHS advantage often disappears quickly.

11. Look to the different pictures comparing random, LHS and LDS. What do you expect regarding LHS and LDS versus number of MC points? *Indeed that a set is LHS and not random can be seen easier for low MC counts, so one can expect already for low n some kind of stabilization in the estimates. For large n the MC results are anyway quite stable, so already random could be accurate enough. For many LDS patterns are visible, these disappear mostly for large n. In rare cases LHS can outperform LDS, but this is not likely for complex cases and large n.*

12. We mentioned that in LDS the inverse discrepancy (giving a count n to achieve a certain discrepancy) rises roughly linearly with the number of dimensions s. However, we also mentioned that to cover and s-dimensional space at all corners we need at least 2^s points. Is this a contradiction? *Not really! If we can achieve a certain discrepancy D we can bound the integration error, but also the function variation V has an impact, and often more complex, high-dimensional problems have a larger V. So over-all the integration error could rise e.g. quadratic (or more).*

13. In our sorted MC example the obtained speed-up was dependent on yield target. Can you explain this? *The speed-up was lower for the lower yield target, and for a low target also normal MC needs less points. In addition if the yield target and the true circuit yield are close together we need really accurate estimations, so the algorithm is tweaked to give that desired accuracy at the cost of a slightly reduced speed.*

Discuss this idea: We know about many different sampling techniques like OFAT, full-factorial and LDS. Each is almost optimum for a certain application. Would it make sense to combine these methods, e.g., for yield analysis or modeling?

Look-up that for yield verification we need to check both environmental range variables x_R and statistical variables x_S! Therefore a direct mix of e.g., full-factorial and LDS makes not much sense.

14. What is bootstrap good for? *The primary application is generation of confidence intervals. BS allows this also for non-normal data and with no further circuit simulations. BS is often part of the internal error checking in advanced statistical methods (such as sigma-scaled sampling SSS, see next subsection).*

15. LDS or LHS can reduce the variance of MC results from run to run. Can this be captured by bootstrap?
    No BS is like doing random sampling from the empirical cdf, from the pure data.

16. In many statistical algorithms there is a trade-off between variance and bias error. Give some examples.
    Using n–1 instead of n reduces the finite-sample bias for the sample variance calculation, but it increases the variance (a bit). In LHS <u>without</u> jittering we get slightly more stable estimates (less variance), but some bias error will be present, especially in estimates like sample yield and kurtosis. With jittered LHS we can reduce the bias, at the expense of (slightly) larger variance.

7

Fast and High-Yield Estimation Techniques

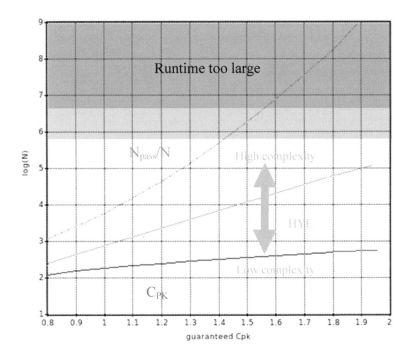

Now, we focus on dedicated high-yield methods giving a further speed-up and we extend the concept of worst-case corners to statistical variables, leading to worst-case distances. The idea of corners (and worst-case corners) and relating the "spec distance" to the yield is quite old. This idea and the use of non random or mixed methods are the base for these more advanced yield verification concepts. These are required because we know that using the C_{PK} comes often with a too large extrapolation risk, especially when addressing high-yield problems.

Here, we will again also talk about methods giving real design <u>insights</u>, e.g., such statistical corners can be related to a certain yield and are perfect for circuit debugging, design tweaks, and optimizations. In addition, there is a strong link to a multi-variate sensitivity and contribution analysis (Chapter 5). With LHS and LDS, we apply to some degree just a nice simple trick, like a lever in mechanics, and it is just useful because many designs have a structure that fits well to these sampling techniques, such as few variables and first-order relationships *dominate*. This works also quite well, just because designers create their circuits in that way that a certain transistor (single or structure) has a *certain* function; and the others should *not* disturb that function! However, as usually with tricks (although indeed with mathematical "backup," remember the Koksma-Hlawka theorem Equation 7.1), the benefit is sometimes limited; so you need further "tricks" like a steam engine, like feedback, or the advanced methods in this chapter.

One powerful method already described was sorted MC, sorting based on modeling to predict circuit performance. However, why not improving further? The simplest technique we already applied intuitively for MC result interpretation was the C_{PK}. We applied a two-parameter fit to the MC data for a normal Gaussian fit and hoped that such Gaussian model would predict well the true yield. Using the C_{PK}, we completely *ignore* that in the real design, there are typically thousands of statistical parameters—although when doing the MC analysis, we *could indeed* gather the information about all these statistical parameters and their impact on the circuit performances! In sorted MC, we already do so, but instead of generating MC samples and applying the quite time-consuming sorting, we could also directly search, e.g., with optimizers (see Chapter 8) for critical combinations of the statistical input parameters! This way we can realize one further key idea for "intelligent" Monte-Carlo: Get tail samples faster than in normal MC.

Are these further advanced statistical methods providing a free lunch? Like for sorted MC, actually for the designer the answer is often yes! However, such gala dinner has to be served by the EDA vendors, and the implementation effort is indeed much larger than for random Monte-Carlo! For instance, MC can treat any kind of statistical distribution, like normal, lognormal, uniform, etc., and also correlations can be included; also, most advanced methods can do so too but whereas in MC, it is enough to have just random number generators for the distributions, you need now much more behind the scenery, like computation functions for pdf, cdf, inverse cdf, and methods such as matrix inversion, eigenvalue analysis, principal component analysis, and optimizers.

For Further Reading:

- Th. Fischer, T. Nirschl, B. Lemaitre, and D. Schmitt-Landsiedel, "Modelling of the parametric yield in decananometer SRAM-Arrays," in *Advances in Radio Science*, no. 4, pp. 281–285, 2006.
- Variation-Aware Design of Custom Integrated Circuits: A Hands-on Field Guide, Trent McConaghy, Springer, 2013.
- Yield optimization via trust regions, A.S.O. Hassan, et al, 2014.
- S. Sun, X. Li, H. Liu, K. Luo and B. Gu, "Fast Statistical Analysis of Rare Circuit Failure Events via Scaled-Sigma Sampling for High-Dimensional Variation Space," in *IEEE Transactions on Computer-Aided Design of Integrated Circuits and Systems*, vol. 34, no. 7, pp. 1096–1109, July 2015.

How to enhance Monte-Carlo? In the MC methods described before we can run the simulations and then simply make an analysis, so we acted like a <u>straight</u> signal chain system. You know that one of the greatest innovations in design is *feedback*. Using sample yield and C_{PK} is like a system with 1-bit ADC (=pass/fail) vs multi-bit ADC. The latter can often do better as we have seen, but actually you know the story about ADC has not stopped at Nyquist-rate ADCs: We have also sigma-delta ADCs, and they use feedback to get higher linearity and lower quantization noise!

So we may ask: How would feedback look in statistical analysis? Mathematically it is just iteration, so we run MC and then do an analysis (e.g., based on sample yield, C_{PK}, C_{GPK} or multivariate analysis), then *feedback* these result to the "sampling" algorithm, which now has to <u>decide</u> which further samples to simulate. Actually numerical experts use quite similar methods (e.g., in WCD), more than mixed-signal or analog designers! Innovation is in circuits, numerics and technology— More than Moore! Like SPICE with its adaptive step size setting, many of such advanced techniques come with clever internal self-calibrating techniques to reduce the user setup effort.

There are many criteria for ADCs such as resolution, area, power consumption, bandwidth, delay, need for accurate elements, design risk, etc., so actually all kinds of ADC (flash, half-flash, SAR, $\sum \Delta$, dual-slope, pipeline, interleaving, etc.) make sense and have their *sweet spot*, so also many numerical techniques makes sense and are good to know; they just differ regarding systematic errors, variance, treatment of nonlinearity, simulation count, etc.! Some are very universal and are often part of bigger analysis (like bootstrap or just random MC). Direct MC gives you $1/\sqrt{n}$ speed or 3 dB/oct, and other algorithms can beat this—often with compromising a few other parameters and with being slightly more complex.

7.1 Worst-Case Distance (WCD) Analysis

With the C_{PK}, we just simply fit a normal Gaussian pdf to the MC result performance data and "hope" for a enough good fit, but with worst-case distances (WCDs), we really aim for truly *finding* the (multi-dimensional) point in the statistical variable space x_S where the spec fail happens with highest probability. For this reason, such methods are not only well-suited to reduce the extrapolation risk, but also give insight to the design behavior, i.e., the detailed dependencies of the design regarding the statistical variables! Actually, this way advanced statistical methods also become a valuable, accurate debugging and design tool. In principle, we can also get rid of confidence intervals because the tolerance of a worst-case search (using a very accurate optimizer) can be usually made much smaller than the standard deviation from the MC sampling process. In addition, such advanced methods can be (much) faster than pure MC, especially for high yields.

For the case of only one statistical variable x (representing, e.g., a threshold voltage V_{TO}), the basic WCD idea is extremely simple: We run a nominal simulation ($x = 0$) and get for example an output voltage of, e.g., 1.0 V. Now, we set x to 1σ and get 1070 mV. This may come because the sigma of x is 1 mV and the gain of the amplifier is 70. Assume the spec limit is 1210 mV, so we could sweep on x till we *hit* the spec violation point. This would happen at 3σ in this case, if the amplifier is linear, and the yield would be now 3σ (as you can check using the C_{PK}). And the method would even work for nonlinear circuits; a differential pair shows some compression on gain (like $V_{out} \sim \tanh(x)$) and so we may need 3.1σ to reach the 1210 mV spec!

To see the full power of WCDs, we could inspect Figure 7.1 showing two independent statistical variables: The design works usually well if we are at nominal parameter settings, so when having no elongations and being at the origin. However, if we change one or both parameters too much, then the design may start to fail. The vector length from origin to the point of interest is related to the probability density and follows for normal Gaussian distributions the $e^{-\beta^2}$ law ($\beta^2 = x_1^2 + x_2^2$). The most critical point is the one having <u>highest</u> fail probability, so it is the <u>shortest</u> distance to the fail boundary and we call it worst-case distance (WCD).

As mentioned, there are similarities to the C_{PK}, and if the statistical parameters follow a normal distribution, we can even use the same formula to calculate the yield from the (normalized) worst-case distance, but in opposite to the C_{PK}, WCD methods can also address problems with non-normal <u>performance</u> distributions (remember the nonlinear amplifier example!).

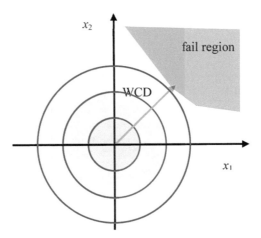

Figure 7.1 General WCD example with two statistical variables $x_S = (x_1, x_2)$.

To treat non-normal performances, the WCD method [Graeb] works *not* directly in the *performance* space (as C_{PK} and generalized C_{PK}), but directly in the original *statistical parameter space* x_S. In this space, we <u>know</u> the pdfs and often they are indeed just normal distributions—if not we could apply a parameter transformation, e.g., to treat also lognormal element distributions.

$$Y \approx \int_{-\infty}^{WCD} e^{-t^2} dt = \frac{1}{2}[1 + \text{erf}(WCD/\sqrt{2})] \qquad (7.1)$$

Like C_{PK} or the generalized C_{PK}, the WCD concept is designed to basically address the <u>partial</u> yield estimation problem only, not the total yield. Usually, we have many specs, so we get a WCD for each. The total yield is usually roughly approximated by the worst-case, *smallest* WCD, so given as min (WCD_i).

From the mathematical view point, the use of the minimum function is a very bad approach, because there would be no sensitivity to the non-dominating WCDs till we reach the equality point. As described in Chapter 5 for multivariate C_{PK}, a better approach is, e.g., to calculate from each WCD the partial yields, find the correlations, calculate the total yield, and then calculating back the overall WCD. For instance, leakage and speed often lead to *competing* specs in digital designs; e.g., the correlation is in the order of −0.8 to −0.95, so that assuming the worst-case where we have to add the yield *losses* is a good approach. However, if we have many redundant specs such as phase margin PM and overshoot or slew-rate SR and rise time, this approach would lead to over-pessimism. Some redundancy often makes sense for difficult specs, like on stability (here you need to get 100% sure), or for informational reasons (maybe 60 Hz PSSR is by far most important, but you get much more design *insights* if you also know about DC (can be, e.g., improved by using a cascode) and high-frequency PSRR (often dominated by capacitors).

On the other hand, there are also some high-yield estimation methods that can take correlations into account, so we will also describe further methods *beyond* classical WCD.

7.1.1 Worst-Case Distance Analysis by Hand

Let us now apply worst-case distances to a simple circuit and discuss the links to typical manual design techniques in detail. Actually, there are *two* major outcomes from such WCD analysis:

1. We can <u>verify</u> the yield efficiently and even for high-yield targets (like 6σ or 1ppb = 1E-9 loss)
2. And we obtain a <u>set</u> of statistical parameters, similar to what is offered in most PDK as process corner, such as slow, fast, nominal, but now we *can* include mismatch and make it correct for *any* performance specification! Also we can set the sigma level according our requirements.

The foundry-provided corners can never fully represent your actual worst-case circuit behavior:

1. You are often interested in a <u>certain</u> yield, but foundry corners are usually setup only for a "standard" yield level like 6σ.
2. If you run a process corner analysis and a MC analysis for process variations, you often find <u>contradictions</u>, especially if your circuit differs a lot from CMOS logic and if your goals are not speed-related!

Figure 7.2 gives a typical example for these problems. The scatter plot shows two performances rise time and phase margin of an op-amp, both usually compete against each other. Usually t_r has an upper spec limit and PM a lower limit. The plot shows the situation for a design done for given foundry corners, and we just pass the specs. However, it could happen that the true situation according to a MC analysis is different, like you may over-design on PM, but under-design on t_{rise}!

Often the situation is even more difficult, like the MC plot is more rotated (this also depends on axis scaling) or much wider or denser or not symmetric. Only for CMOS logic-style circuit speed specs (and when mismatch does not matter), you can expect that MC and foundry corners are in sync!

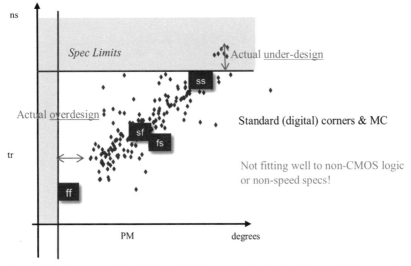

Figure 7.2   Corner result and Monte-Carlo scatter plot for rise time and phase margin.

What the designer needs for analog design are *tailored* corners, specific to each performance, and set according to target yield, as shown in Figure 7.3 Also note: In the performance space, the data are usually non-normal, so we need also a clever way to treat non-normality, and WCD does it.

How would such a statistical corner set look like? Indeed, in simple cases, you can calculate it by hand, for a diff-pair (see Figure 7.4) or current mirror.

If you know that the threshold voltage (for a BJT just the V_{BE} for given collector current I_C) of a single transistor has a sigma of 0.71 mV, you can calculate the total amplifier offset voltage sigma to $\sqrt{2} \cdot 0.71$ mV = 1 mV. If the spec limit is 6 mV, we can accurately expect a yield of 6σ (1 ppb loss) or $C_{PK} = 2.0$.

So what is now the 6σ-WCD? Actually, we need to find the statistical parameter combination that gives us an offset of 6 mV with maximum probability, and that is for sure +3 mV and –3 mV, so each transistor is set to 3 mV/0.71 mV = 4.22σ. Instead of repeating a big MC analysis to obtain the overall statistical behavior, we can focus on the WCD and use this WCD parameter combination alone $x_{WCD} = (+4.22σ, –4.22σ)$. In very old days, designers just inserted a small DC voltage source in series to the transistors to mimic this mismatch behavior manually, so WCD is a near-perfect "substitute" for a big MC analysis.

Modern EDA tools do such WCD analysis not by hand calculation, but numerically. And they can do so also for non-normal performances, and of

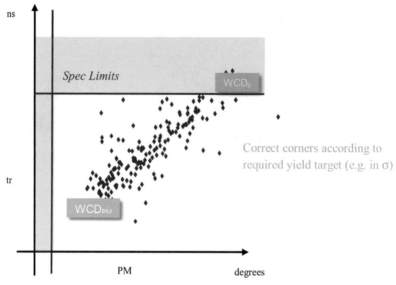

Figure 7.3   Monte-Carlo scatter plot with more accurate worst-case corners.

$$\sigma_1 = 0.71 \text{ mV} \qquad \sigma_2 = 0.71 \text{ mV}$$

Figure 7.4   Differential pair circuit for offset-WCD calculation example.

course for all your transistors in the circuit! For the simple linear 2-variable example, it may take only 15 simulations to find even a 6σ-WCD accurately (no confidence interval, no extrapolation risk—we just know the component sigmas *from the model files* and *exploit* that). This is something normal MC does not. However, for more complex circuits, maybe 1000 simulation points can be regarded as a minimum, but only 1000 points for 6σ verification is a huge improvement; the sample yield confidence interval would only guarantee roughly 3σ, and that with limited confidence!

One good further property of WCDs is that you also can really see <u>which</u> parameters have a strong influence and which one has almost no impact. The latter simple stick to their mean values (so $x_{\text{WCD}} = 0$). Doing a *ranking* on the sigma values, the designer can easily identify <u>which</u> transistors are critical, e.g., he/she can decide to make specific transistors larger to improve on offset. Also, the interpretation for multiple transistor stages is easy, or for the statistical parameters causing the mismatch *within* one transistor. For instance, mismatch may come from mobility mismatch or V_{TO} mismatch—often the 2nd part dominates—and the WCD will give you exact numbers for this. If V_{TO} dominates, the transistor is in "voltage mode" and you can usually better improve it by increasing the width; if mobility dominates, consider an increase in length.

For a two-stage amplifier, the second-stage mismatch is usually less critical, and how much can be also readout from the WCD set. So a WCD analysis is *also* a very useful and very accurate statistical *sensitivity* analysis. It offers even somewhat more information than a contribution analysis can provide.

An interesting question is would the WCD change if we change the design when tweaking the circuit? Yes, unfortunately it would change, but for moderate changes, the WCDs can be quite stable, so the concept is also very useful in combination with optimizers (Chapter 8)!

Note: For WCD search in the statistical variable space x_S, we can actually use optimizers, whereas for circuit optimization, we would optimize the design parameters x_D. Mathematically, there is little difference, often even the same optimization algorithm can be used! In conclusion, optimization is also useful for circuit analysis. If we optimize on range parameters x_R, we could also perform a circuit calibration via optimization.

One key advantage of WCD for yield estimation over plain MC is that—as our example calculation has shown—we can calculate the WCD accurately, i.e., without the usual sampling error, present in all MC results! You just have to know about the distribution parameters of the statistical variables and the simulator has indeed access to them! So, whereas normal MC methods (using sample yield or C_{PK}, etc.) always have a certain inaccuracy (quantified as confidence interval), the WCD concept has in theory no need for such statistical extra margins! On the other hand and as usual: There is no free lunch, and we have also discussed the WCD limitations.

The question is usually *which* errors are more critical—the MC sampling error or the WCD errors? There is no simple answer, but in general, WCD is usually more efficient and more accurate for mildly nonlinear systems (like a robust design) and moderate variable counts (like for typical analog blocks), and especially for high-yield verification (see Table 7.1). Where exactly MC and WCD are head to head is hard to say—maybe 3σ is a good rule of thumb—some later examples can give you a gut feeling.

Table 7.1   Comparison of MC against advanced statistical methods

	Monte-Carlo	High-Yield Analysis
Variants:	Random, LHS, LDS sampling, yield estimation via sample yield of fitting methods such as C_{PK}, sorted MC	Fast k · sigma, SSS, IS
Problem structure:		
Large number of variables	Causing <u>no</u> slowdown on yield verification (slowdown only in sorted MC)	Slowdown, e.g., variable screening needed
Large number of performances	Causing almost <u>no</u> slowdown (slowdown only in sorted MC)	Slowdown, e.g., variable screening needed
Nonlinearity	No slowdown for sample yield, fitting methods may become less accurate (slowdown only in sorted MC)	Slowdown, more iteration needed
Stochastic	No slow down for sample yield, fitting methods may become less accurate (slowdown only in sorted MC)	Not really suited
Systematic error:		
Total yield	None if using sample yield, fitting methods require same approximation as WCD	Roughly approximated (e.g., using min (WCD))
Partial yield	None if using sample yield, some error if model fits not well to real data	Approximated, some error from spec border shape or if multiple fail regions become important
Statistical error (variance):		
Yield	Large for high yields, moderate for fitting methods, easy to reduce in sorted MC	Generally small
Results:		
Histogram, QQ, mean, sigma, etc.	Yes (partially in sorted MC)	Only from initial MC
Yield	Yes	Yes
WC distance corner	Rough approximation	Yes
Effort:		
For moderate yield	Moderate using sample yield, low using fitting techniques	Moderate, low using SSS or fast k · sigma
For high yield	Huge using sample yield, moderate using fitting techniques	Moderate
Parallelization	Possible (only partially for sorted MC)	Limited (e.g., for initial MC part and for gradient calculation in optimization phase)

7.1.2 **Worst-Case Distances for Yield Approximation**

For now, we have not exactly explained in the general case how we can obtain WCDs for our design and what the risks are! For simple problems like our differential pair you can do it by hand, but the key idea for a general numerical solution is to use not plain MC but to combine a short MC run with optimization methods. The optimization target and the concept of WCD can be easily visualized. And this visualization also helps us to understand the limitations of the WCD concept.

If we would have only *one* variable, then there is a one-to-one relation from the statistical variable space to the performance space (although it might be nonlinear), but this is not true anymore for two or more variables. In our example hand calculation for the WCD on a differential pair, we derived the parameter combination $(+4.22\sigma, -4.22\sigma)$ as our 6σ-WCD, but $(+4\sigma, -4.44\sigma)$ or even $(0, +8.44\sigma)$ is giving the <u>same</u> offset voltage—but is <u>not</u> a true WCD! So the optimization criterion is to hit the fail point and to <u>minimize</u> the WCD Euclidian scale $(\sqrt{(x_1^2 + x_2^2)}$ for two variables), because this <u>maximizes</u> the probability following the e^{-x^2} law! The further criterion is on the angle between the WCD vector and the pass-fail border; it must be $90°$ for a true WCD (Figure 7.5, note that the vertical axis represents $-x^2$ to have the fail boundary in the 1st quadrant as in most WCD examples).

In normal MC, we have to accept a certain variance of our estimations, whereas the WCD allows us to accurately finding the <u>point</u> in the statistical variable space x_S where the spec fail happens, and which is closest to the

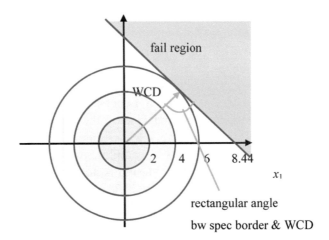

Figure 7.5 6σ-WCD example for linear offset of a differential pair.

origin. However, by doing this, we also accept a certain systematic error, which is related to the shape of the failure region. The WCD is indeed 100% accurate on yield in our linear example for offset voltage, and here the spec border is a straight line. This way the yield integral is easy to manage (remember our hand calculation), but if the design is highly nonlinear, we often end up in a nonlinear spec border and the WCD indicates either a too low (being too pessimistic, leading to over-design) or to small yield (being too optimistic on yield)! We discuss now how the classical WCD analysis works before coming back to a more detailed errors discussion.

7.1.3 Classical WCD Analysis

The classical WCD analysis uses a short MC run and then optimization techniques to find the WCD point efficiently (see Figure 7.6). One key problem is that the initial MC analysis should be large enough to give the optimizer a good starting point. A second key problem is that any optimization becomes slow if too many variables have to be optimized. For luck, most real designs are *dominated* by not so many variables, e.g., a certain performance like offset of PSRR is often dominated by maybe 20 variables, even for a bigger design, featuring maybe more than 1000 statistical variables. So it is a native step to do after the MC analysis a parameter screening step. The optimization step has to be done for each specification, whereas the MC run is usually jointly done for all performances.

Actually only very simple things like the C_{PK} are as simple for 3σ as for 6σ, but luckily also the WCD effort increases only *moderately* with the yield level! This is because actually only the way the optimizer has to go to find the spec border becomes longer for high-yield targets, so maybe we need 200 points for 3σ and 300 for 6σ; the exact ratio depends on the optimizer, on correlations, and on nonlinearity.

Figure 7.7 shows the log file from a commercial design environment of a 3σ WCD run with 8 statistical variables. Of course the initial MC run inside the WCD analysis should be large enough to allow the variable filtering, so for typical blocks having 20 important variables, we should have at least much more than 20 samples, so with 1000 MC samples, you might be already on the conservative side (especially if the design is not too nonlinear). As our LC filter DUT has only 8 variables the WCD algorithm choses a small MC sample count of actually only 18!

Already based on that, a filtering is possible, so the optimization is done only for the six dominating variables. The optimizer has taken three iterations to find the WCD point. For a less nonlinear design, we would need a lower count, so less additional simulations. The WCD length is 3.047σ, which fits well to the C_{GPK} result or to the sample yield of a golden MC run with many points.

Starting High Yield Estimation, Spec limit set to 2.5

Run Monte Carlo analysis

Number of points set according to complexity: n=18

Sample yield evaluation

Max. passband loss: Yield=100%, Sigma to target is 3.685

Run nominal simulation: x_s=0

Variable screening based on the MC results

Number of selected statistical variables : 6 out of 8

Selected: C2, C4, L1, L2, L3, L4

Calculate initial WCD estimate from MC result

C2: -2.5664, C4: -3.6528, L1: -1.1206, L2:-2.259, L3:3.5791, L4: 1.9041

Predicted WCD: β=6.53634

Start optimizer to improve WCD length and direction

WCD spec value: 2.94207 (target 2.5), current WCD: 6.53634

Spec error: 23.58%, gradient direction error: 52.45º

C2: -2.2637, C4:-0.36359, L1: -0.73182, L2:-2.1433, L3:0.084107 , L4:-0.13042

WCD spec value: 2.59292 (target 2.5), current WCD: β=3.22646

Spec error: 4.95617%, gradient direction error: 4.91º

C2:-2.19244 (-2.1924, C4: -0.15205, L1:-0.7696, L2: -1.9637, L3: -0.044206, L4:-0.063202

WCD spec value: 2.5011 (target 2.5), current WCD: β=3.0470

Spec error: 0.058%, gradient direction error: 0.54º

WCD algorithm converged because spec value error and direction error are small enough.

Figure 7.6   Flow for WCD analysis using MC and optimization.

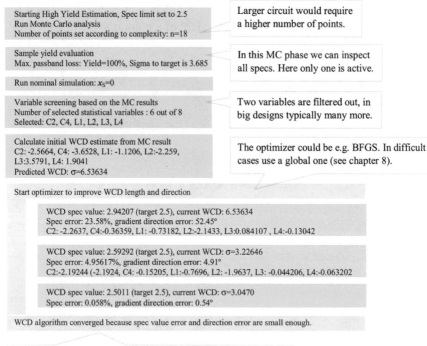

Starting High Yield Estimation, Spec limit set to 2.5
Run Monte Carlo analysis
Number of points set according to complexity: n=18

Larger circuit would require
a higher number of points.

Sample yield evaluation
Max. passband loss: Yield=100%, Sigma to target is 3.685

In this MC phase we can inspect
all specs. Here only one is active.

Run nominal simulation: x_S=0

Variable screening based on the MC results
Number of selected statistical variables : 6 out of 8
Selected: C2, C4, L1, L2, L3, L4

Two variables are filtered out, in
big designs typically many more.

Calculate initial WCD estimate from MC result
C2: -2.5664, C4: -3.6528, L1: -1.1206, L2:-2.259,
L3:3.5791, L4: 1.9041
Predicted WCD: σ=6.53634

The optimizer could be e.g. BFGS. In difficult
cases use a global one (see chapter 8).

Start optimizer to improve WCD length and direction

WCD spec value: 2.94207 (target 2.5), current WCD: 6.53634
Spec error: 23.58%, gradient direction error: 52.45°
C2: -2.2637, C4:-0.36359, L1: -0.73182, L2:-2.1433, L3:0.084107 , L4:-0.13042

WCD spec value: 2.59292 (target 2.5), current WCD: σ=3.22646
Spec error: 4.95617%, gradient direction error: 4.91°
C2:-2.19244 (-2.1924, C4: -0.15205, L1:-0.7696, L2: -1.9637, L3: -0.044206, L4:-0.063202

WCD spec value: 2.5011 (target 2.5), current WCD: σ=3.0470
Spec error: 0.058%, gradient direction error: 0.54°

WCD algorithm converged because spec value error and direction error are small enough.

In such case of only 6 active variables and moderate nonlinearity
WCD finds almost the ideal solution (only small rounding errors).

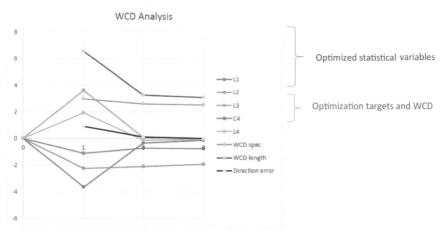

Figure 7.7   Execution of a WCD analysis on a LC bandpass filter with eight statistical variables [Liu2013].

7.1.4 Problems in Worst-Case Distances

To some degree, it looks magic that a single WCD <u>can</u> represent the full design worst-case behavior for a certain performance. We mentioned already some inherent systematic errors of the WCD concept if applying it for yield analysis. On top, there are the numerical errors because the algorithms to find a WCD are not perfect (e.g., stopping tolerance of the internal optimizer).

One basic error appears if you want to calculate the total yield from multiple WCDs, because these are related to partial yield and a certain performance only. If you have a single WCD at 3σ, this is equivalent to a yield loss of 0.135%. If a second WCD on another performance is 6σ, then the total yield is highly dominated by the smaller WCD, so also the total yield is 3σ or the total loss is 0.135%. However, if the second WCD is also 3σ, then the total loss depends significantly on correlation, and the total loss may range from 0.135% to 0.27%! In terms of sigma (or C_{PK}), the error in loss is equivalent up to roughly 10%. We discussed a significant improvement already in Chapter 5 for the multivariate C_{PK}.

However, the problem can appear again in WCD and in a slightly different fashion. A simple example case is offset voltage: Usually, you have two specs on offset like $V_{offset} \leq 6$ mV and $V_{offset} \geq -6$ mV, so we would need to find <u>two</u> WCDs for the offset specs. A native approach to this problem would be combining both specs into one $|V_{offset}| < 6$ mV, but that already causes another problem! This is because a further WCD error appears if the pass region has gaps, and this can happen even for a *single* spec! For a simple spec like $V_{offset} < 6$ mV, there is (at least usually) a single fail region, so the WCD concept fits well, but for a spec like $|V_{offset}| < 6$ mV, there would be (almost fore sure) <u>two</u> fail regions in the statistical variable space, and the yield error would be similar to what we have found for the total yield, i.e., for two fail regions, the error could be $2\times$ in yield loss or roughly 10% at 3σ! Luckily, at higher yields this $2\times$ error would go down both in terms of absolute and relative sigma error, e.g., at 6σ it would be only roughly 0.1σ. So for luck, this type of error becomes quite acceptable if you aim for high-yield targets like 6σ, but on the other hand, typical WCD algorithms do <u>not</u> check well for such type of errors, so often you get no warning on them (Figure 7.8)!

A good approach might be to avoid such kind of result evaluation setup, i.e., just not using functions like $|x|$ or x^2 in specs. However, this is quite a limitation, and functions such as minimum, maximum, and rms are indeed <u>intensively</u> used by analog or digital designers:

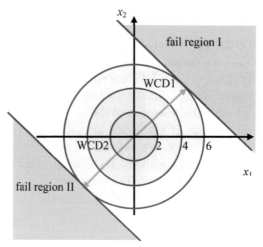

Figure 7.8   WCD plot for a spec like $|V_{\text{offset}}| < 6$ mV.

- When looking to filter characteristics
- When looking to worst-case timing problems
- When using decibel, because $\log(x)$ is actually doing $\log|x|$!
- The circuit itself may create functions like x^2 or $\max(x)$!

Figure 7.9 shows a further example of WCD errors in the two-variable case; comparing the WCD plots, you can see that the yield related to the fail area is

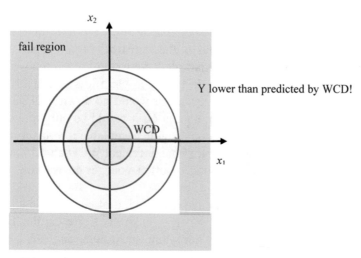

Figure 7.9   WCD plot for performance related to $\max(|V_{\text{offset}}|)$ – like ADC DNL.

different, although the WCD is identical in value! A nice circuit example in which the WCD error is huge and the whole concept becomes almost obsolete is a DAC or flash ADC. In the later, the differential nonlinearity (DNL) depends on the comparator offset voltages. These typically follow a normal distribution, but the ADC DNL depends on the <u>maximum</u> offset voltage among all comparators (like 255 for 8 bits), and the maximum function causes here severe problems. The maximum function is only sensitive to the largest variable, so if WCD finding starts from a certain point in the statistical space, it would adjust the current largest variable to find the DNL spec violation point. And to maximize the joint pdf, the optimizer would set the other comparator mismatch variables to zero! This way the yield estimation becomes completely wrong, and the obtained WCD point has nothing to do regarding what happens in production; actually, you would tweak your design in a highly artificial "corner" point. For an 8-bit flash ADC, the WCD yield error is already in the order of 1σ, so really significant and maybe in the *same* order or above the MC sample yield confidence interval width!

People who overlook that problem and argue that in two dimensions the error is typically well below few percent simply trap into the false assumption that the 2D case is essentially already showing all kind of problems you may have in n dimensions—this is <u>not</u> true, but unfortunately, it is difficult to visualize and hard to understand from pictures!

In general, WCD cannot handle well such stochastic behavior, whereas random MC can even deal with distributions with infinite number of variables, or even if the count of statistical variables is not fixed, but itself a random number. Other examples for stochastic behavior are bit-error-rate BER, effective number of bits ENOB, results from transient noise analysis, etc. In these (quite special) cases, it is better to use no WCD but multiple extreme samples from MC. We discuss this problem later in some more detail in Chapter 9.

You may think, OK if I have no very special digital random circuit problem I can safely use the WCD concept, but actually also classical analog problems can lead to similar problems! For instance, simple linear filters have poles and zeroes, and e.g., a Butterworth filter may show a ripple causing a 1 dB deviation in production. This can happen in many different ways like the peak ripple may appear quite well-distributed over the frequency range like for a Chebyshev filter or it may be at the upper or lower filter edge—this often cannot be captured in a single WCD.

Another systematic WCD error arises from the specification border <u>shape</u>. If it is linear (like for offset voltages from mismatch), WCD is accurate, but

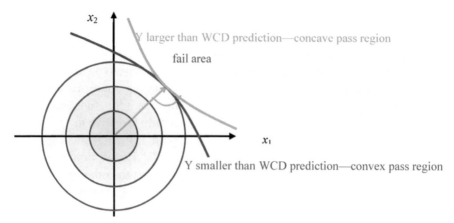

Figure 7.10   WCD yield error in case of nonlinear spec border.

for some *curvature*, the WCD might be too small or too large. The problem is again that this error is difficult to quantify, because when calculating the WCD numerically you do not really aim for checking the spec border shape, as this would slow down the WCD calculation. Some papers report that this type of WCD error is quite small, usually below 1% in terms of σ for high-yield designs. So it is justified to regard the WCD as a kind of "realistic" worst-case, but for sure this is not true for the general case with high nonlinearity and many statistical variables (Figure 7.10).

In principle, we could extend the WCD concept to model also the spec border not only as linear, but as quadratic. However, this would lead to much higher simulation effort, so it is seldom used. Actually, in <u>most</u> cases, there is luckily no need to choose a difficult model because we anyway "start" our modeling already at the most interesting point, which is the spec violation point! This is a clear <u>advantage</u> over response models, usually applied in sorted MC and in contribution analysis. In all the latter, we would indeed need (quite) high-order nonlinear models because in those we usually "start" the modeling at the distribution center! Actually, a linear model around the spec border (as in WCD) can be more accurate than a third-order overall fit.

What can bend spec borders? That is an important question because this makes WCD inaccurate. Obviously circuit nonlinearity is an important factor, but for luck actually there are also several important kinds of nonlinearities that cause *no* such "bending" errors. If we look to a Gaussian circuit response and then apply decibel, we may distort the histogram significantly and the C_{PK} becomes inaccurate. This problem could arise even from two reasons: Of course you get a bias error, and 2nd also the variance may increase significantly if the distribution becomes more long tailed. Such errors are completely uncritical for WCD! Mostly, different in nonlinearities in <u>different</u> statistical variables (like V_{th1} and V_{th2}) makes spec borders nonlinear. At the end of this chapter, we look to some more examples and compare the major different methods.

Our presented examples are still quite linear: So doubling the WCD from 3σ to 6σ is also geometrically doubling the WCD length, and in a linear design, also the spec border would be shifted according to this factor. The first statement is even true for nonlinear designs, because the WCD method works <u>directly</u> in the statistical variable space x_S, but in a nonlinear design, the spec border would <u>not</u> shift by the same amount. Actually, doubling a 2σ-WCD to get a 4σ-WCD is <u>not</u> correct for nonlinear designs, due to arising spec-border nonlinearity, but WCD itself as a method is still acting *correctly* because it really aims for hitting the spec violation (4σ-point in this example). Such nonlinearity (see Figure 7.11) is present in a CMOS inverter chain at low supply voltages: If V_{TO} is close to the supply V_{DD}, the gate overdrive would almost disappear, so that the transistors are not in strong inversion anymore, but in subthreshold

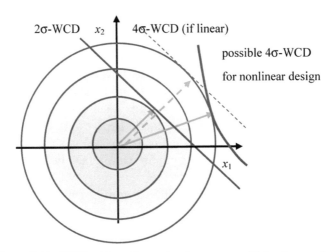

Figure 7.11 WCD with nonlinear circuit response at high-σ in x_1 and x_2.

region, whereas other statistical variables may still have a linear impact, like mobility.

An extreme example, where WCD works almost perfectly but other methods often fail, is a calibrated circuit: think of an oscillator with frequency calibration; such unit has a certain trim range and can "push" back samples with large technology variations into the spec range. However, for too large variations, the calibration would fail, and the histogram would be a big center bar with the good samples plus a few "outliers." Such extremely nonlinear behavior is hard to model, but it is quite manageable if the model is around the calibration point! This does not mean that sorted MC will fail, but the job gets much harder than for WCD. On the other hand, it is also a nice example where manual divide-and-conquer will be successful too, and also just picking the MC worst sample can help enough for circuit understanding and improvements.

In general, WCD effort and accuracy depend <u>highly</u> on the number of variables, mostly on the number of really important variables, and the nonlinearity of the system. The problem is if we filter out too many statistical variables after the initial MC run (see flow in Figure 7.6), we could end up in a severe WCD error! If we filter out too few, the optimization may take much more time than if using the optimum number; so a compromise is needed, and advanced algorithms use clever methods to find a good balance between the effort spent in the initial MC run and in the optimization part.

For a mildly nonlinear system with many statistical variables, we typically assume such behavior: If we move a statistical variable from 1σ to 2σ, we can identify *which* variables are important, and if we have one (or multiple) of these important variables at 5σ, the design may start to fail. So the usual assumption is that the variable ranking we find from low sigma changes (e.g., via OFAT or a small MC run and a contribution analysis) does <u>not</u> change much for higher sigma levels. However, this assumption might be violated in nonlinear designs, e.g., one variable might appear in the denominator and could cause a pole in the performance response, so its sensitivity might be indeed low around 2σ, but $10\times$ larger at 5σ. Unfortunately, it is hard to make sure that such variables with "progressive" sensitivity will not be filtered out. Such variables might be "hidden" within a storm of numerical noise from hundreds of other variables!

For speed reasons, often local optimizers are used to find the WCD, but we are interested in the global optimum. Luckily, most robust designs are quite well-behaved and usually the optimization starting point based on the initial MC run is quite good, so the risk of trapping into a local minimum is small—although not zero. Setting the initial MC count too low can indeed end up in WCD to fail, due to too much filtering or due to local minima.

A last type of WCD error we want to mention is of course that it might be simply not easy to calculate a WCD point numerically with high accuracy (see Figure 7.12). We typically need optimization techniques and we can only achieve a certain accuracy with a given amount of numerical effort. For instance, in our WCD example run log (Figure 6), a small error in angle of only $0.6°$ is reported and a spec value error of 0.057%. In most designs, this level of accuracy is fully acceptable, but in very complex and highly nonlinear designs, it may indeed happen that the WCD optimization fails in finding the worst-case setting, especially for difficult, progressive variables, just due to numerical noise and simulation accuracy limitations. A quite general solution for many of these problems will be discussed in chapter on sigma-scaled sampling SSS.

An interesting question is also how much WCD accuracy we actually need in conjunction with a circuit optimization? This will be discussed in Chapters 8 and 9; we discuss there what could happen if we use too inaccurate WCDs and how much the WCD will change if you (or the optimizer) modifies the design component values.

All in all, the WCD method often gives a huge speed-up for real high-yield verification, but as in several other advanced methods, scalability can be a problem, and it needs to be addressed with clever algorithm extensions—here is some room for improvements.

```
// Define function for circuit performance, e.g. acc. to SSS example eq. 7.1
function f(x: PVektor): Double;
Begin
  f:= 20*abs(x[1])- (sqr(x[2])+sqr(x[3])+sqr(x[4])+sqr(x[5]));
End;
// Define function to be minimized to obtain WCD
function GoalFunction(x: PVektor): double;
begin
 GoalFunction:= sqr(X[1])+sqr(X[2])+sqr(X[3])+sqr(X[4])+sqr(X[5]) +1000*sqr(f(X)-specLimit);
end;
// in main program we call the optimizer, e.g. BFGS (see chapter 8)
// set starting point, make some tests on your own!
x[1]:=1; x[2]:=0; x[3]:=0; x[4]:=0: X[5]:=0;
BFGSN(x, 5, 1e-8, Itmax, fmin);
Writeln("Obtained WCD: ",x[1], x[2], x[3], x[4]. X[5]);
```

Figure 7.12 PASCAL source code for WCD calculation.

7.1.5 Contribution Analysis versus WCD

A very useful analysis has been described in Chapter 5: The (mismatch) contribution analysis MMC; it supports finding the most critical instances on speed or offset, and based on that, you can often quickly improve your design manually! How is this linked to WCD results? For a linear circuit and Gaussian distributions, there is indeed a one-to-one relation, and the parameters with highest contribution are those with largest deviation (in sigma) in WCD! WCD is even a bit more, because it is also valid for non-normal distributions and it has a direct relation to yield. It could happen that the effect of a statistical variable is very nonlinear, but the contribution analysis does not take this nonlinearity very well into account; the contribution based on a kind of average. And even if a nonlinear model is fitted, the mismatch contribution model is typically fit well around the distribution center, so usually the accuracy in the tail regions is limited. In WCD analysis, however, the model is really around the worst-case position and we try to hit the pass-fail boundary accurately. This is possible because in the WCD analysis, the specs are *available* as input, and that is *not* required for MMC analysis. Also WCD has in theory no confidence interval, whereas for an accurate mismatch contribution result you typically need quite many MC points (much more than for stable sample sigma on offset voltage).

So all in all, the WCD analysis results are as useful for designers as the ones from a contribution analysis, but often more accurate in the critical region of the design, especially in non-normal cases. In addition, you can find in the WCD also the <u>direction</u> for the worst-case shifts, so from the signs of each component, you can see in *which way* the individual impacts add-up.

Let us go back to our initial WCD example on a differential pair and compare WCD and mismatch contribution in detail. Having a linear design and Gaussian distributions, the <u>ratio</u> between the statistical variables would <u>not</u> change on the sigma level and the spec setting, so if the 3σ-WCD is $(2.1\sigma_1, 2.1\sigma_2)$ then the 6σ-WCD would just double everything. This corresponds to the fact that the contribution analysis does not depend on specs. In the diff-pair, we have $\sigma_1 = \sigma_2 = \sigma$ and the contribution of each transistor would be of course 50%. For transistors with no impact on offset (like common-mode circuitry or power-down transistors), the contribution would be zero and the WCD entry would also be 0. To find the accurate relation between contribution and WCD entry, we could inspect a slightly more difficult case, like inspecting the total resistance of two resistors in series of different mismatch accuracy. Let $R_1 = R_2 = 10 \ \Omega$ but $\sigma_1 = 1 \ \Omega$ and $\sigma_2 = 3 \ \Omega$. Then, the total R is $R_{\text{tot}} = 20 \ \Omega$

and $\sigma_{tot} = \sqrt{10}\ \Omega = 3.162\ \Omega$, and the mismatch contribution is (94.1%, 5.9%). The 1σ-WCD is a vector giving 3.162 Ω for magnitude in absolute terms or (0.97σ, 0.243σ) as vector (in terms of sigma). In both cases (as well known), the resistor with the larger tolerance clearly *dominates*, and the relation between WCD and MCC is WCD/σ = $\sqrt{\text{contribution}/\%}$. This fits also to the well-known fact that n identical resistors give $\sqrt{n}$ as overall standard deviation.

7.2 Fast k · Sigma Corner Estimation

The described classical worst-case distance method has not only its sweet points, but also its weaknesses. It is very efficient for high-yield cases, but for low-sigma yield targets, the speed-up against pure MC is only moderate. One reason is that the classical WCD flow is designed for an almost accurate WCD calculation, but we have also many applications where a reduced accuracy can be accepted (e.g., in early design stages), so this can be exploited to create faster algorithms. For instance, we can do just a one-dimensional sweep on a scaling parameter instead of a full n-dimensional optimization (see WCD flow in Figure 7.13). This way we cannot correct on the WCD direction anymore, but this would be usually only needed at higher sigma levels. If the model is a quadratic one, then even this sweep part might be skipped [Zhang2009] with

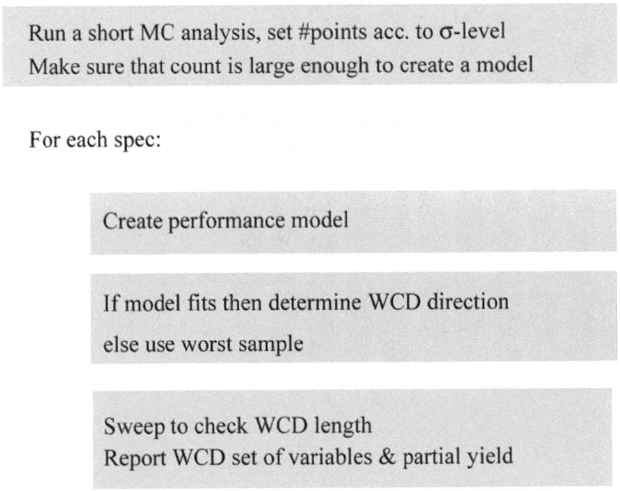

Run a short MC analysis, set #points acc. to σ-level
Make sure that count is large enough to create a model

For each spec:

Create performance model

If model fits then determine WCD direction

else use worst sample

Sweep to check WCD length
Report WCD set of variables & partial yield

Figure 7.13   Flow for fast k · sigma analysis.

minor loss of accuracy (in Chapter 8, we will show that a Newton-optimizer based on a second-order approximation needs in principle no step size control).

The scale factor is checked against the yield prediction, which might be based on kernel densities estimation KDE. KDE is quite popular because it allows to fit almost any kind of distribution, but it has usually the problem that it is well suited for interpolation, but not at all for larger extrapolations. In Figure 7.14, the KDE bandwidth parameter is chosen large enough to get a smooth fit to the true pdf, but the extrapolation capabilities are still very limited. If we would smooth more, then the fit on more complex distributions (like Laplace or Gaussian mixes) would become worse with still only moderate improvements in extrapolation.

Such fast $k \cdot$ sigma technique can be modified in different ways. Actually, the simplest method is just running MC and using the worst sample as statistical corner for further design tweaks. However, this is not very accurate, <u>neither</u> on length, <u>nor</u> on direction! For a 200-point MC run, the worst sample might be 2.7σ or 3.2σ or whatever. However, we know that the CI for sample standard deviation is already quite small for n = 200 (sigma is then 5% only), so if

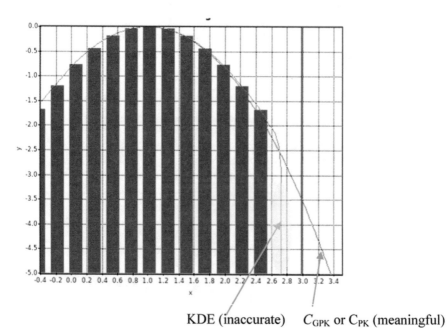

KDE (inaccurate) C_{GPK} or C_{PK} (meaningful)

Figure 7.14   Kernel density fit vs. C_{GPK} fit (normal distribution, averaged MC run, 256 points).

we just *divide* our worst sample by the sample sigma and scale it up (like we get 2.7σ as worst sample, but we aim for 4σ) or down, we could significantly improve on the *length* of our corner sample! The problem is that for non-normal data, the scaling is more difficult, but using the C_{GPK} (Chapter 4) or KDE, it is still feasible. Further improvements are possible by using not only the single worst sample, but e.g., the worst four samples, scale them, and average across them. By doing so, you actually would end up in multivariate methods, as described in Chapter 5.

As a last step of the fast k · sigma flow, we run a simulation sweep for the scale factor as shown in Figure 7.13; this way we double-check our initial estimation (e.g., from KDE or C_{GPK}). All the estimations from the simpler preceding steps might be used as backup for cases where the multi-dimensional modeling fails (e.g., due to too strong nonlinearities) or for internal error checking.

All in all, there are also some similarities for sorted MC with a corner stopping criteria, but fast k · sigma is faster, because in a pure sorted MC, we would have only the sorting speed up, not the speed-up of the direct search step! For small designs, fast k · sigma can give 3σ-corners already with 50 to 200 simulated points. The effort increases moderately with the number of specs due to execution of the sweeps, but for mildly non-normal behavior, each sweep may take only three simulations, and even for strong non-normal data, it would typically take no more than ten steps. Note that fast k · sigma is in most implementations no dedicated high-yield method, for instance if KDE is used for data fitting, the typical verification level is approximately 2σ to 4σ. However, we can expect improvements in the near future, just because designers highly need fast and reliable methods in general, especially when dealing with long simulation times, like for PLL, ADC, RF circuits, etc. In such cases, even a 100-point MC run can take a day, but the sample yield confidence interval would only give a guarantee for roughly 1.75σ, so you need advanced methods even if you only design for 3σ.

7.3 Importance Sampling IS

Monte-Carlo integration is not very efficient in general as we know. It is even very inefficient if we are interested in the distribution *tails* which is often the case in circuit design when aiming for high-sigma corners or for debugging. This is because you need to wait long to get samples in the tail area, so *any* statistic on this has a large variance, so high-yield verification takes many simulations.

340 Fast and High-Yield Estimation Techniques

We discussed several ways to improve it, like LDS or searching the WCD directly via optimizers. However, if we would have already many points in the tail, we would actually have much less problems and would often not need the additional search part! This is the idea of importance sampling (IS). We do not use the original pdf directly, but we apply a variable transform to "lift up" the tail region! This is similar to a normal substitution in manual integral calculations (which you learned in school), so actually we combine analytical methods to improve numerical integration for yield via MC sampling.

IS is quite popular in other fields of engineering, but not so much anymore in circuit design, because this "trick" is hard to apply for problems with *large* number of statistical variables, so you will not find it in many EDA environments currently. Also it is not a true corner-generating method and WCD generation based on pure IS would be usually less accurate and less efficient than the classical WCD search method. However, the idea of IS is still appealing and may receive a comeback! For instance, importance sampling is the core part of the IBM in-house tool "Rambo" [Kuang2012, Joshi2012], which is specific to static memory (SRAM) design. For 4.75σ yield verification, 2,500 simulations are required, giving a reported speed-up of $7360\times$ against random MC. As for WCD, the simulation effort rises only very slowly with the sigma level. Like in parameter estimation, also in memory design, highly tailored tools make sense, because exploiting the problem structure (low variable count, but large variations, nonlinear performances) is almost a prerequisite for maximum efficiency and accuracy. In addition, memory design is a good example that also in digital design analog aspects can be very important.

7.4 Sigma-Scaling Method SSS

A problem in classical WCD analysis is that the number of statistical variables is often very large in typical state-of-the-art blocks. This is usually addressed by parameter screening, but that step also comes with risks. To some degree, scaled-sigma sampling (SSS) [Sun] picks up the idea of importance sampling IS, so it is also aiming for getting extreme tail samples earlier to speed-up yield verification. The idea is to run multiple MC analysis, first a normal MC one and then further MC analysis, but with upscaled sigmas! As desired, this way we immediately get just more tail samples, and the failure rate can be predicted quite accurately, because now we can create a statistic based on more samples as usually available in normal (direct, unscaled) MC!

Like IS also SSS does not really aim for WCDs, and we have still to live with confidence intervals, but SSS has the advantage that we do <u>not</u> need an extra simulation step for each performance spec anymore. The parameter screening step in classical WCD search by optimization is not needed too, just because there is no optimization step; so overall SSS is well-suited for *complex, nonlinear* designs with *many* specs. Further principal advantages are that SSS runs not in trouble if the spec border is nonlinear (Figure 7.15). SSS can also treat correlations among the outputs, and so in principle, it is also suited for total yield estimation.

Another way of looking to SSS is regarding it as a kind of combination of OFAT sweep and Monte-Carlo. We demonstrated that OFAT is a risky method for worst-case finding, but as we sweep in all "directions" given by the MC samples, the OFAT risk went down a lot, practically (Figure 7.16).

Also in SSS, you can think of several refinements like adjusting the number of scaled sigma runs and its ranges adaptively. In principle, there is also no need to run <u>all</u> MC samples again with their upscaled version; you may select just the extreme samples only ("sub set simulation"). For internal yield estimation, we may use the sample yield or uses kernel density estimation KDE, but also other techniques are suitable. The accuracy of SSS is typically checked internally by bootstrap techniques or cross-correlation.

In addition, you may receive further results, like the failure rates. The results for the different upscaling factors can be related into a unified model failure rate with good accuracy (Figure 7.17), but like in other methods,

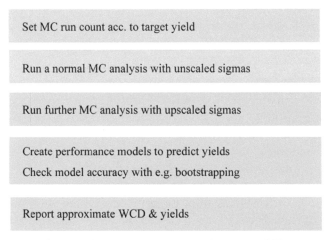

Figure 7.15   Flow for sigma-scaled sampling (SSS).

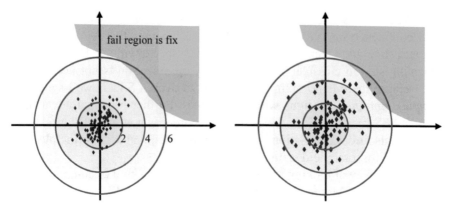

Figure 7.16 SSS with original MC run and with 1.9× upscaling.

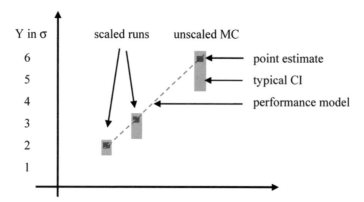

Figure 7.17 SSS yield estimation from three upscaling factors s (linear case, normal variables).

actually at such place, additional assumptions (e.g., which laws of nonlinearity are addressed) can influence the yield estimation.

Following [Sun] we can translate the SSS core part quite quickly to a R program. R features many powerful routines, so without the sampling part, just using the ideal fail rates, the code is very compact (Figure 7.17a).

Figure 7.17 may indicate that SSS is an extrapolation method (like C_{PK}), but it is *not*, because we really *hit* the fail area.

Note that also the yields from the upscaled MC runs have a meaning: Having no functional errors even in the upscaled MC runs is a good indication for a truly robust design. Due to upscaling, there is no real need to increase the

```
require(MASS)

# calculate true yield loss as reference via cdf

# 4 sigma is equivalent to Cpk=1.333 or 99.997%

p.true <- 1-pnorm(4)
# define up-scaling factors
s <- c(2,3,4,5,6)
# calculate up-scaled yield loss (would require circuit simulations)
p.scaled <- 1-pnorm(4/s)
# apply the SSS yield estimation flow
# according to IEEE article via least-square fit
mmat <- p.scaled * cbind(1,log(s),1/s^2)
mmat.i <- ginv(mmat)
models <- mmat.i %*% (p.scaled * log(p.scaled))
# take fit result to get the SSS yield loss result
pest <- exp(models[1,] + models[3,])
# report the theoretical and the SSS results
# as yield loss and in sigma using quantile function
cat("p.est=",pest,"(",-qnorm(pest),"sigma), ptrue=", p.true,"(",-
qnorm(p.true),"sigma).\n")
```

Result:

```
p = 1.878991e-05 ( 4.121877 sigma), ptrue= 3.167124e-05 ( 4 sigma).
```

Figure 7.17a   R code for SSS yield estimation in the Gaussian case (Y = 4σ).

number of MC samples much with yield sigma level from 4σ to 6σ; usually 2,000 to 10,000 samples in total are often enough. This behavior is quite similar to other high yield methods (like WCD) and other methods like C_{PK} or C_{GPK} and in *big contrast* to yield estimation by direct MC and the sample yield. 1000 points can be regarded as a kind of minimum for SSS, because with low sample counts the confidence intervals are still quite wide.

Note that SSS is aiming for yield verification, not for WCDs. This avoids some WCD weaknesses, but makes corner generation *from* SSS less accurate. Corner generation is still possible via multivariate modeling or by picking just the worst sample. In opposite to fast k · sigma flow, there is *no* need for additional sweeps on scaling factors. In principle, we could also use multivariate techniques in SSS for the yield estimations, but it would make SSS less applicable on large circuits. In Figure 7.18, you see a plot derived from the SSS run regarding the yield for op-amp power supply rejection (PSRR). The sample yield drops with the maximum scale factor, of s = 5.82 below 50%, so below 0σ; whereas for low scaling factors, we get 100% sample yield, so actually an infinite sigma, but in the plot we cut at 7σ. In the *gray* area of

Over-all SS confidence interval   Region in which sample yield is noisy and close to 100%

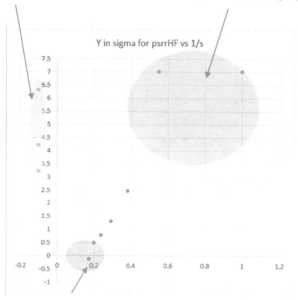

Yield drops (even below 50%=0σ), because scaling factor is large

Figure 7.18   SSS yield estimation plot form (yield in sigma versus 1/s).

low scaling factors, the variance in yield is quite large, so we see quite some randomness in the points.

At some point, here for s = 1.8, the yield drops to a finite effective sigma; and the real behavior is indeed similar to our sketch in Figure 7.17. To see the random variations, the values from a second run (with other MC seed) have been added to the log in parentheses. However, for higher scaling factors, we get really stable results. In addition, the SSS confidence interval is provided. The lower confidence bound is still quite low due to chosen low sample count of n = 1000, but against the CI from sample yield, we get an improvement by roughly 0.6σ; on top of the SSS other benefits (hitting the spec boarder, getting samples in the fail region for debugging, etc.).

In several circuit designs, SSS has outperformed IS [Sun2015]; also an implementation of SSS in R is quite straight forward. Table 7.2 shows the SSS results for a mathematical 6σ 5D-test case, which is quite nonlinear (the C_{PK} is already roughly 3σ too optimistic!). Also the classical WCD method would be quite inaccurate due to multiple fail regions and nonlinear spec shapes. Actually, this test case is difficult for many model-based algorithms,

Table 7.2 6σ test case result using SSS and different number of simulation points

Number of Samples	Estimated Yield	Bootstrap 90% Interval (in σ)	90% Interval from Repeated SSS Runs (in σ)	90% LCB Using Sample Yield*
7000	6.23σ	(5.47, 7.21)	(5.36, 7.15)	3.354σ
14000	6.23σ	(5.63, 6.98)	(5.64, 6.96)	3.54σ
21000	6.21σ	(5.69, 6.86)	(5.51, 6.91)	3.72σ

*Assuming no fails and using Clopper-Pearson limit.

just because obviously a simple linear model would not fit well. On the other hand, it is also a nice example showing that systematic errors could cancel, at least partially: WCD tends to be too *optimistic* on yield estimation if there are *multiple* fail regions, but is too *pessimistic* for concave pass regions (Figure 7.19, some more details you can read in the question and answer Section 7.6).

Synthetic data with function:

$$r = |x_1| - \sum_{i=2}^{5} x_i^2/20 \qquad (7.2)$$

Actually, SSS shows some moderate bias of +0.23σ. For a fair comparison to the sample yield we should subtract this from the LCB for 7000 points, so we would obtain 5.24σ (or 76 ppb loss). If we want to guarantee this with the sample yield method we need roughly 30 million simulation points, so overall

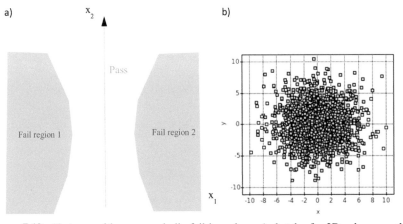

Figure 7.19 Testcase with near-parabolic fail boundary. a) sketch of a 2D sub-space plot, b) SSS run with 3x upscaling.

the SSS speed-up is 4000; it is a beyond the typical C_{GPK} speed-up. The C_{PK} speed-up would be even beyond 200,000, but only available for highly normal distributions (remember our example Student's t vs. Irwin-Hall showing how risky the C_{PK} is). In [Jallepalli2016], a slightly improved version of SSS has been presented, giving less bias and tighter confidence intervals.

As usual, it is not easy to find out algorithm limitations by reading the original papers, but like for WCD, many examples and pictures are only working really well *if* the statistical variables itself follow a normal distribution. If you would try the WCD or SSS method on multivariate *uniform* distributions, then pictures like Figure 7.5 or Figure 7.16 would look quite strange, and indeed, the results will become inaccurate (although *not* completely wrong). However, in conclusion, SSS is a very useful and quite robust method, and even more improvements can be highly expected in the near future, e.g., by picking up the idea of sorted MC, we can improve the corner generation and speed-up the yield estimation further by simulating only the potentially critical samples with upscaled sigmas [Sun2]. Also Bayesian techniques have been included to SSS, giving some further speed-up.

> **Are you confused on high-yield estimation, sample-re-ordering, WCD, fast kσ, sigma-scaling, LDS?** Actually most methods have the same goal, just making MC a faster. The more you *exploit*, like whether the system is dominated by a few variables, the more *speed* (or accuracy) you can get. There is *no* one-fits-all, you can always give test cases where one method beats the other! Random MC is slow but most reliable, and well implemented in almost all design environments. LDS is also a low-risk easy-to-use method, only for highly complex designs and high yields, the speed-up is limited, but we may expect some improvements in the future. The other methods are more advanced and follow the idea of focusing on the worst-case which is in the outer tail regions and usually they *combine* MC and iteration techniques. The goal is usually the same, focus on WC instead of doing a huge MC analysis, just the methods just differ a bit, like in fast kσ only a simple sweep is done, instead of a full optimization.

7.5 Design with Pictures Part Five

With pure MC results, you can do a lot as described in Chapters 3–5; some things are common and an almost immediate output for all the advanced analysis of Chapter 6—like yield reporting. Often histograms are missing as outputs, but you will often get not only higher accuracy but other valuable insights. Let us now inspect our latch comparator as a more difficult example with respect to worst-case distances.

7.5.1 Contribution versus WCD for a Comparator

We mentioned the tight relations between a (mismatch) contribution analysis MMC and worst-case distances WCD, so we made both on mismatch only for our complex latched comparator block. Table 7.3 gives a direct comparison for the (absolute) total offset voltage and the most interesting statistical variables.

Table 7.3 Latched comparator 1.5σ-WCD, worst-sample vs. relative parameter contributions for $|V_{\text{offset}}|$ (major parameters only, full xls file available at River webpage)

Statistical Parameter	MMC Result	Worst Sample	WCD Entry	Comment
COMP0/C0	0.019155	0.822434	0.195323	Parasitic wiring capacitances, C0/C0x are at located 1st stage output
COMP0/C0x	0	−2.06674	−0.245877	
COMP0/N10.pvt	0.00358664	1.53747	0.0114896	
COMP0/N13.pltw	0	−0.33717	0	
COMP0/N13.pu0	0.0118747	1.19616	0	
COMP0/N1.pltw	0	−1.15992	−0.0413625	NMOS for diff-pair bias, minor impact because common-mode transistor
COMP0/N1.pu0	0.00395791	2.5611	−0.00689375	
COMP0/N1.pvt	0	0.886413	−0.0206812	
COMP0/N5.pltw	0.00178266	0.21779	0.126385	Input diff-pair, is dominating the offset
COMP0/N5.pu0	0.00243737	−1.29853	0.0459583	
COMP0/N5.pvt	**0.235319**	1.50624	**0.744525**	
COMP0/N6.pltw	0	0.698224	−0.15396	
COMP0/N6.pu0	0	0.467513	−0.0919167	
COMP0/N6.pvt	**0.248905**	−1.9733	**−0.792781**	
COMP0/N8.pltw	0.00346528	−0.240356	−0.379156	2nd stage NMOS, quite significant impact
COMP0/N8.pu0	0.0310793	0.224258	−0.00689375	

(Continued)

Table 7.3   Continued

Statistical Parameter	MMC Result	Worst Sample	WCD Entry	Comment
COMP0/NM1.pltw	0	0.53319	0.0367667	Input charge cancellation NMOS, negligible impact
COMP0/NM1.pu0	0	−0.239264	0	
COMP0/NM3.pltw	0.00351871	−0.563122	0	NMOS output inverter, negligible impact
COMP0/NM3.pu0	0.0186326	0.848688	0	
COMP0/NM6.pltw	0	−0.559429	0.167748	Reset NMOS transistors
COMP0/NM7.pltw	0	−0.692203	−0.17694	
COMP0/NM7.pvt	0.0037437	−0.166839	−0.137875	
COMP0/NM8.pltw	0	0.606616	−0.027575	Hysteresis compensation NMOS, negligible impact
COMP0/NM8.pu0	0	−1.82869	0	
COMP0/P10.pltw	0	1.17276	−0.027575	
COMP0/P13.pltw	0	−1.09653	0.00229792	
COMP0/P13.pu0	0.00722655	1.06833	0	
COMP0/P1.pltw	0.00374428	0.374136	0.0413625	PMOS load of 1st stage, surprisingly small impact
COMP0/P1.pu0	0.0171627	−0.471267	0	
COMP0/P2.pu0	0.00540636	**−2.28249**	0	
COMP0/P2.pvt	0	**2.86687**	−0.0137875	
COMP0/P5.pvt	0	−0.122529	−0.103406	2nd stage PMOS, significant impact
COMP0/P7.pltw	0.00632563	−0.871859	−0.353879	
COMP0/P7.pu0	0	−1.49746	−0.105704	
COMP0/P7.pvt	0.00732191	1.1915	0.278048	
COMP0/P8.pltw	0.0172066	1.40118	0.326304	
COMP0/PM6.pltw	0.00470468	1.10965	0.00229792	PMOS output inverter, negligible impact
COMP0/PM7.pltw	0.0189103	−0.690735	0	
COMP0/PM7.pu0	0.0100838	−1.34689	0	
R0.R0.pres	0.000722049	−1.10588	0.0436604	Generator resistances, small impact
R3.R0.pres	0.0182012	−0.893763	−0.082725	

For instance, it is nice to see that in MMC some clearly non important instances, (like the output inverter NMOS NM3) still may have some small impact like 3%, whereas WCD is more accurate, giving really zero if no impact on offset exists. On the other hand, there is of course some overhead in WCD, because we made the MMC on the WCD-MC run having 400 random points, and the WCD optimization part took further 602 simulations. Also note, just picking the worst sample (it was #119 giving 37 mV, equivalent to approximately 3.5σ) is even much less accurate compared to both methods. For parameters (such as those related to P10 or P13) with low sensitivity, the worst-sample is typically at random values, like $\pm 1\sigma$.

The spec is nonlinear and the histogram for $|V_{offset}|$ is non-normal, but still both methods work quite well. We also calculated WCD and MMC for V_{offset}, and indeed the results are more stable (r^2 of 0.99 vs. 0.90). The design was not fully optimized (yield only of 86.5%), so the offset was quite large and dominated by the NMOS input diff-pair N5 and N6, as often the V_{TO} mismatch dominates. The overall WCD was 1.501σ, and the two major variables (among 102 in total) give already 0.7445σ and –0.7928σ (so together 1.08σ). This corresponds well to the MMC results.

Actually in the WCD analysis, we manually limited the initial MC count to 400 samples, and for the optimization part, we limited the number of iterations to three; so the optimizer started at roughly 0.3σ for the two dominating variables, till we almost converge to approximately 0.77σ. On the less important variables, the optimizer job was almost the opposite, e.g., for a noncritical transistor N13 in the output latch, the mobility variable was at 0.467σ, but the optimizer pushed it back to zero. In the last iterations, it is also nice to observe that the symmetry of our circuit corresponds well to the symmetry in the WCD entries—on this MMC is less accurate (but we may improve with low effort by using LDS instead of random). The WCD search stops with hitting $V_{offset} = 0.014862$ V (spec limit was 0.015 V) and the gradient direction error was 6.76°. With extending the number of iterations, we could further improve the accuracy down to $\leq 1°$.

Without looking to the other performances, an optimization by hand would be now very easy: Just make the critical transistors N5 and N6 larger, but of course this will impact input capacitance and speed negatively, so a realistic optimization is not so easy.

7.5.2 WCD in a Complex Filter

As mentioned and for luck, many difficulties present for WCD in theoretical mathematical examples will not really occur in typical robust circuit designs.

Note, that from the circuit perspective, a latched comparator is something heavily nonlinear, but from a statistical view point, it behaves quite linear! However, there is no guarantee that only difficult circuits such as DACs or ADCs cause WCD problems. After the quite complex but quite "linear" comparator let us now increase the level of nonlinearity; actually, nonlinearity regarding statistical parameters has nothing to do with circuit linearity! Here we have a real pure analog example on which WCD works quite fine, but just not 100% accurate. As DUT, we chose an eight-element LC bandpass filter.

Note: This filter is also fully integrated to our real-time optimization app. It has eight statistical variables, so the design is still quite tractable (like we could run a big MC analysis as near-golden reference); and of course you can transfer the results also to (much) more complex implementations of that filter as $g_m C$ or an op-amp RC active filter.

App

Typically filters should have a flat passband region, but also have high good stopband attenuation. This usually defines a certain minimum filter *order* and a *number* of filter elements. Once this is chosen, there is some more flexibility on setting filter poles and zeroes, like using a Butterworth or Chebyshev filter. From filter synthesis programs or catalogs you may find that, e.g., an eight-element Butterworth filter fits to your specs, hopefully with some margin. In our example, we focus on a few key specs like the (passband) filter gain ripple in a certain frequency range. By definition, a Butterworth filter has no true ripple (it is maximum flat), but of course we have some attenuation (like 1 dB) at the two corner frequencies of the bandpass. In an MC analysis, the elements will vary (σ set to 2.3% for each element) and the filter total ripple (including the drop at cutoff) will often become larger (like 3 dB), but sometimes it might be even *smaller* (like 0.8 dB). The histogram of the filter ripple is typically significantly skewed, so that the C_{PK} is not suited and other yield analysis methods should be used, like WCD or the C_{GPK}. In general, the latter is much easier to apply, but the WCD can give some further design insights and accurate statistical worst-case corners (helpful for finding the most critical elements and for doing optimizations) (Figure 7.20).

If we run a big MC analysis (e.g., 32K points) and save the waveforms, we can learn more about the circuit behavior. A spec like $(A_{\max} - A_{\min})$ between f_{Lower} and $f_{\text{Upper}} \leq x$ dB can be violated in many *different* ways (like the actual violation in gain might be located at $f_{\text{Lower}}, f_{\text{Upper}}$ or somewhere

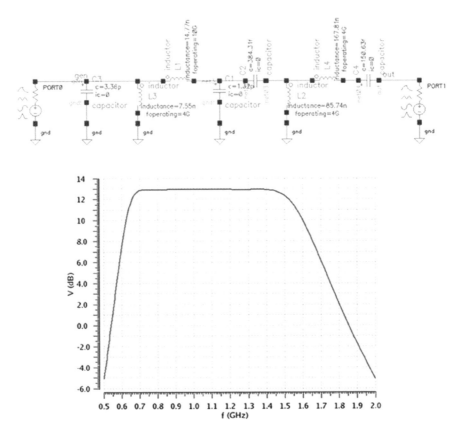

Figure 7.20 LC filter circuit and its nominal Butterworth behavior (note: frequency axis is linear).

in between), but at least ultimately you and your customer are usually only interested in the worst-case, and just avoiding the production of too many bad samples.

On the other hand, inspecting the MC waveforms indeed different kinds of spec violations occur (like at lower or upper passband edge), so what does it mean regarding yield and WCDs? (Figure 7.21).

In many design environments, you can select the most extreme MC samples and save a statistical corner set for this MC point. If you do this for the ten most extreme samples of a big MC analysis, you will usually observe some "clustering." Actually you will typically get *one* cluster regarding critical variables for simple cases like offset voltage, but in our filter (or other complex

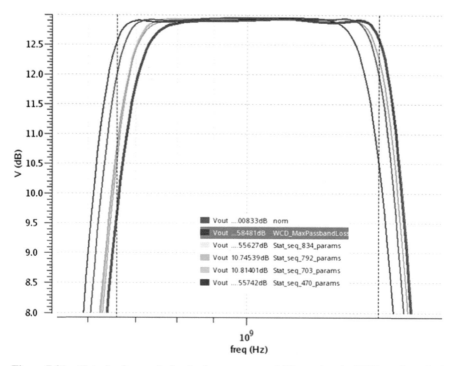

Figure 7.21   Plot of voltage gain for the 4 most extreme MC samples, the WCD, and nominal point with logarithmic frequency axis.

examples), we will indeed have <u>several</u> clusters! What typically helps is that even if you have three clusters, usually one cluster will *dominate*, like in cluster no. 1 there are maybe 6 samples of 10, and in no. 2 and 3 only 2 or less. Of course, most WCD algorithms will find correct WCD point according to the largest cluster. For two balanced clusters, we would have a WCD error of $2\times$ in yield loss, and we run in a situation similar to the problem of partial yield vs. total yield (discussed in Chapter 5) (Figure 7.22).

In Table 7.5, we compare some extreme samples against the WCD. The most extreme MC sample regarding spec violation is #834 and has an overall pdf of 4.772σ (for a multi-variate normal distribution we have $\mathrm{pdf}(\mathbf{x}) \sim \prod e^{-x^2} = e^{-\sum x^2}$), so we use this sigma level for the WCD as well. Actually, the WCD algorithm filtered out two variables (elements C1 and C3) as less important, although both are set to quite extreme values in the worst sample, which is no surprise—just a limitation of the worst-sample picking method! Also the most important variables for WCD (C2 and L2) are only

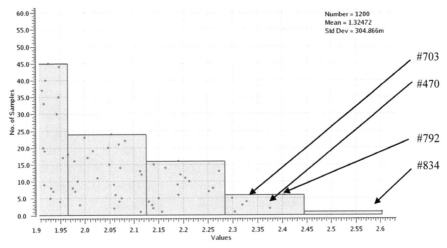

Figure 7.22 Histogram with zoom-in to fail area (for passband loss).

Table 7.4 Four most extreme MC samples out of 1200 and 4.772σ-WCD in comparison

Corner	Seq_834	Seq_792	Seq_470	Seq_703	WCD Max Passband Loss
mismatch:C1	−2.05791	−2.25753	3.16688	0.731617	−
mismatch:C2	−2.77369	−1.93971	1.17822	−1.31317	−3.4336
mismatch:C3	1.03959	−1.01327	−0.33884	−0.13460	−
mismatch:C4	−2.68008	−0.28853	1.76989	−0.65692	−0.23812
mismatch:L1	0.037479	0.175438	1.07004	−0.92655	1.20529
mismatch:L2	−1.12574	−1.70383	0.7316	−2.45219	3.0754
mismatch:L3	−1.13393	−0.089535	0.059932	0.87793	0.069232
mismatch:L4	0.159214	−1.03981	−0.97871	−1.7757	0.098982

Table 7.5 Performances for all points in Figure 7.21 and nominal ($x_S = 0$)

Corner	Max (Passband Loss)	Pass/Fail
Nominal	1.093 dB	pass
Seq_470	2.376 dB	pass
Seq_703	2.329 dB	pass
Seq_792	2.405 dB	Pass
Seq_834	2.604 dB	fail
WCD	3.506 dB	fail

moderately in sync with the worst sample. This indicates that the worst sample is not a very good approximation to the true WCD: Although the joint pdf is the same, the WCD gives a (much) worse performance (see Table 7.4), so the optimizer has really made a good and important job on finding the correct *direction*! Such large deviations between sample joint pdf and performance are quite typical; even for pure normal distributions and linear circuits, they are intrinsic to the <u>multi-dimensional</u> characteristic of the problem! Clear one-to-one relation are only available for one-dimensional cases.

Interestingly, the WCD components are also quite small for L3 and L4, so actually also these could be filtered out maybe, and the decision on which variables will be filtered out depends a bit on chance. In more complex test cases, maybe only 1 to 10% of the variables are really significant, and this helps to apply WCD also to cases with large number of variables. Mathematically, we say the "effective" dimension s_{eff} is typically much lower than the nominal dimension s (= number of statistical variables forming x_S).

Another interesting effect can be found if looking to MC sample #470, giving us <u>another</u> fail mechanism: The WCD and the other extreme samples fail on *upper* cutoff limit, whereas #470 fails at *lower* edge (Figure 7.21) and indeed #470 is to belonging to another cluster, as can be seen in Table 7.4: The entries for C1, C2, C4, L1, L4 differ a lot.

What would help in finding a WCD approximation manually by simple means is averaging among the samples of the dominating cluster; if we do this, we would get the vector (−1.195, −2.01, −0.036, −1.21, −0.24, −1.76, −0.12, −0.89), which is slightly better in sync to the WCD. Further averaging would mean to include also less extreme samples (thus lower sigma values), and those should be upscaled before doing the averaging. Of course averaging can only give moderate direction improvements according to the $\sqrt{n}$ random MC law, and we have to look up carefully to average only *within* the same cluster.

To improve also on the vector length, we could use the generalized C_{PK}. In this example, WCD and C_{GPK} are quite close together (4.772σ—WCD being too optimistic on yield—vs. 3.993σ from C_{GPK} being a bit conservative; its bootstrap lower confidence limit is 3.564σ; this is only 11% below the sample point estimate, so the C_{GPK} accuracy is quite good), whereas the effective sigma for #834 is significantly lower (actually only at 3.006σ using the C_{GPK}), which means that we need to upscale (by 1.5875) it to get a better approximation to a 4.77σ-WCD. This is in synch with our simpler qualitative inspections; and the re-simulation of the upscaled version of #834 gives actually 3.874 dB so it is now really close to the classical WCD! Note: We got this excellent result although we actually used no compute-intensive multivariate methods. So the C_{GPK} can also help to get approximations to worst-case distances. It is also a nice example of a kind of feedback method:

The direct C_{PK} or C_{GPK} application to MC data is a mix of interpolation and extrapolation, leading to some risk, but in this example, we use the yield estimation for a scaling on the statistical parameters, and we re-simulate, so we get this way more information which allows a verification.

Note: The upscaled joint pdf of the approximated WCD from #834 is significantly <u>larger</u> than the one for the true WCD, but having <u>un</u>important variables at nonzero is no problem because they make any way no big impact, and also the worst sample #834 is the sample with most extreme <u>performance</u>, <u>not</u> the one with largest overall sigma, thus the most extreme probability (this sample was actually #980 but giving only 1.445 dB).

How many "fail clusters" exist in the statistical space? Even for a fix design and spec this is not so easy to say, because if you take into account only clusters within 0.5σ there are surely less than if you would increase the spread to 1σ. However, a cluster 1σ below the WCD would usually have only little impact on yield (roughly 5% in terms of sigma). On the other hand, many such clusters could add up—not only theoretically. An extreme example is the maximum function, which is used e.g., in ADC or DAC DNL specs. It could lead to hundreds of clusters (actually even one for each quantization step so $2^{\text{Number of Bits}}$) of even identical sigma level, so ending up in to huge WCD errors! Also our LC filter can be made more critical on WCD error, by a small change in specs or in the nominal design points we will change the yield impact of the clusters, so that two clusters become almost balanced, making WCD more inaccurate. Actually our example looks fully symmetric and centered, but the LC bandpass itself is actually *asymmetric*: the highpass transition is *smoother* than the lowpass edge, leading almost to a single dominating WCD. Matlab (and other math packages) feature basic analysis for such problems, like so-called k-means clustering. Of course, it does not mean that one Matlab function can do the full job, a lot of pre processing is needed, and also further improved algorithms exists. We made such analysis from an MC analysis with 32K points, picking all samples beyond 2.5 dB (giving 3σ). This way we obtained 42 extreme samples, and a two-cluster analysis indicated that 7 belong to the fail area close to sample #470, and the WCD fail area contained 35 samples, so it is really dominating. So here the WCD error on yield loss is approximately +16% (or 0.04σ). This is an excellent WCD result, for using less than 50 simulations. The C_{PK} standard deviation would be 11%, but the data is non-normal, making the normal Gaussian fit 1.2σ too optimistic! The C_{GPK} would be slightly too pessimistic and has a sigma of 4.9% even for 1K samples (LCB = 0.915 or 2.7σ). Therefore here WCD works like a clever manual hand calculation, treating the non-normality very accurately. The only (moderate) problem is that the WCD effort would rise for more complex circuits (e.g., when we want to address an OTA-C filter), whereas complexity has <u>no</u> impact regarding number of simulations for random MC and using the C_{GPK} or the sample yield.

Besides inspecting the gain vs. frequency behavior, we could also inspect the filter behavior in a Smith chart or look for the filter poles and zeroes. Which method gives you most insights depends highly on the circuit, so the filter poles and zeros depend on the element values, but we have <u>no</u> one-to-one component value to pole zer relation; whereas one transforming trace in the Smith chart is indeed highly related to *each* corresponding individual component. For inspecting the filter quality factors or cutoff frequencies you are typically in an intermediate situation, e.g., the Q-factor of depends mainly on component ratios and is quite insensitive to process changes, but very sensitive to mismatch. Unfortunately, a Smith chart is not very useful for most active filters (it might be used of some g_mC or gyrator-based filters, but not really for biquad or Sallen-Key filters). However, as we deal with an LC filter, let us try the Smith chart: A first step should be the inspection of the pure nominal behavior, which is already not so easy to understand for non-RF experts due to high filter order (Figure 7.23).

For the passband cut off frequencies, the chart becomes much more difficult to interpret; the end point gives now $r = 0.42$ (for nominal circuit values) (Figures 7.24–7.26).

Using these corner frequencies, and even entering the component values of an extreme MC sample, the situation will change further. The last impedance point showing the reflection coefficient r is related to the overall gain $|S_{21}| = 1 - |r|^2$, and that is of course impacted by <u>all</u> the eight elements. In the passband, the changes against nominal are not large, so the filter is quite robust, but the situation changes quite dramatically if we select a frequency close to cutoff where the filter shows some ripple.

In such cases, the individual impedance translations add up in a certain extreme way, and for different extreme MC samples the "random walk" in the Smith chart may look quite different (Figure 7.26).

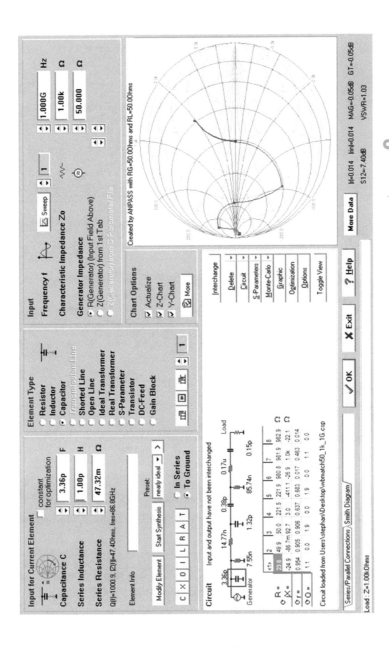

Figure 7.23 Smith chart design program applied to the LC filter (nominal values, center frequency).

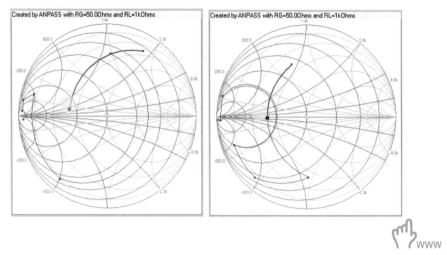

Figure 7.24   Smith chart at passband limits (657 MHz and 1.523 GHz).

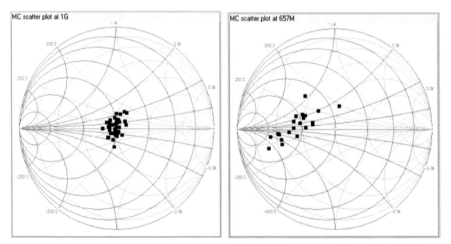

Figure 7.25   Scatter plot for output at 1 GHz (center) and at 657 MHz (more critical spec limit).

So far we found that our methods for WCD and even simply picking the worst sample plus scaling it via C_{GPK} work fine. What about using them for circuit tweaks for performance improvements? In the very simple case of just reducing the statistical variations on all components *simultaneously* no

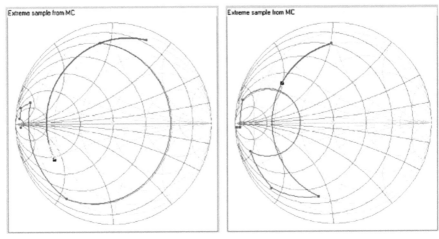

Figure 7.26   Transformation of two extreme MC samples one at 657 MHz and another at 1.523 GHz.

problems will arise, even for nonperfect approximations to WCD. However, if the WCD indicates, e.g., low sensitivity on L_1, and you decide not to optimize it (e.g., to save money or area), you may still end up in a nonoptimum design! Maybe in the (currently) nondominating cluster, L_1 might be indeed a critical variable, indicating <u>another</u> optimization strategy! All in all, a mixed method would be reliable, using multiple "worst"-case distances instead of only one, even for one specification. Typically, the situation will be further relaxed because in real designs you would have anyway *more* specs and according WCDs, so the risk to miss critical fail areas during verification and optimization will be often significantly reduced further.

By the way, why treating an optimization as extra step? Also our MC analysis is helpful; we just should inspect not only the worst-case MC samples, but the <u>best</u> case! Indeed, some of the MC samples have a very good spec margin, even much better than the nominal design! This is possible because "accidently" MC found filter versions close to a Chebyshev filter; and those can indeed give wider bandwidth than a Butterworth filter! Inspect Figure 7.27 for detailed results, e.g., you can see that the wider bandwidth comes with some in-band ripple which might be not desired. In the Smith chart, you would find that such near-Chebyshev filter would also have higher quality factors, which typically lead to larger tolerances in frequency response.

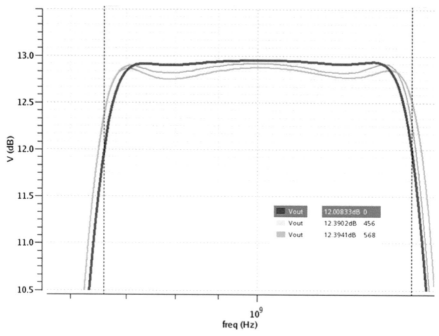

Figure 7.27   Best-case samples from MC analysis and the nominal behavior (red).

7.5.3 Sorted MC in a Complex Filter

Beside so many investigations on WCD, you may ask about MC with sample re-ordering? Indeed it would also work quite nice on the filter test case. If you run our 1200-points MC run with identical setup, but turn-on the sorting, then 50 points will be taken to generate the 8-dimensional model (takes <1 min, would be larger for complex circuits), then the model will be applied to all remaining MC samples (takes <1 min, would increase for higher complexity and also for higher yield levels!) and then the really critical, near-spec samples according to our model will be simulated. Actually, only further 20 samples are needed to decide that the yield is beyond 99.75% with 95% confidence level! The worst sample is simulated 1st, the 2nd worst one as 2nd, etc., so by comparing the sorted MC results, we can also find out if the sorting from the internal model is really *in synch* to the true circuit simulations; Table 7.6 shows that the correspondence is amazing, although not 100% perfect: for instance, sample #703 is critical, but is in an area not detected by the model (leading here to minor errors well below the CI width).

Table 7.6 Full MC results vs. model-based sorted run

MC Point	Max (Passband Loss) in dB	
	Full	Sorted
834	2.604	2.604
792	2.405	2.504
470	2.376	2.376
703	2.329	2.299
540	2.320	2.261
375	2.305	2.252
778	2.299	2.213
1081	2.273	2.195
526	2.261	2.189
468	2.252	2.164
912	2.213	2.157

7.6 Questions and Answers on Advanced Statistical Methods

1. What should I do if something gets wrong in an advanced statistical analysis?
   Try to run a MC analysis and double-check the results, by using the C_{GPK}. Also advanced methods may start with MC internally; this MC run count should not be too small!

2. How accurate is the worst sample in a big MC analysis compared to a true WCD?
   As mentioned the worst sample can vary a lot (even for a fix count like 200 points), like giving 2.7σ or 3.2σ or whatever, but even if you normalize it by the sample standard deviation to 3σ there will be still some errors. These depend mainly on the number of variables, the nonlinearities, the sigma level and the MC count. For moderate sigma levels and only one dimension the difference is very small, just given by the confidence interval of the standard deviation (Gaussian case). For higher dimensions there is mainly a certain angle error, and we can expect quite a good accuracy in the dominating variables, but high randomness in the worst sample for the less important variables. So actually not the total number of variables matters, but also how many variables are really important. To treat non-normal cases you can use the C_{GPK} for normalization instead of using the sample standard deviation!

3. Which method for high-yield verification is best in case of strong nonlinearity but only few statistical variables?
   Often classical worst-case distance methods fit well. They base on MC and optimization techniques. For more complex problems this method is usually only fast if a parameter screening is applied, but it comes with risks and some accuracy degradation.
4. Can I use the WCD if change all my transistors by 20% in W and L?
   If the nonlinearities changes not much yes! With 20% larger W and L the mismatch would reduce by 20%, so the 3σ-WCD for offset would go down from maybe 10 mV to 8 mV. So it looks that the WCD has changed, but actually almost all PDK's are setup in a clever way to reflect the scaling by W and L correctly, because we actually save the statistical variable itself (like 3) and not the offset voltage directly!
5. Can we deal with WCD also in performances with poles? E.g., an oscillator it would give infinite period if it stops to work.
   Yes, WCD is quite robust on such nonlinearities, because it works in the statistical variable space. The C_{PK} shows quite the opposite behavior. The only WCD difficulty might be that the optimization part becomes a bit more difficult.
6. Is it possible to have a normal Gaussian histogram, but still problems to find WCDs?
   In theory yes! You may compose a Gaussian output from multiple Gaussian variables combined in a nonlinear way, like $sign(x_1) \cdot |x_2|$. The sign function is not continuous, which causes classical WCD methods to fail although the output is perfectly Gaussian! The distribution given by $|x_1| - |x_2|$ is less nonlinear (at 1st glance) and actually normal, but WCD fails too; the reason is again nonlinear spec border and multiple fail regions. Luckily this is not typical for good circuit designs!
7. Can we apply the WCD concepts also to lognormal or uniform element distributions?
   In principle yes, and almost all WCD papers include sentences such as "Without loss of generality we can assume that the statistical variables can be described by independent normal distributions." However, the nice pictures often presented as well would look quite strange, e.g., the lines of constant probability would be no circles for a uniform distribution! We mentioned that WCD becomes slower if we need to treat many statistical variables, but in case of non-normal variables the CLT would help us again, beside the many uniform variables the

overall behavior would usually be quite similar to the normal Gaussian behavior.

8. Can we apply WCDs to discrete probability density functions?
   In principle, but optimizer has a hard job in such cases. For luck there is usually no need for this in circuit design.

9. How would a classical WCD algorithm work on a completely "circular" spec border (see Table 7.6)?
   The MC run would give a certain starting point for the WCD optimizer, but all fail points have an equal distance to the origin the WCD direction (the angle) would not matter. So a gradient optimizer would have no reason to change it, so the final reported angle would depend on the initial MC, so on chance.

10. Check-out the scaled-sigma chapter: If we compare the unscaled MC run with an upscaled run using the same seed value, can we expect that the most extreme samples in the first one are identical to the second one?
    No, not completely, if e.g., the different statistical variables behave much different, like odd vs even order nonlinearity, it could easily happen that the ranking would not be the same. However SSS could still work fine.

11. Discuss the differences in effort, accuracy, design insights between WCD and the simpler C_{PK} or C_{GPK} methods! Which kind of non-normality is difficult for each technique? What is the impact of design complexity? What can happen if we skip the MC part in WCD calculation and start an optimizer directly from the origin?
    WCD is a multivariate technique so its speed degrade for complex cases. If e.g., the origine is a local minimum, then a local optimizer would not start! WCD has the big advantage that it offers statistical corners, and this way also sensitivity information.

12. Figure 7.19 shows a five-dimensional test case. Try to interpret the 2D plot and to calculate a WCD.
    Actually the 2D plot is a strong simplification, because there is <u>no</u> hard fail boundary in any 2D projection! The simplest attempt is to regard the sum term as a small, almost constant value, which would lead to a linear spec boarder at both sides. Looking more to the details would show that the true spec boarder looks more like a parabola, as sketched in Figure 7.18. In addition, the sum of squares is actually leading to a chi2 distribution with degree of freedom equal to 4, which is quite

manageable. So we can indeed easily reduce the 5D problem to a 2D problem! The WCD is even easier to calculate, because we just need to find the <u>shortest</u> way to the fail boundary; and for this we need $x_2 = 0$. If we would <u>add</u> the sum of square instead of subtraction, we would get convex fail boarders, and the testcase would look more similar to the memory example in Figure 7.30.

7.7 Summary of Advanced Statistical Analysis for Yield

We presented methods to evaluate Monte-Carlo simulation results with focus on yield, by using the sample yield, the C_{PK} and the generalized C_{PK}. Also multivariate analysis has its value, giving insights to sensitivities and trade-offs. All these are often pure MC <u>post</u>-processing, whereas for real advanced analysis, we really "drive" the simulator according to the results of a short initial MC run.

In the Subsections 7.7.1 and 7.7.2, we present now again a small "show-down" like we did for pure worst-case corners, but actually statistical problems are so difficult in circuit design, that clever manual methods are hardly practical. However, still the person on front of the computer matters too. A "golden reference", like full-factorial for corner analysis, is much harder to provide for statistical problems. Making Monte-Carlo "golden" requires a huge number of samples for the verification of higher yields.

One example for an adaptive statistical method is sorted MC; it gets a speed-up by avoiding simulations with high chance to pass specs anyway. It is based on a multivariate model and it can give a significant speed-up. It has quite a low risk, e.g., if the model is bad it would typically only observe a degradation of the speed-up by sample re-ordering. In the extreme case, it may snap back to simple MC without sorting. As in other advanced methods, a speed-up (against random MC and the sample yield) is typically only possible for higher yield targets (like $>2.5\sigma$ to 3σ) and the speed depends on further factors like number of statistical variables, number of specs, degree of nonlinearity, correlations, etc.; for most methods, 1000 points are a kind of minimum (single spec, small block, simple statistical models). In most advanced methods, the speed depends not much on sigma level, for pure linear case in WCD just the optimization part takes a few more points; for a linear circuit and moderate complexity, the effort would be almost flat, i.e., the number of required simulations would increase just by few percent, e.g., from 3σ to 7σ (Figure 7.28). The WCD concept comes with some slightly stronger

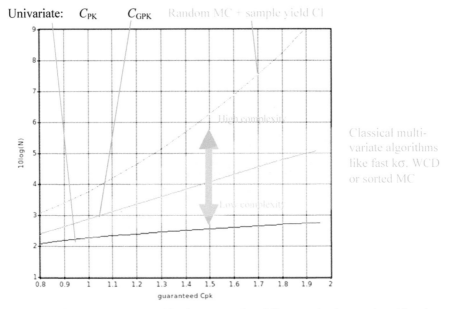

Figure 7.28 Typical yield verification count for different univariate and multi-variate methods.

assumptions, like that there is a single failure region or at least a dominating region plus making assumptions on the spec border at WCD point.

When there are multiple outputs, MC naturally accounts for overlapping or disjunctive failure regions, but almost all advanced methods like sorted MC and IS or WCD focus on partial yield estimation only. As demonstrated, using $\min(Y_{\text{partial}})$ is only a rough approximation for the total yield; it is better to exploit knowledge about correlations to obtain a better estimate.

As mentioned, beside many differences, one key idea of modern statistical yield verification is the generation of <u>statistical corners</u> and especially worst-case distances, because that gives further insights and the option for yield optimization.

The plots in Figure 7.29 depict different approaches that all have their application for good reasons in modern custom IC design environments. Usually, a combination is used, especially for high-σ targets.

7.7.1 Different Methods on Difficult Mathematical Cases

Circuits can be difficult for *many* reasons, such as complexity, correlations, or nonlinearity; let us now first focus on low-dimensional problems and later

Table 7.7 Comparison on typically required number of fully simulated points for different yield levels for different statistical analysis

Analysis	#Points for 3σ	#Points for 6σ	Scalability	Comment
MC & picking worst sample	500	100,000,000	Applicable to extreme complex problems, because no need for multi-variate modeling	Accuracy very limited but quite independent on linearity and number of variables
MC & picking worst sample + scaling	250	1000, but significant extrapolation risk		Accuracy good on length but limit on direction, quite independent on linearity and number of variables, scaling could be obtained via C_{GPK}
Fast $k \cdot \sigma$	200+3 · #specs	Not yet commercially available		Slightly depending on linearity & variable and spec count
Sorted MC for yield verification	100–2000	200 … 5000	Critical, because accuracy depends on multi-variate modeling	Depending on linearity & variable and spec count, internally many more points need to be simulated using the performance models
Worst-case distance (WCD)	100–2000	200–3000	Critical, because variable screening becomes more difficult for complex designs	Depending on linearity & variable and spec count
Sigma-scaled sampling (SSS)	1000–3000	2000 … 10,000	Applicable to extreme complex problems, no need for multi-variate modeling	Slightly depending on linearity

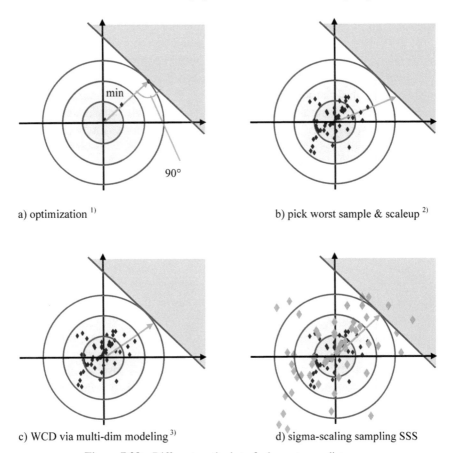

a) optimization [1]

b) pick worst sample & scaleup [2]

c) WCD via multi-dim modeling [3]

d) sigma-scaling sampling SSS

Figure 7.29 Different methods to find worst-case distances.

[1] The starting point for optimization is usually obtained from another method, leading to the classical WCD search method.
[2] Upscaling is usually needed for higher sigmas, like beyond 3σ.
[3] The modeling allows a more stable estimation compared to worst-sample method; it includes usually also some kind of upscaling; a variant is the fast $k \cdot$ sigma flow.

extend to higher complexity. Some of these problems look "too special," but actually each has also some strong circuit design connection! Circuit design is simply sometimes difficult; analog designers indeed stress numerical algorithms more than others, although not always. So in real designs with some complexity, indecd algorithms can fail for quite similar reasons. In many cases, it is unfortunately often harder to find out <u>why</u> problems appear from time to time. Regard these tricky examples as "forerunners," as precursors for real circuit design problems. In compact mathematical examples, the problem is more often already in the method, not so much in the tool implementation. So this is no tool benchmark, but clearly a method comparison. In theory, one could even create more "critical" test cases, but in circuit design, we have indeed almost always the case that the design is in-spec at nominal ($x_S = 0$); and being not far from the origin, the circuits is also typically in spec. So we have a *large*, usually dominating *pass area* in the coordinate system center part. Other cases would be no robust designs, and of course, high-yield methods are not needed to "prove" a low yield. For instance, already the WCD or SSS first MC part would show this!

To detect problems, the algorithms have usually some built-in internal error criteria, which might be also formulated in different ways, e.g., as rms or maximum error. In new generation, EDA tools luckily the quality of error checking is quite high already.

Actually, you can easily find circuits in which one method works better than the other, and you can also find examples where even all existing algorithms work much worse than their inventors think! For luck, you are sitting in the designer's seat and, in most modern environments, you can decide what to use and what your priorities are: more on design or on verification, more on speed or on coverage, more on high yields or moderate yields, big designs, nonlinear designs, etc. You should know now quite well the difficulties, the trade-offs and that there is simply no free lunch! You should not get foolish anymore!

Table 7.13 gives an overview for the systematical and statistical errors of some major yield estimation methods and different simple, low-dimensional test cases. In all of them, we generate non-normal output data (which is typical for nonlinear circuits), but all statistical input variables are purely Gaussian (like most variables in PDKs). Note that this setup favors SSS and WCD to some degree, whereas the sample yield, C_{PK}, and C_{GPK} do not depend on this at all.

The (true) WCD—so the distance from origin to the shortest fail boundary point—is 3.35σ (0.04% loss or true $C_{PK} = 1.167$) for all test cases, and the WCD estimation itself has been tweaked for high accuracy; actually in this small testbench, WCD is already statistically accurate by using a few hundred simulation points. Let us start our little benchmark with a discussion of the error of WCD regarding yield estimation. In the multivariate normal case, WCD is an accurate method, but in our more difficult examples, WCD has a bias error arising from spec border nonlinearity (see Figure 7.10), and that might be surprisingly large unfortunately. In opposite to WCD, the sample yield has no bias at all, but a significant variance. The 95% lower confidence bound for a 3.35σ yield and n = 10,000 is approximately 99.88% (0.12% loss or true C_{PK} = 1.1013 or 3.04σ) and the standard error is approximately 150 ppm (0.015%)—so overall the sample yield error is in the order of 10% in terms of sigma, which is a quite low value thanks to the high MC count.

As we know, the sample yield is accurate to this value for both normal <u>and</u> non-normal cases, but WCD has bias errors listed in the table, and these are up to 0.65σ! To defend the WCD concept, we should mention that for many simpler non-normal cases, also the WCD would be correct. So on the one hand, the listed WCD errors are a kind of worst-case for most practical circuit design applications, but on the other hand, with more statistical variables, you could even create worse examples, in which WCD would fail almost completely (e.g., we mention the flash ADC DNL example already).

The C_{PK} is the third method we inspect: Its bias errors are usually larger than the WCD errors, which is expected for non-normal data anyway. Overall, the C_{GPK} as fourth method behaves best in such mid-yield scenarios—but as C_{PK}, it cannot give accurate statistical corners as direct output.

This is also a limitation for the sample yield method, having also a good ranking in the table. However, if we would reduce the circuit simulation effort, e.g., from 10,000 to 1000 points, the ranking among the different methods would change significantly (due to $\sqrt{n}$ law for standard error and confidence interval), because the CI width in terms of sigma would grow a lot for the sample yield, but not that much for the other methods, and almost not at all for WCD!

We also run sorted MC as fifth method (with yield-based auto-stop on these examples); the speed-up from sorting depends on nonlinearity. On the overall yield, the reported speed-up was $3.5\times$ and the auto-stop occurs after 265 simulated points (out of 950). If we would only run sorted MC on the one-dimensional pure normal case and same yield level, sorted MC would be (of course) faster, using only 50 points instead of 265 (speed-up $20\times$). Only WCD has a near-zero CI width, but WCD-internal the bias error is very hard to quantify in complex design cases.

The sorted MC auto-stop can be also done based on reaching statistical corners (instead of yield confidence interval). This allows to compare the statistical corners from classical high-yield estimation WCD and sorted MC. Interestingly, sorted MC can indeed find almost the same samples as worst samples as if you would run the full MC (without sorting) and doing the sorting (ranking) manually after the simulation. So the sorting based on the multi-dimensional model is done quite well, using only 110 simulated points in total (for $\geq 99.65\%$ corners). On the one hand, this is no surprise because in the simple examples, we have just to deal with two statistical variables; on the other hand, the nonlinearities are quite strong.

Besides classical WCD, we could also have used SSS as 6th method, but look up the sweet spot of SSS are cases with many statistical variables and many specs! Interestingly, the SSS yield results are quite different from the classical algorithm results. SSS uses more points by default; 2000 are chosen manually to stay close to more complex realistic circuit designs. Classical optimization-based WCD has to run MC only *one* times, not several times with multiple scaling factors. Actually, reported yields from SSS were always too pessimistic (even for the normal distribution), and the absolute value of the bias error is typically comparable to the classical WCD (but in WCD, they tend to be too optimistic, not too pessimistic). In further examples with higher yield targets (like 5.5σ), the SSS bias went down significantly (but in these examples not the one for classical WCD). The reported SSS confidence interval is quite large (it would of course reduce for n = 10000 instead of 2000). Also note that in [Jallepalli2016], some improvements in SSS have been described which are (probably) not yet part of commercial products.

According to Table 7.8, there is always at least one method that works fine, and the C_{GPK} method is overall the "best" one. For very high-yield targets (like 5σ instead of 3.3σ), the dedicated high-yield methods would gain some ground as the C_{GPK} is based on extrapolation (at least in the simple form applied here). One example showing this effect would be normal data with *cut* at a certain sigma level. If the cut is a 3σ and the spec is at 4σ, the yield would be 100%, but the C_{GPK} would "only" give approximately 4.5σ. The C_{PK} is even more pessimistic giving 4σ only; to get 4.5σ from C_{PK}, we would need to cut at 1.5σ! This sounds bad, but using the sample yield you would need to run a huge MC analysis with approximately one million points (for 95% confidence). So here the winner would be (often) WCD, but in advance it is not easy to know which method is best for general circuit design!

7.7.2 Different Methods on Circuits

Looking to the math examples, the big arising question is of course how "general" the results are. Well known and for sure: For lower MC counts or higher yield targets, the variance errors would grow, especially for using the sample yield this is a <u>severe</u> problem, which is a well-known fact and as mentioned essentially the *driver* for inventing all the advanced methods! For higher *complexity* the WCD, fast k · σ and sorted MC would *lose* speed,

Table 7.8 Errors for different yield estimation methods for difficult 2D cases ($n = 10000$, WCD = 3.35, 95% confidence interval used)

Specification & Characteristic	Method	Bias Error in σ	Estimate-LCB in σ	Comment
V_{off} < USL, normal Gaussian, one variable, linear spec border	Sample yield	0.0	0.3	99.96% or 3.35σ
	C_{PK}	0.0	0.05	C_{PK} = 1.12, Good method
	C_{GPK}	0.0	0.15	C_{GPK} = 1.11, good but C_{PK} preferable
	Classical WCD	0.0	0	Good method
	SSS	0.35	0.27	99.86% or 3σ
$|V_{off}|$ < USL, non-normal, one variable, two fail regions with linear border	Sample yield	0.0	0.3	99.92% or 3.2σ
	C_{PK}	0.9	0.06	C_{PK} = 1.41, bias far too large
	C_{GPK}	<0.05	0.12	C_{GPK} = 1.04, good method
	Classical WCD	0.1	0	Bias just acceptable
	SSS	0.2	0.3	99.88% or 3.05σ
max(V_{off1}, V_{off2}) < USL, 2 variables, non-normal, two fail regions with linear border, skew is 0.142	Sample yield	0.0	0.3	99.92% or 3.1σ
	C_{PK}	0.27	0.05	C_{PK} = 1.12, bias quite large
	C_{GPK}	<0.01	0.15	C_{GPK} = 1.04, good method
	Classical WCD	0.26	0	Bias quite large
	SSS	0.39	0.25	99.66% or 2.7σ
max($|V_{off1}|$, $|V_{off2}|$) < USL, 2 variables, non-normal, box spec border, skew is 0.7	Sample yield	0.0	0.25	99.85% or 2.97σ
	C_{PK}	0.72	0.06	C_{PK} = 1.23, bias far too large
	C_{GPK}	<0.02	0.09	C_{GPK} = 0.98, good method
	Classical WCD	0.38	0	Bias quite large
	SSS	0.3	0.22	99.6% or 2.67σ
$V_{off1}^2 + V_{off2}^2$ < USL, 2 variables, non-normal, circular spec-border, skew is 0.63	Sample yield	0.0	0.2	99.61% or 2.7σ
	C_{PK}	0.51	<0.02	C_{PK} = 1.07, bias quite large
	C_{GPK}	<0.02	0.08	C_{GPK} = 0.9, good method
	Classical WCD	0.65	0	Bias far too large
	SSS	0.25	0.28	99.28% or 2.45σ

unfortunately. However, as WCD needs often only few hundred simulations, there is quite some margin for higher complexity up to a typical block level. Also, all methods could be combined with LDS to get a moderate additional speed-up, at least for not too complex cases (see Chapter 6 on advanced Monte-Carlo sampling schemes).

As mentioned, a big benefit for WCD is getting a true worst-case direction, but many of the mathematical benchmark examples are too difficult, and actually only in the simplest case, the direction is well defined! If we would pick the worst sample among the 10,000 MC points, we would typically also get a useful statistical corner, suited for optimizations. For less nonlinear examples (like op-amps, bandgaps or comparators, etc.) the WCD behavior would improve significantly, making it (together with scaling the worst-sample) a very attractive method again [Bűrmen].

Let us now inspect some real-world circuits. A major change we can expect is that for complex problems not only the method counts, but also the implementation (e.g., [McConaghy2013, Zhang2016]). For instance, quite many numerical algorithms come with a few "magic" numbers, like "first run a MC analysis with 50 points"? 50 might be good in eight of ten examples, but combining the problems of the two critical test cases needs maybe 250 points (or more)! Actually, sometimes the EDA vendors are in a dilemma: On the one hand, they want to provide a fool-proof tool, with little setup effort, but on the other hand, playing with such settings can indeed give a somewhat better compromise between accuracy, robustness, and speed. So like in a simulator, also the designer's skill has some impact.

In [Gu], a memory cell example is given; actually, the example is simplified, because statistical problems with hundreds of variables are hard to visualize, but if you compare the fail region in Figure 7.30, you will observe indeed some similarity to our mathematical examples, and the memory cell is something in between the Gaussian case and the "circular" case (on which WCD is significantly inaccurate). Actually, the spec border is approximately L-shaped, leading to a too optimistic yield estimation by WCD. Luckily, the error is (by far) not as extreme as for the mathematical examples, because the "second worst" distance point (at the "nose" in Figure 7.30) has not the same length as the real WCD, maybe there is a difference of roughly one sigma; and this reduces the WCD bias error to quite an well-acceptable level (like from 0.2σ down to 0.02σ). In addition, the graphical plot tends to give a (far) too pessimistic <u>impression</u>: The outer fail *areas* are large (actually infinite!), but do not contribute much to yield loss and error on it! This is due to the e^{-x^2}

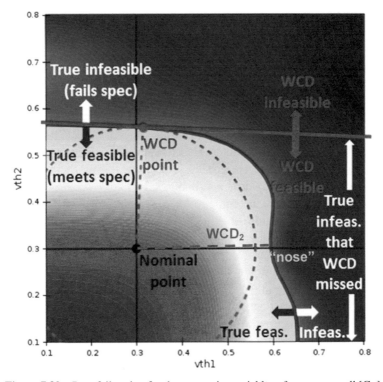

Figure 7.30 Pass-fail region for the two major variables of a memory cell [Gu].

pdf law of the Gaussian distribution, which has to be integrated. For large $|x|$, the contribution to the integral becomes almost negligible.

Let us modify the example figure and inspect another difficult situation: Assume that the second WC point at the "nose" would dominate, i.e., WCD_2 becoming shorter than the initial drawn WCD. Imagine further that now this nose would be very thin and peaky in direction to the origin: In such cases, the "peak" of the "nose" would be hard to find for an optimizer, so the optimization may fail and would not find this global minimum but typically stopping at a local minimum still pointing to the initial WCD, having *larger* length in our modified example, so this WCD would indicate a *too large* yield! Actually, the optimizer may not work fully reliable and the obtained WCD may jump (e.g., if we tweak some WCD-internal parameters or the circuit) like the WC corner set in our example on the CMOS inverter.

For luck, these kinds of problems tend to be quite artificial for most circuit designs. For instance, in the same reference, also a ring oscillator is analyzed in detail, but without showing such problems and having even very linear spec borders, so very small WCD bias. There are also mechanisms which would cancel out the WCD error to some degree: If the optimizer would find the thin nose accurately, we would obtain the WCD correctly, but the yield estimate from it would be too pessimistic, due to the (locally) concave fail border. If on the other hand, the optimizer would find the original WCD (now second worst) the yield estimation would be too optimistic, and for intermediate results, the yield prediction would be even quite accurate.

One other interesting characteristic of Figure 7.30 is also that there are no fails for small *vth1* and *vth2*! This is because the plot shows only one performance (access time). For a full block verification, you have further checks (such as leakage current power consumption) and further WCDs, in classical WCD all require a dedicated optimization run with additional simulation points to find the accurate WCD points via optimization. This is some overhead compared to C_{GPK} or SSS.

Interestingly, for a circuit *optimization* on yield, we need to combine different specs to formulate an overall best *compromise* in some way. Doing so for WCD is no good idea: If you would re-organize the testbench to check many or even all performances in a *combined* spec like $x \leq 0$ with $x = \max(0, I_{DD} - I_{max})/$uA $+ \max(0, t_{access} - access_{max})/$ns, then the fail region would deviate more and more from the WCD assumption and becoming extremely difficult! So it is better to avoid such "spec tricks," they do not fit, neither to many nice fast yield methods, nor to the goal of getting wishful design insights! Unfortunately, you cannot always avoid difficulties, e.g., the SNDR of an ADC might be impacted by several effects such as mismatch, integral nonlinearity, and thermal noise, which could lead to difficult fail regions and worst-case conditions. It is best to *complement* such complex specs with more specific performance checks which measure effects more individually.

Of course, what we state is related to year 2016 state-of-the-art design tools; in research several enhancements have been already proposed (e.g., nonlinear surface sampling in [Gu] or just you as user take the pilot seat as usual and, e.g., run WCD, but double-check the results with C_{GPK} or k-means cluster analysis). We can highly expect that in the near future, advanced and *more adaptive* statistical analysis will be almost as easy to handle as standard Monte-Carlo, but offering both higher accuracy, speed, *and* better reliability, e.g., methods which inspect the spec border and the yield volume in more detail. The idea of such boundary search methods is to find not only one distinct worst-case point, but the whole failure boundary [Gu] or at least all the dominating parts [Roos]. Looking at the fail boundary instead of the full yield integral looks promising regarding efficiency. However, again, it is difficult to make such methods working for complex cases; the speed-up looks dramatic when going from a 2D integral to a piece wise linear approximation (enabling the use of analytical formulas), but in n dimensions, we would only reduce the effort from 100 dimensions to still 99 dimensions. A somewhat more detailed outlook will be presented in Chapter 11.

Too much hype on "variation-aware design"? Sometimes you can get the impression that there is a kind of "war" for the best yield estimation method or something like a "high-sigma showdown" (cite from a newsletter I received every day!) —So one company promotes WCD as the Holy Grail, whereas another pushes their version of sorted MC as the High-Sigma Monte-Carlo (google for "deepchip muneda solido cadence worst-case distances"). There are many papers around promoting huge improvement rates like "100× over Monte-Carlo," but now you know many alternatives to pure random Monte-Carlo, just counting fails and using Clopper-Pearson confidence intervals, so can you really *trust* such marketing promises?

A good recommendation is clearly *Audiatur et altera pars* – Listen to the other side; and being not afraid to ask; yield estimation is about math, so don't worry and stop only if you are really satisfied.

An interesting comparison is looking to the progress in ADC design [Murmann], in 15 years the performance has been improved by 60×, but the pure CMOS scaling has provided only a 10× gain in speed and power. So roughly 6× has been found by other means like circuit innovations, more careful design tweaking, etc., but partially maybe also by "paper tuning," like using selected parts, ignoring power consumption of supporting blocks like reference generators, buffers, etc. Actually [Murmann], we can expect also a saturation in performance in roughly 2027. There are of course also some clear limits in computation power and numerical algorithms, not only in ADCs! Paper tuning in statistics? One clear example is only comparing the *number* of circuit simulations. Often indeed simulation time dominates, but the circuit simulations in MC can run fully in parallel, which is not possible for WCD or sorted MC. On top: Only because MC has little internal simulation time, the circuit simulation dominates, whereas in sorted MC, also the model creation and application takes also significant amount of time.

In the next chapters, we will discuss optimization techniques in detail and we will also link them to statistics. The core idea of making any statistical optimization (a complex topic!) efficient is exploiting the problem structure and focusing on corners—corners as much representative as possible—instead of running MC for large sample counts.

PART IV

Optimization and Advanced Flow Techniques

8

Optimization Techniques for Circuit Design

$$X_{i+1} = x_i - gH^{-1}$$

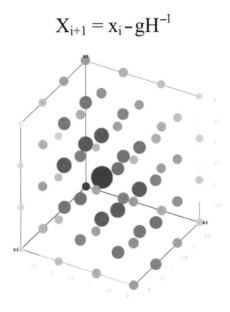

In this chapter, we discuss when optimization is beneficial, and how the most important optimizers work. We show the best practices for setting up an optimization and how to respond in case of problems.

Optimization is the act of obtaining the best results under given circumstances. This sounds simple, but optimizers seem to be special, much more special than simulators, for circuit designers. This is because optimization is almost always something "on top," it comes usually quite short in school and university lessons.

Luckily, optimization can be translated to math easily; quite often, there is something to minimize (like noise figure) or to maximize (like bandwidth, ENOB). Also hitting a certain performance target y_o as accurate as possible can be described easily, e.g., minimize (f) with $f = (p(x) - y_o)^2$. If you have a <u>minimization</u> program, then you can also solve all the other problems, e.g., minimize f is the same as maximize $(-f)$. Therefore, optimization is simply just minimization. It also does <u>not</u> matter how the goal function is calculated; there could be a circuit simulator involved, or instead, you may speed-up by using hard-coded simplified design equations. Using a circuit simulator is just more flexible, because the extra work to derive the circuit performances is quite low compared to a *full* hand calculation.

The formulation of the goal function can be simple in many cases, but it might be more difficult sometimes. It is also a key point, because the result of the optimization and also the difficulty highly depend on goal definition. Key targets are time-to-market, costs, yield, or electrical performances—but some goal might be tough to capture and to balance against other targets. Examples also show that optimization is quite hard to standardize. Optimizers need to be more flexible than circuit simulators. Later, you can always use, e.g., modified nodal analysis and the element equations to formulate the system behaviour; then, you can solve it via Newton–Raphson (NR). Another classic method for solving equations is bi-section, bracketing the solution. The equivalent for optimization and minimum search needs three points for bracketing, and the best method is the so-called golden section search method. However, actually we can create much faster methods with little effort.

The simplest function with a minimum is a quadratic function $f = ax^2 + bx + c(a > 0)$, and the minimum is the apex, where $f' = df/dx = 2ax + b = 0$. Indeed, many ideas to minimize any general function can be borrowed from the ideas to minimize a parabola! For instance, we can calculate f if we have 3 points or if we have the function value plus the first and the second derivative. This way we can calculate the coefficients a, b, and c and solve $f' = 0$ for x leading to $x_{opt} = -b/2a$. Using matrices we can also extend this scheme easily to more than one parameter.

Already in the one-dimensional case, several things can prevent us to find a solution; for example, if a is 0 or negative, then we will simply have no minimum. In multiple dimensions, some more things can go wrong, but luckily almost all tricky things can be well explained if we look to the two-dimensional case $x = (x_1, x_2)$. Actually, you only have to deal with nothing more complex than a 2×2 matrix to understand almost all optimization schemes!

Because optimization is highly related to quadratic functions, the simulation effort typically rises in a quadratic way with the *number* of parameters to optimize, whereas the number of goals or the allowed step-size matters much less. For n = 10 parameters, you can expect that a good optimizer gives you a significantly improved circuit after, e.g., 100–200 simulation points. Of course, it depends on how good the quadratic approximation is and whether there is indeed enough room for improvement.

If you would tweak ten parameters manually just by sweeps, the effort would even grow much faster, namely exponentially! So although optimizers have no a priori knowledge about the design, they can be still very efficient.

Optimization can be applied in many ways like on system level or on block level. It can be done in numerical ways or—often easier—if you have analytical formulas for the performances. Some classical optimization results are quite popular:

- How many inverters and which driver strengths do you need for optimum speed driving a large load capacitance?
- What should be the receiver bandwidth in relation to the signal bandwidth for optimum signal transfer?

Optimization is also useful as subalgorithm in other problems, e.g., for obtaining WCDs (Chapter 7). If you have a little optimizer in C or Pascal, then you can also write your little design programs. We put a program ANPASS to this book, to optimize many kinds of RF-matching networks. Internally, a BFGS optimizer is used, and also a global optimization example is included, which is solved via simulated annealing (Figure 8.1).

Remember the chapter on transistor biasing and sizing; you may put the different approaches like g_m over I_D in a difficult flow chart, but actually optimization is almost the most native approach to solve this problem. An optimizer can easily act as transistor sizer for W and L based on given technology parameters, drain current, V_{DS}, etc., plus targets, e.g., on g_m, f_T or noise behaviour (Figures 8.2 and 8.3).

8.1 When to Use What?

Look at Table 8.1 to get an overview on different optimizers.

For Further Reading:
There is a lot of literature available on optimization, starting in the late 70s also for circuit design. A lot of material is related to RF design and model fitting—here optimization is a highly desirable method, and being used intensively. Note searching for yield or statistical optimization gives not that many hits, also search e.g., for reliability optimization.

- McKeown, An Introduction to Unconstrained Optimization, ISBN-10: 0750300256
- M. J. Box, A comparison of several current optimization methods, Computer Journal No. 9, 1966—really a good benchmark!

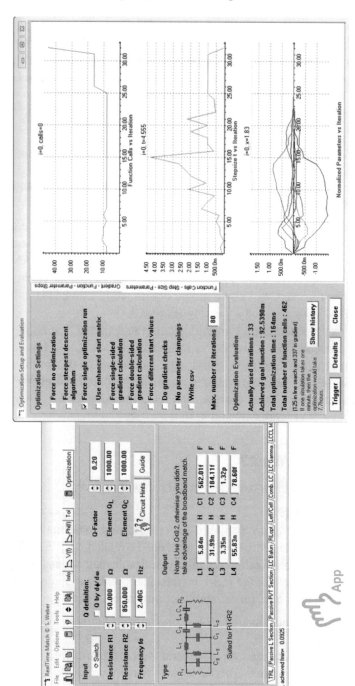

Figure 8.1   Real-time match to design networks with built-in optimizer.

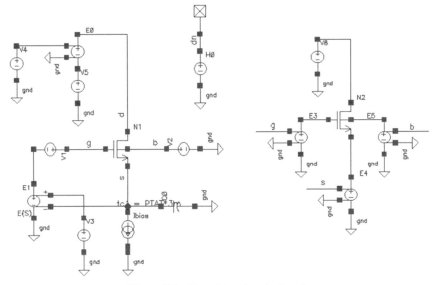

Figure 8.2 Transistor sizer testbench.

Testbench setup for output expressions
e.g. g_m=OP(N1.gm), V_{noise}=sqrt(integ(VN)), etc.

Define spec targets and optimizer settings (like number of
Iterations, gradient limit, etc.). Choose a start point
dnd define which parameters to optimize.

Run simulations and calculate over-all goal function
from spec deviations

Let the optimizer decide how to set the parameters
and when to stop

e.g. decide for further optimizations or accept result.

Figure 8.3 Transistor sizing via optimization.

Table 8.1   When to use what—local vs. global

Optimization	Algorithms	Limitations	Applications
Local optimizers (will only find the next local minimum), typically a fix algorithm, like	Coordinate search	Very slow	Use it only if the number of variables is very low and minor correlations, useful if the gradient evaluation is difficult
	Hill-climbing based, e.g., Nelder-Mead	Usually slow in the case of many variables	Use it only if the number of variables is moderate, useful if the gradient evaluation is difficult
	Steepest-gradient	Goal function behavior should be smooth, usually slow	Never use it, as there are better gradient methods!
	Conjugate gradient	Goal function behavior should be smooth, moderate speed	Use it only if number of variables is moderate
	Quasi-Newton	Goal function behavior should be smooth	Generally a good choice
	Newton	Goal function behavior should be smooth	Make only sense if the effort for calculating the 2nd order derivatives is low enough.
Global optimization, typically multiple techniques will be combined	For example evolutionary algorithms, simulated annealing or using local optimizers with different starting points	Really finding the global optimum requires more time, so if you are sure the local and global optimum is identical will need typically more runtime.	Use it if you are not able to provide a good starting point and if your problem is highly nonlinear

- J. W. Bandler, An automatic decomposition approach to optimization of large microwave systems, Microwave theory & techniques, No. 12, 1987—good ideas to enable hierarchical optimization of complex systems.
- Comparing Results of 31 Algorithms from the Black-Box Optimization Benchmarking BBOB-2009, Nikolaus Hansen et al.!

- D. Agnew, Efficient use of the hessian matrix for circuit optimization, Circuit & systems, no. 8, 1978—good article showing how valuable the Hessian matrix is, with good examples.
- An Evolutionary Algorithm-Based Approach to Automated Design of Analog and RF Circuits Using Adaptive Normalized Cost Functions, Abhishek Somani et al.—explaining goal function graphs.
- A Random and Pseudo-Gradient Approach for Analog Circuit Sizing with Non-Uniformly Discretized Parameters, Michael Pehl, Tobias Massier, Helmut Graeb, and Ulf Schlichtmann.
- The Sizing Rules Method for CMOS and Bipolar Analog Integrated Circuit Synthesis, Tobias Massier, Helmut Graeb, and Ulf Schlichtmann, IEEE Transactions on Computer-Aided Design of Integrated Circuits and Systems, vol. 27, no. 12, Dec 2008.

8.2 Introduction to Optimization; When to Optimize?

A few years ago, I was asked "We need an optimizer demo. Our customer has problems with VCO phase noise, can you do it?"

I actually run first a periodic noise simulation to check the circuit—and the noise summary told me most of the noise comes from the bias part and not from the VCO! So in this case, a more "directed" solution was much more efficient than running "blindly" an optimization. Another example would be an LC oscillator which operates so well that the phase noise is already close to the theoretical minimum, given by the Leeson formula, which relates the phase noise to the Q factor of the oscillator tank circuit and the power consumption.

Often, it is indeed possible to achieve the desired performance with some hand calculations and purely manual parameter tweakings or you may construct the circuit step by step. Also in such cases, an optimization can be unnecessary, although some improvements are usually always desirable, e.g., a smaller layout area, lower noise, better PSRR, or lower-power consumption. So sometimes even for circuits "in spec" optimization can be useful. Also very often such manual design only leads to full spec achievements at most corners, but not all of them. In this case, e.g., a worst-case corner or yield optimization makes sense, but a nominal optimization is not necessary. Of course, advanced numerical techniques, like a contribution analysis or an automated WC corner set search, are very helpful also for those "non-optimization sweet-spot types" of design problems.

Over-optimization at typical may also lead to nonoptimum results over corners! Quite obvious is this, e.g., in LNA design, if you only optimize on noise and gain, without taking bandwidth, and production tolerances into account. The result of a pure nominal optimization could have a high gain, but to small bandwidth and too bad tolerance behavior. Another typical problem appears often in linear amplifiers like op-amps. For good PSSR or gain at nominal conditions, you would need to go quite close to the limits regarding saturation voltages (e.g., using long transistor to minimize current mismatch), but at certain conditions like slow MOS corner combined with minimum supply voltage we can easily end up in headroom problems for current sources or the amplifying transistors. A third example could be stability, even a perfect 90° phase margin PM at typical conditions cannot guarantee a good phase margin over corners, e.g., including large variations in load impedances. In conclusion, be careful with pure nominal optimizations, the best thing is probably that you can setup it up and run it quickly; and switching to more realistic scenarios (e.g., inclusion of environmental or statistical corners) is usually very easy.

Beside these facts, there are also some prejudices about optimization. Of course, people *like* to learn about circuits and are happy when (or if?) finding out *how* a circuit can be improved; e.g., why it was bad to use a minimum L for a certain transistor? On the other hand, it is simply not true to think of optimization as a black box algorithm; measures like sensitivity information are actually *also* used internally by optimizers, and they often report this "somewhere." So as the contribution analysis is an almost free lunch for getting sensitivity information regarding statistical parameters, you can also get a nice sensitivity report regarding design parameters—after the optimization, without further simulation runs.

A real need for optimization occurs if the manual design fails due to difficult specs. In this case, the question is often whether the specs for chosen topology can ever be achieved in the given technology or whether we need to change the circuit topology (problems: new topologies can come with further design risks and major changes are time-consuming too) or to relax the specs. Often the latter is possible, but with impacts on the system design.

Optimization is also a good choice for real high-performance designs and for RF designs, where usually the device limitations have very severe impacts. RF designs are also good candidates, because here you seldom have near-ideal elements to which simple design formulas fit well; and often changing *one* parameter like transistor width or bias current can influence *many* performances, like gain, input impedance, noise, linearity, stability, and area. For humans, treating many relationships is difficult, but software can do! Table 8.2 gives an overview on manual sizing versus optimization,

Table 8.2 Comparison of manual sizing vs. optimization

Manual Sizing Is Easy, As Long As	Examples and Comments
Symbolic small-signal approximations are sufficiently accurate	If big parts of the designs can be treated well via small-signal equivalent circuits (e.g., amplifiers), then MOS-square law fits well (like in older technologies).
There are no tough trade-offs between multiple difficult specs	Bandgap without area hard area requirements
Process variation and mismatch can be compensated by safety margins and structural solutions (feedback, symmetry, calibration, etc.)	Op-amp amplifier design at low frequencies.
There are only few design variables, best with a 1:1 relationship to specs or at least clear step-by-step sizing instruction	Low-order filters or simple amplifiers
Manual Sizing Becomes Difficult If	Examples and Comments
Your design is not as easy as an analog textbook example	Many "tricky" designs are harder to size than most classical circuits.
Large impact of second-order effects, parasitics, etc.	RF PA, high-Q filters, etc.
Specs on highly nonlinear performances with no good or difficult symbolic estimate	e.g., filter ripple or ADC DNL, frequency compensation for 3- or 4-stage op-amp
Impact of PVT variation and mismatch is so large that it adding them up leads to over-pessimism	Low-voltage ultra-deep sub-um designs, circuits for wide temperature ranges
Tight specs, multiple trade-offs	High-performance and/or low-power designs, e.g., RF, ADC, DAC
Many design variables	Many circuits like VGAs and PLLs. Consider an iterated hierarchical approach.
One parameter would already change many performances	RF designs, like LNA, PA, but also filters.

e.g., op-amp design is often easy by hand, but for subtasks (like finding a perfect frequency compensation or to improve the common feedback part) optimization can be very helpful still. Also having prepared an optimization, by setting up specs and making a parametrization, manual tweaks and sensitivity analysis can be made much easier!

Actually, all these examples show that also optimization is not a push-button solution for bad designers, it is interactive, like manual design, but just often faster and being able to go deeper. Figure 8.4 gives an overview on different optimization techniques and scenarios. Of course, there are

Accuracy

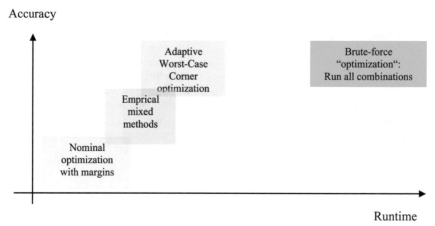

Figure 8.4   Optimization accuracy–speed trade-offs.

also several mixed optimization scenarios possible, like instead of adding performance safety margins we may use "expected" worst-case corners, or instead of using more reliable adaptive WC finders we may use the faster OFAT method, at least in the sizing loop.

Such ideas can give some further speed-up, but without double checks it typically ends up in lower accuracy and harder to estimate risks.

8.2.1 Optimization Pre-requisites and Limitations

In analog design, including RF and mixed signal, there is almost no automated synthesis available, so designers define the circuit topology based on experience plus exploiting basic transistor behavior (e.g., gain can be set by g_m or g_{DS}, which could be exploited for a variable gain stage) and then they have to determine the component values like transistor width and resistor values. Since many years parameter optimizers can do the second job, at least if the problem is not too complex, not too nonlinear.

Optimization is done by tweaking the parameters—just as the designer would make a little change in the schematic, e.g., on transistor length—and by keeping an eye on the circuit performances. A first prerequisite for optimization is a fully automated verification setup (typically based on simulations) with specs. And the second prerequisite is the definition of the parameters to be optimized, e.g., their ranges and step sizes.

In principle, also a circuit <u>topology</u> optimization is possible, but practically not for complex circuits. One reason is that the mathematical problems become much higher than for pure parameter optimization, so that runtimes would become impractical for real-world systems. The runtime is typically <u>highly</u> dominated by circuit simulations, just to obtain the *performance* for each of the tweaked circuits, whereas the <u>internal</u> optimizer runtime is much lower (like < 1 second). A successful optimization of 10 parameters usually requires the simulation of roughly 500 design points, and this means optimization can be 500 times more compute-intensive than the pure verification!

Obviously, a designer has the advantage of having experience, but any pure manual process of changing parameters, re-running the simulations, deciding which solution is best is time-consuming too, so automated optimizers can remove the burden of doing that stupid task; with an optimizer the designer has just to decide <u>which</u> parameters to tweak and in which range, plus giving criteria on circuit performance.

The availability of optimization does not mean that anybody will become immediately a good analog circuit designer, because still the topology choice, the starting values, the circuit goal setting, the testbench setup, etc., based on classical skills like experience and circuit understanding.

So one may wonder: Can optimization beat manual design? Indeed, there are optimization sweet spots and also difficult optimization problems. Mathematically or circuit-wise you can construct cases to "prove" almost anything, like "current optimizers are far too bad for difficult problems" or "optimization is $100\times$ faster." So often the question is different: Can I apply optimization in my current design case to my advantage? *Often* the answer is yes, so when it might be "no"?

The main indicators for optimization difficulties are as follows:

- The specifications are not complete! This can cause the optimizer to provide a design that is not suited.
- A starting point is hard to provide, maybe it is even impossible to make such circuit in given technology, or there are even physical reasons against it (e.g., you cannot achieve simultaneous power match on a two-port with stability factor below unity).
- The problem is too complex, too nonlinear, and numerically too noisy, respectively. If simulation accuracy is only 0.5% due to numerical noise, then it is difficult to obtain accurate gradient information for an optimizer (Table 8.3).

To understand why these points are critical, the designer should understand how optimizers work. Actually different kinds of optimizers are available, and there is no single best algorithm.

Table 8.3   Inputs and outputs for an optimization

What	Why	Comment
Inputs		
Models	Prerequisite for verification	Usually from foundry
Testbench and Conditions	Prerequisite for verification	Usually already done early
Performance evaluation	Otherwise the optimizer would not know about the circuit	Usually already done for automating MC and corner runs
Specifications	To set optimization targets	Do not overlook one! Often weights can be applied
Optimization options	To select and control optimizer	Might be optimizer-specific
Parameter setup	Otherwise the optimizer would not know which parameters to tweak	e.g., range of parameters, step size
Starting point	Needed for all local optimizers	Usually the schematic values
Outputs		
Optimized circuit parameters	This is the major output	Backannotate to your schematic to make the improvement permanent
Optimizer log file	To check in case of problems	Giving, e.g., a history, reason for termination
Sensitivity results	Usually a by-product	Often as table and with plots
Plots, e.g., on parameters and performances	To check optimization progress and for understanding	Often not all waves will be saved to reduce amount of data
Performance models	Support for quick manual tweaks	Are often created internally anyway, but not always the designer has access to them.

The mathematical *input* for an optimizer is typically only the set of parameters and the so-called goal function, so it does <u>not</u> matter much if the performance results are obtained from one testbench or multiple ones, from a schematic simulation or if it already includes layout parasitics. The only difference is that post-layout simulations simply take longer due to larger netlist content. The opposite way would be running no SPICE-like circuit simulations at all, and using e.g., symbolic, hand or tool-generated, performance formulas instead (as our RealTime apps do). This pushes time-consuming circuit simulations completely *out of the sizing loop*, and due to the large achievable speed-up it was very popular in the past. Nowadays, "SPICE-in-the-loop" is usually preferred because symbolic solutions take much time for preparation, are less flexible and are often not accurate enough, anymore.

8.2.2 Classifications of Optimization

There are different kinds of optimization algorithms for the task of finding the optimum. Such optimum can be either at the ends of an interval, or somewhere in the middle. Two major categories are global and local optimization (see Figure 8.5). Global optimizers *can* solve problems with multiple local optima and can work even if the starting point is bad. Local optimizers require usually a good enough starting point (like point 2 in Figure 8.5) from manual design techniques, but are often much faster. For a local optimizer, we usually want *monotonuous* improvements; i.e., the next best-accepted point should be really better than the previous one. However, for a global optimizer it could be required to accept a nonmonotonic behavior to get out of local minima (see Figure 8.5 for starting at point 1).

Note: For histograms, the *mode* is the value that appears *most often* in a set of data, so it is the point of maximum probability density. Many distributions (like the uniform or normal one) have only *one* mode and are called unimodal, but mixed distributions are often *multimodal*. We can regard functions with multiple minimums as multimodal too.

The simplest type of optimizer can typically already deal with multiple parameters but treats the performances in a single real-valued goal function. Usually multiple competing goals exist like maximize (S_{21}) till $S_{21} > 10$, maximize (BW) till BW >1 GHz, minimize (NF) to the possible optimum. So the most integrated design solutions combine the different individual goals into a single-goal function like $f = w_1(S - 21) - w_2(BW) + w_3(NF)$. Usually this is even done automatically and the user just has to set the weights.

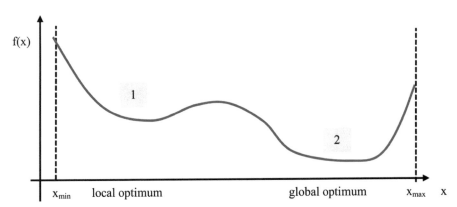

Figure 8.5 Local versus global minimum search.

Optimizers able to treat the individual performances directly are also available, but require even more runtime.

A further classification can be made by looking to the *type* of variables. Usually we are interested in optimization of real-valued parameters, like transistor width or capacitor value. However, in several cases also integer parameters are of interest (like for the number of transistors in parallel, e.g., in a bandgap circuit). Interestingly, integer or even mixed optimizations can be very hard. For instance, in the case of real-valued parameters, the gradient can be often calculated and used for deciding in quite a reliable way in which search direction the optimization should progress. However, integer problems tend to be much more nonlinear and no true gradient information is available. Mainly for mildly nonlinear problems existing methods can be quite easily extended [Pehl] by using finite differences also for integers.

Here is a simple but difficult integer optimization problem:

Imagine a gear system for which you want to obtain certain transfer ratio, but of course the number of teeth for each wheel must be an integer. A well-known example is this for finding a certain ratio:

$$\text{Min:} \ f = (1/6.931 - x_1 x_2/(x_3 x_4))^2 \ \text{with} \ 12 < x_i < 60 \qquad (8.1)$$
$$\text{A good set is } 19, 16, 43, 49 \text{ giving } f = 2.7 \cdot 10^{-12}$$

This optimization problem has no ideal optimum ($f = 0$) and it has multiple local optima! For fix value of the other x and sweeping only one x, f is quite smooth, but sweeping in other directions shows that f is having significant steps. Therefore, this kind of function needs efficient global optimization algorithms to be optimized. In principle, this optimization can be done in brute-force style by simple evaluating all $48^4 = 5308416$ combinations, but this is not efficient.

What we have presented so far is called unconstrained optimization. Often real problems come with further restrictions to define a solution; e.g., all widths and lengths need to be positive (leading to inequality constraints) or for some parameters you may want to force a certain ratio (leading to equality constraints). The solution of such constrained optimization problems is mathematically a bit more difficult, but in circuit design this often causes only minor headache anyway; e.g., often both kinds of constraint can be often removed by a simple transformation or by substitution of variables; e.g., by setting $R_1(x_1) = R_{\min} + \Delta R \cdot (1 + \tanh(x_1))/2$, you can force physical values in a certain range ($R_{\min}, R_{\min} + \Delta R$).

For these reasons, we focus on unconstraint optimization in this book. On the other hand, it shows well that also the user has some responsibilities to setup the problem to let it fit well to the available algorithms!

From the user's perspective, they are also further kinds of optimizations. As mentioned, some designs are so difficult that you may first need a nominal optimization, then, e.g., a corner optimization and last a statistical optimization. Luckily, all optimizers usually allow, in principle, the treatment of such different "optimization scenarios," like optimization on a single corner or at multiple corners, or yield optimizations including Monte Carlo runs (Table 8.4).

Table 8.4 Different types of optimization problems

No.	Type of Problem	Example	Comment
1	Real variables, single-goal function	Many (BFGS, conjugate gradient, but also nongradient methods like Nelder-Mead)	Simplest type, but often used
2	Real variables, single-goal function and constraints	Optimizers based on quadratic programming	You may use #1 and transformations or "penalty" functions which give an increase in goal function when constraints gets violated
3	Mixed real-integer variables	Few, e.g., using a gradient optimizer and approximate gradients by finite differences also for the integer parameters	Optimizers which are natively acting as global optimizer can often be used directly.
4	Multiple goal functions	Pareto optimization (e.g., weighting based or constraint based)	You may use #1 and adjust the weights to get different solutions of the Pareto front.
5	Statistical variables included	Yield optimization, e.g., using WCDs	If combined with other problems, this is the most complex task
6	Surrogate-based optimization (using meta model)	Often applied in difficult cases, like yield optimization or 3D-FEM designs [Bo Liu2011].	If the goal function evaluation is very noisy or very time-consuming it makes sense to create a surrogate model, and to optimize on this instead.

Nowadays, several commercial design cockpits features all these and also other advanced techniques like adaptive-set multi-corner optimization with built-in worst-case corner selection (sizing over adaptive corner sets). We will come back to such complex scenarios in Chapter 9.

In a corner or MC analysis we perform a design (or variable) space exploration (DSE), this has similarities, but also differences to design automation (DO); Table 8.5 gives an overview and Figure 8.6 give pictures for a two-dimensional example.

Note: If we perform a DSE, we can create (approximated) performance models $f(x)$. And instead of running further circuit simulations in the optimization loop, we could just use such the much simpler models for goal function evaluations. This is called surrogate-based optimization (SBO), and if you have already performed a design analysis, it is an elegant way to speed-up optimizations! Sometimes this method is also used to make optimization feasible (e.g., using a gradient optimizer although the original simulations are noisy, or performing a global optimization with a split of surrogate and direct optimization). For instance, in [Bo Liu2011] surrogate-based optimization is compared to direct techniques for on-chip inductor and transformer design. Both designs have only four parameters, but the EM simulations takes hours to run. Due to SBO roughly 75% of the optimization time could be saved.

Table 8.5   Design optimization versus exploration

Design Optimization (DO)	Design Space Exploration (DSE)
Converging-iterative process; aims for the optimum design.	Diverging, sometimes iterative process. Aims for characterizing the design, e.g., for modeling or worst-case finding.
Runtime for well-behaved problems typically quadratic on the number of variables.	Runtime could rise exponentially with the number of variables (using full-factorial method).
Optimization has two distinct parts; formulate the problem (e.g., as goal function) and converge to the solution.	Once we know the design space, a better design solution can then be found e.g., through surrogate-based optimization.
Depends on a well-posed problem formulation (starting point, tolerances, etc).	We do not need a well formulated problem; only the performance evaluation is required.
Moderate effort, e.g., quadratic	Larger effort, like quadratic to exponential
Can be done stand-alone, but picking up results from DSE can improve convergence and speed significantly	Can be done standalone. Simpler sensitivity results might be also derived from an optimization run (but hardly more, because DO does not aim for space filling).

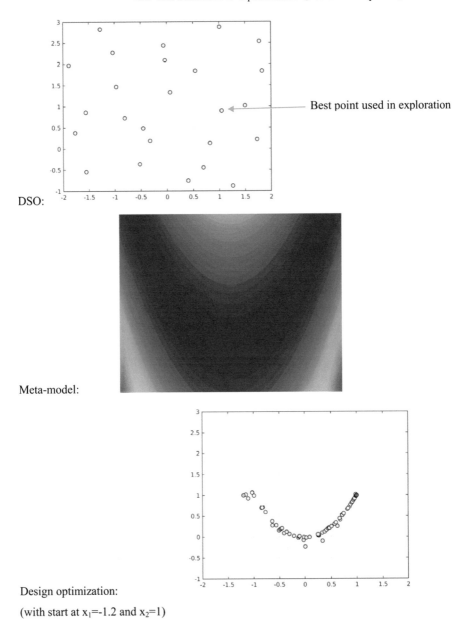

DSO:

Meta-model:

Design optimization:

(with start at x_1=-1.2 and x_2=1)

Figure 8.6 Typical samples setting for an optimization versus for design space evaluation, and a meta-model contour plot derived from the space evaluation.

Global optimization has generally some more elements of space *exploration*, whereas local optimizers try to find the shortest way to the optimum, ignoring huge parts of the variable space. So in the bottom picture of Figure 8.6 you see the points a typical local optimization algorithm would take. In SBO we would sample the space (e.g., using LDS), create a *meta-model*, and then we could e.g., run a first optimizer from the best-suited LDS point using now the meta-model, i.e., the model would gives us a kind of *emulator* for the true design. And optionally we can double check or refine our solution (and the meta-model) further with direct simulations with a second optimizer.

8.3 How Successful Optimizers Work

A good optimizer should be robust <u>and</u> fast, because the result should not depend much on the starting point and circuit simulations are very time consuming. Why should we aim for "robustness"? Simply because real-world optimization goal functions may not behave like simple quadratic function with a unique optimum! You may have to deal with strong nonlinearities (like exponential functions) or even discontinuities (like absolute value function, minimum function, and reciprocal function) or multiple optimums (like $x^3 - x$) or very flat optimum (like x^4)!

Unfortunately, making an optimizer robust comes usually with some speed penalty; e.g., to avoid trapping into a local minimum, the optimizer has to apply also larger parameter steps, but such bigger steps have usually a lower chance of being successful than well-directed steps based on gradient information.

Beside the internal optimization algorithm structure, *also* the optimization setup by the designer has an important impact on optimization speed! And when talking about speed, obviously one important question is: Is there a theoretical limit on optimization speed?

One key point is the number of parameters, not so much the number of steps for each parameter, because good optimizers have an adaptive step control algorithm. For efficient optimization, it is best to minimize the number of variable parameters, e.g., by exploiting matchings, like W(N1) = W(N2).

Often user gets direct support in the schematic entry tool for definition of 1:1 or 1:x matching (Figure 8.7). Such assistants can also help on managing different optimization parameter sets, to do comparisons, e.g., between current optimization setup and last-saved schematic.

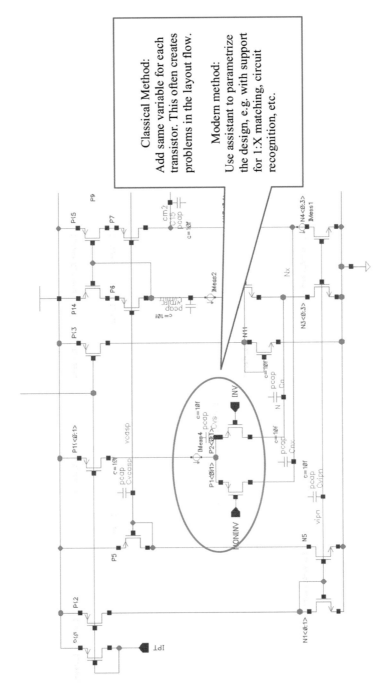

Classical Method:
Add same variable for each transistor. This often creates problems in the layout flow.

Modern method:
Use assistant to parametrize the design, e.g. with support for 1:X matching, circuit recognition, etc.

Figure 8.7 Parameter assistant for defining matching on selected parameters for two transistors.

It is best to focus on the major parameters with largest impact on performance, but also make sure not to optimize too few parameters, because this would limit the optimizer too much to obtain a good solution. The performance goals should vary smoothly regarding the parameters, so it is often best to optimize on the minimum average or rms and not on the minimum (or maximum) of a certain performance, because the minimum function is like the absolute value function, a nondifferentiable function. If the goal is to hit a single value like $V_{\text{out}} = 1$V, then terms like $(V_{\text{out}} - V_{\text{outwanted}})^2$ would be typically used.

People often think that their problem is something *special*. Many things in optimization can be indeed special, but on the other hand, internally most optimizers work in quite similar way (Figure 8.8).

Probably, our starting point x_0 is usually not the optimum x_{opt}, but a $\Delta x = x_{\text{opt}} - x_0$ exists, so the key is how to determine this Δx. Usually we want to achieve at least a certain reduction in f.

8.3.1 Newton and Quasi-Newton

Taking a brute-force grid-based strategy to find a minimum of a n-dimensional goal function $f(x_1, x_2, \ldots, x_n)$, the number of points would be in the order of 10^n (e.g., for taking 10 values for each x_i). For a small optimization with $n = 5$, this would mean already 100,000 points or more, which is by far too much for most nontrivial circuits.

Set parameters $x_D = (x_i)$, e.g., acc. to schematic values

Consider to setup parameter ranges too

Get back a goal function f back
(for circuit design, this requires simulations in the background
and an automated result evaluation and specs)

Set the next parameter set $x_{i+1} = x_i + \Delta x$
using some strategy defining a suited Δx.

Figure 8.8   General flow chart for an optimizer.

A better way is to remember basic math and looking for the point where the gradient $g = df/dx$ becomes 0. That leads to a set of equations to be solved iteratively and the order of points needed is often roughly n^2 which is obviously <u>much</u> lower than 10^n.

So let us follow this powerful approach, and let us approximate the function f with a Taylor series, like $f(x + \Delta x) = f_o + g\Delta x$. That speeds up well, because it allows optimization along a search path with direction s with biggest (local) improvement (Gradient optimizers) by setting $s = -g$.

Obviously using a second-order approximation, $f(x + \Delta x) = f_o + g\Delta x + \frac{1}{2}\Delta x H\Delta x$ is even better, for two reasons: First, the second-order approximation is valid over a wider area, then the first-order one. Secondly, it is easy to <u>directly</u> calculate the minimum of a second-order polynomial (Newton's method); by using the second derivative, we can obtain not only a good search direction, but also the search step *length*.

The basic idea of Newton's method is to (iteratively) find the point $x + \Delta x$ with $g = 0$ and approximating the function f by its second-order Taylor series. In one dimension, a parabola can be defined by 3 points or by one point + first + second order derivative. Newton's method is doing the later in all n dimensions; in this case, the second derivative is the Hessian matrix $\mathbf{H} = d^2 f / dx_{ij}^2$:

Newton step:

$$g(x + \Delta x) = g(x) + H(x)\Delta x = 0$$
$$\Rightarrow \Delta x = -gH^{-1} \Rightarrow x_{opt} = x_S - gH^{-1} \qquad (8.2)$$

Example: assume $f = 2x_1^2 - 2x_1x_2 + x_2^2 + x_3^2$—the optimizer does not know this formula!

Start point: all $x_{iS} = 1$

Basic math gives:

$$g = (4x_1 - 2x_2, -2x_1 + 2x_2, 2x_3), \text{ so } g(x_S) = (2, 0, 2) \qquad (8.3)$$

and a constant Hessian matrix

$$H = \begin{pmatrix} 4 & -2 & 0 \\ -2 & 2 & 0 \\ 0 & 0 & 2 \end{pmatrix}$$

$\Rightarrow -gH^{-1} = (-1, -1, -1) \Rightarrow x_{opt} = 0$—which is only easy to see if you know the formula!

As we can see, the gradient g is a vector with dimension n and it collects the n first (partial) derivate $\delta f/\delta x_{Di}$). H is just a matrix collecting the 2nd (partial) derivatives and is of dimension n × n.

Note: We get the <u>exact</u> result in just one <u>single</u> step (once we know g and $\mathbf{H}$)! For pure quadratic functions, this is possible for any <u>arbitrary</u> starting point, which means that the Newton method has a <u>proven</u> quadratic convergence: All kinds of problems with quadratic behavior can be solved with a simulation effort of approximately n^2 points! For more nonlinear functions, more steps are needed, but often not that many (about 5 to 10 iterations).

So for near-quadratic examples the Newton algorithm is an extremely powerful algorithm. Interestingly, the gradient component on x_2 is zero in our example, but the Newton step is still correcting this variable as well, as required—although a simple linear sensitivity analysis would not indicate this! This is one reason why Newton's method often highly outperforms simpler optimization algorithms, like searching along the steepest descent direction $s = -g$ (instead of $-g\mathbf{H}^{-1}$). Steepest gradient is often taking quite short steps and needs therefore many more iterations (compared to Figure 8.9). Often steepest gradient is not much faster than even simpler algorithms using no derivatives at all (e.g., just bracketing the optimum).

Remark: As well-known g can be approximated with finite differences $\Delta f/\Delta x$. That is also possible for the Hessian matrix: $\mathbf{H} = f_{xy} = \delta^2 f/\delta x \mathrm{d}y \approx \Delta^2 f/\Delta x \Delta y$. Gradient calculation has an effort of approximately n simulations, whereas $\mathbf{H}$ calculation requires roughly n^2 simulations. The exact value depends on the method, e.g., if you use single-sided gradients $g_{SSG} = (f(x + \Delta x) - f(x))/\Delta x$ or the more accurate double-sided gradient approximation $g_{DSG} = (f(x + \Delta x) - f(x - \Delta x))/2\Delta x$.

As mentioned, usually the simulation part to obtain $f(x)$ is the most time consuming, and the internal optimizer calculations take typically less than seconds; e.g., even the matrix inversion to get $\mathbf{H}^{-1}$ from $\mathbf{H}$ takes not much time at all, because $\mathbf{H}$ is much smaller than the Jacobian matrix used within circuit simulations.

To some degree, most advanced optimization algorithms use the Newton idea. A very important class is so-called quasi-Newton optimizer (e.g., the Davidon-Fletcher-Powell DFP and the Broydon-Fletcher-Goldfarb-Shanno BFGS methods, both named according to names of their inventors). They avoid the need to calculate the Hessian matrix upfront, before being able to

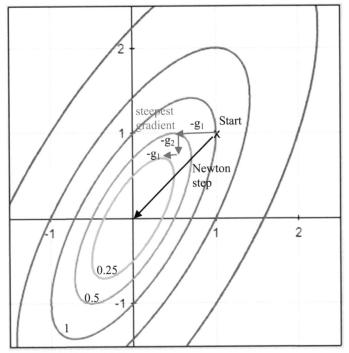

Figure 8.9   Contour plot showing the optimization of $2x_1^2 - 2x_1x_2 + x_2^2$.

make to the parameter step in the direction of the minimum. Instead, they perform a minimum search along the search direction(s) and compose the inverse Hessian matrix iteratively (actually they do this directly without really doing a matrix inversion). This looks like a disadvantage because of course the linear minimum searches also need some circuit simulations to decide on step length, but there are also some advantages. For instance, the Newton method may fail in cases where using $s = -g$ would at least lead to an improved design! Calculating g and $\mathbf{H}$ from the starting point then doing a potentially big step is a riskier strategy, more impacted by higher-order derivatives than just making smaller steps (Figure 8.10).

8.3.2 Parameter Setup Hints and Stopping Criteria

The user has to define which parameters should be tweaked for the circuit optimization. Usually it is a subset of all possible variables x_D. In principle, we can even optimize on range parameters x_R; then, we would use the optimizer

Set parameters $x_D = (x_i)$, set start matrix (e.g., $\mathbf{iH} = \mathbf{I}$)

> Calculate goal function f
>
> Calculate gradient g (usually by finite differences)
>
> Update matrix $\mathbf{iH}$ using g result
>
> Calculate search direction s from iH and g
>
> Do a linear step-size search $x_{i+1} = x_i + ts$

Check if $|g|$ is small enough to stop

Apply other stopping criteria

Figure 8.10 Flow chart for a gradient-based, quasi-Newton optimizer like BFGS.

not for circuit sizing but for a kind of calibration; e.g., we may find the best bias current or the values for digital trim bus to hit a certain performance (like filter cut-off frequencies).

Usually, designers have a good gut feeling about which parameters should be optimized. In addition, you can run a sensitivity analysis, e.g., to get the mismatch contributions of each instance and decide in a parameter screening step for which parameters an optimization makes sense.

For instance, you may want that all Ws, Ls, and Rs are positive and in a meaningful range suited for layout. Some optimizers require to setup only parameter ranges, but others may also need a step size. The need to stay on a grid might be a requirement also for other reasons like layout restrictions. Usually having a grid decreases the optimizer performance a bit—especially for gradient-based optimizers. For too large grid steps Δx, the optimizer may tend to stop *too early*! That is an important to know: typically a user would think that if we use, e.g., $\Delta x = 0.01$ grid, then the optimizer finds the best x within 10^{-2} accuracy—but, that is usually too optimistic.

Sometimes designers claim that optimization results are not well suited for layout reasons. One problem is often that some goals are simply not set up, such as the block area; another reason could be the style of parameterization. For transistors, you typically have the option to tweak either the finger width or the finger count, or the m-factor (plus gate length). However, for applications at moderate frequencies there is little difference between two transistors having, e.g., 8 fingers, $wf = 5$ μm and m = 2 or 2 fingers, $wf = 2.5$ μm and m = 16. If you only need a local optimization and expect that your starting value is close to the optimum, it is better to set up only wf as real-valued parameter to be optimized. This makes the optimization much faster, and if wf becomes too large you can still consider to select, e.g., a higher m-factor after the optimization.

One general question is also whether we should optimize on all parameters just present in the design or only on a subset of really important ones. The latter is faster, but a bit more risky. A gradient optimizer would simply not shift parameters having no impact on performance, so regarding the optimization result there are no disadvantages, only on speed. Internally, clever optimizers could focus first on the most important variables, which would speed-up the optimization on many variables, with quite low internal effort. So the user would not gain much by treating the optimization variables set x_D adaptively.

Another part for an optimization setup is to define when to stop. Usually you can select and combine different criteria like:

- Stop when all performances are in spec;
- Stop if a time limit is reached, like 60 minutes;
- Stop after a certain number of simulated design points, like 1000; and
- Stop when there is no improvement in, e.g., the 200 last points.

In addition, most optimizers have also internal stopping criteria, e.g., if the gradient or the parameter step sizes become too small. It is best to look for the optimizer log file to get more information!

Note: Stopping when all performances are in spec looks meaningful, but quite often it makes sense to continue a bit, e.g., to make the block area or power consumption even lower.

Here the optimizer is doing a gradient calculation (Figure 8.11).

8.3.3 Hill Climbing Techniques and Global Optimization

We can also get inspiration for solving optimization problems from other areas than a quadratic function; e.g., we can look to the problem of finding the highest (or lowest) point (x, y) in a landscape. These techniques are called "hill climbing," and one quick approach for finding the highest point could be moving just in the direction of steepest gradient, e.g., till we get no improvement anymore. At this point, we can recalculate the gradient and repeat our step length search.

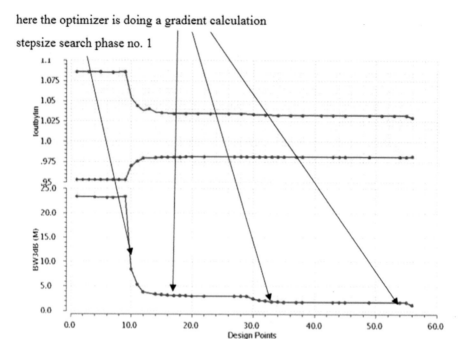

here the optimizer is doing a gradient calculation

stepsize search phase no. 1

Figure 8.11   Typical optimization-run with BFGS optimizer (3 performances vs. points).

A less efficient but even simpler method would be just walking in x-direction to the highest point, then changing the direction to y and repeating the local search again and again till the height differences become very small, indicating convergence.

Not only water would take a way along the gradient directions, but also other mechanisms in nature work like an optimizer, e.g., crystal building (here energy is minimized) or evolution. Some of these are also suited to address the global optimization problem directly. For instance, simulated annealing works fine on most combinatorial optimization problems like the traveling salesman problem (check out our RealTime app for matching networks). It has no real internal memory or model building, but as it allows up-hill steps randomly it can solve problems with local minima. The runtime is typically in the order of $(An + Bn^2) \cdot \log(n)$, and usually still much better than a brute-force approach (often requiring n! simulations).

An alternative way to form a global optimizer is using a local optimizer (e.g., Newton's method or steepest gradient) with <u>multiple</u> starting points [Shutao Li].

Table 8.6 Some popular global optimization methods

Type	Algorithm	Comment
Local optimizer on multiple start points	Check sensitivity and create a starting grid at least for the major parameters, then run, e.g., BFGS	Goal function should be smooth enough to let local optimizers work efficiently
Genetic optimizers	Treat the parameters as genes. Create a start generation and emulate evolution (mutation, inheritance)	One tricky part is setting the number of childs in each generation. Using too many or having too much mutation slows down the optimization.
Simulated annealing	Works like hill-climbing but with some randomness on top to allow also up-hill steps to get rid of local minimums. The closer you are to the final solution, the less randomness.	The tricky part is to control the randomness. If it is too large you optimization will be slow. If it is too small you may still stop at a local optimum.

One interesting point is to compare all these different methods like simple coordinate search, steepest gradient, Newton's parabola method, and a genetic optimizer (using mutation, crossover and selection to create new generations). Actually many such benchmarks have been done.

Note: It seems that global optimizers (Table 8.6) are superceeding local optimizers, but beside longer runtime in smooth, near-quadratic problems there are also few negative side effects. For instance, if we parameterize elements with almost zero sensitivity, then a local optimizer would usually stick to your (hopefully meaningful set) of start parameters, but a global optimizer *may* shift them a lot without having a benefit. In a similar situation you would be, if your specification setup would not be fully complete, such as a bandgap to be optimized without area limit. A global optimizer may adjust all parameters far away to get just some improvements—but the achieved set of parameters might be not satisfying (like far to large area).

Is There a Best Optimizer in General? For quadratic problems the Newton or quasi-Newton methods are almost perfect; no significant improvements are really possible. However, an optimizer that always works as fast as possible and always finds the global optimum is much harder to construct!

In principle, we would require that a good optimizer would <u>use</u> all information it has to decide on the next steps. However, there is also a problem if doing so, because circuit problems can be so nonlinear that it makes even sense to ignore from time-to-time too old information.

All optimization problems are to some degree similar, but can be also be much different in some details. Regarding local optimizers and smooth problems (continuous parameters and near-quadratic relations between goals and parameters), the BFGS algorithm as derivative of the Newton algorithm is for most problems a very good choice. It is well-proven, fast and reliable. For such problems, the number of points is in the order of n^2, i.e., if we double the number of parameters we need usually $4\times$ more points. If the problems are more nonlinear, e.g., because the starting point is not a good one, then the number of points will increase, e.g., to typically $10n^2$. So for a typical nontrivial optimization task with 10 variables, we can expect a solution in 1000 points.

If we would do a plain n-dimensional sweep instead (brute-force optimization) we would require, e.g., 10^{10} points! In both cases, getting a solution would not mean that the circuit is in spec, it only means that we can be quite sure that the optimizers reach a (local) minimum. If the problem has few multiple minima, it is best to use multiple different starting points for the BFGS run. If the problem has indeed many multiple minima, it is best to directly use a *global* optimizer.

8.3.4 Do Real-World Circuit Designs Have Local Minima?

The short answer is "'Yes," but it really depends on the circuit problem and starting point if a local minimum can disturb the optimization progress or not. Other problems such as small gradients, parameters at the edge of their range (like $L = L_{\min}$), nonsmooth/nonparabolic goal functions (such as minimum or maximum), or parameter redundancy can cause even more difficulties.

Example #1: In a fourth-order filter composed of two second-order filters in series, there can be easily two solutions giving *exactly* the same gain vs. frequency; which one is found may depend on chance or starting point! The noise figure, input impedance, or nonlinear behavior might be slightly different, so that one solution could be (slightly) better than the other one—looking at the full set of specs. This way one optimum becomes only local, but it can be still preferably found by a local optimizer unfortunately, if the starting point is bad. For a fourth-order LC ladder filter usually a single solution is optimum already from pure gain vs. f specs, but if, e.g., a sixth-order filter should be optimized to give a gain possible to achieve with already fourth order, then it could happen that one L and C values are "optimized" to zero— an example of parameter redundancy, but it could also happen that only one parameter becomes zero and maybe two inductors lie in series—and only the sum of both becomes well defined.

Example #2: An ideal bandgap design depends mainly on area and resistor ratios, so the optimum W/L might come out of the optimization due to restrictions on supply voltage, whereas W·L can be at any value. If mismatch comes into the game, then certainly W·L matters, and a certain minimum area is required for achieving a low output voltage tolerance. For many other real circuit problems, such situation will not really happen, because typical constraints (like parameter ranges), more specs, etc., restrict the design space more and more, so leading more and more to the desired global solution (Figure 8.12).

Example #3: If the circuit uses on-chip inductors and if we introduce the number of turns as optimization parameter, then the quasi-integer nature can introduce local minima too. It is better to optimize on the inductance L directly (e.g., assuming a fix quality factor) and later looking for the physical implementation (number of turns, diameter, width, metal layers, etc., fitting to optimized L and assumed Q). Similar effects can occur in transistors, e.g., if a model binning boarder is crossed.

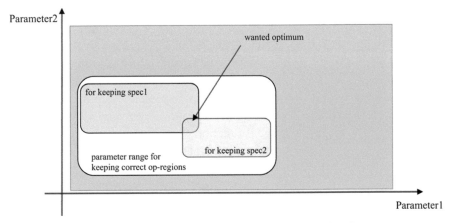

Figure 8.12 Typical behavior of optimization goal functions.

Example #4: A buffer amplifier testbench with a pseudi-random bit-generator (PRBS) has been created to analyze jitter. The main source of jitter was inter symbol interference (ISI), and this was dominated by filter capacitances, bondwire inductance, etc. So a native question is how to minimize the jitter, e.g., by tweaking the capacitances, inductances, etc. One result was no surprise: The capacitance has to be minimized. However, interestingly, a minimum inductance was not best, the optimum was somewhere in the middle of a realistic range for L. So we run different optimizers, and find a solution, but when checking this solution with a big dense sweep, we found that actually our optimizers stopped at a local minimum! There was a second, slightly better jitter minimum, just a bit more away from the starting point. Interestingly also the global optimizer stopped to early, because we just set the number of point limit to rigid. Inspecting now the testbench in more detail we found set our PRBS sequence was quite (too) short (just to save runtime), with more cycles the two optimums smoothed out more and more. Also our initial optimization was done a nominal conditions only (just for speed reasons, and because C and L are not much corner dependant). Going for a true corner optimization, again the two minimums smoothed out. So in conclusion, making things too simple, being too greedy, can cause problems, which luckily disappear when making the problem more realistic. Also when adding futher optimization criteria (like power, noise, area, etc.), we can expect that some local optimums are simply giving too bad performance on these other criteria (leading to a somewhat larger over-all goal function).

8.3.5 Advanced Techniques Beyond (Quasi-)Newton

We mentioned the different kinds of optimizers, and actually for special problems more tailored optimizers could work more efficient. However, let us first inspect if it possible to make a faster optimizer than e.g., BFGS or Newton for pure quadratic problems.

These types of optimizers treat the goal function as a single real-valued function. Some optimizers are designed to optimize on sum of squares (rms error), so if the problem (the goal function f) is of that type we can create faster algorithms by exploiting this special mathematical structure even better. For instance, the Gauss–Newton GN optimizer can optimize the famous Rosenbrock function (also a sum of squares) very efficiently, faster than any other optimizer even. Pure GN cannot guarantee a strictly monotonic minimization of the goal function—so it is quite risky to use it. Therefore using a damped version with stepsize control is standard. The Levenberg-Marquard optimizer is also a kind of damped GN, and for least squares problems it works sometimes faster than BFGS. The pity is that we would loose flexibility in defining the goal function with such special optimizers. However, for model parameter fitting applications Levenberg-Marquard (LM) is actually a kind of standard optimizer, because e.g., for fitting of transistor or package parameters we natively have to minimize the rms error between model and measurement results.

Note: You may ask what is so special in the Rosenbrock function, that is is so much harder to optimize than a similar looking 2D quadratic function. Indeed the Rosenbrock function looks like a *steep* narrow valley with small down-gradient. At the River webpage you can download a spreadsheet example rosenbrock.xls a contour plot which is visualizing this behavior nicely. The narrow valley dictates the downhill way to go, and because the gradient is small, only very small deviations are allowed, so that over-all we need many small steps.

A further aspect in advanced optimization is large-scale optimization, indeed if the number of parameters is large (like $\gg 100$), then classical quasi-Newton optimizers are not best-suited. This is because often large optimization problems tend to be sparse, e.g., the Hessian matrix contains many entries close to zero, and unfortunately BFGS is *not* preserving the sparsity. For circuit design this is luckily seldom a problem.

A more important problem related to optimization is how we treat constraints like $x_{D1} = L(N1) \geq L_{min} = 180$ nm. As mentioned, for such simple linear inequality constraints we can apply (even automatically) a parameter transformation, and we still end up in a fast and reliable optimization. For non-linear relations this method becomes sometimes more difficult, and actually we can ask is optimization always *needed*, and is it really always "minimization"? Indeed, going back to Chapter 2.1 on transistor sizing, most design goals are defined as constraints (not as minimization or maximization), e.g., $f_T \geq f_{Tgoal}$ and $V_{noise} \leq V_{noiseTarget}$. Remember our example calculation for offset voltage to obtain width and length for a given W/L ratio. In this found that for a given technology a certain W and L (or larger values) can guarantee a certain maximum offset standard standard deviation. If the offset spec is relaxed we can maybe even use minimum-L transistors, so that the constraint on L becomes <u>active</u>. For current mirrors often the opposite is the case, because we anyway need long transistor for good current matching, high output impedance, etc. So here the minimum-L constraint would be <u>inactive</u>.

Note, that functions like $f = x_1^3 + x_2$ would have *no* global optimum for the *unconstrained* case, but adding constraints like $g_1: x_1 \leq x_2$ and $g_2: x_2 \leq 1$ can define a problem with a clearly defined global optimum. Sometimes already the constraints restrict the <u>feasible region</u> so much that they are more important than the minimization part.

In general performances, or design measures like area, could appear as objective or in constraints. *Mathematically* constraints would be *hard* design targets, whereas the minimization target would be a *soft* target. It may look, that often designers need no full optimization. In addition, optimizers, dedicately designed for *constrained* minimization problems (see Table 8.2), can work more efficient than e.g., BFGS. However, if we *only* look to constraints, then the price we have to pay is that we would get a *feasible* design solution, but maybe not the *optimum* solution. For instance, maybe a PSRR of 120 dB is possible, but the constraint (just the spec) was only 80 dB. Often this is acceptable, but if a better performance is indeed possible, we could obviously improve our chip design further, and maybe we could relax the spec on other blocks (which could e.g., lead to small chip area or lower power consumption) or on other performances (like CMRR). *Hard* and *soft* are indeed quite fuzzy terms, e.g., of course, the efficiency is a very important target in almost all high-power designs, and also in extremely low-power systems, so from the customer side $\eta \geq 90\%$ would be a *hard* spec. However, as designer you often want to over-fullfil such critical targets to some degree, and truly optimizing η. And interestingly, this would require mathematically to treat η as a *soft* spec!

All this flexibility is possible with constrained optimization, which works a bit like a combination of e.g., quasi-Newton *and* our hand calculation for constraints! A very good circuit example can be found at [Allen2008] for matching network design. Mathematically a classical general approach is using so-called Langrange multipliers λ, one for each constraint. The calculation effort is not only depending on the number of variables n, but to n plus the number of constraints n_c.

An alternative method is to "transform" a constrained optimization problem into an unconstrained problem (and using a well-known standard *unconstrained* optimizer). For instance, we can just make the goal function f very large *if* we *violate* a constraint. This technique is called *penalty function* method. Unfortunately it works sometimes not so well: If we increase f too much, the problem is not well-behaved, e.g., having highly different eigenvalues (see Section 8.5.1), or being not close-to-quadratic anymore; and if we increase f only a bit, we may still have some constraint violations.

To avoid this weights are typically introduced, and only in simple cases it is possible to set these penalty weights in a meaningful way upfront. To avoid this problem an adaptive weighting procedure can be implemented. If e.g., one constraint is still violated obviously its weight needs to be increased. Unfortunately such weight change will change the goal function, which could slow down the optimization loop, e.g., because the Hessian matrix would also change, more than in comparable truly uncontrained optimization cases.

So if possible it makes much sense to *reduce* the number of constraints, or making sure that they are anyway fulfilled, e.g., to force $g_1\colon x_1^2 + x_2^2 = 1$ and $g_2\colon x_2 \geq 0$, we can solve this equation, and replace everywhere x_2 by sqrt$(1 - x_1^2)$. This way both the number of variables would be reduced, and also the number of constraints, so the optimization would become much easier! For an inequality constraint like $x_1^2 + x_2^2 \leq 1$ one clever way could be using $x_1 = r \cdot \sin(\varphi)$ and $x_2 = r \cdot \cos(j)$, giving $r \cdot \sin^2(\varphi) + r \cdot \cos^2(\varphi) = r \leq 1$. And with $r = \frac{1}{2} + \frac{1}{2}\cos(\rho)$ we would have translated the complete problem into a simpler *unconstrained* optimization task.

These examples also show again that the problem definition and testbench design has quite a <u>significant</u> impact on how good any optimizer <u>could</u> work, i.e., obtaining a really feasible solution in acceptable time. With a bad testbench you sometimes "debug" effects, which simply do not exist. Two classical examples are bandgap start-up circuit or is testing a Schmitt-trigger only in pure DC simulations. Some topologies have multiple loops and can give you glitches in DC sweeps, although in a transient (and reality) analysis everything might be clean.

Another aspect is that from the pure math view point hard and soft constraints are very different, but for a designer the specs might be not so hard. For instance, of course your receiver must work in a WLAN system defined by an IEEE norm, but the block specs might be still a bit fuzzy, so some "specs" might be violated; they might be specs, but no hard constraints in the mathematical sense. Here the idea of spec or performance margins could jump in, and they would find a mathematical place e.g., as weighting factors in the optimizer goal function (and not as hard contstraints).

8.4 How to Support Optimization

An optimization usually requires much more simulation effort than a pure verification, because many different design parameterizations must be simulated till the optimizer reaches really an optimized version. In a transistor-level testbench and running simulation, you have many options to decide how to check stability, like via transient analysis or with an AC loop gain analysis. For optimization, it is preferable to choose the one which is fastest and smoothest.

In DC or transient analysis, you as user can help the simulator with a good testbench setup or by providing initial conditions, tweaking the options carefully, etc. Similarly and even more this is true for optimizations! We mentioned already that a certain minimum accuracy is required to enable advanced analysis, and this accuracy is a bit higher than the minimum for pure verification. Another aspect is to make the testbench realistic using realistic bias generators, correct load and source impedances, maybe package models, etc. The more realistic the testbench is, the higher the chance that the optimization result is exactly what you are looking for.

Testbench setups can be also quite different, so you may create one single big testbench, or follow a more modular concept. So the questions arises on how should we organize the testbenches for an efficient optimization? In older days, many environments can only run *one* testbench at a time. Therefore, designer often tried to make one big testbench which can check many or even all performances. This usually also reduces the time for netlisting, simulator start-up, etc. However, there are also disadvantages: such complex testbenches may become too hard to maintain and to debug. In addition, some optimization environments can *exploit* a more modular testbench structure, e.g., by not running all performances for optimization points which are already bad in other performances. Such split is also better for optimizations over worst-case corners. Also you can often run multiple testbenches in parallel, so faster than one big testbench. All in all, a compromise is usually best.

8.4.1 Goal Definition

In pure analytical calculations, a goal function like $f_1 = x_1^2 + 0.001x_2^2$ is as easy to minimize as $f_2 = x_1^2 + x_2^2$—but numerically f_1 is usually harder to treat. The optimization environment has to deal with highly different parameters—varying by orders of magnitude, and also the performances can have highly different numbers, like pF vs. MOhms. For this reason, the optimizer core is usually isolated from the direct design values and deals internally only with <u>normalized</u> values. For this reason, it is not good to let the optimizer start from $x_i = 0$ and giving no parameter range and no grid. Preferably, use realistic parameter ranges and also realistic spec targets. Of course you aim for noise figure zero, but that makes the job for an optimizer more difficult—it is better to set a realistic goal like 0.2 dB. Also do not to overlook essential specs, it is preferable to use detailed specification templates or look to good spec examples, e.g., for commercial products or coming from IP tools (Chapter 9).

We already mention that most optimizers work best if the goal function can be well approximated by a quadratic function. However, it is often not so obvious which kind of function your goal function follows. For instance in filter or matching network or amplifier design often *maximum flat passband* response is desired. To define "what" an optimum is, *different* goal functions f can be defined. Their characteristics can have significant impact on optimizer performance. For instance, f should:

1. be always defined (bad $1/x$)
2. be continuous (bad $\text{sign}(x)$)
3. be continuous in df/dx (bad $|x|$)
4. be not too nonlinear (bad $\exp(50x)$)
5. close to optimum it should be well approximated by a second-order polynominal (bad x^4)
6. have no local optima (bad $\sin(x)$)

An optimizer may stop too early if the goal function is too flat, but also using functions like $|x|$ or square root or maximum—"somewhere" in the goal function—can lead to problems. If we take, e.g., a six-parameter quadratic function and place square root on top or square, it will lead to a $|x|$ or x^4-like minimum. Both modifications make optimization harder, and BFGS takes in both cases roughly 30 iterations instead of 8 iterations; i.e., we have to accept a $3.5\times$ slowdown.

A native definition—often found in datasheets—is "maximum deviation is ± 0.5 dB." That so-called max-norm $L\infty$ leads to nondifferentiable functions, in opposite to using sum of squares (L_2 norm or rms) or the average. Another well-known norm is L_1 (sum of absolute deviations), which leads to robust parameter results. However, also L_1-norm is more difficult to optimize than L_2 norm, and for *circuit* optimization it is usually of lower interest.

There are algorithms available to optimize well also for max-norm or L_1-norm-like problems. However, in general it is not clear if the problem to be optimized is indeed of that special type.

Often there is also a compromise needed between good optimization capabilities and easy definition; e.g., the max-norm allows simple goal function definition from *user side*, and the customer is usually on the safe side if he has the guarantee that the *maximum* error is below a certain limit.

Using a weighted-average (as an alternative) could lead to the same optimum, but numerically faster—at the expense that good weights are not known *upfront*. For instance, for typical filters getting flat response is more difficult at the transition(s) from passband to stopband, so larger weights are needed there.

Also the readout of the bandwidth in a testbench from AC simulation can cause optimization problems. Figure 8.13 shows such bandwidth evaluation; we made a sweep on frequency with 101 and 501 steps and on top we swept the BW tuning parameter. With the dense f-sweep we get a perfectly linear relationship (red curve), but with too few AC points we generate numerical noise, which could stop an optimization quite early. In this case, the DUT was a Butterworth bandpass, with more difficult filters (like those having a ripple) the effects might be even stronger.

Of course, similar noise is also critical in transient simulations for overshoot or settling time. So sometimes it is not so much the DUT that is hard to optimize; sometimes the devil is in the *testbench* setup; and optimization is natively more sensitive to testbench inaccuracies. In a separate pdf and in our

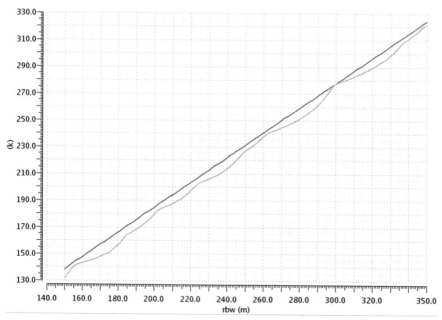

Figure 8.13   1 dB-bandwidth in kHz result of a trimmable g_mC-filter (sweep on tuning parameter) from two different AC simulations.

RealTime app, you can do some experiments on such issues (e.g., selecting either single- or double-sided gradients).

8.4.2 Worst-Case Corners and Worst-Case Distance in Optimization

As mentioned pure nominal optimization is often not needed, but a corner optimization is often highly desirable. One problem is that the WCD and also the WCC will <u>change</u> as we change the design essentially during an optimization! But luckily for moderate changes both are often be quite stable. Let us pick up the differential pair example in Chapter 7 on WCD.

The true 6σ WCD for offset of our diff-pair design was at $(+4.22\sigma, -4.22\sigma)$, but maybe our algorithm is too inaccurate and delivers, e.g., $(+4\sigma, -4.44\sigma)$. This point has a certain error in angle but it gives the <u>same</u> offset voltage; it is not true WCD because the joint pdf is a bit lower. So what does it mean for the design task on minimizing offset voltage? If we would make the transistors $4\times$ larger, the offset would decrease by $2\times$. This is correct in full MC and also in the WCD analysis, leading to $(+2.11\sigma, -2.11\sigma)$. For the approximated WCD, we would get $(+2\sigma, -2.22\sigma)$, i.e., both corners give us the <u>same</u> total offset. In conclusion, our sizing would still work quite independently whether we use the true or an approximated WCD! So if the WCD gets less accurate because we basically change the design during the optimization it would not cause big optimization problems; i.e., the WCD and also WCC concept are quite robust. This is true at least in simple cases. For complex cases with essential nonlinearity or many more variables too bad WCC or WCC settings will indeed slow down the optimization process or the optimizer stops earlier then for using a true WCD. In conclusion, you need to update the WCC and WCD from time to time. We discuss such adaptive corner optimization for yield improvements in Chapters 9 and 10.

As mentioned, more variables or more nonlinearity can cause problems in the WCD method, but even if we have many variables which model the MOS transistor mismatch (maybe one or two in old PDKs, but maybe many more in 14 nm CMOS), an approximated WCD can be still very helpful, because what counts is usually the effective offset voltage shift, e.g., composed from threshold voltages, mobility, t_{ox}, L, and W variations—to which weighting ratio does not matter so much. If you increase the transistor area, you would anyway shift all of them in sync, so that an optimization would still work pretty well.

Imagine we get an approximated WCD directly from a MC analysis (like picking the worst sample), and then we tweak the circuit on this statistical corner. If we would run again the "same" MC analysis (same seed and number of points, no change in topology, only in sizes), we would typically get almost the same analysis with just the improved circuit—and the statistical errors in MC (which are normally significant due to $1/\sqrt{n}$ law) would usually not add up significantly (due to same MC seed). One reason why this relation could be destroyed is stochastic behavior, like having a random generator or if doing a transient noise analysis, but in most cases analog circuits are quite well behaved by construction.

However, besides this good news, there are *also* problems! For example, if we look to a two-stage amplifier and its offset, then the 1st stage contributes to offset and also the 2nd stage, but usually the 1st one dominates. If we rely on a too bad WCD, being correct on the total overall offset voltage, but not on the WCD *direction*, then we may have problems at some point of WCD inaccuracy, due to the multi-dimensional characteristic of designs: If in true 3σ-WCD the offset error is 2.5 mV from 1st stage and 0.5 mV from 2nd stage, and if our approximation gives 2.75 mV and 0.25 mV, we would have still little problems, but if the direction error is large, like having 3.5 mV and (–0.5 mV) a numerical optimization could be easily in trouble! You as designer would see the overall offset +3 mV and you would think "OK, I need to improve the offset, and the 1st stage dominates, so let us make the area, e.g., 4× bigger"—you would often not touch the 2nd stage (because you are lazy?), and actually it works out—you get already the desired improvement because the 3.5 mV 1st stage impact would go down to 1.75 mV, so overall we get 1.75–0.5 mV $= 1.25$ mV, which is maybe even *beyond* expectations! However, an optimizer could also recognize that if we decrease the area in the 2nd stage, the offset would go there from –0.5 to –1 mV, and the overall offset would improve accordingly. However, overall like when double-checking the result in a new big MC analysis this numerical "optimization" would not give a better yield on offset, no lower standard deviation. So the optimizer may improve on the selected statistical corner from 3σ to 5σ but the real yield improvement is only from 3σ to 4σ—just due to relying on a too bad WCD!

How often you may find such bad WCD approximations via worst-sample depends on several factors, you get into trouble if using a too small MC count (like only 50–200 points) or if the sensitivities are too different (like 2nd stage has only 10% impact—instead of 33% in our example). In MC example runs with 100-points on such 2-stage amplifier, such severe direction error in the worst samples on $|V_{offset}|$ was present for roughly 11% of the runs, so this happens not so seldom! It would even increase when having more statistical variables. And the error rate would only reduce slowly by using higher MC counts.

We also mentioned that in some special cases the WCD method can completely fail or it can be very compute-intensive to find them; here a reliable workaround is using <u>multiple</u> such approximated WCDs from scaled worst samples (requiring much fewer simulations than a full WCD search in complex designs). As shown, picking one bad worse sample can lead to <u>under</u>-optimization, but using a (not too small) set quite can safely prevent this—at the expense of more simulations (and thus runtime) <u>in</u> the optimization loop. Actually you would typically over-optimize a little bit, like with the true WCD the optimizer would hit the same specs (e.g., on offset) and yield targets with 5% less area.

As demonstrated earlier, if the optimization really would work significantly, suboptimal depends also on the parameters you choose for optimization. If we would only optimize for a global scaler variable in transistor width of both 1st and 2nd stages, then all kinds of numerical optimization would still work fine, even with such bad WCD. Further techniques with the goal to help on optimization can be quite systematically applied and are often called sizing rules.

8.4.3 Sizing Rules

One reason why a WCD or WCC with moderate accuracy is still usefull for optimization is that designers typically follow certain sizing rules almost intuitively. Benchmarks show that if your optimization starting point is bad, optimizers will quite often fail and will simply not find an optimum solution. One problem is often that the goal setup is simply incomplete. For instance, for an op-amp you may set up *many* goals like BW, I_{DD}, DC Gain, V_{offset}, PM, SR, noise. But *still* this may not define a "good" circuit! For instance, it could happen that even some transistors are not in their desired operating regions like "saturation" (usually for active transistors and current sources) or "ohmic region" (e.g., in a triode-region common-mode feedback circuit). This could cause a bad PSRR or CMRR, but if both are not in our set of goals you may end up in a real bad design! Maybe also the DC gain will be bad by such bad op-region, but often the sensitivity on V_{DS} to DC gain A_{DC} is quite low—compared to the sensitivity to CMRR or PSRR! Sometimes also the specs might be simply too relaxed, like even with op-region violation

$A_{DC} = 60$ dB is possible, just because the op-amp topology could even give 100 dB!

Everybody knows that in a diff-pair we should have identical transistors, so it is native to only parametrize <u>one</u> transistor as master and to let the second transistor follow as slave. This speeds up the optimization due to using less variables, although in principle also the optimizer could find by itself that both transistors have to match! However, even if V_{offset} is among our goals this is surely not a good idea to do so—it is better to follow the sizing rules you know for diff-pairs, current mirrors, etc.—it also helps to be able to make a good layout after the optimization!

In benchmarks, it has been shown that auxiliary goals like "for M11 we want $V_{DS} - V_{DSsat} > 20$ mV" are indeed helpful. There are even circuit recognition tools (Figure 8.14) that identify typical circuit structures, and many PDKs provide built-in support for them [TowervJazz]. If a finder recognized a certain structure or characteristic, we can use this information e.g., for layout purposes, e.g., large instances (typically output transistors or capacitances) should be placed early in the layout process, or if a symmetry

Diff-pair   Low-voltage cascoded mirror   Big devices (layout first)

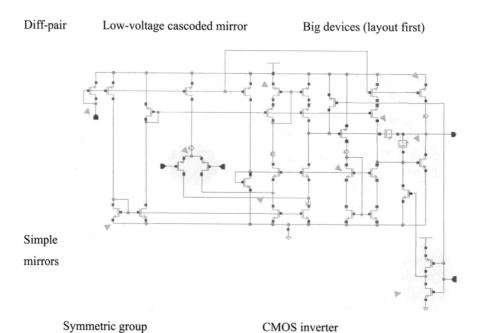

Simple

mirrors

Symmetric group   CMOS inverter

Figure 8.14   Finder application for our 3-stage op-amp circuit.

axis is found, we can translate this also layout, either manually or by using constraints.

Based on the found structures design goals can be set even automatically from a template [Massier]. If, for instance, a differential pair or current mirror is recognized, then an auxilliary goal like $V_{DS} > V_{DSsat}$ could be created. This technique is helpful in front-end optimizations, e.g., if no good starting point is available. If you already have a good starting point (and a complete spec setup), the auxilliary goals will usually give no contribution to the goal function and are not necessary (but still a good backup, if e.g., during a global optimization something gets wrong).

Note: Auxilliary goals like $V_{DS} - V_{DSsat} > x$ or $V_{GS} - V_{TO} > x$ can be easily generated automatically and the user can also adjust the safety margins. On the other hand, we can easily create such saftey margin in a slightly different way, like by extending our corner spread on supply a bit; e.g., if the customer spec is $V_{DD} = 3.0-3.6$ V, then you could just internally optimize across 2.7–3.9 V—this method requires actually less effort and has the advantage that you would be able to tell the customer "Yes we can meet the specs at 3 V, and we test it even down to 2.7 V." Using arbitrary margins in the auxilliary goals could end up in the problem that some constraints are too hard or others still too restricting. Also a full automatic setup is quite difficult to obtain, often it leads to "false errors" and also checking many such goals in a transient simulation may lead to a big overhead. So the best method would be a combination of both, like optimizing over a slightly extended range (on suppy, temperature, etc.) plus monotoring the auxilliary rules.

Keeping sizing rules is important for successful optimizations. What also helps to keep them and to get a good overview during optimizations is to optimize not only at the worst-case corners, but also to include at least one corner that you know very well (like nominal) or that gives many insights (like min V_{DD})! If you do something wrong in the parametrization (like creating an asymmetry in a diff-pair), you can often see this immediately this way.

8.4.4 Optimization Shortcuts

The concept of having a goal function is quite native, but indeed if we would know its structure in more detail we could obtain some valuable speed-up. In simple optimizers, the simulator and the optimizer core are highly decoupled, which is surely an advantage for flexibility, but more "communication" could be beneficial. Some simple methods are indeed often implemented in design environments and can speed-up significantly especially for global optimization. One simple example is "conditional simulation" (Figure 8.15).

A variant would be to simulate first those performances that require only short simulation times (like DC performances and operating points), and doing the compute-intensive simulations only if really a good design point has found (at least according to the 1st simulation group results).

One way to further speed-up optimization is exploiting hierarchical information, in the design itself or with respect to performances. For instance, we may not use a single-goal function combining all performances; instead, we may treat the individual performance goals separately and introduce a ranking of them. This way the optimization effort rises according to n^2 might be reduced. Of course, doing such spec ranking (in analogy to typical manual approaches, like optimize first on DC operating point, gain and maybe bandwidth, later include also supply current and noise, etc., see Section 2.1) or doing similar for parameters comes with some effort and risks. For instance, if the users decides on ranking upfront, but having wrong assumptions we may slow down the optimization. And if we let the optimizer decide it needs to run additional simulations to collect such information, e.g., at least from time to time a full-gradient calculation makes sense to double-check if the ranking still makes sense. Several such techniques have been already suggested like so-called hierarchical cost graph or target cascading [Somani]—for problems with a fix structure (like model parameter extraction), the idea is used since many years.

Divide simulations into two groups: A pass at x_i or B fail at x_i

Get new point x_{i+1} from the optimizer

Simulate group B (those who fail at x_i)

If results better than x_i then simulate also group B
Else Discard x_{i+1}

Figure 8.15   Flow for conditional simulation.

8.5 Design with Pictures Part Six

Basic optimization techniques are easy to visualize with 2D or 3D plots, but for more variables this becomes much harder [Kammara], e.g., for f: $|\mathbb{R}^3 \to |\mathbb{R}$ giving quadruples you may end up in 3D color plots (Figure 8.16), and numerical techniques are usually required to get a full understanding.

As mentioned, usually the goal function f to be optimized behaves close to the optimum as a quadratic function featuring the Hessian matrix $\mathbf{H}$, and this gives ellipsoidal contour plots. Let us inspect quadratic functions in more detail before we treat in part six with our latched comparator circuit example. Actually we will do there even a statistical yield optimization, because for a comparator a pure nominal optimization (without mismatch) makes seldom sense.

8.5.1 Deeper Dive on Quadratic Problems

The most interesting characteristics of such ellipses—being the "contour lines" of a quadratic function—are the *principal directions*. Having correlations,

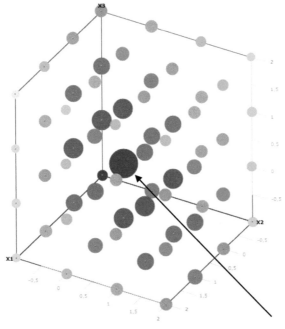

minimum at (0,0,0)

Figure 8.16 Visualization of $f = 2x_1^2 - 2x_1x_2 + x_2^2 + x_3^2$ (dark and big balloons stand for low f).

these directions are usually <u>not</u> simply the coordinate axis! Optimization becomes more difficult if strong correlations are present (because then no independent step-by-step parameter optimization is efficient) and if the different variables have highly different sensitivities. An analysis to characterize this is a so-called eigenvalue analysis for $\mathbf{H}$, and actually BFGS uses the (inverse) $\mathbf{H}$ matrix internally for finding the best-suited optimization directions! For pure quadratic problems, the Hessian matrix is fixed, so it really contains almost all information we need to understand the optimization.

Let us went back to our 3D quadratic example for introducing optimization, for $f = 2x_1^2 - 2x_1x_2 + x_2^2 + x_3^2$ the Hessian matrix was:

$$H = \begin{pmatrix} 4 & -2 & 0 \\ -2 & 2 & 0 \\ 0 & 0 & 2 \end{pmatrix}$$

Generally, the eigenvalues of a matrix A are defined by $\mathbf{Ax} = \lambda\mathbf{x}$. In our case, the eigenvalues are all real due to the symmetry of $\mathbf{H}$. For difficult designs, the ratio between the smallest and the largest eigenvalues can be very large, indicating highly different sensitivities and second-order derivates. Numerically it can even happen that some eigenvalues are negative; in such case, BFGS has to modify the Newton search direction, because the quadratic approximation has actually no minimum at all!

The eigenvalues λ_i and eigenvectors of $\mathbf{H}$ are:

$$\lambda_1 = 0.7639320225: \quad (0.618033988749895;\ 1;\ 0)^{\mathrm{T}}$$
$$\lambda_2 = 2: \quad (0;\ 0;\ 1)^{\mathrm{T}}$$
$$\lambda_3 = 5.2360679775: \quad (-1.618033988749895;\ 1;\ 0)^{\mathrm{T}}$$

Note: All distinct eigenvectors of a symmetric matrix are perpendicular. The sum of eigenvalues is equal to the sum of diagonal elements. In the appendix we list some internet online tools for such tasks. In addition you can you e.g., the R environment with command `eigen(cbind(c(4,-2,0), c(-2,2,0),c(0,0,2)))`.

From a step along the coordinates, we can approximate the (partial) derivatives, and if we do a step along the largest eigenvector we would find the directional derivative in this direction, which would be here the largest (quadratic) sensitivity. So an eigenvalue analysis is highly related to optimization; if the maximum over minimum eigenvalue *ratio* (so-called condition number) is *large*, then we have parameters with highly different sensitivity, and often in complex optimizations several parameters have only a minor impact; i.e., we may disable them to speed-up the optimization. If the condition number is large, also simpler optimizers like steepest descent slow down significantly, even if the problem is quadratic! However, for BFGS large condition numbers are only a mild numerical problem, but of course if an eigenvalue analysis is done the results could be used to give the user a warning, and often by re-scaling the variables significant improvements are possible. For instance, it makes little sense to optimize a parameter from e.g., x_i from 1 fA to 1 mA; here it is much better to use $x'_i = \log_{10}(\text{xi}/1\text{nA})$.

The eigenvalues are also related to the optimization accuracy for the parameters to be optimized. At the optimum, the goal function is almost flat we have $f \approx f_0 + f_{xx} \cdot x^2$, so accepting a small Δf like 0.01 leads to parameter inaccuracy of $\Delta x = \sqrt{(2\Delta f/f_{xx})}$, for small f_{xx} this might lead to large uncertainties. To capture the whole situation, you have to look for derivates *not* only along the coordinate axis but along the eigenvalues! The existence of small eigenvalues indicates that the parameters are <u>not</u> really well defined—even if all the f_{xx} itself might be large.

Doing the calculation with worst-case eigenvalue instead of f_{xx} leads in real circuit often to large Δx, i.e., to get accurate parameters via optimization you need really a very high accuracy on f!

One further aspect is also interesting: From these eigenvalue analysis results and inspection of **H**, you can see that variable x_3 is not correlated with the others. So the optimizer can minimize f with a plain x_3-sweep, and the minimum found from this is also the absolute minimum—this is <u>not</u> true for the other variables! We can also exploit this and just set x_3 to the optimum $x_3 = 0$ and reduce the complexity to two variables only, which allows a visualization easier to interpret.

$f_{\text{red}} = 2x_1^2 - 2x_1 x_2 + x_2^2$ and the Hessian matrix is now:

$$H_{\text{red}} = \begin{pmatrix} 4 & -2 \\ -2 & 2 \end{pmatrix}$$

The eigenvalues and eigenvectors of H_{red} are now:
$\lambda_1 = 0.7639320225$: $(0.618033988749895; 1)^{\text{T}}$
$\lambda_2 = 5.2360679775$: $(-1.618033988749895; 1)^{\text{T}}$

Note: In the appendix you can find a web-based tools for this task.

In a contour plot, you can interpret the eigenvalues and vectors. The only little tricky thing is that in most references on quadratic functions and ellipses you typically find a form $f = xAx$, but for optimization it is more convenient to use the Hessian matrix and $f = \frac{1}{2}xHx$. So actually the ellipse and contour plot interpretation is usually done according to $A = \frac{1}{2}H$.

$$A = \begin{pmatrix} 2 & -1 \\ -1 & 1 \end{pmatrix}$$

In the 2D case $Ax = \lambda x$ leads to quadratic equations, so also a solution by hand is possible; the eigenvalues are now just halved, whereas the vectors are scaling-invariant:

$$\lambda_1 = 0.381967, \sqrt{\lambda_1} = 0.61803 =: \quad (0.618033988749895; 1)^T \quad (8.4)$$

$$\lambda_2 = 2.618034, \sqrt{\lambda_2} = 1.61803: \quad (-1.618033988749895; 1)^T \quad (8.5)$$

The rotation angle is $\alpha = \arctan(1.618/1.0) = 58°$.

In the appendix, we give a real LC circuit example with 8 parameters, and you can run analyses on this with the RealTime apps available to the book (the results of the eigenvalue analysis are presented in the log file opt.txt of the matching network app). It can happen that at the start point of an optimization the behavior is highly nonquadratic and the quadratic approximation may even have negative eigenvalues. One classical BFGS solution is to switch back to steepest gradient and another one is to shift the eigenvalues (thus to modify the Hessian matrix) (Figure 8.17).

How can we influence eigenvalues? In this chapter we analyzed difficult optimization problems mathematically, but how can we exploit this? Indeed some options exist. The very first option is to scale the variables, so that their values are e.g., not too far from unity, but often this is done anyway internally, by the optimization environment. A next step would be to also normalize the sensitivities. However, this is more difficult, because natively the parameter sensitivities can be quite different. So the next option would be to try to influence the curvature, the second order derivatives in a positive way, like reducing the non-diagonal elements! Such mixed terms occur often, e.g., to reduce offset designers would increase the area $A = W \cdot L$, but to keep e.g., the saturation voltage we should keep $k = W/L$ constant for our change. Indeed, if we introduce A and k as parameters, the optimization would often run faster! Similar methods are possible for filters; here we should use e.g., ω_0 and Q, instead of the element values.

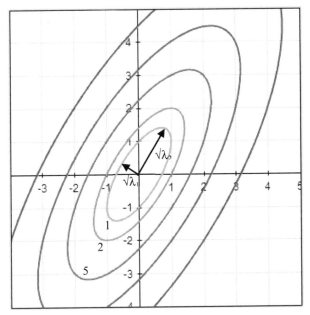

Figure 8.17 Contour plot of our 2D quadratic function and eigenvalues.

8.5.2 Men versus Machine? Construction versus Optimization?

All in all, advanced optimizers use techniques an experienced designer would also apply for circuit sizing, like the concept of sensitivities and even correlations. Of course, other advanced tasks are much harder to automate, like debugging, testbench extensions, circuit modifications—for this and many other tasks, designers also need fantasy and intelligence.

The *a priori* design insights a good designer usually has lead to the situation that an *initial* design can be usually made much quicker *manually*. However, at some point, if many variables, goals, and corners need to be treated, the systematic algorithmic work of an optimizer and its high accuracy pays out: Imagine you sit in a laboratory and you have to trim two pots for wanted output at 4 mA and 20 mA input. Often if you trim one pot, you usually have to readjust back also the 2nd pot; i.e., there is some correlation between them—which makes the trimming harder. A human designer can recognize such influences, but only to some degree and he/she would probably fail to apply a clever strategy also for five or ten of such interacting parameters! Good optimizers such as the BFGS algorithm have no such limitations, and thus, the more complex the cases are, the more advantages for BFGS. Step by step it collects more information on the design by running more and more simulations and "composes" an efficient strategy. The latter is mainly in the gradients $\mathbf{g}$ and Hessian matrix H and in the optimization algorithm itself (like conjugate gradient, BFGS, DFP, or Newton).

In the initial design phase, designers typically apply direct solution methods, execute small parameter sweeps, and use no optimization. For instance, if you have to design a voltage divider with several taps, you can use Ohm's law and the voltage divider rule $-V_{out} = V_{in}R_2/(R_1 + R_2)$. This becomes obviously a bit more difficult if you have many output taps and maybe also nonfix ΔV in between each pair of taps. However, with some effort you can often solve the set of design equations step by step. For instance, the output easiest to calculate is usually the one above ground because it is only impacted by one resistor and the divider current. Having this you can move to the 2nd tap, 3rd tap, etc., till you are done. Usually the number of equations n is equal to the number of unknowns x, and the calculation effort is in the order of $n = x$.

We can solve the problem also by formulating it as an optimization, but the effort would be typically in the order of n^2! So we have some price to pay for our "laziness," but one advantage is often that optimization setups are usually much easier to extend for further goals (like on noise or area or output impedance) or more complex circuits (like loading the R-divider with a diode)!

Another example can give further insights: What about the results of a mismatch contribution analysis and using it for a manual re-sizing vs. an optimization at WCD? We already explained the close relationship between contributions and WCD; in both, we can find out which instances impact the offset mostly. So if I need to reduce the offset by 1.4× the design could increase the area of the important elements (like two transistor pairs) by 2× with keeping the W/L ratio for transistors to maintain g_m, V_{Dsat}. This way the designer needs just <u>one</u> re-simulation using the WCD to double check the improved sizing! A gradient optimizer, however, would need to calculate the gradient according to all design parameters that the user defines. So actually also the optimizer speed depends to some degree on user know-how. This way an optimizer can find out the direction with best improvement rate with maybe nine simulations. Next usually a step-size search is done, which needs roughly five simulations. This sequence of about fourteen simulations needs to be done several times, till the design is in spec, so overall we may need about

50 simulations—and more if we simply ignore the mismatch contribution results and just optimize on many more (unimportant) parameters.

Ultimately the manual design and the optimizer could end up in almost the same result, but the designer was much more efficient regarding number of simulations. This was possible by exploiting pretty much know-how:

1. You exploit the ranking provided by the contribution analysis
2. You know that an area increase helps to minimize mismatch
3. You exploit the Pelgrom's law by applying the factor of two for W and L
4. You know about basic sizing rules, like if keeping W/L many transistor characteristics remain almost constant (like V_{Dsat} and g_m) to make sure that our sizing has no big negative side effects (beside larger area and larger capacitance)

Step by step also the optimizer is gathering this information (e.g., by calculating gradients and composing the inverse Hessian matrix) and exploits this (applying the Newton step). In manual design, you may need some iterations, and also the optimizer is doing so in his internal step-size setting algorithm.

Again, the manual approach becomes more difficult when you need to start looking to other performances in fight with offset (like bandwidth, stability, area, power consumption). Typically designers are able to include maybe one, two, or three more specs, but at some point and especially if the circuit or the sizing rules become difficult (e.g., because in your technology the MOS transistors do not follow the simple MOS square law model), other methods are helpful—and that is often optimization, or asking a colleague or using another circuit or SPICE monkeying or catching an enlighting idea during sleep.

8.6 Questions and Answers

1. How many simulation points are typically required for the optimization of ten parameters?
   This depends on the nonlinearity of your problem! If the quadratic fit done inside many optimizers fits well you can expect good progress in $n^2 = 10^2 = 100$ points. Of course, the optimizer also needs enough freedom to be able to optimize, so you should optimize the right parameters in the right range. To identify the best parameters consider to run a sensitivity or contribution analysis.

App

2. Can you expect to get the same final solution when starting from significantly different reference points?
   Only sometimes this is realistic! If there is one such unique solution it could happen, at least with acceptable accuracy. However, if you have many goals like performance < limit, and the optimizer is set to stop at "All spec met," then actually it would be possible to be even better—just by chance and for sure dependent on starting values. For extreme case, it could happen that a local optimizer stops too early at a local minimum. Then better use a global optimizer.

3. What should I do if something gets wrong in an optimization?
   This can happen for many reasons, best inspect the log file. Also sweep some parameters manually, does the circuit respond to such sweeps, are you able to improve by hand? Also inspect your goal setup: Avoid zero as spec limit, zero as parameter values, and strong nonlinearities in specs (like output which can only take binary results).

4. You modify your optimization setup, instead of 10 parameters, you now optimize on 20 parameters. What do you expect in runtime?
   Typically the optimization effort rises quadratic, but maybe 10 a parameter optimization was giving not enough freedom, so it stops too early. So maybe it is really worth to spend this time. For global optimizers, the increase in runtime might be also beyond the quadratic law. The actual number of points depends also on the nonlinearity of the optimization problem.

5. Setup an optimization and run it. Often you will observe that the optimizer is changing some parameters not much, at least at the beginning? What do you expect e.g., regarding parameters with low and higher sensitivity?
   Typically optimizers do not "touch" unimportant parameters, having low sensitivity. Instead optimizers typically optimize first the major parameters. This is actually similar to the manual design process. Here you typically also look to the most important parameters and effects!

   For instance, in a PA design the primary goals like power and efficiency are mostly related to the output stage. So you design first the transistor stage connected to the load, then go "back" to the driver stages, step-by-step. For an LNA it is vice versa: Here the input specs are more critical like noise figure, so you start with the stage directly connected to the antenna, then looking to the other stages.

6. We can use optimizers for circuit sizing, and what else is possible?
   You may use it also for performing a calibration or even an optimization in which a calibration is applied. Of course, it does not mean that optimization is the fastest technique for calibration problems. Optimizers are also key parts for other methods like for finding WC distances or WC corners. In addition, many optimizers provide many further results, e.g., in the log file or in special results tables; one such by-product is often a sensitivity analysis!.

7. You want to reduce the sigma of the offset voltage (or improve on other performances highly impacted by statistical parameters). Instead of doing a big MC analysis within the optimization loop, you want to exploit the idea of statistical corners and worst-case distances. However, because you are not so familiar with WCD methods you just run MC and pick the worst-sample and optimize on this corner. What are the problems?
   This idea could work quite well, especially for low-yield targets like 3σ, because then MC effort is moderate and WCD methods would be hardly much faster. However, such MC worst-samples are seldom accurate on direction. Of course, the true WCD has the largest probability, so also the worst-sample should be not far away, but actually your "protection" on limiting the direction error is quite small, just the little higher probability of the true WCD vs. errorness WCD. So it is realistic that in 5 to 15% of such MC analysis, the worst-sample has such big direction error against the true WCD!

8. Can I use my circuit simulator as optimizer?
   Yes, indeed in simple cases you can do so. For instance, you may remember how old analog computers work. Then create a testbench reproducing this optimization problem (e.g., by using VerilogA) and run a transient analysis in the simulator.

9. Which kind of optimizer is realized in Figure 8.18?
   The idea is following steepest gradient, but actually real software implementation of steepest gradient work slightly different! In the schematic the gradient g is calculated <u>continuously</u>, but in software we usually calculate g and then do a step-size search with fix direction $s = -g$. Calculating g all the time (so even after a very small step Δx) would usually take (far) too much time!

10. If you can program your optimization goal function e.g., in Excel, then you can use the built-in solver for this task.
Try to run it for yourself or use the file rosenbrock.xls from the River webpage (last tab sheet).

11. In Section 8.3.4 we gave sample circuit examples with local minima, but optimization can be also used for maximum likelihood estimations or worst-case distance search. What about local minimums in these applications?
Indeed for symmetric WCD problems and starting at the origin, we would run into problems, because often the origin would be a local optimum. In MLE for a triangular distribution the goal function looks also not like a nice quadratic function at all.

8.7 Summary: Why Optimization Was So Hard?

We mentioned already several problems and indeed you may ask why is simulation available since almost 50 years—able to find thousands of unknowns

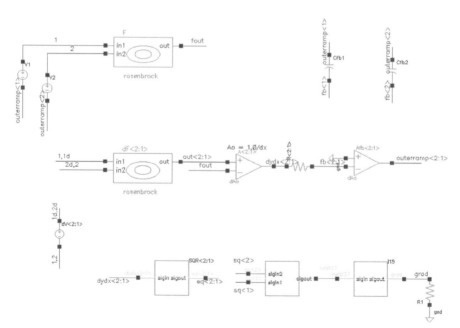

Figure 8.18 Testbench acting as "analog computer" for a two-parameter optimization (goal function defined with VerilogA block).

quite efficiently—whereas optimization of real complex circuits—maybe
featuring 30 parameters of major interest—is still regarded challenging?

There are several reasons:

- Getting the equations to solve for simulation is quite a regular task due
  to Kirchhoff's Law and the element equations.
  *In optimization there is no real standard method for this! So historically
  optimization has been indeed applied successfully on problems for goal
  formulation was easy, e.g., by using least-square criteria for modeling
  purposes!*
- In circuit simulation, the element equations standardized and even the
  partial derivatives are usually fully available to enable Newton–Raphson
  method (NR).
  *The gradients for an optimization are harder to provide (requiring further
  simulations, sensitive to rounding errors, etc.). For yield optimization,
  the goal function and its gradient are even very difficult to calculate!*
- The equations for simulation and optimization can be both very nonlinear.
  However, in circuit simulation you know this upfront. For instance, a BJT
  has an exponential characteristic, which can cause Newton–Raphson to
  fail, but luckily you can implement dedicated supporting algorithms to
  get rid of such difficulties.
  *This is more difficult for optimization, basically you as the user has
  sometimes to pamper up your design and <u>also</u> your optimization!*
- In circuit simulation, you may fight with multiple possible operating
  points, and you may apply few nodesets or initial conditions to force the
  desired state.
  *You typically know well <u>where</u> to apply nodeset; good candidates are
  latches, start-up circuits, etc. In optimization, you may try different
  starting points to get rid of local minima, but this is not so easy to do;
  e.g., a too bad starting point may stop the optimization completely.*
- Optimization is something on top of simulation, so if the optimizer takes
  a set of parameters which make the simulation difficult (e.g., circuit
  becomes unstable), then this immediately effects the optimization.
  *So optimization requires usually very careful numerical implementations.
  For instance, the gradient calculation is much more difficult in optimization! In circuit simulators, the derivatives are typically hard-coded, so
  very accurate.*
- For circuit simulation, the equation to solve is something like $f(x) = 0$,
  whereas for optimization it is $f'(x) = 0$.

This sounds so similar but in reality Newton–Raphson often takes only few iterations (like 5) to find a solution, whereas an optimizer (e.g., Newton's method) often needs more iterations, even for much lower number of parameters. And also one iteration itself takes more runtime because almost always full circuit simulations are required. One Newton iteration would require simulations in the order of n^2.

- A big question is whether we can formulate mathematically what a good design is!

This is a key point for design automation. We showed how to translate a datasheet into a minimization problem, but the weighting factors might be a bit arbitrary. Also "soft" factors are difficult to include. Surely a good circuit is one which cannot be improved! This can be checked in a designer review, but is hard to program—and the same problem occurs also in layout automating tools!

Table 8.7   Do's and Don'ts for optimization

Do's	Don'ts	Comment
Define all performance specs which matter	Be lazy and forget specs	If your specs are incomplete the optimized design may not fit, e.g., regarding layout
Make sure that performances vary smoothly with circuit parameters	Make life difficult for an optimizer and define binary outputs, like Hi or Lo!	Optimizer run faster with smooth outputs
Choose a suited optimizer	Stick to an unsuited optimizer	Sometimes you should really try multiple, global optimizers are usually more robust but slower on smooth problems
Exploit what you have find a good starting point	Waste time in using a bad starting point	
Define parameters and ranges carefully	Do not optimize all or too few parameters	For too many variables the optimization becomes slow, but maybe you can do it in an overnight run
In difficult cases optimize 1st with ideal elements or just an input stage	Optimize a big system without having good sub-blocks	Divide-and-conquer can save a lot of time also in optimization, check if ever a certain target can be achieved, like NF < 0.9 dB at 5 GHz or BW > 1 GHz in a matching network from 50 Ω to 1 kΩ

The key points to make optimization tractable in spite of high complexity are working in a systematic way using some key speed-up techniques:

- Reduce complexity on statistical and environmental variables by using statistical and environmental corners.
- Focus on the important parameters, but do not overlook parameters to optimize. For this, check design on sensitivities. Look-up that often the optimization effort rises quadratic with number of parameters. Set up parameter ranges to achieve meaningful values also suited for layout.

- Define realistic goals; the optimizer will collect them into an overall goal function. The user can typically set weights to tune the optimization further. Sometimes users have anyway a good feeling on how to set the sets, like 0.1 dB in noise figure as critical as 1 dB in gain, but sometimes this is not the case. In the future, we can expect even more powerful algorithms, like Pareto optimizers, which create full sets of optimized designs (so-called "Pareto front") with different optimization trade-offs (more in the last book chapter).
- Optimization has beneficial by-products like giving a sensitivity report after the optimization. This helps for manual tweaks or for doing the next optimization. Optimization is no push-button solution! (Table 8.7).

9

Advanced Front-End Design Methods

Now we put the discussed techniques together, but we still focus on front-end design topics; e.g., to enable yield optimization, we combine optimizers and overall worst-case finders. We also discuss options for further analysis methods for higher speed, more automation, and more design insights. You typically cannot find built-in features addressing all these advanced methods, but most environments offer powerful scripting capabilities.

Let us pick up the front-end design flow proposal from Chapter 2 and inspect which methods are available for different difficult tasks and which can be merged to an even more advanced analysis. Figure 9.1 shows a merger ("Improve Yield") for the worst-case search and the circuit optimization. Such option is available in advanced design environments, and it makes sense because the worst-case corners (either for deterministic range variables or statistical variables, or for both) may *change* during the optimization.

We have to solve truly complex, very difficult mathematical problems to make yield optimization feasible, but beside this, <u>as usual</u>, you should first get

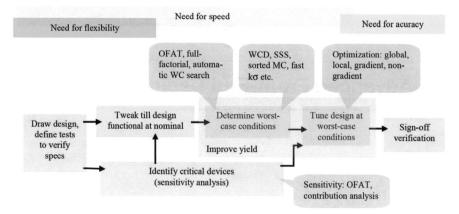

Figure 9.1   Circuit design flow chart and methods.

an overview on the status of your design before deciding for an optimization. Doing an optimization at typical conditions is sometimes waste of time, even if the design might be improved significantly (like you can increase the op-amp open-loop gain from 60 to 100 dB) and easily by an optimizer (like in an hour), because it may still show severe problems under specific conditions (like gain drops to –10 dB at minimum supply)! It may happen that a design optimized at typical conditions still shows the same problems, sometimes even in a <u>more</u> severe way! Better anticipate, identify, and solve the urgent problems first before fighting for the last decibel on other things. Beside this, many less difficult design problems can be still (almost) solved with classical divide-and-conquer.

Often it makes sense to optimize up front on already-known (or easier to find) difficult conditions. We could first do separately a WCD analysis and a WCC analysis and look which kind of variables, statistical or deterministic range variables, causes more pain (Figure 9.2). Then we should optimize on the more problematic case—before applying a full flow, while taking both range parameters and statistical parameters into account. Such classical divide-and-conquer has the advantage to be faster at the beginning, and the methods for that are available out-of-the-box in many modern design environments. Note, in some cases, we may have problems: Yield improvement by combining optimization and corner finding might be difficult, e.g., if the direct combination of statistical WCD and worst-case corners does *not* represent the full overall worst-cases or if the WCDs give no accurate yield estimation.

Prepare design, create testbenches, set specs, specify desired yield

Set design parameters x_D

A Run a WCD analysis at (single) expected WC conditions

B Run WCC at a single overall critical compromise WCD

If more fails in A Optimize x_D on WCD
Else Optimize x_D on WCC

Enter overall yield optimization, e.g., based on overall WC

Figure 9.2   Fast preparing optimizations to provide a good starting point.

9.1 Task-Driven Adaptive Statistical Analysis

A good statistical analysis is pretty much more complex than a SPICE transient analysis. Yield analysis or even a full statistical analysis (e.g., also addressing corners and contributions) is more like fast-SPICE or mixed-signal simulation, in which also many clever speed-up techniques are integrated. In Chapters 5, 6 and 7, we described many useful statistical techniques, but actually no single one is as universal as random Monte Carlo, and MC is not well-suited, e.g., for high-sigma corner generation or yield optimization. If we connect different techniques, we can compensate weaknesses—by still having a significant speed and accuracy advantage over pure MC (rando, LDS, etc.). In principle, we just need to arrange the manual statistical analysis steps (Figure 5.2). Some decisions are done based on yield target, because some methods are natively better suited for high-yield estimations than others. Also inside each sub-task there is room for improvement. For instance, we can speed-up the MC parts by applying LDS or e.g., optimized LHS. This works best if the number of important statistical variable is not too large. For larger circuits it makes sense to apply the better sampling scheme at east to these important variables. You may find these iteratively, or from experiences. Usually mismatch is dominated by the threshold voltage parameters, and for filters or current generators also process variables like sheet resistance are important.

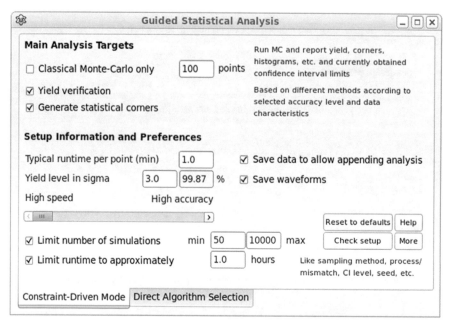

Figure 9.3   Example for a task-driven user interface.

One problem in pure algorithm <u>selection</u> is to decide on the crossover point; because seldom we have a clear separation, usually there is a smooth transition and overlaps. So, instead of *selecting* a certain "best-suited" method, we can also regard the different methods as a continuum: WCD is clearly optimization based, but if we aim for statistical corners with SSS or sorted MC the act not much different, just the type of optimizer is different (like gradient vs random), is also the type of information used and the target differs (slightly). Also almost all advanced methods start with pure MC anyway, and some extensions like creating a multivariate model (as in sorted MC), up-scaling sigmas (SSS), or starting a direct search for WCD are not really contradicting. Imagine we start with MC and classical WCD search, maybe supported by SSS to improve on the starting point for the optimization part. If WCD converges, we may double-check the results (e.g., via C_{GPK} and pure SSS, or via sorted MC) and stop, but if WCD looks inaccurate (e.g., due to numerical problems or nonlinear fail boundary) or too slow (due to too many variables), then the results could be still used, e.g., for creating or refining a multidimensional model and applying sorted Monte Carlo or (if we need maximum speed) by we could just rescale the current worst samples, maybe with additional cluster analysis and averaging for accuracy improvements. So we obtain a good solution even in the extreme case where no WCD even exists (typically only for a certain difficult circuits and output measures) by switching back to basic MC plus using (a set of scaled) statistical extreme samples—actually this can be very efficient and often accurate enough.

This way we end up in a very robust, reliable, and still efficient algorithm, able to deal with all kinds of nonlinearity and complexity.

Note in this analysis we still focus on statistical variables only; i.e., we run it at nominal conditions or an expected critical corner. For a true yield optimization (next sub-chapter), we would need the full WC corner combination, and we need to update these corners from time to time. In this context, the optimum corner generation method might also depend on the currently *achieved* yield level.

9.2 Yield Optimization and Overall Worst-Case Search

The simple divide-and-conquer approach of pure corner optimization plus optimization at pure statistical corners is non-optimum in the general case, and especially for more difficult designs. We actually underdesign, because the true overall WC would be not captured explicitly; and we could end up in "oscillations." An example could be an op-amp critical on offset due to mismatch but being more critical on other performances regarding PVT corners. If we would first optimize on corners (as proposed in Figure 9.2), e.g., to improve on speed, we may make the offset even worse! And vice versa tweaking for lower offset could worsen the speed unfortunately. A certain improvement could be to "take an eye" to the non-optimized performances, e.g., just by constraining the size of the input transistors. With design know-how, small experiments and sensitivity investigations, designers are usually able to anticipate such problems, but such approach comes with the risk of over-constraining, leading still to a non-optimum final design. In addition, it becomes difficult for treating more than a few specs. For uncritical designs, you may solve many problems this way, but ultimately a true overall worst-case optimization is the more elegant way because it treats many problems directly.

The difficult yield optimization case is having a design with significant nonlinearity, so that the simple OFAT method, of just directly combining WCC and WCD and doing no further checks or refinements, often would fail. The extreme opposite way would be switching back to (an almost) full-factorial method; like doing a (bigger) MC analysis at all corners inside the

Prepare design, create testbenches, set specs, specify desired yield

Set design parameters x_D

Run a (huge) Monte Carlo yield analysis for each of all the conditions set by the deterministic range parameters x_R

If yield < target yield go for another design point $x_D = x_D + \Delta x_D$ and test this on yield

Figure 9.4   Brute-force design via MC and all corners inside the optimization loop.

optimization loop (Figure 9.4), here we would address all kinds of variables *simultaneously* and we have no such "oscillation" problems, but the speed for design would be hopelessly slow.

Can math help us again? We formulated the yield and optimization mathematically, so what is yield optimization? We are looking for the design defined by $x_D = x_D^*$ (being in the valid range of design parameters) which is minimizing our performance goal function f with keeping the yield loss (fail probability $p_{fail} = 1-Y$) below our target yield loss (e.g., 6σ). And we need to do that for all environmental conditions defined by the x_R ranges.

$$x_D^* = \arg \min f(\mathbf{x}) \text{ subject to } p_{fail}(\mathbf{x}_D, \mathbf{x}_R) < p_{failTarget} \qquad (9.1)$$

Often, we may have to include further constraints. One not so nice property of Equation (9.1) is that the WC corners do not appear! These would be the quantiles, coming from the inverse cdf, e.g., at 6σ we would need to calculate $q = \text{cdf}^{-1}(1-10^{-9})$, and we would get the spec setting for this situation. Again the key problem is that the yield integral and the cdf is hard to evaluate, because the yield integral fail regions are only defined implicitely; and sampling efforts time-consuming circuit simulations. As the quantile function is monotonic we can reformulate yield optimization as:

$$x_D^* = \arg \min f(\mathbf{x}) \text{ subject to } q(\mathbf{x}_D, \mathbf{x}_R) < q_{spec} \qquad (9.2)$$

Using full corner sets and MC in the design tweaking phase is not only highly inefficient, but it has also the disadvantage that the MC analysis with its confidence intervals would make an optimization very difficult. In MC, it can happen that a worse design gives a (slightly) better yield just by chance!

If we want to optimize on the standard deviation σ of the offset voltage, then we could use the sample standard deviation s in the optimization goal function—but s depends on chance (the more the lower the MC count)! So an optimization goal function based on s (or any other MC results) would present quite noisy data to the optimizer, which could prevent an efficient and accurate optimization (especially if using a fast gradient optimizer like BFGS).

At best, the brute-force method might be used as "golden" reference for benchmarks. [Brayton1980] and [Javid2007] propose geometrical approaches and advanced optimization techniques to solve the problem more efficiently. Actually, some moderate speed improvements are possible without losing much of generality; e.g., we could do the MC analysis at all corners by first using only a moderate MC count (like 50, and using the C_{PK}), and then we could extend it only for the worst deterministic corner combination(s) for each spec, e.g., to 500 (for a sample yield around 3σ–4σ and using the C_{GPK}, or more points plus using the sample yield). Another approach would be remembering, e.g., the latin hypercube idea: We can indeed generate LH sets which address <u>both</u> statistical variables and our environmental range parameters, but typically also the LH Monte-Carlo variations would be still critical for an optimization (and of course hardly suited for high-yield optimizations).

Therefore, and for speed reasons, we propose design tweaking based on overall worst-case corners, because this way we can substitute the full mix of "MC plus corners" by a compact set of overall corners which treat both statistical and deterministic variables. To some degree this is still a kind of (iterative) divide-and-conquer approach with some limitations, but it is a realistic approach which can be followed in modern design environments. At the end of the chapter, we also present some algorithms, which go beyond this and *merge* the optimization and variable space evaluation part.

9.2.1 Methods for Overall Worst-Case Search

We already presented reliable and efficient methods for worst-case finding in a set of corners x_R and also for statistical parameters x_S, so what is the best way to find the <u>overall</u> worst-cases among (x_R, x_S)? We mentioned OFAT is often not reliable enough, and full factorial is far too slow. In addition, we should also get a feeling for how accurate do we need to find the overall worst-cases. Some algorithms are optimized for finding the true worst-case very accurately, but may require quite a lot of runtime (like classical WCD analysis applied to very big designs), whereas others may exploit some extra assumptions and are more optimized for speed (like fast k·sigma flow or the C_{PK}). If you are in the sign-off phase, you should better use the first kind of algorithm to avoid any risk—if feasible. However, if you are in the design tweaking phase and if your specs are not fully confirmed, better choose a fast algorithm.

9.2.1.1 Example and heuristics for overall worst-case search

To understand the requirements and the general problems for finding the combined worst-case on both range and statistical variables, let us start with a mixed example, with two statistical variables representing the threshold voltage variations (often a dominating effect in analog design) and two range parameters temperature T and supply voltage V_{DD}. Let us pick up the CMOS inverter of Chapter 2, but let us now use no process corners like SS and FF but treat the process variations as statistical variables on thresholds V_{TOn} and V_{TOp} (which is also a more realistic assumption). Let us again focus first on delay and on one-parameter sweeps.

Usually the WCC on speed is maxT (due to mobility degradation) and minV_{DD}, but if V_{DDmin} is very small, like for ultra-low-power applications, it could be that the gate-overdrive becomes very small, leading to a strong increase in on-resistance, especially at <u>low</u> temperatures, because $|V_{TO}|$ has a negative TC. If we now tweak the design, it could happen that the WCC is <u>either</u> at T_{min} (as shown in Figure 9.5a) or at T_{max} (as usual for moderate and large V_{DD}). As we have seen in Chapter 2, the WCC can actually even "jump," during the optimization, which makes the job for the optimizer very hard! To prevent this, a meaningful workaround would be not only to focus on the worst-case but just to simply use more corners combinations, like second worst combination or a critical setting in opposite to the found WCC—at least for critical variables, such as temperature. This can be understood in more detail looking to full-factorial corner and to the MC results. Figure 9.5b to 9.5d display the varying responses over a range of conditions and variables (sometimes called shmoo plots).

The detection of this "oscillation" condition is quite easy here: The final sweep (done on the expected most critical variable, here it was temperature T) has given another WC than a sweep around nominal conditions.

The "jumping" worst-case the inverter example might be regarded as a bit special, but actually similar problems can happen in many designs, and also the testbench and corner setup have an impact. If we have circuits like a bandgap or variable gain amplifier VGA, we may include temperature T or tuning voltage sweeps directly into the tests, so having them not directly as corner variables anymore. This way the whole verification might be faster, but exactly that clever arrangement could also cause such jumping worst-cases; e.g., the VGA linearity might be impacted at different tuning voltages by completely

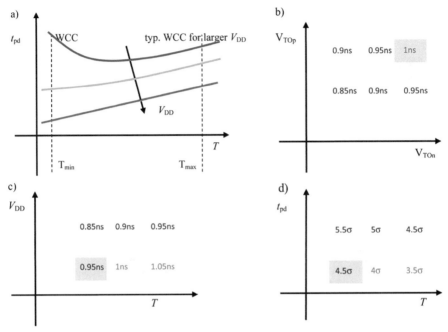

Figure 9.5 Principle inverter delay behavior: a) Temperature behavior for different V_{DD}, b) Threshold voltage impact ($V_{DD} = V_{DDnom}$, $T = T_{nom}$), c) for VT corners (process at SS), d) Monte-Carlo yield vs VT corners.

different circuit mechanisms! Or the bandgap may show difficult behavior at maximum temperature and special process variable combinations, but is sensitive to quite different variables at other temperatures. The good thing is that having such examples in mind typically designers can identify such problems quite quickly, even more when using "dirty" design or testbench "tricks." For instance, due to channel length modulation a voltage regulator has usually a finite power supply rejection PSRR like 1 mV/1 V (60 dB); you may improve this easily by just adding a little negative compensation voltage of –1 mV/V, e.g., via a resistor divider. However, it is likely that the compensation is not perfect, so you may get –0.1 mV/V (80 dB). This sounds uncritical, but exactly these "too clever" methods will often cause larger, more nonlinear variations, and difficult WC behaviors!

To find the overall worst-cases efficiently, we can actually borrow a lot of ideas we applied for pure statistical WC distances and WC corner finding. A starting point, we may remember the OFAT method, would be just to "compose" an overall worst-case corner set according to Figure 9.6 [Schwencker].

Prepare design, create testbenches, set specs, specify desired yield

Set design parameters x_D

Run a WCC analysis for all the conditions x_R

Run WCD at each WCC (sigma acc. to yield target) to get overall WCC

If design fails at overall WCC decide for another design point $x_D = x_D + \Delta x_D$

Figure 9.6   Design flow proposal with WCC and WCD in the optimization loop [Schwenker].

In many design environments, you can just follow this approach by hand; step by step, Figure 9.7 is showing an example successfully executed in WiCkeD^{TM} from MunEDA for a memory cell design.

Would this user approach work in general? First searching for WCC then for WCD, so doing a kind of ordered OFAT? Or should we better do the statistical analysis <u>first</u>? The answer is quite easy: For symmetric circuits, a corner analysis would give no offset voltage (maybe also infinite CMRR, PSRR, HD3, etc.)—so any WC corner search would completely fail! So here it

- Assume global variation of e.g. 2 σ
- Consider only global parameters (vthn, tox, xl ...) (1)
- Find parameter set for 2 σ Worst Case Point (2)
- Now consider only local parameters (dvth, mobility) (3)
- Based on global parameters find Worst Case Distance for a given Specification of a performance (e.g. Write Level, Iread) (4)
- Find the performance for local variations representing the array size (5)

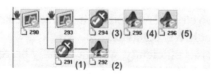

Figure 9.7   Stepwise manual worst-case finding in a commercial design environment (Courtesy of MunEDA).

is obviously much better to run first a statistical analysis for statistical corners, like WCD. In this case, we can obtain a WCD for offset voltage (and other performances), and in the subsequent WC corner search we would find out the WC appears at maximum temperature, and maybe maximum (or minimum) supply (due to channel length modulation).

A second (strong?) argument against inspecting first environmental corner is simple: Doing so would be in contradiction to what happens in reality! Of course, first the fabrication needs to be done (setting the statistical parameters); then, the devices have to be tested at the environmental corners. Then finding the first case has a better chance if we use the bad production samples!

This leads to a flow with "improved" order (Figure 9.8). A third argument to treat statistics first is that often mismatches (and process variations) are very critical, so it makes sense to care first for the first-order effects (and in many *analog* designs, mismatch is a major effect!); this makes sure that the design will not behave in an artificial ideal way. In the questions and answers section we also discuss a more mathematical "proof".

If such "well-ordered" OFAT-oriented method would work *in general* is again a more difficult problem! We can expect that if there are strong correlations between statistical parameters and deterministic ones, also such enhanced "well-ordered" OFAT approach could fail. Luckily "fail" might be not so critical anymore because it would just mean you get a certain combined WC corner, which is unfortunately not as extreme as the true WC corner. So if you add by hand some small safety-margin, you may accept the risk even for sign-off. In addition, you can improve the flow by adding expected WC corners manually. Because in the proposed corner-based flows the sizing

Prepare design, create testbenches, set specs, specify desired yield

Set design parameters x_D

Run a WCD analysis at nom. conditions

Run WCC at each WCD to get overall WCC

If design fails at overall WCC decide for another design point $x_D = x_D + \Delta x_D$

Figure 9.8 Improved design flow with WCD first and then WCD.

is done in a fast loop without full MC analysis, it is best to double-check the optimization result in a succeeding separate verification. For efficiency, such split between a fast sizing loop and a final verification makes generally sense.

Let us check the flow (inspired by the op-amp example) in our inverter example (Figure 8.10) and try to solve the WCD problem manually: As mentioned, the delay-WC on V_{DD} and process is almost trivial (even OFAT would have no problems), but the temperature characteristic is (quite often) more difficult, because the mobility drops at higher temperatures (so the MOS on-resistance increases), but the threshold voltage also gets smaller there. The later leads to a <u>larger</u> gate-voltage overdrive, and thus lower on-resistance. These two effects can cancel or reverse the overall TC. As a result, for very low supplies the threshold voltage effect can dominate and the WC can shift from maximum temperature to minimum temperature.

Mathematically we can model the problem as followed:

$$t_{pd} = kT/(V_{DD} - V_{TOp} - V_{TOn})$$
$$\text{with } V_{TO}(T) = V_{TO}(T_{nom}) - 2 \text{ mV/K} \qquad (9.3)$$

Note: This is a simplified description, just to make the difficult statistical optimization problem tracktable. On the other hand, it is quite meaningful and even a bit more nonlinear than the true relationship (having no pole due to subthreshold onset in real CMOS transistors).

As already mentioned, it is often better to treat first the statistical variables, e.g., in a WCD analysis for 3σ as yield target. In the inverter case, the threshold matters and we would obtain $V_{TOn} = V_{TOnnom} + 3\sigma/\sqrt{2}$ and $V_{TOp} = V_{TOpnom} + 3\sigma/\sqrt{2}$.

Now we need to find the WC among the normal corner variables, and we can do it via preordered stepwise OFAT (see Chapter 2). So next we can sweep V_{DD} and would find for sure that V_{DDmin} is the WC for delay. And last we would sweep T (with $V_{DD} = V_{DDmin}$) and get the desired overall WCC set. As already mentioned, this way a clever applied OFAT method would still give the correct solution, very efficiently. However, how can we generalize this and how can we avoid "oscillations"?

As mentioned, difficult variables (with strong and nonlinear impact) should be treated with quite dense sweeps, and—as we described in Chapter 2—they should be treated in <u>late</u> sweeps (if using OFAT, for full-factorial there is anyway no problem, but it is very slow method), because they decide mostly on the overall WC—whereas the more linear or less sensitive variables could be set earlier in the WC search. This is also in synch on how we treat statistical corners; we treat them first, and indeed most process parameters like sheet resistance luckily anyway only vary by maybe ± 15 to 20%, also mismatch effects rarely change the bias point significantly. In such ranges, the design should of course behave well—often in opposite to temperature behavior. In the inverter example actually also the process behavior can be regarded as critical or dominant, but it still does not matter whether you sweep first V_{DD} or on process.

You think inverter speed or op-amp offset are special, or maybe the other way round: Too simple to be representative? Actually, everything in custom design can be regarded as special, but there are also quite many *common* problems; e.g., many analog blocks show similar difficulties: At *low* supplies, most circuits become challenging due to transistor saturation, so especially at slow corner (SS) the DC gain may drop significantly, but usually not at fast corner (FF). This is normal behavior most designers anticipate anyway. However, if you choose for that reason quite short values for the transistor lengths to lower V_{Dsat}, then usually also the FF corner *may* show a somewhat lower DC gain. So sometimes at V_{DDnom} or V_{DDmax} the FF corner might be the worst-case on DC gain, whereas at V_{DDmin} it is usually still the expected SS corner. In our flow proposal, we sweep first on simple variables like V_{DD} and almost for sure we get V_{DDmin} as WC for DC gain, so doing later the sweep on process corners would give indeed the correct overall WC combination. Unfortunately, this approach becomes more difficult if we want to treat process variations as statistical variables, because we also proposed to vary statistical parameters first—to treat mismatch there is actually no alternative, as we have shown! Solving this conflict is possible in several ways; one obvious solution would be to split the MC run into two, and running the MC process analysis after setting of the less critical variables, as proposed. And a second solution would be of course to run the MC simulation already at an expected WC, which is luckily SS, and this is easy to anticipate.

Note: To some degree, good designs are usually also mathematically "better behaving," which means that bad designs are usually harder to optimize over WC corners. That stresses actually the importance of applying well-known manual best practices and sizing rules! Avoid dirty tricks (like hoping that the TC of a threshold voltage and a sheet resistance would cancel) and try to be on the conservative side!

Prepare design, create testbenches, set specs, specify desired yield

A MC analysis and a WCD search (at expected WC conditions if available)
In case of functional problems consider to reduce WCD-sigma
For outputs with extremely high yield we may need no accurate WCD

B Run WCC at each WCD to get overall WCC
Stepwise preordered OFAT/automatic is most efficient
Start with expected WCC & less sensitive, linear behaving parameters

Figure 9.9 Heuristic preordered OFAT for overall worst-case corner searches.

Summary and further possible improvements:

- For overall WC finding focus first on statistical variables then on deterministic range parameters. In the later, you may apply adaptive WCC methods or full-factorial but stepwise preordered OFAT search maximizes the speed and comes with little risk because the sensitivity and nonlinearity of the range and statistical variables is taken into account during setup.
- Optionally, we could iterate; i.e., after WCD and WCC, we could go back to WCD (and WCC).
- Remembering the design pampering idea, we can also adjust the yield target over the optimization loop, like starting with 3σ-WCDs then moving to the actual yield target like 5σ. One advantage in doing so is that finding a low-σ WCD is possible with lower number of simulations!
- As mentioned WCDs can only address the partial yield to optimize on total yield, you can use the approach explained in Chapter 4 for finding the overall C_{PK} based on correlations or via blocking-min. This optional step takes only very short time and is usually only needed if you need high accuracy and if many goals exist (because here min(WCD) can lead to too optimistic total yield estimation).

9.2.1.2 Worst-case corner effort reduction methods

Using corners instead of MC runs provides a big speed-up, but further speed-up in conjunction with optimization is still desirable, and it is often possible by reducing the number of overall WC corners. In principle, we would need one overall WCC for each spec, at least, so the number of WCCs can be quite large (like twenty or more). Therefore, a reduction, exploiting correlations, is desirable, at least for the optimization phase. Often this also helps to really understand a complex design, just because too many corner cases are extremely hard to manage.

Indeed, we can often merge environmental WC corners in x_R (due to spec redundancies), but can we do so also for WCDs and the overall WCC? Yes and no! Merging in range parameter corners is often easy because for simplicity we treat each variable with <u>discrete</u> values, like V_{DDmin}, V_{DDnom} and V_{DDmax}, but for WCDs we work on <u>continuous</u> variables x_S. For designs with a minimum count of specs, there is a low chance that we can merge WCDs, but the more specs we have the more redundancy can be expected, surely, e.g., in specs like power supply rejection: PSRR(DC) > 100 dB, PSRR(50 Hz) > 80 dB, PSRR(1 MHz) > 50 dB (or similarly for rise time and bandwidth, phase margin and overshoot, DNL, and effective number of bits).

Corner merging could be even done adaptively during the optimization; e.g., if design is far from the optimum point we can apply more corner mergings, to optimize faster. For an optimum verification coverage in late stages, we can stop on merging and also extend the set with "second-worst" corners.

The merging might be done based on correlations coefficients between the outputs, which may come from a parametric model fit or from nonparametric ranking methods (Chapter 5). As mentioned, at the beginning of an optimization or for low-yield targets, the worst-sample method can give an acceptable approximation to the true WCD. So we could rank the MC sample points for each spec directly: For instance, MC sample #109 may have rank number 1 (= worst-case sample in this MC analysis) on V_{offset}, rank 4 in THD and 3 in PSRR—so overall average rank is (1+4+3)/3 = 3. We can do this also for the other MC points, and pick as wanted "compromise WC" corner the one with lowest average rank.

An alternative method would be inspecting the (nonparametric) correlation coefficients among the outputs, which reach from $c = -1$ to $+1$. Based on that we could merge statistical corners which have a correlation $|c| \geq 0.9$. Note that this method is a bit riskier, because most correlation measures take the whole dataset into account to get a low variance in the correlation estimates. However, for worst-case inspections it could be better to focus on tail samples, because actually the correlation might be impacted by nonlinearities, which could cause that the correlation across tail samples is different to a correlation obtained from the all samples.

Besides pure merging we can also use other speed-up techniques; e.g., if one spec is totally uncritical we may simply skip simulating its related WCD. This makes sense at least if such performances have a long simulation time and if the WCD performance on this spec is similar to the nominal performance. In addition, we may generate no WCD, e.g., if the standard deviation of this performance is very small (leading to a large C_{GPK}), and similarly we can do for WCCs. Actually, a designer works in the same way: He/she runs sweeps and focuses the work on things that matter, not wasting time on things that work anyway.

After the WC mergings, we would ideally end up with a moderately large set of merged worst-case corners. And the number of corners in this set would be the number of fighting spec groups! We could do it also do vice versa: Given we want six balanced near-WCC (compare to Figure 1.7), so we could adjust the ranking or correlation criteria till we reach that number. Or we may set the correlation limits to a value giving us, e.g., a corner count reduction by 50%.

For low-sigma designs and at the beginning of a circuit optimization, it could also make sense to reduce the effort inside the search for statistical corners: Doing a full WCD search for each spec at an expected WCC is usually nice to have. Unfortunately there is no single WC condition for all specs and doing a MC analysis or a full WCD analysis on all these different conditions would be a significant effort. Therefore, a MC or WCD analysis at nominal conditions or only at the expected most-critical overall WC makes sense. For analog circuits, this is usually the combination of minimum supply and slow MOS corner, because here many key performances (like distortion, output power, PSRR, CMRR, gain) become critical, whereas on performances with almost opposite behavior (such as breakdown or leakage) often some overdesign is usually easier to apply. Of course, the design should be robust enough to not fail completely on other specs like stability.

9.2.2 Fast Full-Yield Optimization with Heuristics

Let us now pick up the overall worst-case corner yield optimization concept (Figure 9.10) and extend it with the discussed searching heuristics. We already mentioned that design is "pampering," so for final verification we need to set the yield target according to our desired production yield; this might be a high-yield target like 4.5 or 6σ. However, in the design tweaking phase the circuit may show bad behavior for such extreme statistical corners. Maybe the circuit is not even functional at such high-sigma WCD, because it may cause a strong nonlinear behavior and big problems for an optimizer (maybe also for circuit simulators and output evaluations). Therefore it could make sense to increase the yield target step by step during the optimization. In addition, the WCCs and WCDs may <u>change</u> during the optimization (next subchapter).

Prepare design, create testbenches, set specs, specify desired yield

Set design parameters x_D

> Adaptive search for overall WC corners for both deterministic range parameters x_R and statistical parameters x_S for given yield target

Let optimizer change design point $x_D = x_D + \Delta x_D$ to minimize goal function representing the spec violations

If no improvements anymore go back for new WC corner creation

Figure 9.10 Concept of fast yield optimization based on overall worst-case corners.

Also note: Ultimately, the sigma level should be set (finally) according to your <u>desired</u> total yield. Indeed, maybe only for <u>some</u> specs a high yield target like 6σ is really needed, whereas for power consumption 3σ might be regarded as good enough. However, a <u>consistent</u> yield spec is surely less confusing; having some "uncritical" specs at low sigma level lead to nice and tight specs, but actually you foolish yourself and add room for misunderstandings. Better make clear statements and demonstrate consistent spec limits!

Actually the key concept for fast yield optimization is quite straight forward (Figure 9.11), but we can plug-in quite many advanced mathematical and heuristic methods:

1. Using expected worst-case corners, e.g., at the beginning or for speed-up the searches.
2. Apply fast stepwise preordered OFAT at the beginning but later go for deep adaptive search methods, optionally denser sweeps, tighter internal tolerance settings.
3. Apply corner mergings in case of strong correlations, especially in case of long simulation times, many specs and in early optimization stages.
4. Consider to skip WCD generation for performances with large $C_{GPK,}$ or to skip totally uncritical specs and variables in next iterations.
5. For reliable convergence and better monitoring includes expected WCC (like V_{DDmin} + SS) and nominal corner (without mismatch).

Prepare design, create testbenches, set specs, specify desired final yield

Set current yield target according to to final yield and design status
Set design parameters x_D

Find set of overall WC corners for current yield target, if functional
problems then lower it otherwise increase if needed it till we reach final yield

Optimize design based on performance goal function

If no improvements anymore go back for new WC corner creation

Figure 9.11   Yield optimization with overall worst-case corners and adaptive yield target.

6. Reuse performance models created for WC search, instead of generating them always from scratch, reuse internal optimizer data (e.g., inverse Hessian if using quasi-Newton).
7. In late stages, check also for total yield (based on correlation and partial yields) to update WCD sigma levels of the individual specs.
8. Extend the pampering idea if there is too little progress, like consider to relax specs on power or area.

Which mix is best depended on many things like how time-consuming the circuit simulations are, how the testbench structure looks like, how complex and nonlinear the optimization goal function becomes, and how much compute power and simulation licenses are available. In our chapter "Design in Pictures Six," we give some examples.

9.2.2.1 How the worst-cases may change during an optimization

We mentioned that from time to time during the optimization, an update on the worst-case corner set is required. If we do not, we may slow down the optimization, but how much actually depends on several parameters, and of course some slow-down might be fully acceptable, because we would save simulations for worst-case finding. In the simplest WCD cases, like for offset on a differential pair, even no update is needed, but we cannot expect this in more difficult cases. So let us inspect a typical analog optimization, like a two-stage amplifier and looking again to offset and, e.g., area (or speed) as conflicting goals. One advantage is that we can directly calculate the WCD by Pelgrom's square-root law. Assume a gain of 2 (6 dB) of the 1st stage, so here the matching matters two times more than for the 2nd stage. For instance, for the starting point of the optimization, let us assume an area of 1 um² (for each stage) and a matching constant of 1 mV/um (related to a diff-pair). We get an output offset sigma of 2 mV from 1st stage, and 1 mV from the 1st stage, so in total we see $\sqrt{5}$ mV = 2.236 mV. If the spec is at 4.472 mV, we would get a 2σ yield only (C_{PK} = 0.667).

To improve on offset and yield, we should obviously increase the area, and best with focus on the 1st stage. So let us increase the area by $4\times$, to get an output offset of 1 mV from 1st state, and now 1.414 mV in total. It is not unrealistic that the optimizer would stop here (e.g., because the area limit is reached or because bandwidth is now significantly lower). The partial yield on offset is now $4.472/1.414\sigma = 3.163\sigma$ (C_{PK} = 1.054) which could be regarded as a significant improvement! Look up, this is an accurate analysis, but actually we have not used the WCD concept, so what does it mean for a WCD-based optimization?

For simplicity, let us assume 3.163σ is also the target yield, so it would be native to use this sigma level for the WCD! We know that the 1σ-WCD gives us 2.236 mV at the starting point of the optimization, so the 3.163σ-WCD gives 7.072 mV. For this we need now the setting for the mismatch variable at each of the two amplifier stages, which are 2.83σ and 1.41σ; this gives at the output indeed 2.83 mV·2+1.41 mV = 7.07 mV.

Looking now to the optimized result with doubled input area, we would get for the WCD an output offset of 4.26 mV (which is slightly too optimistic), not exactly 4.472 mV. This small mismatch to our initial accurate hand calculation is because the circuit and the WCD have been *changed during the optimization*.

It is interesting to check (Figure 9.12) how large the WCD change really is, e.g., regarding the angle or the component ratio of the initial WCD (before optimization it was a ratio of 2) and the one for the optimized circuit (here the ratio is 2.27/1.575 = 1.414).

The angle change is 18° which looks maybe not that small, but the error in WCD magnitude is related to $1-\cos(\Delta\varphi)$ which is still small (5%).

The scenario looks simple, but it can be easily extended, e.g., maybe our optimization setup on parameters gives the optimizer more room for improvements. In this case, it is realistic that maybe also the 1st stage voltage gain increases (e.g., from 2 to 20) during the optimization, so that this way

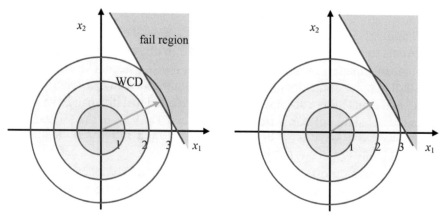

Figure 9.12   WCD for 2-stage amplifier before and after 1st stage area optimization.

the 2nd stage offset has much less impact. Of course, also here the WCD situation would change during optimization, just in another way (the spec boarder angle would change on first stage gain: with higher gain the second stage offset would become less critical).

Also for constant gain, further scenarios are possible; like of course, instead of changing only the 1st stage area, we could apply an even more aggressive area increase on it, but compensating the total area increase by making the 2nd stage area smaller. Or we may increase the area of both stages by some amount. In most cases, the WCD would change a bit and in a slightly *different* way.

What we also see is that the WCD changes can limit the optimization progress a bit, so a native workaround would be to overoptimize a bit by spec tightening. This would lead to some (small) over-design, which indeed could be probably only reduced further by *updating* the WCD <u>often</u> enough. And this is of course exactly what advanced yield optimizers will do.

How much will normal WC corners change? Actually there is no real difference, and of course also classical VT corners may change; and process variations can be anyway often treated with either process corners or statistical models! Some corners might be very stable, because the optimizer just has no option to change the design significantly on leakage current corner behavior or supply current behavior. However, this is no guarantee: If the TC of a bias current can be optimized, then a constant current vs T may give a significantly different behavior than a PTAT biasing. With constant current usually the phase margin PM becomes critical at low temperatures, whereas for PTAT bias it might be almost temperature-independent.

9.2.3 Advanced Yield and Surrogate-Based Optimization

A too heuristic method often leaves a bad feeling, and indeed we mentioned some case, where a unique, well-defined WCD simply does not exist. As described, the sizing on over-all WC corners is a very efficient method, applicable to *most* circuits, but what about the real difficult cases and even more adaptive methods?

We mentioned already that e.g., LHS could be used to treat not only the statistical variables x_S, but also the environmental range variables x_R. And of course, we could even include the design variables x_D; variables are just numbers. Simply sampling across $x = (x_S, x_R, x_D)^T$ would be similar to brute-force design style; and applying LHS or LDS would be not much better. However, indeed there are also interesting *adaptive* methods available in research, acting in more general way e.g., compared to sorted MC or WCD. In addition, there are methods which combine to some degree space exploration methods, multi-variate modeling and optimization.

Let us start with the latter; if we apply a space exploration, we can just take the best goal function point x_D as optimization result. If we do the exploration with standard random MC we would work like a pure random optimizer, and with an effordable sample count, we would at best come somewhat nearby the true optimum design solution. An improvement would be to generate a multi-variate model (called meta-model, because it is based on simulations) regarding x_D and to run further optimization steps (e.g., via BFGS). This way the model would provide some interpolation, and we could come much closer to the solution. The good thing is that for these optimization steps there is *no* need to run true time-consuming circuit simulations. We just need to evaluate the model, and this is orders of magnitudes faster.

This basic idea of using a model as surrogate is called surrogate-based optimization (SBO). The advantage is a huge speed-up in optimization, at least if we exclude the model creation part. In addition, the modeling part can help to make the optimization easier, e.g., we can create a smoothed model, although e.g., the original simulation suffer from some numerical noise. In yield optimization this point is important, and SBO is a quite native choice, because the natural unbiased estimator for the yield, the sample yield from a MC analysis, is indeed quite "noisy"!

So optimization e.g., based on the C_{PK} or C_{GPK} or WCD is a kind of surrogate-based optimization, at least in a wider sense.

In addition, we can use the SBO idea iteratively. Many such approaches are similar to a step-wise "MC-across-corners" approach: Start with a (relatively) small sampling set $x = x_1$ (e.g., n = 500 points) and then extend the sampling for the most interesting regions, with focus on the variables with largest and most nonlinear design impact [Shan2011]. This way the optimization (for WCD or sizing) is not a separated part anymore; and e.g., compared to sorted MC the internal simulation effort is reduced too.

In [Wang2003] a method is described which picks up latin hypercube sampling and adaptive response surface modeling methods, plus the use of simulated annealing for global optimization. This way the method can adapt itself to the complete problem structure efficiently, ending up in a quite acceptable number of circuit simulations, even for highly nonlinear problems with multiple local minima.

So you may ask, how can I use it? Unfortunately, commercial implementations are still missing. And also in research the problem of high-yield estimation is often excluded.

9.3 Connecting Design Methods

More advanced analysis methods can be created by linking several basic analyses (like Monte Carlo and sensitivity or optimization and corner finding). This can be done manually or by scripts; the later makes sense for standard tasks often required during design and verification.

Typically, designers try to solve the problems directly, e.g., by improving the schematic or (if necessary) the testbenches or even go back to system design plus considering spec changes. These tasks are extremely hard to automize, and a fully automated approach for circuit design might be either very limited (like to very special circuit classes) or very complex. In Figure 9.4, we excluded the debugging and setup parts, and in general it is probably very hard to have more automation, because it is complicated to provide all the many inputs up front, and to forsee many decisions (like which parameters to optimize, in which range, with which weights).

On the other hand, there are many further design tasks to be solved and there are also many references discussing really highly advanced topics, like yield optimization including stress-induced aging effects [Pan2012], or optimization including a topology selection step. In commercial tools, this is partially possible too by creating user-defined scripts.

To some degree, this means "back to the roots," because (roughly in the 90ies) before graphical user interfaces become popular, designers use scripts quite intensively, and many simulators have built-in analysis capabilities too. For instance, the PSpice^{TM} plotting tool Probe^{TM} has a great macro-recorder to just do the evaluation one time manually, and having it immediately automated for all succeeding runs. Also some older commercial (like Aptivia^{TM} from Antrim, bought by Cadence Design Systems in 2002) or some in-house graphical design tools offer similar features, plus control blocks like loops or conditional execution.

User scripts are often in competition with built-in features of the design environment; in some aspects, built-in functions are preferable. For instance, optimization is not only for improving a design, but it could also provide detailed sensitivity information—almost "for free." Something similar is possible by running a contribution analysis based on existing MC results. Actually such "by-products" are highly desirable and unfortunately harder to obtain via user-scripts.

A second advantage for built-in methods is that most advanced methods work with a kind of internal model or memory; e.g., the BFGS optimizer uses the inverse Hessian matrix, or classical WCD uses parameter screening methods, and also the starting and endpoints for the WCD optimization step could be reused. Reusing and/or extending this memory could speed-up the execution significantly; e.g., if we have already executed a sensitivity analysis, why should the optimization start "from scratch," without taking earlier simulation results into account? In fix problems like circuit simulation or MOS transistor parameter estimation (e.g., for a BSIM model), such reuse methods can be applied easily, but unfortunately in circuit design there is also a risk that the design is subject to changes (like adding a cascode, changing a resistor), so such reused information might be unfortunately too old, so maybe misleading.

What makes highly sense is putting obtained worst-case corners in a database. Those corner combinations which appear very often should be put, e.g., in a corner template so that all designers in a team can quickly check their circuits at known difficult conditions. Some combinations are of course quite trivial, like the worst-case on leakage current (being at maximum supply and temperature), but others may give indeed new insights into your designs. Also of course a worst-case corner search could reuse older worst-case results— even if the design has changed a bit, the fundamental trade-offs will usually not change much. High-yield techniques (like sorted MC or WCD) might be improved in similar ways; e.g., information on fail boundaries and critical variables could be reused or even specified up front (designers often know well about critical nets or instances, like from design tweaks, intuition, or earlier sensitivity analysis).

Data mining and "big data" are hot topics, maybe also for future IC design. We can expect more in that direction and surprising results. Designers often

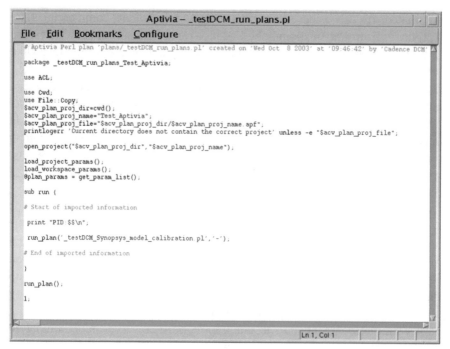

Figure 9.13   Scripts as integral part of circuit design, also in graphical environments [Klevbrink2005].

exploit their knowledge of the circuit, and often for a certain task (like change in a capacitance) there is no need to run the *whole* verification again from scratch. This idea could be applied in many ways, like for speed-up of post-layout analysis [Wang] or for speeding up optimizations. The math behind such techniques, like Bayesian statistics [Jaynes2003], is quite well founded; just commercial EDA implementations are missing.

In this subsection, we discuss some further advanced front-end design and analysis tasks. One already established link was putting worst-case search and sizing by optimization together in an iteration loop for yield improvement. This is one key step in automated variation-aware design. What about linking other typical design tasks? Or applying reuse techniques between analyses?

9.3.1 Script-Supported Design

Design flows based on optimizers are executable in many modern design environments, further automation—even a kind of synthesis—is possible by capturing design steps in a script [Crossley2013]. Such synthesis scripts may, e.g., even include design calculations, checking for conditions, construction steps or even a topology selection. At the extreme, the designers output would be such script instead of just a single-sized circuit! Maybe this will become a future fun part in the analog designers work, but currently there is only little infrastructure supporting it, e.g., in making it technology-independent, offering a GUI, and an object-oriented circuit programming language. Of course, such scripts have to be created, maintained, etc.— usually all these have to be done by hand. For sure, such scripts are (much) harder to read then schematics, unfortunately. We pick up such IP topics in Chapter 10.

In general, often even simple design-supporting scripts, which just collect simple tasks like documentation and datasheet creation, can make the designers work easier. As mentioned, such script-based approach is always a bit in competition with methods anyway built-in to design environments. Datasheet creation for a design review is a good example. If simulation data are already available, the datasheet creation task is nothing else than data collection, e.g., in a HTML file. In many design environments, you can create simple scripts, which run, e.g., a nominal simulation and a corner analysis up-front and then create an HTML datasheet. A native environment extension would be just to offer different kind of datasheet creation, with running up front some standard analysis (like sweeps on supply and temperature, corners, Monte Carlo); plus supporting further options, e.g., on including schematics, testbenches, setting the level of details in compliance tables.

In the past, scripting comes often with learning about special programming languages like SKILL$^{\circledR}$, Perl or Tcl, but nowadays most environments offer also graphical script creation (see Figure 9.13, design description for sensor front-end and $\sum \Delta$ ADC at [Osman2016]).

9.3.2 The Split Monte Carlo Method

Designers often prefer MC with mismatch only or process only, because this gives more design insights compared to a MC run with all statistical variables. This technique is also often used in memory design, so can this idea be exploited in general to speed-up e.g., the WCD search? Or are there other advantages?

A typical situation is this: Some performances are strongly impacted by mismatch, but not much by (global) process variations, like offset voltage, maybe also PSSR and CMRR. Of course, there are also performances with

Table 9.1   MC results for mismatch only and process only of a typical analog block

	Mismatch MM	Process P
σV_{offset}	10 mV	1 mV
σI_{Leak}	<1%	20%
σI_{DD}	5%	5%

almost the opposite behavior, like leakage current, delay or filter cut-off frequency, and several outputs are usually impacted from both, like supply current I_{DD}. Table 9.1 gives an overview on such a typical design.

How can we combine the results or compose WCDs from such results? Actually the larger variance counts much more, so is dominating the WCD by far, only for "balanced" outputs like I_{DD} we really have a mix and for 50–50% a 3σ-WCD would be composed (assuming normal Gaussian behavior) as 3σ-$\text{WCD}_{\text{MM}}/\sqrt{2} + 3\sigma\text{-WCD}_{\text{P}}/\sqrt{2}$! However, we should notice that in difficult cases the method may fail; e.g., offset or PSRR might be perfect in a MC process-only analysis, but not for mismatch, so the WC samples on process can never be found correctly!

One further problem is that such two MC analyses just take two times more simulation time theoretically, so the major advantage is the WCD speed-up from having a lower count of variables, offering slightly higher accuracy, and lower internal runtimes. In random MC, the speed disadvantage for doing two runs is usually still significant, whereas with LDS the negative impact of high dimensionality could be reduced with the split method.

What designers also often do is that they speed-up the MC–MM run by only doing a fast DC analysis, e.g., for offset, I_{DD} and DC PSRR, so they use their knowledge that, e.g., the WCD on MM for leakage (or, e.g., phase margin, slew-rate) often anyway does not really matter!

To let a tool follow this manual approach, a kind of plan is needed; e.g., in this it can be decided in which way the WCD (and, e.g., the closely related contributions) has to be found. Modern analog design environments allow the execution of such verification plans.

So all-in-all, this split MC method is useful for design insight and also for finding small effects, and even further splits may make sense, like on wiring parasitics (as we have done for our latch comparator) or for only chip-external components (e.g., to decide how much tolerance can be accepted in these elements).

How to combine statistical results? It is well known, and we mentioned it already, that for calculating an over-all sigma we have to add up quadratically. The same approach is also good for combining statistical and bias errors into an overall rms error:

$$\varepsilon_{tot}^2 = \sigma^2 + \varepsilon_{bias}^2$$

Now let us extend the scenario a bit. Imagine you have two results (e.g., from MC or from a laboratory measurement), and you know the result should be the same, but actually both results differ due to statistical variations. How can we combine the two individual results into one hopefully more accurate overall result? If we would have the same accuracy for the two measured results, it would be native to take the arithmetic mean (average) as overall estimate! And the overall standard deviation would go down by $\sqrt{2}$. This method is correct if we have no correlations and Gaussian distributions, and the idea can be extended ending up in a general theory! To some degree, it is even possible to treat your designers a priori know as information with a certain tolerance, and to combine this with simulation results following certain model assumptions. Or we can combine results from a previous simulation run (like an older big MC analysis) and a new one (maybe a shorter one), instead of making one big analysis from scratch, without information reuse.

9.3.3 The Eye Opener

Maybe the most important thing for you as designer is to understand "where" the design currently is, plus "where" and how you can improve—with respect to statistical variables, design variables, and corners.

RF PA engineers do not only use small-signal S-parameters but also large-signal S-parameters. With sensitivities we can do almost the same. Sensitivities $S = \Delta y/\Delta x$ are useful for design understanding, and they are also a native starting point for WCC finding.

The best way to get the design status and sensitivity information is to run OFAT sweeps (with 3 or more points) on corner and design variables to get spread of all performances, e.g., for temperature, V_{DD}, C_L. In addition you should run two short MC analyses—one on mismatch, one process only—to get, e.g., the $\pm 3\sigma$ spread for all performances. A table with sorting option can give you a perfect overview, e.g., to check which performance improvements are needed and also how you can improve, e.g., by reducing mismatch or more by improving on PSRR (reducing V_{DD}-spread) or by changing the bias concept (PTAT, constant current, replica biasing, calibration, etc.) for lower temperature coefficients, etc.

Such performance analyzer could be even extended by creating performance models from these "large-signal" sensitivity analysis results, supporting what-if analysis or, e.g., giving warnings in case of strong nonlinear behavior (like bandgap output voltage having a significant curvature) or non-normal data where the split MC method may fail. A further output easy to create would be, e.g., to show at which (extrapolated or interpolated) parameter value the design would start to fail on each spec. So you would, e.g., see that over –40 to +100°C the design fails, but using interpolation methods you may get that it works within 0 to 80°C. Or if the design is fine the targeted environmental conditions, you would, e.g., get that the design starts to fail at +150°C. By doing mixed sweeps or a full WC corner search (as described in Chapter 2) also correlations and the impact of the other parameters could be included.

Most design environments support automatic datasheet generation, a worst-case corner analysis and as by-product you can often get also sensitivity tables, correlation plots, regression results, etc. Check-out the documentation, often just clicking a special menu item is needed. With scripting features you can extended and automate such analysis further. Figure 9.14 gives an example: The contribution table reported that the dominating variables on V_{offset} are V_{CC} (24% contribution) and temperature (69%). So it is native to plot them for more circuit understanding. Indeed the correlation coefficients are large (beyond 0.5) and the relationships are quite linear (as it should be for a robust design).

One outcome could be also how wrong you would treat the design if you would only look to OFAT results and if you would ignore correlations. If you see difficulties, then you are often in a situation where circuit optimization makes sense.

9.3.4 The Spec Inverter

Spec negotiations can be hard, and it is good to have concrete numbers and more than a gut feeling. Usually we think specification are given, but actually many block specs are "soft," and a designer should be aware which requirements are really "a must," and which are more a kind of recommendation or guideline (e.g., the current consumption for a block which is anyway consuming not much power or the area of an anyway small block, used only once in the chip). In principle, you can run, e.g., a MC analysis with specs set to current knowledge and obtain the sample yield—but you can also adjust the specs and check how the yield would change! Using a less quantized estimator

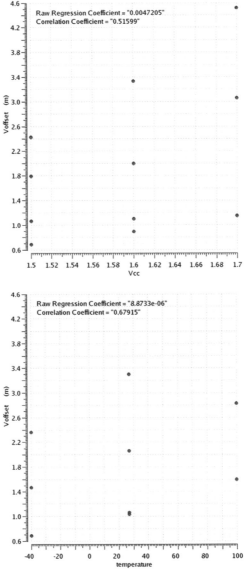

Figure 9.14 Sensitivity plots obtained from our 3-stage CMOS op-amp WC corner analysis.

like the C_{PK} (in case of normal data) or the generalized C_{PK} (also valid for non-normal data) you can even find the specs quite accurately (and even if sample yield is 100%). Such spec-yield trade-off analysis could be useful

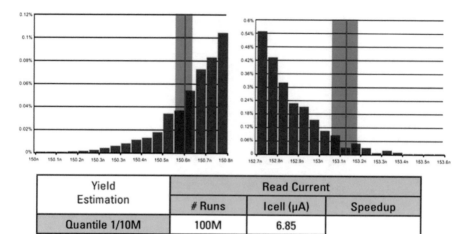

Yield Estimation	Read Current		
	# Runs	Icell (µA)	Speedup
Quantile 1/10M	100M	6.85	
VARIATION MANAGER	20k	6.84	5,000 X

Figure 9.15   Silvaco's high-sigma performance limit analysis [SilvacoVM].

Note: The blue bars represent the confidence interval.

for discussions with system designers, quality experts, and customers. Most statistical analysis packages support such yield vs spec analysis, so once you have exported the MC results, e.g., as comma-separated value files (CSV) it is mathematically quite a simple task; we have to use the inverse cdf (percentile function).

Some design environments have already some built-in features for this task (Figure 9.15).

9.3.5 The Automatic Optimization Parametrizer

In many cases, we can assume that for any given circuit structure designers know anyway quite well *which* components need to be tweaked, but there is some risk that some parameters are overlooked, especially in new circuits. So even if a designer follows many best practices for an optimization setup, the design result could still depend significantly on the designer's expertise. So instead of purely relying on the designer's decisions which parameters to optimize, we may run systematically a full sensitivity analysis (e.g., a mismatch contribution analysis) and optimize the top-10 parameters with biggest impact. Unfortunately, there are also some key problems: These top-n components do <u>not</u> really lead automatically to a good optimization setup! For symmetric circuits, we can expect a parameterization exploiting the symmetry is much better than one just directly optimizing on the top-n components! Usually the designer has such and further constraints, like on current mirror ratios. Also it can be almost trivial that, e.g., the transistor length has a significant impact for speed reasons, but the best transistor size might be anyway the minimum length (e.g., if this transistor is not critical regarding mismatch or voltage breakdown), so there is no need for optimization on it.

Another problem is that such mismatch-based optimization setup would be to often exclude critical transistors which are large, thus having only small mismatch. For instance, a big output transistor or a switch could be still critical, e.g., due to absolute tolerances. To include them it makes sense to use the mismatch results again, but to normalize on $\sqrt{area}$. This way can obtain a second ranking table and find the optimization "candidates" which would be overlooked in a non-weighted contribution analysis.

Many of such rules might be collected e.g., as constraints (see Chapter 10). In EDA tools you can also use structure-based constraint finder or "circuit prospector" (Figure 9.16 [Dennison2010]), and even fully automated methods are available [Eick2011].

Ultimately all these together could be arranged to obtain an almost fully automated or at least highly assisted parameterization setup for optimization. This automation makes sense depends of course on how familiar the designer is with applying the parameterization just quickly by hand—and how much support he/she gets anyway from the environment, e.g., via assistants. In addition, when doing such automatic parameterization, we could in principle easily pass *further* information to the optimizer, like which parameters are the most sensitive ones. If many variables have to be optimized, the optimizer could focus first on these parameters and speed-up the optimization compared to an optimization on (too) many parameters.

9.3.6 The Circuit Terminator

Most analyses presented in the book are for <u>design</u> purposes, but if you are only interested to find design weaknesses, e.g., for pure verification or for a design review, you can organize, e.g., WCD and WCC search in a slightly different way—with focus on "breaking" design "efficiently"! Such analysis might be part of a big regression run, and the people triggering it might be not the same who will fix the circuits.

We can pick-up many WCC search ideas, like running OFAT sweeps, but for instance, instead of terminating the search loop when <u>all</u> WC combinations have been found we may already stop once we found the <u>first</u> (usually most

Define default parameters to be optimized for each device type
(like w and l for transistors)

Prepare design, create testbenches, and automate performance extraction

Analysis circuit structures and apply constraints
(like $w_1 = w_2$ and $l_1 = l_2$ for a diff-pair)

Run mismatch contribution analysis

Add most sensitive instances to list of optimization devices

Scale-up contributions acc. to area to identify critical large devices

Add again most sensitive instances to list of optimization devices

Match parameters by applying constraints in structures like current mirrors

Display list of parameters to be optimized

Allow manual edits, setup specs, run optimization

Figure 9.16   Flow for automated circuit parameterization.

critical) worst-case and spec violation! Of course, in the design phase on a non-optimized circuit such analysis may stop early, maybe already at nominal conditions, but for sign-off it can be very helpful, also for intermediate debuggings.

We can even extend the whole idea in other direction: If the design passes the specs even at all WC corners, we may extended the range parameter limits further, e.g., the temperature range till we reach the "break" point. Or we can speed-up our search by starting from an expected WC or giving a ranking on parameters to let the search start with the most critical ones, like temperature or process corners.

9.4 Design with Pictures Seven

As we have treated worst-case distances and corners, and also optimization in detail, let us now combine both for a yield optimization. Doing, e.g., a pure nominal optimization for a comparator makes little sense, because without inclusion of mismatch important characteristics (like offset) would not be addressed—and also not their trade-offs with other targets (like speed)!

Instead of doing a MC analysis at all corners in the circuit sizing loop, we should use overall worst-case corners to treat both range and statistical variables simultanously. So let us start with examples on this subtask.

9.4.1 Overall Worst-Case Search in Action

Using our CMOS inverter becomes a bit boring; we extend it now a bit and apply the flow described in Figure 9.19. One aspect we want to show is the corner merging, so we keep the design itself simple enough to be able to understand all nonlinear effects but we need to make the variable and spec setup already quite complex and realistic enough.

Assume we need to design a clock buffer using two CMOS inverters; we have specs for these performances:

- Speed-related: t_{pdLH}, t_{pdHL}, t_r, t_f
- Input-related: C_{in}, $V_{th} > V_{min}$, $V_{th} < V_{max}$
- Output-related: R_{outH}, R_{outL}
- Transfer-related: DC Gain
- Noise-related: jitter
- Power: leakage current, dynamic average current
- Other design goals: area, target yield, etc.—do not depend on x_R

The corner variables x_R could be, e.g., process corners, temperature, supply voltage, load capacitance, and generator impedance.

Due to given specs, we can assume in the extreme case one worst-case corner for each of the specs (15 in total), but we can also expect some strong correlations across the specs, like among the speed-related specs, but also, e.g., C_{in} and area could be redundant specs. For a single inverter circuit, the correlation would be even close to 100%, but for a more complex design this is not so clear. For instance, in an op-amp highly optimized for low offset the input stage transistor area may dominate, so both specs would be still highly correlated, but in an op-amp designed for high-output drive capability the correlation may drop to an unimportant level.

We can expect also several other strong correlations, e.g., if our speed specs are very challenging, we should use minimum length transistors, and this way there would be a strong positive correlation, e.g., between leakage current and C_{in} (also for area and R_{out}). In analog circuits, we have, e.g., usually multiple speed-related specs such as rise time, settling time, slew-rate, and small-signal bandwidth; and also here we can expect strong correlations, so there is a good chance for some WC mergings.

Such corner mergings are usually already done by default in many design environments, so the output shown in Figure 8.21 shows overall 8 WC corners although we have 13 specification limits. In this simple voltage buffer circuit and for the moderate range parameter variations also the OFAT method works quite good (only one difference to full-factorial) and it needs only 13 simulation points, whereas full-factorial requires 405 points. One "big" joint corner is for rise/fall time, Lo-to-Hi/Hi-to-Lo delay, and R_{outH}/R_{outL}.

We can also modify our setup a bit; e.g., if we only have temperature, supply voltage, and process as corners, we could even merge our WC set further. We end up in now only seven merged corners.

On the other hand, if we use, e.g., more temperature steps and have a circuit with difficult temperature characteristic, we may end up in more overall corners and cannot merge much anymore.

Running a MC mismatch analysis on this buffer would show that mainly the DC gain and the offset voltage would vary significantly, but both actually still significantly less than within the pure corner analysis. For instance, the corner spread on relative threshold is 17%, whereas $\pm 3\sigma$ is only 2%! However, gain and threshold are not much correlated, so we should not merge them.

Note: For more advanced technologies and very small transistors we can expect that the standard deviation from mismatch will grow, so that can reach the regions of the process (corner) variations.

9.4.2 Comparator Yield Optimization

For our latched comparator, we already inspected WCDs and mismatch contributions. Both can already help a lot for setting up an optimization. This is usually more reliable than just only making an educated guess on which parameters we should optimize; it is of course also more efficient than just optimizing everything. One example is the output CMOS latch part: Without sensitivity analysis, many designers would have overlooked it and would not optimize the latch transistors at all—ending up in a non-optimum circuit. Our DUT is also a nice example of combining construction and optimization, because based on the sensitivity and a bit trial-and-error we actually extended the circuit a bit to give the optimizer enough freedom to really end up in a good circuit. In more complex circuits like a multi-stage op-amps or in a high-precision bandgap, even cleverer construction techniques (like adding a cascode stage or introduce a clever frequency compensation scheme) can often nicely complement optimization.

On the other hand, our DUT is still quite compact, a more complex circuit would have also more elements where optimization is simply an over-kill (like for some bias parts or shut-down or mode selection transistors). In opposite to our LC band-pass filter, we should optimize many transistors in synch, just for symmetry reasons! This makes sure to keep important sizing rules, reduces the number of parameters, and increases the optimization speed. In some cases, it is a bit arbitrary how much matching between different transistors we should really apply: In the output latch, we may match not only for symmetry, but also, e.g., across the lower and upper transistors of the series PMOS transistors. This is usually better for layout reasons and should still allow a near-optimum circuit and it can give further optimization speed.

As we have already inspected the comparator statistical behavior, let us also inspect, e.g., temperature and supply behavior and run a worst-case corner analysis on them, best on top of the statistical corner results to get the *overall* worst-cases (Table 9.2).

Note: A further range parameter x_R could be the input common-mode voltage V_{cm}. Indeed, the optimum circuit would depend on V_{cm}, but on the other hand this makes most sense if you would have a clear application in mind, like using a fix reference voltage of, e.g., 0.6 V$\pm$5% or using $V_{DD}/2$ as reference or whatever (like selling an universal comparator with wide-range inputs)—and the principle methodology would not change anyway. In fact, in defining such details and in setting up good testbenches is quite a lot of work to actually prepare manual circuit tweaks and optimization! The testbench setup can have also big impact on optimization results, e.g., in our comparator testbench actually the output rising edge matters, but the falling edge not, so if we optimize the NMOS and PMOS width of the comparator the PMOS will become quite large to drive the load capacitance fast enough, whereas the NMOS becomes smaller and smaller during the optimization to decrease its input capacitance! If this is undesired, we may either extend the testbench or link the NMOS and PMOS width to get

Table 9.2   Latched comparator over-all WC corner results for V_{offset} from a statistical corner

V_{CC}/V	Process	Temperature/C	Pass/Fail	V_{offset}/V
Corner results:				
1.4	**FS**	**−40**	**fail**	**0.0212**
1.4	FS	27	fail	0.0200
1.4	FF	−40	fail	0.0200
1.4	FF	27	fail	0.0190
1.4	FS	100	fail	0.0189
1.4	NN	−40	fail	0.0186
1.4	FF	100	fail	0.0177
1.4	**NN**	**27**	**fail**	**0.0174**
1.2	FS	−40	fail	0.0169
1.2	FF	−40	fail	0.0163
1.4	SS	−40	fail	0.0162
1.4	NN	100	fail	0.0162
1.2	FS	27	fail	0.0161
1.2	FF	27	pass	0.0154
1.2	FS	100	pass	0.0153
1.4	SS	27	pass	0.0149
...	...	...	...	...
1	SS	27	pass	0.0084
1	SS	100	pass	0.0081
1	SF	27	pass	0.0080
1	**SF**	**100**	**pass**	**0.0075**
Relative contribution:				
76%	9%	15%	–	–

a usual width ratio between both transistors. A similar effect happens on the first stage PMOS load: These transistors are not that critical, so they become quite small during optimization (which minimizes their impact on decision speed). On the other hand, these transistors have a reset function, and for a moderate clock period of 5 ns it is indeed possible to make these transistors quite small, but if you want to use later a faster clock (or a lower supply) the comparator might become completely non-functional! So you should really use the worst-case also on clock frequency for the optimization (maybe even with 20% margin) or even treat f_{clk} as further range parameter. Of course, such "findings" from the optimization can also trigger manual circuit modifications; e.g., we may optimize the timing for the NMOS diff-pair and the PMOS load by having a separate clock driver for each. This higher complexity could make sense if you cannot fulfill the specs but have some margin on area.

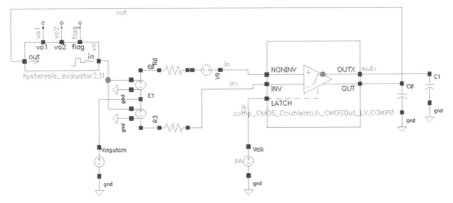

Figure 9.17 Latched comparator testbench.

Spec setting is critical too: It could easily happen that from system design highly contradicting specs are setup, e.g., a low-input capacitance is desirable, but having too hard specs on this may lead to too large offsets. Also speed, power, and noise are usually in competition—and it could happen that your circuit topology or even your technology can never fulfill all specs simultaneously. For high-performance designs some trial-and-error is hard to avoid.

Another aspect is, e.g., to include the surrounding circuitry correctly, like load capacitances and generator impedances; without the latter we would not cover kick-back effects correctly and with $R_G = 0$ we may design a comparators with far too large input capacitance and charge kick-back! It can easily happen that you end up in multiple testbenches to really cover all effects, so for an op-amp you may want to have a testbench for closed-loop and one for open-loop analysis.

In opposite to simpler examples (CMOS buffer and op-amp peaking) for the latched comparator, both statistical variables and range parameters are quite important. To get an offset at all we need to create first a statistical corner regarding mismatch, then we can run a WC corner search for supply, temperature and process corner. Table 9.2 shows that indeed the PSRR is the main problem, because V_{CC} has the largest contribution. Compared the nominal conditions the deterministic corners give an increase in offset from 17.4 mV to 21.2 mV, so roughly 20%. This is a good relation for comparators. For op-amps which are often less speed-critical, we could maybe optimize a bit more (e.g., to improve on PSRR), and the op-amps the offset is often more critical at *higher* temperatures. However, for our comparator it looks different, the WC is at minimum temperature; also the FS corner is most critical. This could be an indication that the circuit shows some weaknesses at slow PMOS, and an explanation is that the first stage PMOS load might be critical. So it can make much sense to inspect all the results which are now available almost for free can lead to further design insights. For instance, the speed and power consumption performances are much more impacted by PVT corners, than by mismatch; e.g., the WC delay, at the combination minimum V_{DD}, maximum T and SS, is 81% larger than the nominal delay. Actually the speed WC combination is no surprise at all. The supply dominates here, which is also typical for CMOS-style circuit. However, these details depend on how *large* the environmental ranges are. Improvements by pure element sizing are difficult for comparators, so here we can only just make the performance better by spending more power (see Figure 1.16c).

9.4.3 What to Do after an Optimization?

Of course, you can start the next one (e.g., with some changes in the weighting factors), or you might be happy and move to layout, but actually beside improving circuits most optimizers give you also interesting circuit design insights like a sensitivity report—just almost for free, i.e., without additional time-consuming simulations!

Also the need for further optimizations is typical: A complete spec is often essential and it is quite easy to overlook something, e.g., without spec on clock input capacitance the optimizer may make the related transistor very wide for high speed, but this could become inconvenient for system integration due to high clock load. Often it is not easy to really set *all* specs meaningfully, so typically some system design, trials, and calculations are required before becoming confident on specs and design partitioning—and after a pure block optimization further system fine-tuning might be required, like a second optimization with DUT and neighboring blocks.

Sensitivities and contributions are often a free by-product from an optimization run. Actually we can expect zero sensitivity to the overall goal function f for an optimum circuit, so the linear sensitivity might be not of the highest interest anymore. However, of course it is also interesting to check the curvature which gives you a feeling how quickly deviations from the optimum lead to performance degradations. Although the gradient to the goal function should be close to zero; still the individual performances may have a clearly nonzero linear sensitivity! So a detailed sensitivity report could help to understand the design or for doing further optimizations. For example, the sensitivity report may show very low sensitivity to some performances, which could mean that maybe some important parameters are just not yet setup for optimization. Figure 9.20 shows some examples from our latched comparator example; a global corner optimization has been executed, but we stopped it interactively to understand the design a bit better. Also table reports are available. For normal environmental corners (like nominal) the offset voltage is almost zero, so also the sensitivities are small and hard to extract (r^2 small), but for the large-offset statistical corner can find the expected linear dependencies. For the total gate capacitance the behavior is different: It depends almost only on one transistor and is almost corner-independent.

The sensitivity results can also support quick manual tweaks, e.g., if one parameter just has a high impact on a certain important performance you want to adjust, just do a manual sweep on the identified parameter to find a good compromise by hand.

Lookup: This does not mean that the normal sensitivity analysis, e.g., based on mismatch contribution or OFAT would be waste of time! Of course the sense report after an optimization can only be related to the parameters which are part of the optimization setup—whereas the other methods can even identify "unexpected" important parameters!

We already mentioned that for really hitting the optimum with high accuracy we really need a very high accuracy on the goal function f. A small deviation Δf can lead to quite different sets of optimized parameters. This is important when using different starting points, e.g., we checked our optimized parameters against another optimization run using just minimum length and minimum width transistors as starting point. Indeed, the lower the sensitivity, the less accurate the optimizer accuracy—which is fitting to theoretical investigations and is also not really a problem for circuit design, just showing that the design is quite robust, as it should.

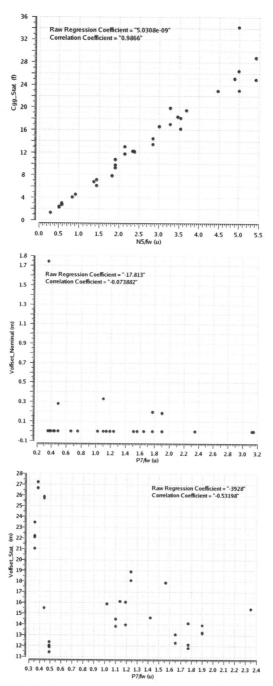

Figure 9.18   Typical sensitivity plots as by-product of an optimization.

9.4.4 Optimization with Inaccurate Worst-Case Distances

For a perfect yield optimization true WCDs are helpful, but what happens, e.g., during an offset optimization, if we use a *worst-sample* instead? We run a global optimization using the true WCD in the goal function, but simulate both statistical corners. This way we can look to the offset voltage of *both* samples *during* the optimization. Actually there are some differences, but generally if the optimizer makes a parameter change, both true WCD and the chosen worst sample go up and down *quite* in synch, but not always.

The small differences are also present in the *optimization results* (Figure 9.19): At start, the worst sample was 10% worse than the WCD, so for having the same spec the optimizer puts more effort in improving on worst sample. So at the end the worst sample was even <u>less</u> extreme than the WCD by roughly 10%. If there would be no direction error in the worst sample, then both samples would be improved almost in synch, so the change from +10% to −10% is the result of the worst-sample *direction error*, and an indication how much accuracy you may lose by using it instead of the true WCD! In many cases, it could be acceptable, because a design with $\sigma = 5$ mV is not so much better than one with 5.5 mV. So in this specific example run it has made not so much difference whether you put more effort in WCD search or in the optimization.

When dealing with many variables such direction errors are hard to visualize, so Figure 9.20 presents a simpler 2D case, two variables x_1 and x_2 are present in the input domain. Let us also inspect also two output performances f_1 and f_2; here the scatter plot might be not elliptic, but distorted. This is because in our example f_2 has an exponential characteristic (which would cause problems if we use the C_{PK}). In our example f_2 depends only on x_2, so the ideal WCD has only an entry for x_2.

Note that the worst-sample is not exactly 3σ, so we need to scale it a bit, but the direction error of roughly 30 degrees cannot be corrected so easily; it would require e.g., multi-dimensional techniques or more samples and some averaging.

Also note, *during* the optimization the performance value of, e.g., a 3σ-WCD would change of course, as we want a performance and yield improvement, so also the direction will change a bit. How much depends, e.g., on how bad your starting design is, if e.g., the input transistors are far too small at the beginning, the optimizer would make them larger and their

WCD entry would lower. If already the initial design has a good balance, the optimizer would, e.g., increase several offset-critical transistors in synch, and there would be only a small WCD direction change.

Begin of the optimization

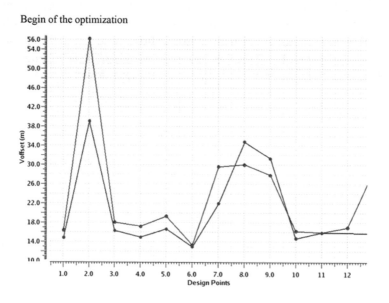

End of the optimization

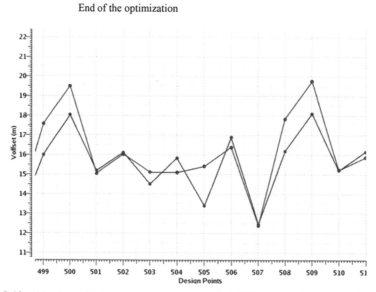

Figure 9.19   Absolute offset vs. optimization points for WCD (red) and worst sample (blue).

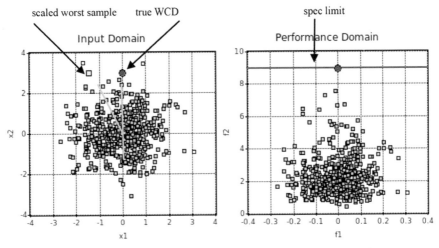

Figure 9.20 WCD versus approximation from scaled worst samples.

9.5 Questions and Answers

1. Discuss these two methods to find the overall worst-cases: 1st approach: find the worst-case (environmental) corners, then at each found corner run a worst-case distance analysis, 2nd approach: Run WCD first, then for each WCD execute a worst-case corner search.

Both approaches have their limitations, but due to the special structure of many circuits, the "statistics first" method works much better. For instance, the nominal offset voltage of an amplifier is often (near) zero. It depends usually strongly on mismatch, i.e., it is following the difference of two statistical variables Δx_S. On top of that there are of course also other effects like process variations or temperature, etc. These effects might be linear or quadratic, or whatever. However, all these have (almost) <u>no</u> impact if $\Delta x_S = 0$ (as in a pure corner analysis). So offset fol-low typically a $_1(k_1+x_{R1}+x_{R1}{}^2\ldots)\cdot(k_1+x_{R2}+x_{R2}{}^2\ldots)\ldots\cdot\Delta x_S$; and whatever you do with $\Delta x_S = 0$ there is <u>no way</u> to make inferences on the environmental effects, because this product is zero! Of course, for some other performances like phase margin or bandwidth the dependency would not be proportional to Dx_S, but e.g., to $(k + \Delta x_S)$, maybe even with $k \gg \Delta x_S$, so that also a "corners first" flow could work.

Another nice examples is looking to offset, but for a bad op-amp with large systematic offset (like +5 mV, e.g., caused by V_{DS} asymmetries) and bad PSRR (like −10 mV/V, e.g., due to use short-channel transistors and no cascodes). Assume a supply range of 2 V to 4.5 V with V_{DDnom} = 2.5 V. If the mismatch is small (like ±4 mV), then a spec like $|V_{off}|<$ 10mV would be critical only for one side, here +10 mV. Taking only this spec side into account would be risky, because e.g., with a +2 V increase in V_{DD} we would shift the offset down by 20 mV, so to roughly −15 mV, so that now the lower spec side is more critical. In conclusion, it is no good idea to use too greedy search methods in such cases, or if e.g., strong quadratic effects are present.

2. Imagine an optimization like this: Execute an MC analysis, find yield, and maximize the yield, e.g., with the BFGS algorithm. Which problems will pop-up, and how can you solve them?

   The MC yield estimation has some uncertainty, like random noise. This causes big problems for any gradient-based optimizer. To reduce the noise we would need a very high MC count, so the overall optimization becomes very slow.

3. How can you calculate the required spec setting for a certain yield, e.g., based on a MC run with 200 points? What to do in case of having no fails? Or if you want a very high yield like 99.99%?

   Consider using the sample yield or the C_{PK} formula. What are the advantages of the C_{PK}, and what are the limitations?

10

The Fully Assisted Variation-Aware Design Flow

Here we extend our view on design and address further custom design flow aspects, like the treatment of layout influences, or intellectual property (IP) creation and management. We discuss design environment aspects and how to arrange all techniques for an assisted, constraint-aware, semi-automated flow, starting with a target datasheet and ending up in an optimized circuit with layout that is "ready-for-production."

Still full analog circuit synthesis through commercial tools is almost science fiction. Even advanced research projects concentrate on special topics like maybe op-amps, filters, bandgaps, certain types of ADC/DAC, or very regular and well-tested blocks. So the aim for EDA vendors is usually to support a good mix of methods, among which the designer has still to choose. So users apply different methods according to the current design status, like exploration phase, sizing loop, or sign-off verification. In this chapter, we look at the overall flow and remaining aspects, like treatment of IP and the transition from front-end to layout. More circuit reuse and use of advanced tools, which are able to treat layout effects very early, are reality. Of course, this short chapter can only give an introduction and an overview, but a book on variation-aware design would be hardly complete without it.

For Further Reading:
Some references go even beyond what we discuss in the book (like topology optimization or Pareto optimization), because we treat techniques really available in commercial design environments.

- ITRS 2011 Analog EDA Challenges and Approaches, Graeb, invited paper.
- Design for Manufacturability and Statistical Design: A Constructive Approach, Michael Orshansky, Sani Nassif, Duane Bonin.

- J. Crossley, A. Puggelli, H.-P. Le, B. Yang, R. Nancollas, K. Jung, L. Kong, N. Narevsky, Y. Lu, N. Sutardja, E. J. An, A. L. and Sangiovanni-Vincentelli, E. Alon, *BAG: A Designer-Oriented Integrated Framework for the Development of AMS Circuit Generators*, IEEE/ACM International Conference on Computer-Aided Design (ICCAD'2013), Nov 2013, pp. 74–81.
- Ramy Iskander and Marie-Minerve Louërat, and A. Kaiser, *Hierarchical Sizing and Biasing of Analog Firm Intellectual Properties*, Integration, the VLSI Journal, vol. 46, no. 2, pp. 172–188, 2013.
- (Invited Designer Track Paper), J. Crossley, A. Puggelli, H.-P. Le, B. Yang, R. Nancollas, K. Jung, L. Kong, N. Narevsky, Y. Lu, N. Sutardja, E. J. An, A. L. Sangiovanni-Vincentelli, E. Alon, Department of Electrical Engineering and Computer Science, University of California, Berkeley
- Yield Model Characterization For Analog Integrated, Circuit Using Pareto-Optimal Surface, Sawal Ali, Reuben Wilcock, Peter Wilson, Andrew Brown, 2008, IEEE.

10.1 IP Reuse and Design Support

Classical custom design environments are centered around schematic, simulation setup, and plotting tools. However, in a new chip, often much more than 50% of the blocks are reused, just because the easiest way to save design time and to reduce risks is reusing what you have. A high reuse is usually possible for behavioral models, because these are almost by definition technology independent, and also there are widely accepted language standards like Verilog-A and Verilog-AMS. This way system design becomes quite easy, because you can often reuse models and just combine them for your design.

Such models allow also a fast testbench setup, and also the testbenches with its stimuli, loads, and performance equations can be often reused efficiently. A bit later, models pin-compatible to your circuit blocks can also help a lot to speed-up top-level verification. Often, it is desirable to have even different levels of models, like a simple one for first-order effects (e.g., excluding loading effects and dynamic power consumption), maximum simulation speed, and a more complex one, e.g., for subsystem performance trade-off investigations. The earlier you have such model, the more beneficial!

Besides modeling tools, also software for entering specifications and for design management (data mining, status checking, etc.) is helpful.

10.1.1 IP Tools

Intellectual property (IP) is quite a general term, so an IP tool can be almost everything. In this subsection, the focus is on IP management and reuse, whereas we address classical design supporting tools (Smith chart, filter programs, etc.) addressed later—although there are some links. Both often require the entering of specifications, so sometimes you can find combined tools.

Direct circuit reuse is often difficult, because circuits are complex and technology dependent; e.g., for going down in supply voltage, significant circuit modifications are often required. On the other hand, having a certain starting point for design is always an advantage, and the access is usually easy: just pick a circuit library and a circuit similar to the one you need to design. For some kind of blocks, the options for reuse might be manifold, like for bandgaps, op-amps, OTAs, and LNAs, because those are needed in almost all kinds of analogs or RF designs. Here a good naming convention or even some kind of IP managing tool makes sense (besides good documentation). This could also act as a starting point for the design in general. For instance, once you specify the block's function (like op-amp, LNA, bandgap, and comparator), the environmental range parameters, and the performance specs, you can easily automate the testbench creation and the documentation. In addition, you may create a behavioral model just by giving the specs, e.g., at nominal or worst-case conditions, with few clicks. Such tools could be used to track the design status, like "Ready for first design review" and "Layout available for post-layout simulation."

This way you can generate quickly a kind of "executable spec," and you would have always something that works! The IP managing tool may help you in finding exactly what you need or you simply do it from scratch—but with some tool support (e.g. at least getting just a symbol according to the pins you defined in the spec).

For op-amp design, this could be a detailed specification template, a calculator for noise performance, matching, frequency compensation, different op-amp circuits, etc.—or maybe even a step-by-step instruction or a whole circuit synthesis script (Figures 10.1 and 10.2). This means not only the circuit could be reused, but also the design strategy for it. Whether this is a key advantage or not depends of course also on how similar your existing IP blocks are: If existing IP is close to your specs, then often already simple scaling rules (like "double all width" to double the output power at same supply voltage) or sometimes even direct optimization could lead to the final circuit quickly (like for using a 800 MHz LNA at 950 MHz).

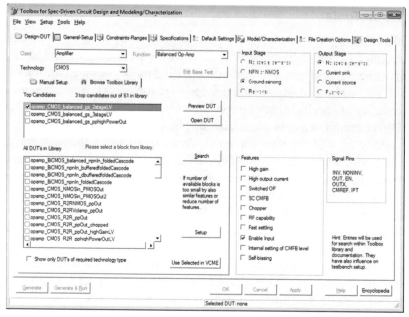

Figure 10.1   User interface of IP selection tool (prototype).

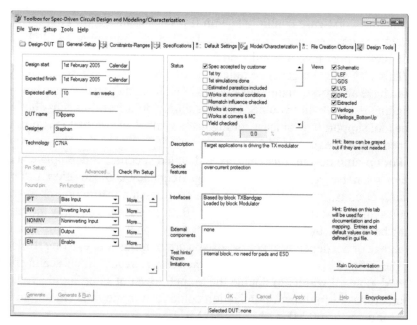

Figure 10.2   Block spec entries to generate documentation, testbench, symbol, and a pin-compatible model (Prototype).

10.1.2 Technology-Independent Design

One way to make the complex task of analog design more efficient is IP reuse. However, a limiting factor is that often you might be able to find a "similar" design, but often not with the same specs and in the same technology. For minor spec changes, we may apply optimization methods directly, but for bigger changes or even for major technology changes, more tool support is desirable, e.g., via migration scripts. Sometimes even an advanced support is already available from foundry-side (e.g. XFAB). For instance, in [Boos2011] you can find an embedded migration frame work which offers structure-recognition for an initial sizing, based on look-up tables for the transistor IV characteristics, and constraints for the voltage stacks (similar to Figure 2.3). Figure 10.3 shows an example setup; the output is just the sized circuit, in which the transistors work in the intended voltage and current ranges, also matching specs are included. This way a robust starting point for further performance optimizations is achieved.

Analog design is never technology-independent; at best, you can get support for technology migration and for design methods based on few key characteristics. For instance, a classical way for a bipolar op-amp design from scratch (and for synthesis even) is to start with the overall gain requirement; finding the number of required stages according to beta, early voltage, and load resistance. Then you can set noise and input bias constraints to define bias currents.

To support an almost technology-independent design, the most elegant way would be, if the semiconductor foundries would offer a kind of really universal process development kit (PDK). Unfortunately, this can be hardly expected between competing foundries, and it could also cause a severe overhead. Some universities have created experimental PDKs for technologies which even not yet exist, e.g., to check how to do designs with such future technologies. In addition, there is also an initiative on so-called interoperable PDKs, but actually this stops much earlier, being still far away from true technology independence! For this reason, most EDA environments give only some support for setting up technology migration scripts quickly. Usually a mapping for components and parameters is needed. And in addition, simple calculations are possible during the transfer. Migrating from 180 nm to 90 nm makes sense in logic blocks to simply scale all lengths and widths by $0.5\times$. This would treat the saturation voltages and drive strength quite meaningful, and we would even take advantage of the higher speed in 90 nm. However, mismatch and flicker noise would suffer significantly for such $4\times$ shrink in area, so this aggressive scaling is often not suited for classical analog blocks.

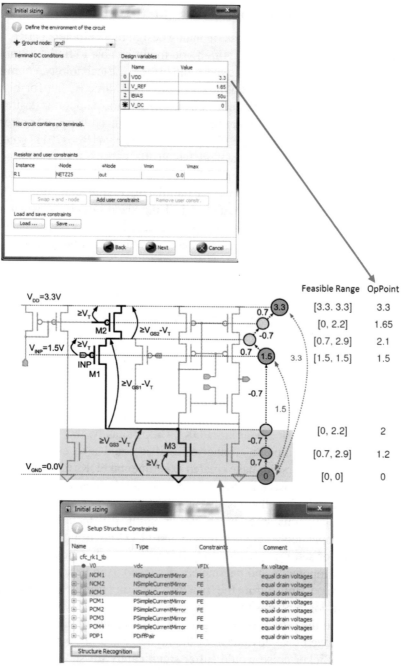

Figure 10.3   Sizing tool based on exploring transistor stacks and structures.

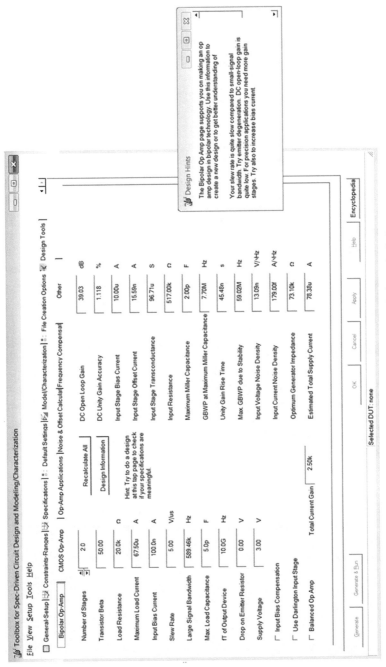

Figure 10.4   Bipolar op-amp design calculations.

If the scaling rules become more complex or if a lot of tweaking is needed, then at some point even full script-based design (not only migration) becomes an option. Full script-based design comes with even more difficulties, e.g., layout generation is hard too. Even if the circuit topology is fixed, a layout construction up-front is not easy, because for an op-amp, the best values of widths and lengths depend a lot on how much you optimize for noise (leading for big input transistors) or output power (output stage will dominate the total area).

The traditional manual flow fails often to incorporate complex data that is readily available, e.g. in the technology process design kit (PDK), like the detailed transistor models. The tool *ID-Xplore* (from *Intento Design*) supports automatic constraint-driven sizing, based on so-called bipartite graphs (Figure 10.6), is offering a design acceleration. The design entering is schematic and constraint-based, and basically technology-independent, only the calculated sizing solutions are of course technology-related. For a technology migration, the user would just need to redirect the PDK paths. Different trade-offs can be selectedand an automatic back annotation of transistor size data to the schematic is available; allowing a smooth integration to any traditional flow.

Design Support. Among the many topics we mention, there are always both: "hot" and "also-ran" topics. I remember in my first design project, a simple thing as the (full-chip) LVS check took days and real expert know-how, although all our blocks itself *were* already clean! Some short circuits were very hard to find, and the debug tool looked like coming from the Flintstones.

Another wall-flower is usually the undo feature. If an undo exists, then often the capabilities are quite limited, so you might switch back to an older parameter setting in a simulation setup, but this will not help much if the underlying circuit has been modified already. Also simply making a backup in complex environments can also cause problems. Often it is a challenge to let other experts (like those from hotline) re-run a certain configuration of testbench views, tool versions, etc.

In fact being good at the "also-ran" features is often quite important for convenient work! This is one big part the major EDA vendors need to do—besides fancy stuff like Pareto optimization, circuit synthesis, or whatever the next "cool" topic is—and over the years, they were quite successful on this.

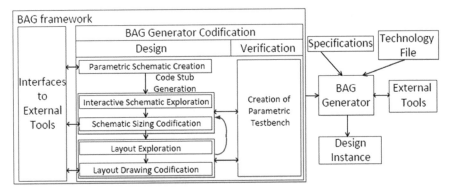

Figure 10.5 Advanced script-based analog circuit design [Crossley2013].

10.2 Cockpit Number One: Augmented Schematic

Of course, one key element in analog design was, is, and will be the schematic entry, and another one is the circuit simulator (actually multiple, like for analog blocks and systems and one for mixed analog–digital systems) and the related setup and result evaluation tools (for analysis selection, result plotting, printing, cross-probing, and backannotation to schematic, etc.). This will not change, but becomes more and more augmented. Opposite to digital design in analog, a schematic has simply so many advantages that it will remain:

1. A well-organized schematic is easy to pick up for other designers, for re-use, getting design ideas, or for making a design review.
2. You can *enter* your design quickly, often with efficient partial copy and paste.
3. You can arrange the elements to reflect your *intentions* or the circuit operation, like the signal flow direction.
4. You can *prepare* the physical implementation, e.g., *guiding* the layout for symmetry and by grouping elements.
5. Many layout tools offer a kind of "place as in schematic" and give support for schematic-driven layout (via flight wires and cross-probing).
6. In modern environments, you can augment it with performance graphs and tables, for *documentation* and debugging.
7. Your schematic editor is also your tool for *navigation* across the whole design.
8. It is a perfect debugging tool, e.g., by using *backannotation* for g_m, I_D, V_{Dsat}, of transistors or by highlighting all transistors in saturation.

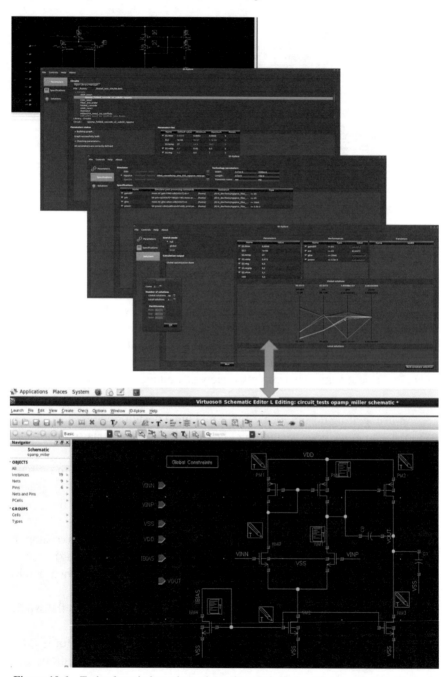

Figure 10.6   Technology-independent schematic-centric design (Courtesy Intento Design).

9. A schematic is often a superior representation of a certain function. For instance, filters can be described by poles and zeroes or by coefficients, but a ladder network has much better behavior with respect to tolerances and it is well – linked to implementation—without becoming inflexible.
10. Drawing a nice schematic is often leading to new *ideas* for improvements. Thus, well-established standard functions can be often plug-in easily, like a little trim-DAC or simply a power-down switch.
11. Schematics allow an easy *mix* of techniques, like analog and digital, real PDK components, and idealized models. Often additional flexibility is possible with a configuration (or hierarchy) editor.
12. Even highly advanced analog synthesis tools offer typically a schematic as output, because this way the designer gets an immediate impression whether the design result is trustworthy [McConaghy2011].

In conclusion, a schematic editor is a very important tool, and analog designers almost think in terms of schematic structures (like diff-pairs, transconductors, and control loops) and performance curves (like filter characteristics).

Until now, we focused on describing key techniques to solve specific problems. Each step often contains many substeps, and the overall flow is usually iterative—the more unknowns, the more iteration. Actually *many* steps in the overall flow *may* take *much* time, but experience, application of hand calculations, and reuse can shorten some steps a lot (like reuse of a spec or circuit topology or testbench or behavioral model, or experience in the estimation of parasitics or in setting MOS widths and lengths). Usually only some parts are "hard" and need most time—although it is difficult to know in advance *which* ones.

We have not talked much about top-down vs bottom-up! Ultimately most custom designs and design flows are still much more bottom-up and quite tightly bound to foundry PDK and technology, but there is no clear separation anyway, because for any top-down approach, you need many good sub-blocks like VCO, PFD, filters, and counters for a PLL frequency synthesizer—or just transistors, resistors, and capacitors. For instance, many environments claim years to enable "spec-driven" design, but actually you have to compose all things up front like DUT, testbench, and measurements, and once you have done all this details, you can set specifications. In conclusion, almost all designs are done in a kind of "meet-in-the-middle" style [Chen]. In addition, also the concept of "constraints" leads to a very pragmatic design style which is neither top-down or bottom-up (see next subsection).

Most tool solutions address sub-problems and have to jump-in to the general flow. So they need to be flexible for enabling a (highly) assisted flow. A key problem is that high-automation methods often do not support well "flexibility." It is often easier to create a construction (or placement) tool that follows a *fix* built-in strategy, compared to the one that is able to also pick up an intermediate *manually* created solution (e.g. for some critical parts)! In conclusion, there is yet no perfect efficient flow. What helps us regarding front-end design is that we can assume that at least our target is a robust design, which works well at least at nominal conditions and having (due to application of design best-practices, sizing rules, etc.) relatively smooth responses to conditional changes. For such "realistic" designs, we can define an assisted flow that works quite efficient in almost all real design cases.

Actually combining all different methods into one environment is a huge challenge. For instance, many reported analog circuit synthesis tools are set up quite differently to the usual analog flow, e.g., with respect to testbench setup or regarding extensions on the block-under-design—unfortunately.

In addition, there is also a strong pressure from technology and market side: Tools simply need to be able to manage the higher block complexity, higher transistor counts, etc. Thus, further improvements in the algorithm details are required; e.g., EDA vendors need to make sure that statistical methods, like the mismatch contribution analysis, will not only work for typical block in 45 nm, but also for advanced Fin-FET-based designs, which have usually many more transistors, so many more statistical mismatch variables.

The decision whether tool assistance or IP re-use makes sense is up to the designer, and the decision depends highly on the following:

- The designer's skills (e.g., circuit understanding helps for design tweaks and optimization setup). This includes tool usage; e.g. a sensitivity analysis can support debugging and can give valuable insights for further decisions on design next steps!
- The circuit status (for a good manual design, we need no nominal optimization and can directly go to size over corners)
- Tool capabilities (is the optimizer able to improve the circuit, e.g., even with a bad starting point)
- Available compute power and time (the more you have, the less tricks you need, but realistically difficult problems can seldom be solved in brute-force style, better use the day for manual tweaks and get understanding and run automated parts overnight)

10.2.1 Designing with Constraints

A general big problem preventing full automation in many cases is that the general custom IC design flow has a huge number of inputs, and it is very difficult to provide them all up front (like accurate specs for all blocks). Actually, something more is needed than pure target datasheet-oriented design. A best-practice design requires also inputs hard to provide up front and better entered as so-called constraints. Making decisions step-by-step is anyway a natural method and compliant to many classical construction methods. For these reasons, many modern design environments allow the definition of constraints at many levels, and they can nicely state the designer's *intentions*. This way, any team member can define constraints to improve teamwork, documentation, and design quality (see Figure 10.7 as an example for a CMOS PDK and design environment with built-in constraint support). A major advantage is their flexibility compared to setting all goals up front, and often tools can pick them up or check them automatically.

Constraints can work top-down (like "layout should be symmetrical" or "high proximity required"—e.g., set in schematic entry or floorplanner) or bottom-up (like "clock lines should not cross this sensitive area"—e.g., set in layout editor). This way, the specialist in each tool can set up constraints in the best way.

We mentioned that it is quite typical that some specifications are not really written. Also the specs in a datasheet may not have equal strictness. In addition, customers and designers simply assume that besides the written specs also general design rules and best practices (like using common-centroid layout for critical parts) will be applied. Constraints complement the hard specs written in a datasheet and strict design rules like those on layer distances, minimum and maximum width (DRC), electromigration (Table 10.1).

Table 10.1 Overview on design inputs

What	Source	Comment
Design rules	Foundry	In most cases, you can run automatic checks, e.g., w.r.t DRC, and electromigration.
Datasheet	Customer/Designer	See Chapter 1. Typically you need to "translate" the content into testbenches, simulator setups, etc.
Constraints	Designer, Layouter	Define internal requirements, e.g., to assure properties not possible to simulate.
Soft factors	Manifold	Hints and design strategies from books, manual checks, inputs during a design reviews and discussions, etc.

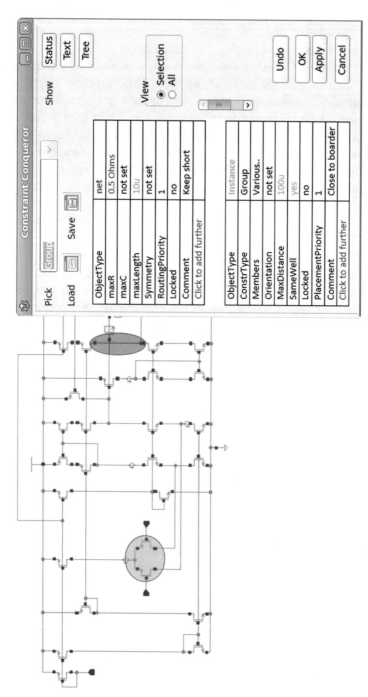

Figure 10.7 Constraint setting on our op-amp as a good option to capture design intentions (artificial UI shown).

Typically, constraints will be set step by step during the design (see Figure 10.7); they are often layout related, used to improve team communication and just to not forget anything important or helpful. For instance, a constraint on symmetry makes sense for a critical differential pair, or sometimes you want the same orientation of two block instantiations in a design to improve matching. Here the definition of a constraint makes sense, because often you simply cannot simulate what would happen if a temperature gradient causes on offset voltage, so better try your best "by construction."

A certain problem is to standardize constraints, defining a meaningful set without overhead and gaps, implementing constraint checkers, etc. So setting up constraints looks easy, but to let the tools pick up them correctly (e.g., checking if a constraint is fulfilled or driving an automation tool to keep the constraint) is a hard task for the EDA vendors, and there is no standard set of constraints for analog design so far. Also, it could easily happen that we end up in too many constraints, or even contradicting constraints (like wanting symmetry *and* same orientation, wanting to keep *too many* wires shorter than 10 um). For that reason also a kind of severity classification is often required for constraints; and also on that some thinking is needed to get an over-all priority, because constraints can be anywhere in the design hierarchy.

As often, in analog design, the devil is in many details. The state-of-the-art for design environments is that designers are able to enter already many kinds of constraints at different levels (either manually or by having assistants for it) and that some tools pick them up for checking them or to allow an autoplacement or routing based on it. What is typically missing yet is that all the (front-end and back-end) tools make efficient use of constraints, and also, a mix of manual and constraint-driven design is often difficult to realize. For instance, some auto-placers indeed pick up many constraints, but they cannot start from a meaningful manual placement.

As many of these limitations are already identified, we can expect more and more improvements in the near future—no-one-fits-all solutions, but embedded pragmatic improvements to make also layout a bit a fun part of design, e.g. comfortable wire routers with high flexibility, bus, and advanced power routing capabilities. Such semiautomated methods have the advantage of direct graphical feedback, and the user can still decide in which sequence he/she wants to progress.

10.2.2 Design Tools

One starting point for design could be IP reuse, and also in an IP management system (or in a design and verification environment linked to it), we may get

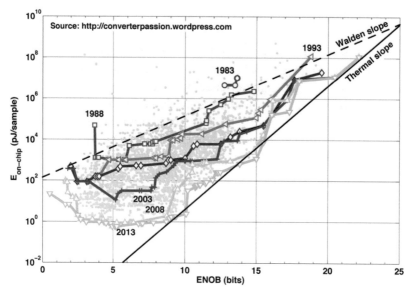

Figure 10.8   ADC FOMs and trend [converterpassion.wordpress.com].

access to some design tools, like for topology selection hints (e.g., which ADC to use according to technology and requirements on speed, power, and area) or for doing basic calculations like getting the expected power consumption from a FOM (Figure 10.8).

Several such tools are usually available embedded in your circuit design environment, like a calculator (with built-in functions for all the classical analog performances like rise time, phase margin, THD, FFT) or a basic Smith chart. Some tools may directly come from your EDA vendor, and others might be created by your CAD department (Figure 10.11 showing a transistor sizer, to find the best-suited width to obtain a certain g_m for given length, V_{GS} overdrive, etc.), 3rd party vendors, or experienced designers. Typical examples are MATLAB® toolboxes (e.g., for PLL design), but there are also many tools available free of charge like Excel® sheets, calculators for two ports, and for dealing with linear equivalent circuits, S-parameters, or programs and catalogs for all kind of filters. Actually, there is a whole bunch of such free or almost tools, and even special simulators for symbolic circuit analysis are available, such as Symbolic SPICE (SSpice). Two good Smith chart programs are CSmith and WinSmith, and two great tool collections are AppCAD and AdLabPlus (Figure 10.9).

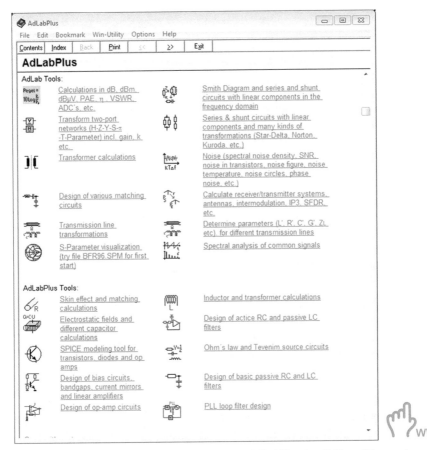

Figure 10.9 Free circuit design tool collection AdLabPlus (available at River webpage).

10.3 Cockpit Number Two: Variation-Aware Driver Seat

Besides the design entry, the analysis setup tool connected to it is the core element in the designer's everyday work. In modern custom IC design environments, you are able to address almost all kinds of problems. Beyond the basic circuit analysis (DC, AC, transient, noise, sensitivities, etc.), you can run corner and Monte-Carlos analysis, plus most of the mentioned advanced methods (like different types of optimizations and worst-case searches). In addition, you can check for layout effects, like running post-layout analysis with wiring or even substrate parasitics included. Verification is a key task and a complex one, and also IP aspects and verification plans are tightly linked.

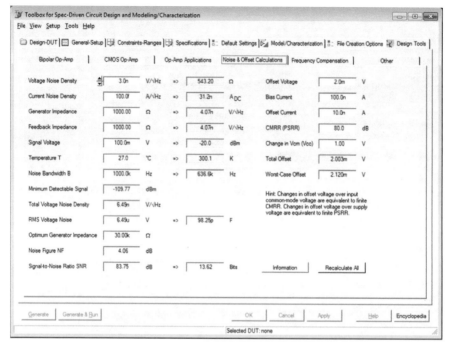

Figure 10.10 Design calculations embedded in IP managing tool—here for noise calculations.

Decisions and tool requirements depend highly on how far the design is from being finished—or at least from moving to the layout phase. Optimization—manually or automated—comes before sign-off verification. And at the beginning, we typically need to pamper our design, and there is no need to really *accurately* find the worst-case for each performance. So actually designers spent a lot of time not only in schematics, but also in driving their simulations—hopefully not too much by pure "SPICE monkeying." Some simulators already offer advanced built-in features like optimization or mismatch analysis, but usually more flexibility is possible in doing these more complex tasks in the design environment and by driving the simulator from a graphical user interface. This way it is usually easier to support multiple testbenches, easy maintenance, comfortable interactive debugging, very complex results evaluations, etc. (and actually also multiple simulators).

A verification engineer has almost the opposite job: He should "hate" the design and try to break it; and he/she has no need to sweep design parameters

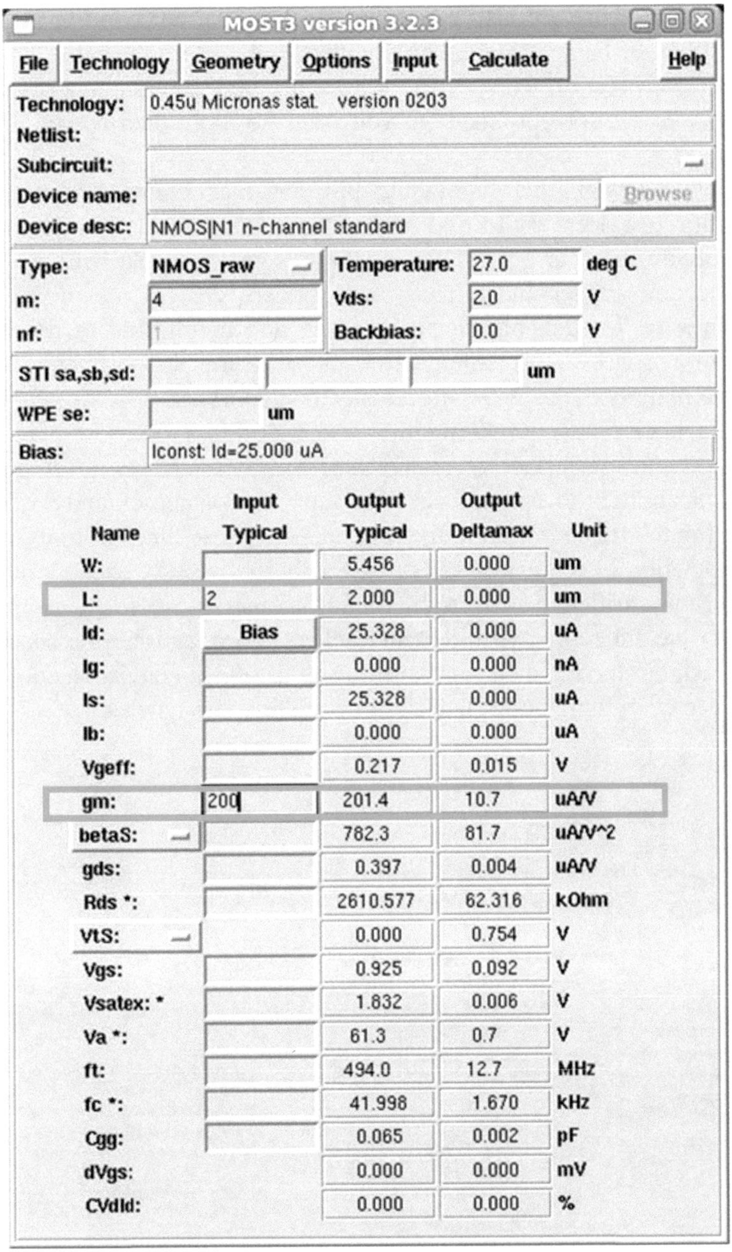

Figure 10.11 Micronas transistor sizing tool (MOST3© Micronas, Courtesy of Micronas Germany).

x_D. However, there are also many similarities, like both optimization and verification can be speeded up by knowing the WC corners. In real-world designs, a verification worst-case corner setup may contain almost hundred corners, whereas for optimization, you want no more than maybe a dozen corners.

In digital design, there is often a split: designers create the system, and verification engineers try to find bugs. This is good for design quality, and more specialization is a trend for analog design too—also because design teams become bigger and bigger; most modern designs are mixed-signal chips anyway. In such big chips, planning and monitoring of design and verification progress is a challenge too, and some tool support is now indeed available in modern design environments to support true spec-driven design, from the beginning and during the whole design process. Figure 10.12 sketches the user interface of such a tool; the main features are import and export capabilities to spreadsheet programs and databases and a GUI for connecting the requirements to the testbenches. These kinds of tools support full automation of regression runs (e.g., daily and weekly execution plans), reporting on verification coverage, found inconsistencies (e.g., due to spec changes), etc. Of course, requirement trackers and spreadsheets are also used in the past, but having a clear connection to the design environment helps a lot to keep everything up-to-date and to avoid inconsistencies.

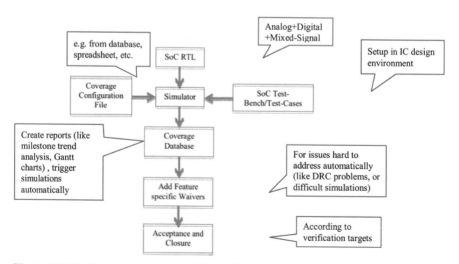

Figure 10.12 Verification tools enabling chip-level top-down design and verification [Venkatakrishnan2014].

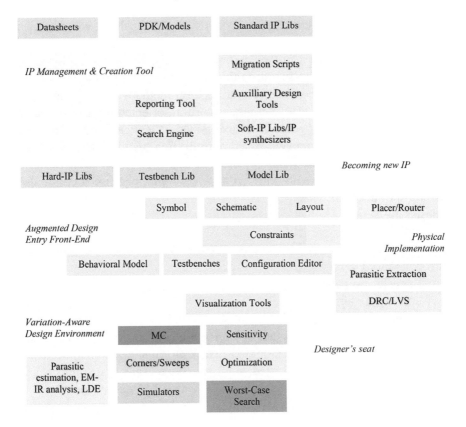

Figure 10.13 Major elements of an augmented variation-aware design environment.

In Figure 10.13, we show most of the tools required for variation-aware custom IC design; only few special ones are sometimes kept out (like EM solvers or MEMS simulators), and also some basic tools like netlisters or auxiliary tools like netlist reducers or a file versioning system.

10.3.1 Task-Driven Design Flow

In Chapter 2, we describe the typical manual (custom IC) circuit design flow. The designer picks the required tools and collects data for his design decisions. However, it is not so clear which method is the best for a dedicated task like finding critical devices. Almost always you have different options, and often a dedicated analysis has no single output; for instance, sensitivities are often also a by-product of a statistical analysis or an optimization.

In Chapter 8, we describe a task-driven statistical analysis, and we use it for complex yield optimization scenarios. If clear specs exist and if the design is at least functional, then further automation and more speed-up by reuse is possible, and there is no real need that the designer himself decides for a certain statistical, corner search or optimization method—a more data- and design-oriented setup instead of a pure algorithm-oriented setup would be a desirable complement in general. As mentioned, "classical" algorithms like worst-case distances or gradient-based optimization may fail, although they are very efficient in many other cases. Such task-oriented or even "design-oriented" setup would be not only an analysis setup, because it could also include optimizations, sensitivity analysis or modeling tasks, so being actually a designing setup! Mathematically this is not completely new, we would just combine the techniques for space exploration (modeling, sensitivity, worst-case corner, etc.) with design optimization (direct optimization, surrogate-based optimization [Bo Liu2011, Moustapha2016]). All this is DACE, design and analysis of computer experiments.

Actually, modern EDA environments offer already quite a lot of support for such integral approach, like for statistical yield and corner analysis, or for switching the design representation from pre-layout to estimated parasitics and to full post-layout with a single switch. Unfortunately, sometimes user-interfaces only look task-oriented, e.g., users can select run modes like "Optimization" or "Sensitivity Analysis," but usually still designers have to select specific methods (like local optimization by BFGS or sensitivity analysis via OFAT sweeps), and users can only run one task at the time. In the future, we

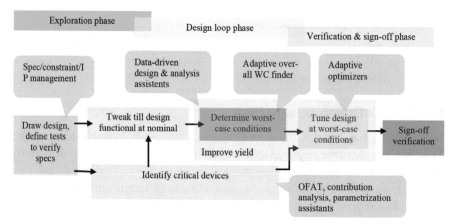

Figure 10.14   Flow chart and algorithms for circuit design.

can surely expect more task orientation and automation; with scripts, users can already do now quite a lot (e.g., via oceanXL in Cadence Virtuoso®). However, for really "merging" e.g., space exploration and optimization, common data formats are needed to transfer information like parameter setups and model coefficients. In such scenario, the designer and the environment would be the true ruler about all variables and functions; and the schematic entry or a hierarchy editor would act like (very comfortable) property editors. And the simulator with the plotting tools would act as debugger. All this is not complete science fiction, and similar to digital and software design flows.

10.3.2 Physical Aspects and Sign-Off

In older design environments, there is often a split between front-end design and layout (being of course a key part of design). In bigger companies, often different people work on the schematics and on the layouts, because learning the complex tools takes some time, and using them blindly and efficiently takes even more time. However, for dealing with advanced technology problems, there is also a trend that both fields, front-end and back-end, will be treated by joint or highly connected tools, e.g., with extensions in both schematic and layout editor. So actually, the variation-aware design cockpit gets extended and will also cover variations due to many kinds of layout effects!

The physical implementation and actually the layout are often regarded as something that "follows automatically" after system and front-end design, but actually this is not really true. Packaging and floor planning should be done ASAP! In addition, circuits for which advanced statistical and optimization techniques make most sense are also often very sensitive to the physical implementation and parasitics. So experienced designers have the layout aspects in their mind, like "Is a certain inductance value and quality factor realizable in the given technology?" or "Can we effort the chip area for filter capacitors required to implement a certain filter with certain noise properties?"

Modern PDKs include detailed MOS models to cover even the surrounding *neighborhood* of a transistor to its electrical characteristics. This means such layout-dependent effects (LDE) can be simulated nowadays, and it works actually in a similar way to the inclusion of normal RC wiring parasitics. If e.g., a well (e.g., from another transistor or a guard ring) is close (e.g., distance d = 1um) to a transistor (like P1), then the electrical behavior (like threshold voltage $V_{TO}(P1)$) changes a bit compared to the default value (e.g., assuming

Given: Design, testbenches, constraints, corners (range parameters)

Design specs & constraints: *From user directly or requirement system*
Target yield? Confidence level?
Preferences?
       Which specs are critical?
Stopping criteria?
       Accuracy for yield, corners, etc., Runtime, Number of simulations, etc.

Design characteristics:
Complexity? If not specified switch to auto
Nonlinearity? If not specified switch to auto

Desired outcomes for variable space evaluation:
Parameter setup for x_D, x_S, x_R
Want sensitivities, contributions or performance models
       If yes, than setup to which parameters, import from corner setup
Want statistical corners?
       If yes, than setup to which parameters
Want deterministic corners?
       If yes, than setup to which parameters
Want combined corners?
       If yes, than setup to which parameters
Yield, C_{PK}, C_{GPK}, mean, sigma?
Histograms, quantile plot?
Performance correlations?
Parameter correlations, mismatch contribution?

Design optimization:
       If yes, then parameter, starting point & goal setup
       Take results from variable space evaluation into account

Figure 10.15   Task and design driven setup.

$d \gg 10um$). In modern PDKs the function $V_{TO}(d)$ is part of the models, so once the layout positions are known (at least the placement, not necessarily the routing) we can run a more LDE-correct circuit simulation (regarding V_{TO} and other parameters). Physically such LDE changes are mainly the result of mechanical stress during the fabrication. The extraction of the LDE model parameters is done by the foundry based on test chip measurements and 3D field solvers. As usual, the normal designer is just using these complex models in his circuit simulator, in the background.

Doing such analysis late, i.e., when you have the full block layout, and recognizing in post-layout simulations that the block does not work as intended anymore, could mean that a lot of time and work has been "wasted"! Especially for high-speed and high-accuracy circuits, it would be a very bad approach to check temperature and supply effects in very high detail, just because it is so easy, and to overlook that already the wiring capacitances degrade the speed or stability by 30%.

For this reason, modern PDKs and advanced custom IC design environments even support doing an early analysis, already on partial layouts (so-called "electrically-aware design" EAD) [White], or by adding at least estimated (not necessarily layout-based) parasitics. Using such tools (e.g. the EAD option in Cadence® Virtuoso® Analog Design Environment and Virtuoso® Layout Editor), the designer could quickly make a layout for the most critical parts and get at least for this part all the layout parasitics. In such flow, the simulator netlist is created from the pure schematic netlist plus a *stitched-in* part of the partial layout with extracted parasitics and layout-dependent effects (LDE). With wiring (or even substrate parasitics), the netlist becomes often much larger, and also LDE can extend the netlist a lot, because for maximum accuracy, even each transistor finger has to be treated individually.

Note: In true RF or microwave design environments, such parasitics awareness is standard since many years, but in these, the approach is often easier to implement, because RF designers are skilled to work almost directly in the layout view—instead of using schematics. In addition, RF designs have usually a much lower complexity.

Corners in Parasitics. In the very old days designers were happy to have models just for the nominal conditions and for the typical average behavior. Nowadays having corner models and MC parameters for process and mismatch is standard. However, also the chip interconnections have not only a performance impact, they can also vary significantly. Changes in the different oxide and metal layer thicknesses lead to variations of series resistances and wiring capacitances, so it could makes sense to work with corners as maxR, minR, maxC, and minC. However, like in transistors also some correlations are present, e.g., it is less likely to have at the same time both very thin metal and oxide layers. So using designing for the combination Rmax, Cmax could lead to some over-design. Therefore, again fix mixed corner combinations are standard in most advanced technologies; and ultimately also a statistical analysis for parasitics could make sense. To keep the level of intra-die variations at an acceptable level strict layout rules need to be maintained, e.g., regarding layer filling. For this task usually metal fill run sets are available, but of course such metal fill elements can impact the circuit performance, and there are places (e.g., near on-chip inductors) where the filling is not desirable.

Once a layout (partial or full) is available, also further checks will be typically applied like design-rule checks (DRC) or an electromigration and voltage drop analysis (IR drop). Of course, the full production yield is not only determined by what you can observe in front-end MC simulations, catastrophic failures due to layout effects will be on top, so also layout optimizations in that direction, like using double vias can be important too, especially in advanced technologies and high-reliability applications [Yu2016]. For this also so-called Design-for-Yield, Manufacturing or Reliability tools are available too.

10.3.2.1 Advanced layout techniques
Like EDA vendors try to optimize the front-end flow, they also created many tools to do high-quality placement and routing, e.g., assisted or even automatically, based on constraints. Also this area is a field of continuous improvements; perfect automating tools are not really available now—neither in commercial tools nor in research. Also here composition and optimization techniques are meaningful attempts, and the problems to solve are quite complex and nonlinear as well.

One problem for automated tools is that you often end up in "eat-all-or-nothing". A schematic sized automatically, but ending up in a non-optimum circuit might be "repaired" sometimes, at least by an expert. A layout proposal from an auto-placer is often really difficult to improve, e.g., because the proposal is often a very dense layout, so that it could be often easier just to make everything by hand—almost as usual. Figure 10.16 gives an example for an automatically generated placement based on user-defined constraints. Unfortunately some parts of the layout look not acceptable, which could be at least partly explained by some potentially missing constraints.

On the other hand, automated tools can give you at least a reference point— much better than nothing, but usually not as good as the hints an experienced designer can give. One trend helps: Newest technologies have so complex

Plus internal state data (like simulation raw data, optimizer Hessian matrix, ranking on parameter importance for WCD, tracker for setup and design changes, etc.)

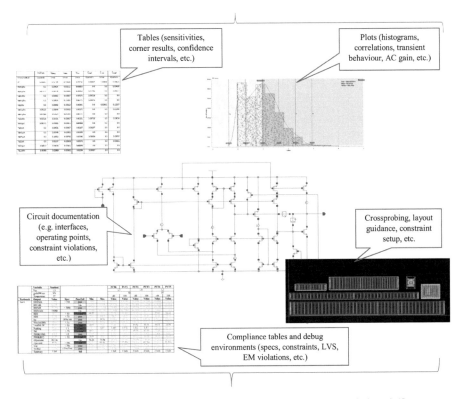

Tables (sensitivities, corner results, confidence intervals, etc.)

Plots (histograms, correlations, transient behaviour, AC gain, etc.)

Circuit documentation (e.g. interfaces, operating points, constraint violations, etc.)

Crossprobing, layout guidance, constraint setup, etc.

Compliance tables and debug environments (specs, constraints, LVS, EM violations, etc.)

collected in a modular document (status documentation, html, xml, png, pdf), reporting to project monitoring tools (Gantt chart, milestones, verification coverage)

Figure 10.16 Comprehensive design driven analysis outputs.

design rules, e.g., regarding densities and orientation, that a kind of "Lego" approach makes sense. This leads to a layout in a kind of "matrix" style, which is easier to automate; and the potential loss regarding performance or area is often quite acceptable.

In addition, also fully automatic constraint generation is possible nowadays [Eick2011], often even generating more constraints than designers typically find. This leads, together with an enhanced set for constraint options plus supporting hierarchy and constraint priorities, to significant improvements in auto-placers. Figure 10.17 shows the fully automatically constraint and synthesized layout of an amplifier, which has indeed only very minor

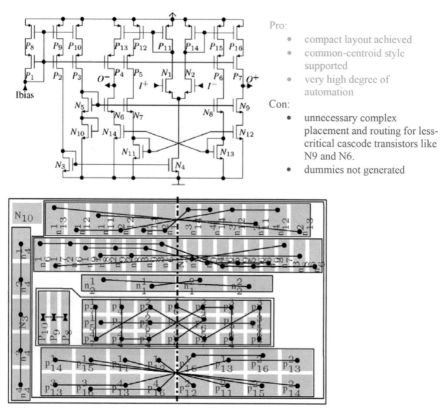

Pro:
- compact layout achieved
- common-centroid style supported
- very high degree of automation

Con:
- unnecessary complex placement and routing for less-critical cascode transistors like N9 and N6.
- dummies not generated

Figure 10.17 State-of-the-art synthesized layout with automatic constraint generation [Eick2011].

weaknesses. Comparing this layout to an older version, e.g., a weakness on placement for N10,11,13 is now gone. Of course, also manually created layouts are usually not perfect, and actually usually a good compromise is enough and a much more realistic goal.

10.3.2.2 Parasitic analysis

The layout influences circuit performances often significantly, like maybe 10 to 30% on speed and bandwidth, so indirectly also on power or more difficult measures like stability or gain flatness. So it is important to include layout parasitics, like wiring capacitances and resistances ASAP to your simulations (Figure 10.18), sometimes even self- and mutual inductances matter, and for

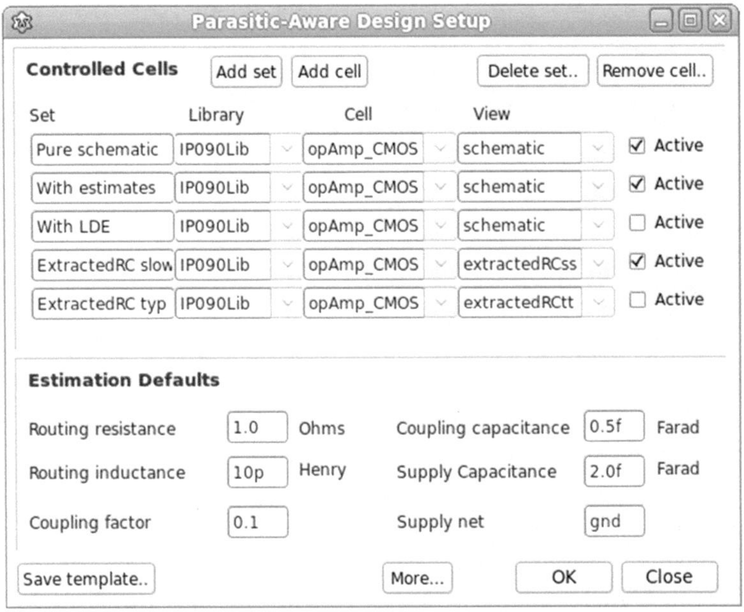

Figure 10.18 Generic parasitic and layout-aware design user interface.

crosstalk and RF behavior, even the substrate characteristics. The standard technique is just to check the layout netlist of the circuit against the schematic source on correctness doing an LVS run and to run a parasitic extraction tool to calculate the wiring parasitics from layout geometries and technology data (like wire and isolation layer thickness, and oxide permittivity). However, such post-layout simulation alone gives no direct design *insights* beyond the pure performance shifts. Designer need to know *which* parasitics are most critical and *where* a real good layout is a must!

The mismatch contribution analysis is a very powerful method; we have seen that statistical methods are amazingly efficient in finding the most sensitive parameters even among thousands of parameters! Once you have obtained estimates for your wiring capacitances at each net (either by pure estimation through transmission line formulas or through direct parasitics extraction on the layout), you can easily run a mismatch contribution analysis also for these parasitics elements and find the most critical nets quickly, e.g., w.r.t bandwidth, stability, and delay. This is a perfect guide for doing a suited layout, and the effort is just a MC analysis on a speed-critical performance; in analog designs, it could be often a fast AC analysis.

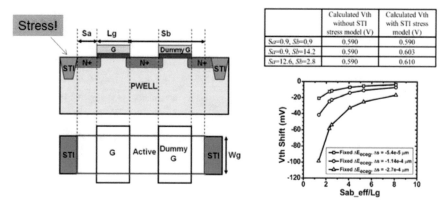

Figure 10.19   LDE from shallow trench isolation (STI) in HV MOS transistors of an older 0.35-um technology [Wei].

In addition, of course also parasitic tables and the classical backannotation of parasitics into the schematic are helpful; especially if you compare the actual parasitics with your earlier estimates. This way you can often learn a lot and transfer also your findings to other blocks.

10.3.2.3 Layout-dependent effects LDE

Wiring parasitics are not the only important reason for deviations between a schematic-based simulation and reality. In addition, you may need to analyze also for package influences, substrate effects, and so-called layout-dependent effects (LDEs). LDE is becoming more and more critical, especially in new advanced technologies nodes, because in these, the transistor parameters like V_{TO} or mobility μ depend quite significantly on the transistor neighborhood, like distance to wells. Interestingly, early papers on LDE are related to even 0.35-um technologies, but in these, the effects are typically quite manageable by classical "over-design," e.g., just keeping enough safety margin to other wells, or only relevant for special structures (like power transistors or extremely matching-critical instances).

In 28 nm or below, a detailed early LDE analysis, even on partial layouts, is usually possible, so there is much less need for over-design or working just blindly. Actually it is a good idea that *both* front-end designers and layouters care for LDE. The designers are interested for *block* performance impacts and they want to set constraints for the layouters (Figure 10.20). The layouters are more interested in making the layout of *individual transistors* correct, and according to constraints in the layout-centric flow, there is no real need for a

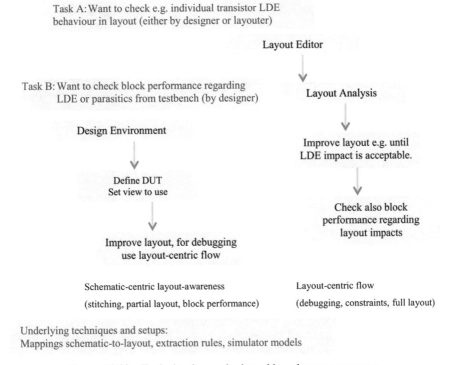

Figure 10.20 Typical tasks required to address layout-awareness.

block testbench, and an LDE analysis tool users can just analyze each transistor individually (and automatically in the background) and check for unexpected changes in threshold voltage V_{TO}, saturation current $_{Dsat}$ or on-resistance R_{on}, etc.

Advanced custom IC environments offer several assistants to give the designer a quick feedback to potential LDE problems in the circuit, even suggestions on *how* to reduce the different layout effects (like well-proximity effect WPE and poly spacing effect PSE). This way the layouter can quickly inspect different layout improvements, even graphically; and he/she can chose the best compromise between accuracy, area, and routability. Figure 10.21 shows such a plot for 20nm NMOS transistors placed in different ways. The manually optimized layout has variations below 2mV in V_{TO} between each transistor finger, whereas the compact layout gives a 13mV worst-case spread. Of course, also the full block-level behavior can be simulated with the correct LDE parameters, even if only a partial layout is available.

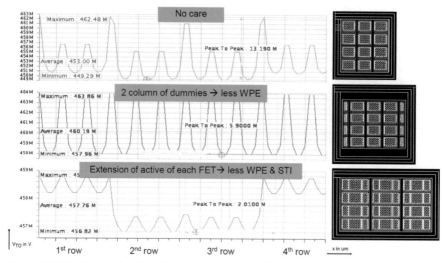

Figure 10.21   Result plot from layout-aware design flow (Courtesy ST Microelectronics, France).

10.3.2.4 Post-layout speed-up techniques

One general problem in all post-layout analysis is that the simulation times can be much larger, like $2\times$ to $30\times$ – with full RCLK extraction and FinFET technology even up to $100\times$! So, parasitic reduction techniques will become more and more essential. Many simulators have built-in parasitic RC reduction, but speed-up is usually limited, and more techniques are needed to be still able to simulate complex RF circuits in FinFET technologies. Some reduction techniques are also available in the parasitic extraction tool itself, to merge fingers in MOS transistors or to filter out very small capacitors and resistors, but in the future, we can expect more techniques, which combine extraction, accuracy constraints, and simulation.

Post-layout simulations and design come also with several special questions and techniques. For instance, it is quite native that the performances get a "shift" after the layout, so we may want to apply optimization also on the post-layout netlist. This sounds simple as the simulator and the optimizer work anyway just with numbers, but unfortunately, the post-layout netlist may differ in <u>structure</u> from the one generated from the schematic, but the designer wants of course still to work with the parameterization, which he/she has used in the schematic! In addition, the new optimized values should be sent back to get an optimized schematic and layout. This is actually possible, but causes a lot of programming work at the EDA vendors to make this possible and to let the flow run smoothly and in an easy-to-understand way!

The mentioned performance shift can be found by just comparing the pre- and post-layout simulation results, but the designer typically also wants to understand and know which parasitic elements are causing which effect. This kind of sensitivity analysis is far from trivial! For instance, many (like dozens or even hundreds) capacitances might be present at each net, and the sensitivity to the total capacitance is giving not always the full picture, like it might be accurate enough for pure loading or speed effects, but not for difficult analog effects like cross-talk, stability, changes in filter poles, and zeroes. For instance, in a latched comparator, a capacitance asymmetry of few % can create a very significant offset voltage—which is simply not present at all in pre-layout simulations. Again a mix of techniques is needed like estimating parasitics ASAP, e.g., by doing a partial layout quickly for the most critical parts and of course just by knowing what matters in which kind of circuit.

Also new and amazing statistical techniques can help to manage complexity in post-layout: They pick up the idea of having a performance shift. Just assuming there is a pure shift and adding x decibel to the pre-layout performance is a risky method. However, this idea of "borrowing" information can be automated and further improved [Wang, Sun] by using so-called Bayesian techniques [Jaynes2003] for *model fusion*. So in advanced fully automated tools, multiple performance shifting models can be applied and verified to "transfer" all the accurate pre-layout results we already have (like MC performance results, and sensitivities) by running just a small number of post-layout simulations. For instance, the tool could combine the results from a big pre-layout MC analysis with 1000 points and a 40-point post-layout MC run into a post-layout MC result and yield estimation almost as accurate as a 1000 point post-layout MC analysis, with low internal computing time and of course highly reduced overall runtimes.

10.4 Summary

Many algorithms for a highly automated circuit design flow exist already. However, it is a big effort to put all the elements together, and some are highly technology-related, so it is tough for an EDA vendor to provide them. Due to software complexity, it is also a big challenge to implement such environment with high user friendliness (and low number of bugs), but actually such design environments with very high degree of assistance exist already.

How much automation we will ever get? Prediction is very difficult, especially about the future! IP reuse to enable a kind of Lego system as we have in digital is a hot topic, but what about real new designs? We have seen that more adaptive methods are a clear trend, but will they be ever foolproof? At some point always some expert know-how and tool training is needed. For instance, a latch simulation can give three different solutions, and the designer has to manage them correctly circuit-wise by using a reset or simulator-wise via initial conditions.

Sometimes, it is easier to make tape-out or you just have to really see the bug in your laboratory, because what you may see is so strange that you simply would have not trusted your simulations! Systematic work, problem anticipation, and simply experience will be always the top methods for engineers. However, sometimes, you just *have to* remember "OMG, an input-stage in 709-style can show a phase-reversal in case of exceeding the common-mode range." or "Upps, my 1st IC design needs a focus-ion-beam (FIB) modification before going to customer, because I made a last-minute change without detailed re-simulation."

Business as usual: Some mistakes can be avoided by experience, some by anticipation, some by detailed verification, some by automated methods, and some bugs will usually present in your first silicon samples. However, due to increasing complexity, you should be good in all these areas, according to the Olympic Motto: Citius–Altius–Fortius.

10.5 Design with Pictures Eight

Let us now go back to our power amplifier from "Design with Pictures One". A PA operating in class AB was addressed, with a tuned driver. RF-PAs in CMOS are not easy to design, due to substrate issues and numerous loss mechanisms when compared with other technologies. For instance, GaAs or SiGe offer better performance, but are usually more expensive, so mainly for high-volume, low-cost designs in CMOS make sense; and having a good optimization flow can really help. We placed this PA example in Chapter 10, because we discuss here also layout effects; and an RF PA is a perfect example where optimization makes sense, and even *post-layout optimization*.

In order to achieve multiple specifications, multi-objective optimization automation is required. Adding a driver and respective input matching network, we redefine the parameters to be used along the optimization procedure. An initial solution should be provided as described, based on previous analyses, so that the optimization tool can search in nearby regions to the

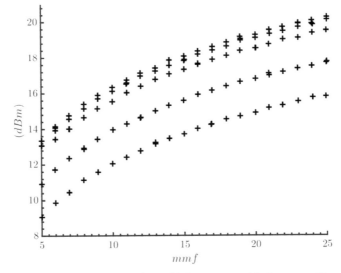

Figure 10.22   Finger sweep for multiple corners and its impact on P_{1dB}.

reference point, hopefully determining a set of parameters that meets all the criteria. Some parameters can be initially swept to get an idea how they behave among all set of corners and to establish an adequate range of values for the optimization. For instance, in Figure 10.22 the number of fingers of the transistors in the output stage is swept to see the minimum at which the P_{1dB} specification is respected. In sync with the basic PA theory we can improve the output power by lowering the on resistance and having just more gate fingers.

BFGS can be employed to provide a local optimization with the reference point early mentioned, and then we can run a global optimization to check if there are other interesting solutions not considered before. Moreover, during the optimization process there might be some useful information along a significant number of iterations in which some correlation can be seen between parameters. Figure 10.23 shows the values for three specifications along several iterations in a local optimization process. Unfortunately the graph is not so easy to interpret, e.g., we see and expect no high correlation between gain and output related performances like OIP_3 and P_{1dB}. The two output measure should be somewhat correlated, but a better statistical analysis should show this in a better way.

For instance, we can directly obtain a table for the correlation factors as depicted in Table 10.2. Such tables are available for the correlation between

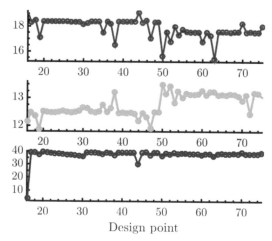

Figure 10.23   Example of (a) local optimization with iteration points for OIP$_3$ (top), P$_{1dB}$ (middle) and gain (bottom).

component parameters and performance metrics, or between different performances. For instance, forward gain $|S_{21}|$ and input reflection $|S_{11}|$ are highly correlated because e.g., a bad input match immediately causes a lower gain. The strongest correlation between stability factor k and any other performance is to $|S_{21}|$, which is also no surprise, because a high gain comes often unfortunately with a low stability factor.

However, some results in the table provided directly are hard to interpret, e.g. we readout a negative correlation for intercept point and compression point. Actually, some such non-intuitive results can be expected, because the table based on optimization data points, not e.g. on MC data (as usual). And the output correlations simply depend also on which parameters you vary! If we e.g. look only to variations in the matching network, we would observe no correlation between DC current and output power, but if we e.g. vary the transistor widths we would obtain a significant positive correlation. Further "error" sources are that e.g. a local optimizer s not designed to cover the whole variable space (see Figure 8.6), also nonlinearities can lead to results which are harder to interpret.

The local BFGS optimizer has taken about 300 simulation points, whereas about three times more simulations are required to meet all specifications when using global optimizers. These optimizations can be done for all corners at the same time, or identifying the worst-case to help improving speed. Figure 10.24 shows MC results for three major specifications,

Table 10.2 Correlation between performance parameters

	Average	$|S_{21}|$	$|S_{11}|$	OIP_3	OP_{1dB}	X_i	R_i	k_{min}
$|S_{21}|$	0.90728	1	0.94398	0.90145	-0.97989	-0.9388	0.69867	-0.88819
$|S_{11}|$	0.88369	0.94398	1	0.84544	-0.90017	-0.99265	-0.99265	-0.75203
OIP_3	0.85292	0.90145	0.84544	1	-0.86925	-0.80482	0.80308	-0.74638
OP_{1dB}	0.87585	-0.97989	-0.90017	-0.86925	1	0.90669	-0.59962	0.87532
X_i	0.86818	-0.9388	-0.99265	-0.80482	0.90669	1	-0.66812	0.7662
R_i	0.71391	0.69867	0.75156	0.80308	-0.59962	-0.66812	1	-0.47633
k_{min}	0.78635	-0.88819	-0.75203	-0.74638	0.87532	0.7662	-0.47633	1

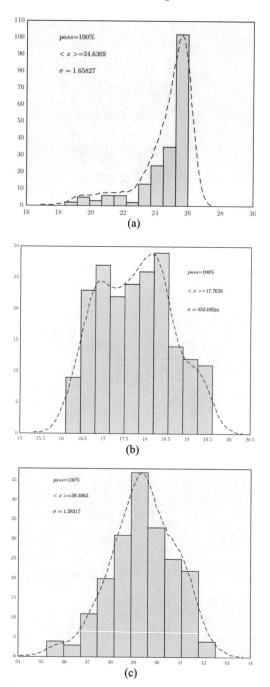

Figure 10.24   MC simulation results after optimization for (a) OIP$_3$, (b) P$_{1dB}$, (c) gain.

namely OIP$_3$, P$_{1dB}$, and gain. These results represent a considerable yield improvement over the first manual design earlier presented.

An interesting graphical representation to see is the one depicted in Figure 10.25. It gives some insight about the MC dispersion of results in specific parameters relatively to corners SS and FF.

Table 10.3 shows a comparison between results from local and global optimizations with some significant differences on the parameter values. As mentioned the differences can be quite significant, which is no real problem, but just a consequence that an optimum is typically quite flat, just indicating robustness and (relatively) low sensitivities.

Following the schematic level simulations, post layout simulations can take place, with parasitics included. At this phase, slight deviations are

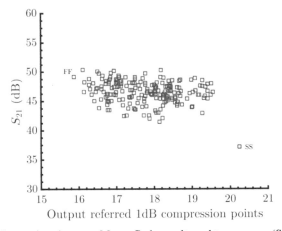

Figure 10.25 Comparison between Monte Carlo results and two corners (SS and FF) for (a) |S$_{21}$| vs. P$_{1dB}$.

Table 10.3 Comparison between results from optimization

Parameter	Local Opt.	Global Opt.	Unit
$V_{b,pa}$	597.4	659.0	mV
$V_{b,drv}$	743.0	758.2	mV
L_{drv}	8	2	nH
mcs	26	22	–
C_{m1}	22.63	18.42	pF
C_{m2}	15.26	13.68	pF
L_m	1.711	1.789	nH
mcg	21	21	–
mdrv	5	9	–

Table 10.4   Correction in the parameters from optimization in extracted view

Parameter	$V_{b,pa}$	$V_{b,drv}$	L_{drv}	C_{m1}	C_{m2}	L_m	η_{drain}
Correction	−2%	+1%	+0.5%	+0.8%	+3%	+2%	−1.5%

introduced due to resistive and capacitive parasitic elements caused by the interconnects. To compensate for this, one could go back to schematic and make adequate changes, change the layout accordingly, and perform new postlayout simulations. However, an interesting alternative is to perform a (local) optimization *directly* on the extracted cell view. As such, for a given number of fingers and multiplication factors, e.g., the widths of the transistors can be re-adjusted to accommodate performance variations, and also other parameters as well, such as the matching network elements. After a postlayout (extracted cell) optimization new parameter values can be obtained to satisfy all corners – Table 10.4 summarizes some results after optimization overall corners. Actually there are no big differences, i.e., the parameter changes are in the same order of magnitude as the usual production tolerances. However, there is no clear limit if post-layout optimization is a must or not. For instance, in a 25 dBm RF PA at 5 GHz in SiGe <u>most</u> of the design time was even spent in post-layout state! Sometimes, it can even happen that you need to change e.g., your matching network topology to address the impedance changes from layout effects. So in such cases, parasitic extraction as accurate as possible, post-layout simulation, and optimization often even saved a re-design.

11

Conclusion and Outlook

Our book's focus is not only the state-of-the-art of variation-aware design, but also technologies which have become more and more complex and the available techniques for circuit designs as well. We give now an overview on what can be expected in near future on our core topics like optimization and statistics, but we also take a more general look to (front-end) circuit design environments and related topics. Indeed one bigger trend is to connect all these tasks more and more, because also mathematically they belong to each other and they can benefit from each other.

The problems in modern custom IC design are not due to a single law (like Pelgroms' law, or Moore's law or the end of it), single parameter (like going down to 7 nm Gate length), or single technique (having FinFETs and mixed-signal). So we can expect that statistics on re-designs (Figure 11.1) will not change so much in the future. Regarding variations, there is really a bunch of problems and solutions, and we provided many examples, the underlying key theory, and a lot of background information. Designers need to find and will find new clever circuits that can work even at 0.7 V supply, subsystems that are accurate enough, although the process tolerances become more and more severe, and they will deal with growing transistor counts, growing number

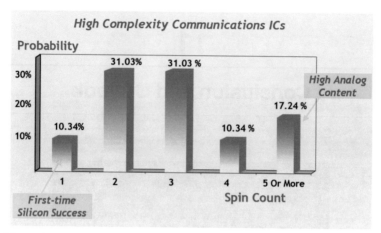

Figure 11.1   Re-spin statistic for large communication IC design [DAC 2002].

of variables to enable multi-mode, multi-system, multi-SoC design with low power, and for low costs; all this almost without expensive high-accuracy elements.

Obviously, there is still and always a gap in engineers thinking and wishes compared to what design tools can really offer, but the key for successfully managing all significant variations is to be able to pair a suited variation-aware methodology with the current design situation. Sometimes even a bit more, just to organize and partion your work in such a way that all tasks become feasible. In the book, we show the different options (with their prerequisites, advantages, and limitations), and in modern custom IC environments, designers have comfortable access to them. Many methods come with the typical variance in the results involved with statistical samples, some come without these inaccuracies, but rely on additional assumptions, so as usual there is no free lunch. Designers who are aware of this can clearly work more efficiently and accurately.

Looking to the math behind design problems, there are many useful proven theorems, but actually still many *unproven* propositions also! In this book, we showed that statistics can be treated by many different techniques and not only by Monte-Carlo. Many of such advanced techniques have faster convergence, but some still degrade significantly if applied in a straight forward way to really complex problems. We also demonstrated the advantages of using optimizers for circuit sizing, also here global convergence and efficiency is dependent on how good the designer managed the setup, and such setups can be complex. So we provided guidelines for analyses regarding corners and statistics, and for optimization.

Solving the specific design problems step-by-step is the focus of current state-of-the-art mathematical investigations, and the integration to real commercial EDA tools is always a bit behind. There are also many funny parts, in creating faster, highly adaptive, more "intelligent" algorithms, to get higher speed with acceptable risks; and also the user-interface part is essential, for general acceptance and for robustness, like having some protection against non-optimum setups.

For further reading:
Here is a list of literature on advanced topics, like analog synthesis or hierarchical optimization. However, look up that often you have to search for other keywords than you think of, e.g., there is a lot of material about multidisciplinary design optimization (MDO), but much less on hierarchical circuit optimization.

- Fast Statistical Analysis of Rare Circuit Failure Events via Subset Simulation in High-Dimensional Variation Space, Shupeng Sun and Xin Li, . . .
- H. Graeb, *ITRS 2011 Analog EDA Challenges and Approaches*, in Design, Automation Test in Europe Conference Exhibition (DATE'2012), Dresden, Mar. 2012, pp. 1150–1155.
- J. Crossley, A. Puggelli, H.-P. Le, B. Yang, R. Nancollas, K. Jung, L. Kong, N. Narevsky, Y. Lu, N. Sutardja, E. J. An, A. L. and Sangiovanni-Vincentelli, E. Alon, *BAG: A Designer-Oriented Integrated Framework for the Development of AMS Circuit Generators*, IEEE/ACM International Conference on Computer-Aided Design (ICCAD'2013), Nov. 2013, pp. 74–81.
- Ramy Iskander and Marie-Minerve Louërat, and A. Kaiser, *Hierarchical Sizing and Biasing of Analog Firm Intellectual Properties*, Integration, the VLSI Journal, Vol. 46, No. 2, pp. 172–188, 2013.
- Automation of Analog IC Layout – Challenges and Solutions, Juergen Scheible, Jens Lienig, ISPD'15, March 29–April 1, 2015, Monterey, CA, USA.
- IEEE Transactions on Evolutionary Computation, Vol. 15, No. 4, August 2011 557, Trustworthy Genetic Programming-Based Synthesis of Analog Circuit Topologies Using Hierarchical Domain-Specific Building Blocks, Trent McConaghy, Member, Pieter Palmers, Michiel Steyaert, Georges G. E. Gielen.
- Analog Layout Synthesis – Recent Advances in Topological Approaches, H. Graeb, F. Balasa, R. Castro-Lopez, Y.-W. Chang, F. V. Fernandez, P.-H. Lin, M. Strasser, Date 2009.

- Multidisciplinary Design Optimization: A Survey of Architectures, J. R. R. A. Martins, Andrew B. Lambey, American Institute of Aeronautics and Astronautics.
- Framework for sequential approximate optimization, J. H. Jacobs, L. F. P. Etman, F. van Keulen, and J. E. Rooda, Structured Multidisciplinary Optimization 27, pages 384–400 (2004).

11.1 Advances in Corner Analysis and Modeling

The worst-case corner analysis was the first advanced method we discussed, and we found that reliable and efficient solutions are already available, so what can we expect in the future?

- Surely improvements e.g., on step-size control (to avoid "blind spots" not well-covered by the model); this makes sense for variables with potentially strong nonlinear impact, like temperature.
- Further new algorithms: adaptive mathematical algorithms can do tricky things like fully *combining* heuristics, designer's *a priori* knowledge, earlier corner simulation results, and advanced sampling and modeling methods. Exploiting existing information is indeed possible, e.g., by applying so-called Bayesian techniques, coming from the statistical field [Jaynes].
- No 100% parallel execution is possible for adaptive methods, but some over-all runtime speed-up is indeed possible if the algorithms will be optimized for at least some parallelization. Actually, modern EDA software is usually at least partially optimized on this already, but there is often room for improvement.

Beside specific improvements, we can see that almost all our book topics are *highly* connected to one key technique: <u>Modeling</u>! WC corner search accuracy checks, optimization strategies, the worst-case distance method, etc. – all this depends on modeling [Martens2008], but note, one general problem remains: Whatever we do, like spline interpolation or Gaussian process modeling, we apply almost by definition a model which is not at all 100% correct. This way, e.g., in WC corner analysis, also the model parameter estimation part becomes fuzzy; so from the pure math view point we do here something ugly. So again, actually only with somewhat more information regarding the system under investigation we can improve further. That is often the reason why dedicated algorithms can outperform general, easy-to-use methods. A dedicated method might be e.g., a circuit-specific design script which takes e.g., corner variations directly into account.

A cool thing is that today's general methods (GPM, machine learning, etc.) can be quite easily modified for this, e.g., in GPM we could plug-in an analytical model (e.g., exponential function or a certain multiplicative law) for the general trend, and remaining fitting part by the Gaussian functions becomes easier (so-called *universal* kriging). Here we are close to the state-of-the-art in math!

We authors remember that many engineers where enthusiastic on deriving formulas for circuits because programs like Mathematica® allow *symbolic* analysis, so why using only numerical simulators like SPICE? Symbolic *circuit analysis* would be indeed *one* attempt to improve modeling by using true theoretically founded functional equations (e.g., based on small-signal equivalent circuits, transistor equations, etc.). Another approach is to use a big set of such typical "circuit design" functions (like polynomials, rational functions, e^x, $\sqrt{x}$, min(x), etc.), and then to perform a *symbolic regression* (SR), ending up in using the best-suited formula for each performance. Actually it seems that this second method is much easier to apply [Guerra-Gomez2015], because it natively fits to numerical circuit simulations; and there is no need to do a real circuit and performance-specific symbolic analysis. Magically also such fitting methods would indeed often deliver meaningful functional results, like risetime behaves reciprocal to the bias current, and maybe linear with temperature, like $t_{\text{rise}} = 3.267\,(1 + T/5.4\text{e}4)/ib$ (*ib* in uA, T in °C, and t_{rise} in ns). The only advantage of the fully symbolic method would be that the constant(s) could be also related to other parameters (like Miller capacitance C_m, TCs of resistors, or current mirror ratios). With pure fitting we can only include the variables which are tweaked. In a WC corner analysis and an related performance model these variables are typically only the corner variables x_R, but in principle we could also extend the performance models to other parameters, e.g., those from x_D (for modeling and design purposes, including optimization) or even x_S (for yield analysis, statistical corner creation, etc.). Of course, such symbolic fitting could not only deliver one solution, usually multiple output formulas are available. The more complex ones just typically give a somewhat better fit, whereas the simpler ones are easier to interpret.

Note that such performance models could be used for any fix circuit topology in design scripts by solving the performance expressions for the design variables, but we can also use them directly in a ultra-fast *surrogate-based* global optimization. However, optimization is another Subsection 11.3, so let us stay for a moment for more basic topics like verification and statistics. In all these topics we can definitely expect more modeling-based, more sensitivity-driven design techniques in the near future.

11.2 Advances in Verification and Statistics

Regarding verification, there is a lot of news from the digital and software domain. Here one focus is finding bugs. You may run so-called *directed* tests to find such problems "as usual", or you could also run *randomized* tests! It is proven that this style of verification helps to find design mistakes quite efficiently, especially bugs difficult to anticipate! So one trend is picking up the verification concepts from these domains, like coverage-driven verification or even formal verification. The latter is still at the research level for analog applications, whereas coverage-driven, assertion-based techniques may quickly swap over from the digital area to mixed-signal and analog design. The main problem is the definition of functional analog assertions, it is a very complex task to describe analog input and output waveforms accurately, including their relationships. So at the moment, it is the best complementing concept for analog applications.

Disproving that the design is correct is quite an easy task for automated tools, but a positive proof is more difficult. With respect to verification, finding the *worst-case* parameter set consisting of both deterministic and statistical parameters is generally a key task. Finding the worst-case for *one* type of variable (either statistical or deterministic) can be regarded as solved, and it is available in many modern EDA tools. We can just recommend to use these methods, because they give a good balance between efficiency and accuracy, and are not difficult to setup anymore. However, putting deterministic and statistical variables *together* (see Chapter 9) could end up in a really complex and highly nonlinear problem.

In addition, especially the problem of increasing complexity of both models and circuits make the handling statistical variables more and more difficult. In old PDKs (like ≥ 130 nm), mismatch has been modeled by usually just one or two variables; in modern processes, it could be a dozen, and on top of that, the mismatch impacts are growing too due to shrinking geometries—even in digital circuits, mismatch effects have become severe! Also the number of transistors per block and per subsystems increased, so overall the same problem becomes easily $20\times$ more complex in new designs.

All in all, we could apply advanced model techniques also to the extended variable space $(\mathbf{x}_R, \mathbf{x}_S)$; the algorithms exist, mainly managing the complexity is the key challenge.

This higher general complexity problem is often preventing the user of brute-force methods like running Monte-Carlo over all corners; and often we have on top higher quality requirements. Having a verification gap is no good idea in contexts such as medicine or autonomous cars.

One way to improve dealing with complexity is to exploit hierarchy; in optimizations, designers often do so. Also in statistical analysis, we can exploit the hierarchical structure of our design. When analyzing correlations, you will typically find out that variables in the same block have stronger correlations than variables in different blocks. In addition, the correlations or WCDs will reflect the system structure, e.g., design symmetries or repeated structures (e.g., in ADCs). This way we can predict the approximate structure of the correlation matrix much faster directly from the circuit topology, avoiding too many time-consuming simulations. Several of our discussed techniques, like low-discrepancy sampling LDS, have weaknesses [Dick2014], but with exploiting hierarchy, we can work around some limitations. For instance, in analog low-power designs, usually the threshold variations are most important, and if we use for these high-quality low-discrepancy points, we can reduce the variance in our MC results significantly.

A big trend in statistics is also the application of <u>robust</u> methods, able to deal with high nonlinearities and outliers. Indeed, we have seen that in some cases already the idea that a worst-case corner can "represent" well the worst-case behavior of a circuit is not valid at all! A fall-back solution is here, e.g., using multiple MC worse samples. So the corner set idea could be still used to speed-up design, but the corner finding becomes a bit more complex—and some speed loss is sometime unavoidable in favor for robustness.

In many cases, we demonstrated that even if using advanced numerical techniques, still the user knowledge has significant impact, e.g., an optimization becomes easier with good starting point. Actually exploiting "*a priori*" knowledge is also possible for statistical analyses! Most people have been taught that probability is either something defined by axioms or is just "frequency of occurrence", but a third way is to regard probabilities as "limited knowledge". Taking experimental data in to account (like MC results) we can obtain a more precise "*a posterio*" knowledge (e.g., about variances, model parameters, or percentiles). Such approaches are called Bayesian techniques, according to Thomas Bayes (1761–1701). Such methods can lead to better results regarding speed and accuracy than assuming "nothing" and purely relying on data only. Note, that defining probability as "frequency" is not so native as it seems, and it creates many further questions, e.g., what is defining the frequency or is that frequency really a constant? So what "probability" really "is" is quite a *philosophical* question! However, in MC we have usually anyway already translated a whole physical problem into a mathematical model, so that we just have to find an elegant solution, within the model, and its limitations. However, indeed advanced Bayesian techniques have reached nowadays pretty much popularity, just because they often lead, following a systematic flow, to good results. Actually the whole way of getting more knowledge about our universe is about getting more precise knowledge from experimental results, and existing know-how.

A nice example [Wang] for Bayesian techniques is reusing pre-layout simulation results to improve on the typically very time-consuming post-layout analysis, e.g., by creating a Bayesian scaled-sigma analysis (BSSS). Bayesian techniques are also popular in the field of decision making, learning, and artificial intelligence (AI). Actually quite many papers pick up such ideas also for circuit design, but we personally expect there are much more fruitful applications for such techniques—especially in a commercial sense, and regarding generally available software tools. In SSS we have not reached the limits at all, e.g., subset simulation extends the idea for further speed improvements [Sun], e.g., we can run MC ones, and then sample again around the critical regions found in the first run, even repeatedly. However, doing this we would lose the advantage of being independent on the number of specifications.

Random, LHS, LDS, and now Super Sampling? Again, it is very interesting to see how many overlap exists between things with not much in common at first glance. Could super-sampling SS, a very successful method in graphics also help in statistical analyses? Looking to old CRT displays, old flat screens, and then to modern 4K screens the progress is impressive. Actually the 4K looks like an analog dream, with good material almost no artefacts, no "aliasing" effects at sharp edges, etc. This is because of the high definition resolution, but also thanks to SS, not to marketing. The artefacts can be avoided by a clever 2D over-sampling scheme, picking up ideas from techniques we explained (LHS, LDS, Poisson disk sampling). The only pity is that e.g., in MC simulations designers are often not in such comfortable situation, we often have not so many points as we want and as a 4K screen simply has. So for us over-sampling looks often unreachable. However, one method we mentioned could enable it, and this is (meta) modeling, using a fast to evaluate mathematical model, instead of long-running simulations, would enable "over-sampling"; and e.g., in a surrogate-based simulation SS might be picked up. An excellent article on synthetic imaging can be found at [Laine2006].

11.3 Advances in Optimization and Synthesis

Luckily in some aspects the future lies not so much in the dark. Currently most commercial optimization tools can treat multiple specs, many also allow a yield optimization, but usually they actually optimize a single real-valued overall goal function using weights for the different performance targets or similar constructs. If a circuit is highly optimized, then improvement in *one* performance is usually only possible by *relaxing* another spec (see Figure 11.2). If this is the case, then such point is called "Pareto optimum". Mathematically we are interested in solutions which "dominate" others. Imagine we look at two solutions x_1 and x_2, and we have two performances f_1 and f_2 to be minimized. If $f_1(x_1) < f_1(x_2)$ then x_1 is preferable, and if also $f_2(x_1) \leq f_2(x_2)$ then x_1 is indeed *dominating* x_2, and if *no* other solution dominates x_1 we call x_1 a Pareto optimum.

Actually a whole *set* of Pareto-optimal solutions usually exists (so-called *Pareto front*, dark green line in Figure 11.2); and one side of the front we have less good solutions, and the other side is empty (the impossible region, "utopia"); no "better-than-Pareto" solutions are possible. However, current commercial optimizers typically only provide <u>one</u> "best" solution, which is e.g., defined by assigned spec-weighting factors of the over-all goal function $f(x) = \Sigma \, w_i f_i(x)$.

In research, such Pareto optimizers have been already developed over years; and also applied to smaller circuits, but their application lacks due to large compute power requirements when applied to more complex problems. This is not because Pareto optimizers are not powerful, but just due to the

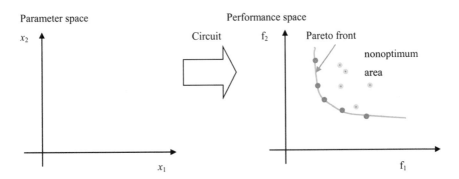

Figure 11.2 Pareto optimization and the relationship between parameter and performance space for 2D case (goal is to minimize f_1 and f_2, e.g., representing risetime and noise figure).

fact that such whole front of solutions is generally requiring more effort, e.g., one meaningful attempt for Pareto optimization is to run many standard optimizations on f just with *different weights*, so the weights would act like a seesaw for the dark green dots in Figure 11.2. Having Pareto optimizers and even Pareto yield optimizers is clearly desirable, because it would remove the tedious task of setting weight factors upfront to the optimization, or even to re-run an optimization multiple times. Once you have found the Pareto front, you can get many different Pareto-optimum solutions—with different trade-offs—in few seconds without further time-consuming circuit simulations. To run a Pareto optimization we would need in principle no spec limits at all, only the direction (upper or lower specs). You can even get trade-off plots like overall yield versus spec limit or total C_{PK} versus bandwidth or power consumption, etc. [Holmes]. The iterative scalar weighting method will not always work well, so in advanced Pareto optimizers often other techniques are also implemented, e.g., converting the Pareto multi-objective problem into a constrained single-target optimization task, just by picking *one* function f out of f, and constraining the *others* to be below a certain ε_i.

Another region of improvement regarding optimization is mixed (real-integer) optimization. The concepts for optimizing a function with real-valued parameters are very old, and advanced variants are dated to the 60s, so here there is little room for improvement. However, when optimizing mixed problems, like the m-factor in a bandgap and the transistor widths and lengths, it is not clear at all how to do it in the best way. If we do it sequentially, we may end up in a non-optimum solution.

Is there a third way, beyond single and multi-objective optimization?
By putting our performance targets into one over-all goal function as weighted sum we can manage the optimization task efficiently by using standard optimizer. However multi-objective optimization would be easier from the user perspective, because there would be no need to select the weights upfront, and we can choose any Pareto optimum quickly from the set of optimization results. However, is this "all" what is possible? What about another intuitive scheme: If we optimize all performances individually, we could directly see what would be a theoretical ultimate optimum, the "utopia". For instance, *without* looking to other specs we can achieve a noise figure of 0.5 dB, and a risetime of 1.2 ns. A natural best point would be the one which is just *closest* to this uptopia $f_u = (0.5\ dB, 1.2\ ns)$. Having this in mind we could actually move away from

the tedious weighting method! However, whatever we do we can only improve *to some degree*! Using the *linear* weighted sum method is just equivalent to the L_1 (or Manhatten) norm regarding $f_0 = (0, 0)$. Using f_u instead of f_0 is somewhat an improvement, and one more improvement is possible. Instead of using the L_1 norm we could also use e.g., the Euclidian norm L_2 or the maximum norm L_∞; and one can prove that actually using the max norm would be the somewhat more correct method, the one which would really allow to hit all Pareto optimums. However, still the weighting is important too, and over-all there is no such "third way". Only nice guidelines to remove the scalarization problems a bit.

11.3.1 Hierarchical Optimization

Optimization is time-consuming, and Pareto optimization is even more complicated, but, in principle, we could just *collect* all testbenches, goal functions and constraints, and run a *single* huge optimization. However, when using classical optimizers, we would need to fully simulate *all* testbenches for *each* optimization point ("all-at-once", see [Martins]). So for industries (automotive, aircraft, etc.) with many optimization tasks more advanced so-called multidisciplinary optimizer structures have been developed.

Some ideas can be also picked up manually; one way to speed-up system optimization is exploiting *hierarchical* information, e.g., by using multiple abstraction levels for simulation and design. It is actually often done for more complex blocks like PLLs, ADCs, or RF transceivers. For instance, on demand designers can switch blocks to a behavioral description to speed-up simulation and doing design tweaks (e.g., on PLL loop filter). Something similar could be even automated in general; for instance, for a full optimization of a critical transistor, you need to tweak the finger length and width, but also the number of fingers and/or the multiplication factor—so you have to set four parameters to define the layout. On the other hand, to the first order, only the total width counts $w_{tot} = wf \cdot m \cdot nf$, and only for few performances the individual full-four-parameter setting has an impact at all (like noise or stability in the RF region). So a clever optimizer should first optimize on L and w_{tot}, and only in a later stage, it should switch to the full set of all four parameters. A similar technique could be also applied for subcircuits like the OTAs in a $g_m C$ filter, the elements of a PLL, or for passive elements like critical capacitors or inductors.

In true multidisciplinary design optimization (MDO) also other, more mathematically inspired, algorithms have been created to avoid the inefficiency of an "all-at-once" approach. In Figure 11.3 there is an example [Martins].

Besides all improvements, pure optimization is not efficient for all kind of problems. It may work out well for migration between foundries having a similar process, but for bigger changes in technology or specs (re-targeting), optimization should be augmented with migration scripts. Several kinds of circuits (like op-amps or bandgaps) natively fit better to *construction* than to optimization. So enhanced, script-based synthesis techniques may lead to very acceptable designs and no need for optimization anymore—at least for non-high-performance applications. Script-based design is often not very flexible, but with a better, more technology-independent *infrastructure*, it may pay off quickly.

On the other hand, also optimizers become better and better. With efficient mixed optimizers and enough compute power, also multiple circuit topologies can be optimized, so that actually also the problem of topology selection and definition can be addressed automatically. In some areas like RF and EM design and simulation, such techniques have been developed and applied already—at least by some universities. In this context, of course also highly hierarchical techniques make sense; an optimizer would almost act like a human designer. An example of a very successful mix of scripting and optimization is transistor model parameter estimation. Here we can extract hundreds of parameters, dealing with many kinds of testbenches, thousands of measurement results and goals. This is possible because such problems have an almost *fixed* problem structure, so spending a high effort to setup a clever fully automated flow makes much sense.

Figure 11.4 gives an example optimization flow transferred to circuit design which might be called multi-level semi-automatic.

11.3.2 Circuit Synthesis

There are different ways of exploiting hierarchy in design; and this helps to address not only transistor modeling, but also true *synthesis* of analog building blocks, like operational amplifiers, bandgaps, etc., also here the specifications are very clear and well-known. One critics on optimization is whether it leads to trustworthy designs. If we even optimize the circuit topology, we can expect that some "synthesized" structures may not behave well regarding process or environmental changes. However, if using classified trustable building blocks (like diff-pairs, different current-mirrors, OTA's, etc.) the problem can be relaxed and could be manageable with use of robust optimizers and enough

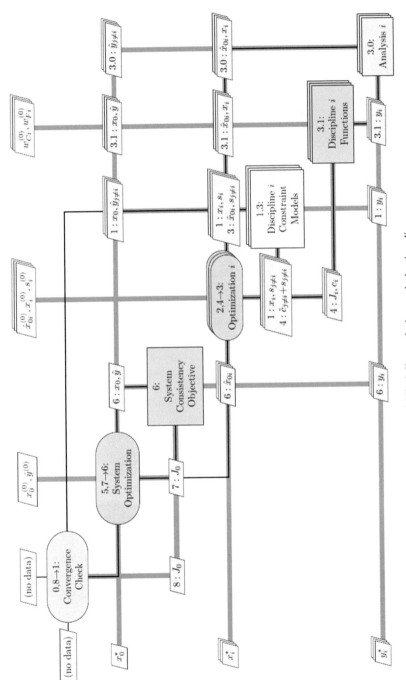

Figure 11.3 Multidisciplinary design optimization diagram.

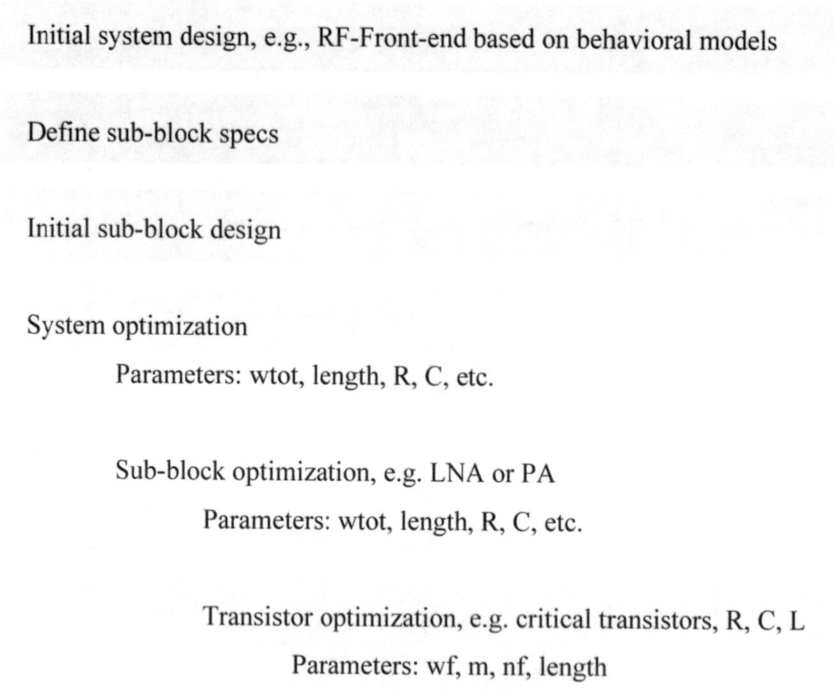

Initial system design, e.g., RF-Front-end based on behavioral models

Define sub-block specs

Initial sub-block design

System optimization
  Parameters: wtot, length, R, C, etc.

  Sub-block optimization, e.g. LNA or PA
    Parameters: wtot, length, R, C, etc.

    Transistor optimization, e.g. critical transistors, R, C, L
      Parameters: wf, m, nf, length

Figure 11.4   Semi-automated hierarchical optimization.

compute power [McConaghy2011]. Maybe we can even expect commercial implementations in the near future. An interesting outcome from such "Pareto synthesizer" would be not only getting *performance* trade-off curves as in Figure 11.2, but also circuit *topology* trade-off information (Figure 11.5).

[McConaghy2011] includes also one comparison to a manual design showing that the optimizer achieved a better performance (Figure 11.6). Actually, six specs have been taken into account, namely DC gain, bandwidth, supply current, area, phase-margin, dynamic range and slew-rate.

Actually, this result is expected, because often human designers have *further* goals in their mind (like offset voltage, noise, power-supply and common-mode rejection, distortion, suitability for layout, etc.) and are often a bit lazy or reluctant in really going to the limits.

A further advanced field is layout-aware design optimization and synthesis [Graeb2009], i.e., the inclusion of layout parasitics and layout *generation* in the sizing loop (controlled by an optimizer). This offers further automation,

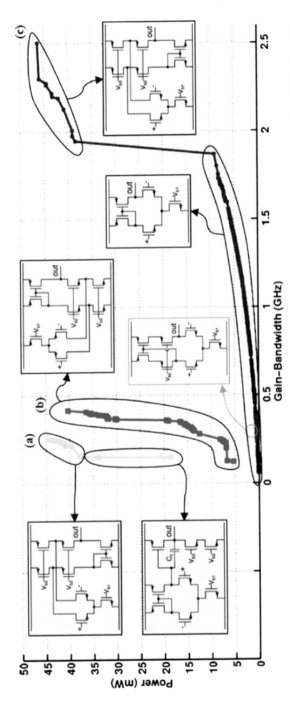

Figure 11.5   Topology synthesis results for different input common-mode voltages and power-speed optimization (a = large V_{cm}, b = V_{cm} close to $V_{DD}/2$, c = large V_{cm}).

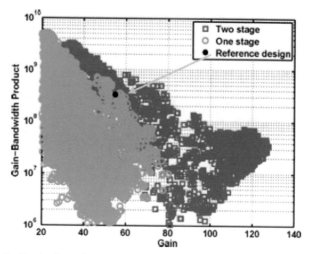

Figure 11.6   Synthesis performance results (2-D cross-section of the 6-D Pareto front).

beyond existing assisted flows, which offer "only" awareness for layout parasitics, design rule violations (e.g., regarding IR drop and electromigration), and for changes of electrical parameters related to layout (so-called layout-dependant effects LDE).

We mentioned, the "fight" between construction and optimization is mostly related to front-end design, but in layout we have a similar situation. Actually parametrizable cells (pcells) are part of all PDK's in the world, and contain sets of layout instructions. However, most pcells are specific to components of lower complexity, like a transistor or resistor. Having augmented pcells e.g., for full op-amps could be one helpful part for really addressing analog synthesis; and indeed powerful object-oriented tools are already available since some years. Unfortunately there is no established standard yet, especially not for dealing efficiently and user-friendly with hierarchy, arrayed structures, constraints, or real complex structures.

11.4 Business Drivers and Trends

Actually both circuit design and EDA software have a bright past and a hopefully bright future. Often the software was actually leading, and it has taken some time to pick up new techniques; and sometimes EDA it was quite behind, at least related to research results.

Einer nur freie die Braut,   *For only one shall win the bride,*

der freier als ich der Gott.   *one freer than I, the God!*

These are the words from the Nordic God Wotan in Richard Wagner's "Ring of the Nibelungs"; and the billion-Dollar EDA industry is in a similar situation as Wotan, having many duties, and limited resources! EDA is also infrastructure, and it has often longer cycles than products, and many more old "stuff" in it. Actually, the SPICE simulator was a huge success, so that even the bugs become an industry standard, but it has still taken some time to convince the design community. So even in the early 1980s, many custom IC designs have been made without circuit simulations, just by thinking, by anticipating, and by analyzing all problems almost manually. In Table 11.1 you can see about the progress in IC design and software, with cordless telephone chips as example. At the end you can see that the "revolution devours its children"! On the other hand, in software many things will still remain, have to remain, just because users want compatibility, and no changes in habits. On the other hand, guess what the product experts for wireless have to do to survive?

What Figure 11.2 also clearly shows, is that one clear direction for design is mixed-signal. And here several techniques from the digital flow swap over, e.g., the use of digital verification and modeling techniques. Actually also digital designers pick up formerly typical analog techniques like self-adaptive voltage scaling. However, for the other direction, the way to go is not so easy. Different options exist, and there is no one-fits-all. For instance, randomized tests can clearly help on verification and debugging, but analog assertions are harder to define; they can become very complex. And we are still far away from enabling analog synthesis: design by scripts is one of the most realistic options.

In most parts of the book, we take the view of an engineer, so actually it does not matter much whether 10,000 people in the world do analog design or less or more—you have to do your job, at best like it and improve yourself! So in our book we looked *not* much at *non*-technical trends. The world is indeed sometimes dramatically and quickly changing, not only technically. The pure number of tape-outs went down because chips become larger and larger—and more expensive. So "Go big, or go home" seems to be a trend! It is difficult for small startups to make a difference, but on the other hand, designing chips can make sense to get competitive advantages for companies never thought of engaging in IC design!

Table 11.1   Cordless telephone technology and EDA software over time

Year	Design	Methods
1985	Plan for a new European digital telephone standard	Corners, MC
	Status: Old analog cordless telephones based on CT1	
	Discrete PA, SAW IF IF filter, LNA, ICs for RX, TX, PLL	
1992	DECT as new digital cordless telephone standard	
1993	1st consumer products	RF simulation, 1st WCD
	Discrete PA, SAW IF IF filter, LNA, ICs for RX (Dual-Heterodyne), TX, CMOS PLL	papers related to circuit optimization
1996	2nd generation telephones	
	Discrete PA, SAW IF filter, Bipolar ICs for LNA/RX (Single Conversion), TX/PLL	
1997	3rd generation	
	Integrated Bipolar PA, BiCMOS Low-IF RX, TX/PLL	
	GAP as DECT standard extension	
2000	4th generation	
	Integrated PA, Low-IF TRX/PLL	
2008	5th generation	Yield estimation by
	130 nm CMOS SoC TRX incl. major digital parts & PA	statistical blockade
2010	No new products, because DECT telephones has become ultra-low-cost, widely displaced by WLAN, Bluetooth, smartphones, software solutions like WhatsApp	
2011	DECT extensions e.g., on ultra-low energy ULE	Early papers on sigma-scaled sampling for yield analysis

All this will certainly have some impacts on which good features and new techniques will be implemented, and what may take much longer.

11.4.1 Design and IP

Even in big growing markets, there are many options to fail and few to become a champion; one example was mobile communication and another is IP. The bigger the chips become, the higher the chance to be able to reuse something;

and the probability rises that you simply cannot do it in your team. For these reasons, digital block reuse is standard since over 40 years—the CMOS gates look 100% the same since a long time, also more complex blocks like flipflops, IO cells, memory, interface standards are highly standardized. On top of that also more and more complex digital *systems* can be *synthesized*. Also most digital systems can be package as IP. For analog blocks, this is not yet true, for many reasons, analog blocks do not only feature NMOS and PMOS transistors, but also special process elements—and they can differ in several ways. The biggest impact on analog designs comes usually from the supply voltage. That dictates how many transistors you can stack and which type of transistors you can use or which design tricks you need.

One further problem in IP (on top of the technical challenges) is making a real business out of it! Mergers and acquisitions are a trend with bounces over the years (e.g., >100 billion $ in 2015), but still some small companies engaged with analog synthesis were unfortunately not bought by the big players and died. They were partially too much ahead.

How many analog designs will be made by IP-providing companies? This will surely have an impact on design tools and methodology, but which kind is not easy to say. Of course, there will be IP companies trying to offer a wide portfolio—having a high need for tools perfectly supporting technology migration. However, there are also IP companies, with focus only on newest deep submicron technologies—which anyway only few foundries can provide. Interestingly, due to high costs the trend for real *immediate* wide use of *newest* technologies has been stopped in the last years. CMOS 350 nm or 180 nm have survived <u>much</u> longer than expected—and will be even present for further twenty years—of course with reduced presence but still used in big "niches" like SMART power, sensors, automotive. All in common is that IP buyers expect robust high-quality designs (best "silicon-proven"), so that also statistical verification methods will be a key element. For migrations to similar technologies or for minor re-targeting, optimization can often be directly applied. For larger changes, some rule-based scalings and mappings should be applied up front in migration scripts to give a better starting point for optimizations. Of course, the more clever migration or even synthesis script or the less challenging the design, the less need for optimization, but if the initial design is done in a systematical spec-driven way, then optimization becomes almost push button.

An ultimate form of "reuse" in digital design is using an FPGA (Field Programmable Gate Array). As technology allows higher and higher integration, the FPGA concept becomes indeed more and more attractive, compared to a full application-specific IC (ASIC). There are also some attempts for *analog* programmable designs. Currently they have only a very small market share, but some analog blocks are indeed so essential that the chance of really using them in the analog FPGA is very high (like for ADCs, DACs, supply monitoring blocks, etc.). On the other hand, there are also many applications (e.g., related to the Internet-of-Things IoT) where analog performance and low power consumption has a dramatic impact on product success.

An alternative method for cost reduction is outsourcing, and it is often done for the layout parts. Indeed, it is worth to think *where* to invest: in tools or directly in work, maybe in India or even all over the world with something like "My eHammer"? To some degree it would be even easier than using this platform for getting a house cleaned, just because ultimately only a *virtual connection* is required.

Besides optimization also construction methods can be extended; e.g., circuit and layout might be codified in scripts to generate highly scalable and portable blocks. This technique is of course in competition with other techniques, like reusing hard IP and apply simple migration scripts. A full script-based flow would require even more changes in habits than using an optimizer on top of what you do anyway. And of course IP protection is a difficult topic too; often hard to align with modeling and documentation requirements.

11.4.2 Computing Trends

In the 1980s, computer software (like SPICE) runs on 16-bit CPUs with Megabytes of memory at tens of MHz, and over the years, we reached the TByte and GHz domain. However, some of these positive developments seem to stop! That a modern computing server is more powerful than its predecessor is nowadays highly related to *parallel* computing and multi-core techniques, but *not* much due to higher clock frequencies. This constraint will definitely have impacts on EDA tools and partially it already has.

With respect to variation-aware design, it has consequences as well, because typically just the older, more brute-force style techniques (like full-factorial corner analysis or random Monte-Carlo) are easier to parallelize! Many implementations of optimizers or advanced statistical methods need to be improved to exploit modern hardware in an optimum way and to adopt them for more and more complex designs and technologies. In the early 1980s, a central computer was standard for a development department.

Later, workstations and PC's became popular; whereas now, working from a PC with connection to the company cloud server is standard. Using encryption techniques, it would be obvious to just go for general cloud computing, at least for challenging compute tasks and for smaller companies. Of course, for some interactive tasks, like manual layouting, but often also for debugging circuits, there is also a general need for low latency.

11.5 Future Analog Design

Up to now we looked to realistic improvements in optimization and statistics, to trends, but of course there are also other topics, needs and ideas to improve the custom design flow. So here is just a list; and many topics are of course connected to variability.

- Actually the hype on IP and FinFETs will *remain*, even FinFETs are not the ultimate solution (e.g., also nano-wire FETs, even in complementary form or vertical FETs, are under investigation), but the driver will be of course digital design. Also the foundries have a big interest to push these topics; it is a self-"runner" if these technologies become mature enough at acceptable costs.
- Even the most advanced EDA tools have gaps, actually designers usually take the decisions, and tools are not that good in decision making. However, in software engineering, decision making and machine learning are hot topics! You as engineer are in the driver seat and you have to formulate your questions before you can expect good answers.
- One key point in successful software design is problem *analysis*, and exploiting everything which is available this could be supported by having a full *constraint-driven* flow. This way design *intentions* can be collected *systematically*, and as tool-readable constraints, this can simplify tool setup, and it can help to avoid overlooking important design aspects. Having a *standard* on constraints would be clearly helpful, but it does not exist yet. From the tool perspective it is of course easier to have specific setups.
- Connecting point tools to a full flow is essential, especially for bringing circuit design ("front-end") and layout much closer together. In ultra-deep sub-micron technologies this is required not only in high-performance analog, but also in digital circuits. Actually, there is already a name for

such techniques, electrically-aware design EAD. It can help to reduce the number of time-consuming iterations during the design flow—and it can also lead to a better understanding of physical implementation effects (like performance impacts of wiring parasitics or proximity effects between neighboring transistors). So what we described in Chapter 10 is only the (impressive) beginning, actually here it looks again that the EDA tools lead.

- Although designers have become more powerful nowadays computer servers than ten years ago, it can easily happen that things which went fine for many years, like doing a post-layout simulation for sign-off, become almost impossible! For instance, if you apply RCLK parasitic extraction on a high-performance RF 14-nm chip, it can easily happen that the number of netlist instances is 30× above older generations! So old topics such as parasitic reduction move much more into the focus; and techniques need to be extended. The whole topic is not only about modeling and simulation runtimes, of course also aspects like variation-awareness and sensitivity matter.

- We mentioned it already, automatic techniques which exploit *hierarchy* will become more important in the future. For instance, a simple flat netlist does not contain the design hierarchy anymore, so using this netlist as the only input might limit speed and accuracy of many algorithms significantly. Implementing efficient divide-and-conquer strategies can help to adopt algorithms (like statistical corner generation) for more and more complex systems [Zou2009].

- Of course, variation-awareness is not only about analog design, also in digital, mixed-signal and system design. It is only realistic to expect in these wider fields much more innovation and tools. However, an first requirement is just to connect IC design to other areas like software, MEMS, printed circuit board (PCB) or package design, e.g., to treat more effects like electromagnetics or self-heating, and to enable multi-disciplinary design, variation-awareness, and optimization.

Have you ever heard this:

Problem too big, market too small!

Unfortunately, analog is often regarded as this! And indeed besides the recent achievements, many presented in this book, one may ask whether there will be a further real breakthrough in the next years? What could that create? And which of the improvements are desirable?

Improvements in Simulators. Actually this topic is not really a core topic for variation-aware designs, but there is some progress! Several of the newer techniques have been implemented since years, and some of them are similar to what we can expect e.g., in statistical tools and optimizers. For instance, matrix inversion is an n^3 process—so time-consuming, but large circuit matrix tend to be sparse, i.e., containing many zeroes. Exploiting that, the rise in simulation time drops typically to $n^{1.5}$. In some places, life is easier for simulators, because they deal with a limited number of device types, and the device models are usually hardcoded. This way, the derivatives and good initial guesses for operating points are available internally—for simulators but not for circuit optimizers. In so-called fast-SPICE simulators, adaptive modeling and partitioning techniques have been implemented very successfully. For less accuracy-critical digital parts, we can use simpler device models, and with event-driven simulation, we can improve further. For regular structures like memories, we can also speed-up by simulating one cell and take this as representative for the other cells.

11.5.1 Enabling the Next Revolution

Inspecting all our topics again, we may ask what is *in common*. And what was the *reason* for success e.g., for the digital design revolution, and for the success of analog simulation? In addition, what would be the ultimate way, *the best way to work*?

- Having always something that works, i.e., lowering risks.
- Knowing what to do, having clear guidance, knowing the current design situation.
- Being able to try new ideas, being flexible!
- Do not waste time with doing almost redundant tasks again and again.

This has many similarities to what designers do anyway in existing design environments, but also software design and object-orientation. Of course, analog design has also these aspects and concepts already, but (historically) there are quite many limiting factors. For instance, you can program some of the described techniques like the C_{PK} or Monte-Carlo with sigma-scaling (SSS) just in very few lines of code in MATLAB or statistical packages such as R, but *translating* it to a real circuit design is a big effort!

In a math package or in advanced programming languages you can treat all numbers or even complex *objects* as you want, almost independent whether something is a number, a vector or a matrix, or a constant, a global variable, or a parameter. Actually having full (read-write) *access* to all variables would be a progress, best in combination with methods to copy or clone variables. In addition we also need good *structures* or categories (like variable belongs to class x_S or to amplifier TOP.VAG.A1, or variable is user-editable or is derived from another variable in back-end flow, etc.). Variables may have to be on a certain grid for manufacturing (like 10 nm), have a certain valid range (like $90\,nm \leq L \leq 10\,um$) or a certain statistic, sometimes even constraints involving other variables, even across the hierarchy (like $OTA1.L_1 = 4 \cdot OTA2.L_2$). Also some categorization based on design priorities makes sense, e.g., variable T is important, should be part of the corner set and being checked on sensitivity by default, or parameter *wf* should be optimized within $\pm50\%$ by default.

In standard design environments the flexibility is much lower, mainly because the different tools offer different capabilities, just to execute *specific* tasks, like circuit simulation, corner analysis, Monte-Carlo, etc. Often the wheel is implemented twice, e.g., the simulator needs transistor models, but also a sizing tool (see Figure 10.10) could use it directly as well; maybe even with different accuracy levels like full-BSIM4 vs quadratic MOS2 vs table-model. Having standards would make maintenance, for technology updates much easier. Interestingly, one IEEE standard, Verilog-A/AMS, offers already quite a nice structure for disciplines, signals, units, etc., so there is not so much to do on top.

In the end, often the EDA *infrastructure* is a clear bottleneck, both the data structures *and* the programming interfaces. The open-access database OA (standard set by SI2 group [SI2]) was a step forward some years ago, but due to the big challenges of CMOS designs below 25 nm (FinFETs, double patterning, variability, etc.) and many others (complexity, system design enablement), this is by far not enough. Actually a large number of auxiliary setup files exist in parallel.

Especially the interfaces between environment and simulators are quite limited, namely the variable setting. Actually the term "simulator" should be treated more general, it should better stand for "solver". Actually in a "variation-aware centric" flow the simulator is only one solver among others. Currently the human designer makes not only the design decisions, he/she is also the only one who "knows" the data. Only he/she has a "good" memory, whereas in most tools these things do not exists or they are very limited. A tool *owning* the data, the variables x, and the simulation results $f(x)$, could perform MC, worst-case searches or optimization, etc. more efficiently, with more re-use. It is quite realistic that future "big data" engines exploit the collected information even better than humans. Such tool, actually the environment, could also "own" the status information (like design is "in-spec at nominal conditions" or "LVS-clean and ready for postlayout verification") and the strategy, and in a circuit synthesis it would also make or propose decisions. At least these advanced topics are quite hard to solve in general with point tools only, but an environment of that complexity needs for sure a good modularity and standards.

Due to the first EDA revolution by SPICE we have quite some standards for models (unfortunately not well-structured) and on netlisting formats, but already for result evaluation there is none (or too many). On its own this is no big problem: designers just have to learn several methods instead of one globally applicable. However, regarding further automation techniques this, and limited tool interfaces, are a clear burden, at least for someone who want to become a power user, or for design environment architects, who need sometimes to go beyond the standard, like

- doing parameter fits (e.g., for process monitoring, see Figure 1.10),
- applying matrix analyses (e.g., for eigenvalues, see Figures 5.16 and 8.16),
- executing special kinds of spectral analyses (for communication systems),
- creating special statistical plots (for risk quantification), etc.

Another requirement is for sure having more automation for these things, like running a verification, and starting an optimizer if needed, or creating a certain well-defined graph and putting it into a document.

A similar problem for designers, but even more for programmers, is netlisting. There are well-defined formats, but depending on design source and simulator capabilities the work can be cumbersome; and in the background tricky things can happen. For instance, changing the *number* of segments of a resistor can lead to a different netlist *structure*, and *multiple* parameters (like segment *length*), not only the segmentation number parameter itself, could change. These modifications are applied by callback functions, and in some cases, like optimization or sensitivity analysis, these changes can cause consistency problems if you forget to trigger these callbacks. Netlisting is also a perfect example showing the limitations of pure tool-centric improvements: Simulators can perform parameter sweeps much faster than any tool on top of the simulator, e.g., because the simulator can directly re-use the results from the previous simulation as starting point for the current one; and in addition there is no file generation overhead. However, callbacks for netlisting or parameter constraints limit the application of simulator-internal capabilities often. So over-all, having "variables everywhere" in a true variation-aware design flow is very challenging.

The challenge is often in the fight between efficiency, flexibility and user-friendliness. For instance, of course, current simulators can do a Monte-Carlo analysis efficiently, and the setup is easy, but having more flexibility, like applying advanced techniques (auto-stop, multivariate modeling, sorting, worst-case distances, optimizations with restart capabilities, etc.), there is unfortunately no standard interface at all! Actually, often analog designers need both: very high efficiency for being able to treat many blocks and complex problems, but they also need to be able to dig into the real details; and following a kind of standard could be also a significant burden regarding speed. For the C_{GPK} or for SSS the implementation effort is quite moderate, because simulators have built-in random number generators, but for other algorithms (like multi-variate modeling or for using more advanced sampling methods) efficient interfaces are often missing, and the workarounds to be taken are often non-optimum regarding runtime and implementation effort. So whatever you do, you often end up in a highly tool and technology specific solution, with limited application range and performance. A more clever solution would be having e.g., more flexible sample generators already in the general design environment, and fast, non-file-based communication channels between the different tools and solvers.

Actually, one way to go would be eliminating tool limitations step-by-step. Indeed internal improvements are usually possible; e.g., often LDS generators are applied to all included statistical variables in a PDK, but often a specific testbench only uses a small subset. This leads to a reduced LDS speed-up, especially in advanced technologies. Also typical environment features like scripting by run plans are clearly helpful. Indeed beside some limitations, plans offer a good compromise regarding user guidance, self-documentation, *and* flexibility!

However, for other issues, and for solving more challenging problems, it might be even easier to pick up directly best-practices from other fields, like math programs or software environments. For instance, math packages or even spreadsheet programs are already good or even perfect for comparing different MC results, whereas EDA environments are seldom good in this, just because this is not really a standard design task. However, limited GUI (graphical user interface) capabilities are often indeed a problem, e.g., different *optimization* results often *need* to be compared, even very carefully, e.g., regarding parameter settings and achieved performances. How else could be decided for the best solution? Of course, one could e.g., extend the MC features in a certain tool, like allowing better comparisons and merging of MC results. This way the user could for instance combine the results from two MC runs, instead of spending much time to run a new even larger analysis from scratch. However, in a space exploration tool *owning* the variables *and* results, these tasks, or the merger of corner analysis results, would be a much more native task, something available out-of-the-box without any real programming effort. In an object-oriented environment such result merger would be as simple as copy paste in a schematic or in a spreadsheet program, or like concatenation in R (just x<-c(run1,run2).

One key problem is surely that in custom design *many* things matter and the programmers of EDA tools can be hardly experts in all fields, like numerics, statistics, analog design, user interfaces, programming interfaces, databases, versioning and issue tracking systems, distributed computing, etc. Of course, also the pure software architecture itself is challenging for a design environment with *full* variation-awareness and even synthesis capabilities. Users need nowadays compute server support, multi-tasking, etc. for convenient work, not always, but frequently. Nobody wants that the plotting window popping up stops all other works, or that a crash due to division-by-zero leads to data losses and a full environment restart, because nobody owns the data.

Interestingly some tasks were often easier in older environments with only minor GUI support or even full setup by text files, e.g., it is often easier to comment out a parameter in a textual optimization setup than doing the same in a graphical environment. This is because in the text file, you can keep your old setup in parallel, as comment, but in a GUI such features often require a lot of programming by the EDA vendor. Here having more object-orientation, powerful property editors (well-aligned with the schematic entry), profilers and debuggers, and "information at your finger-tipps" would help a lot.

Problem too big. ....? It would be indeed a great thing to have clearly more joint work among researchers, between universities and EDA vendors, and more joint company initiatives to find solutions – for a connected world, in which IC design is definitely an essential part. Probably no *more* initiatives are needed, but more *long-term* co-operations. Making connections is not so easy as it looks, e.g., there are standards for postlayout simulation, but beyond netlisting many more features are desirable, like support for design hierarchy, re-use of simulation results (extending a MC run, applying Bayesian techniques, etc.), selective stitching capabilities regarding EAD and LDE or regarding different blocks, sensitivity analysis, backannotation to schematics, cross-probing to layout, global standardized support for constraints, e.g., related to layout style, electromigration and IR drop, treating technology corners and statistics, etc.

Currently, often you can still rely on experience and anticipation, just take the risk or accept a non-optimum design, and tape-out. However, as the challenges are manifold, we can expect many interesting things to come. Indeed, having synthesis in mind, or just only all the great specific IC design techniques, it looks that, to some degree, the "automatic gap" mentioned in our introduction, is by far *not* only in IC *design*, it is in IC design *environments* too. Sometimes existing standard techniques only work to some point, and all further features will only work with a very specific tool set. This is also because custom IC design problems are often much more *complex* than anything people do in classical math or programming environments. Often a big team needs to work efficiently, so the focus on all EDA tools is typically much more on achieving short enough runtimes, not so much really on flexibility, and not that much on variation-awareness or even optimization.

However, do we really have a classical chicken and egg problem, or too big problems? Actually not, because math is compact, powerful, and elegant. There is no huge wild bunch which has to be implemented till anything works. Math in itself is something quite modular, e.g., one more math and optimization-centric environment is presented in [Jacobs2004], see Figure 11.7.

A more recent activity from Sandia is "Dakota" (Design and Analysis Kit for Terascale Applications, https://software.sandia.gov/trac/dakota), an extendable environment for design exploration and optimization. It offers

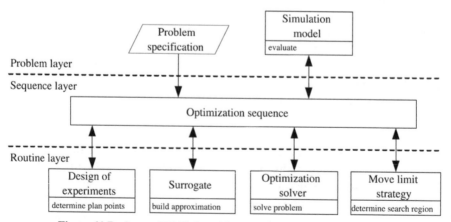

Figure 11.7   Layered UML-based framework for design optimization.

parallel-computing and many built-in algorithms, e.g., including quasi-Newton, Pareto and surrogate-based optimizers, plus a bunch of model generators and sampling methods for design space exploration. Obviously, such platform is also a good attempt for system integration and for documentation or reuse purposes; although of course still the devil can be in the details. For instance, in big designs not only the simulations should run in parallel, sometimes also the pre-processing or the pure result evaluation can take much time.

11.6 Last Words

We expect that many described techniques, being part of a kind of "second EDA revolution", will remain as *core* algorithms, but the design environment and tool vendors will be kept busy just to make sure that all methods work in nowadays complex context.

We think from time to time it is interesting to check how much progress (in research and in productive environments) we really have in analog and mixed-signal EDA environments (Table 11.2), or e.g., in astronautics and space flight (no table). Note, that this table is only according to the author's best knowledge and many features *cannot* be simply represented by *yes* or *no*, e.g., yield optimization becomes more difficult if the environmental variables are included or if the yield level is high. Also layout-awareness can have *many* levels, like fully automated layout generation, inclusion of aging, LDE and electromigration, or (practical, but much simpler) just only treating RC parasitics without updating them in the sizing loop. For instance, already in [Choudhury1993] an efficient method for parasitic-aware constraint-based design has been implemented. There the layout constraints are not only applied in simple terms like $C_p < 2fF$ or $(C_p1 - C_p2)/C_p1 < 5\%$ (see Section 10.3.2), but really circuit *performance*-based, which requires efficient sensitivity calculations (Figure 11.8). In this flow difficult tasks can be fully automated: critical couplings (e.g., from clock nets to an amplifier input) can be *identified*, then they will be *constrained*, and a place and route tool can be driven so that the constraints are kept; even shieldings can be inserted automatically. Topology selection might be a pure selection process (easy to program, but requiring a huge library) or could be trial-and-error (leading to many unnecessary simulations and long runtimes), or may include advanced construction methods.

For commercial application, the users would be typically different from the EDA tool programmers, so many more aspects become important.

Table 11.2   Status on advanced chip design techniques

Optimization Type	1	2	3	4	5
Circuit sizing	X	X	X	X	X
Yield target	X	X	X	X	X
Sizing with layout-awareness	–	X	–	–	X
Pareto optimization	–	–	X	X	X
Topology synthesis	–	–	–	X	X
Availability					
In research	X	X	X	X	X
Commercial	X	X	X	–	–
Comment and examples	No hot topic in research anymore, capturing the yield accurately is the major difficulty. [Javid2007] describes an advanced flow including environmental corners and SGB optimization.	[Graeb2009] offer full layout creation, in commercial tools post-layout netlists can be optimized, but it is assumed that the parasitics will not change much	Available since 2007, compared to #1 just the underlying optimizer is changed	[McConaghy2011], without Pareto yield optimization the topic is quite old	In principle a collection of existing techniques, e.g. [Choudhury1993] addresses the problem of layout parasitics

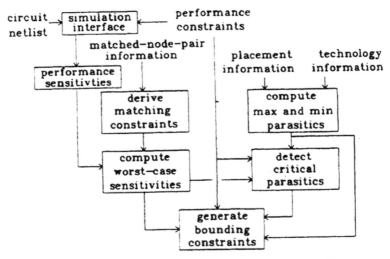

Figure 11.8   PARCAR flow described in [Choudhury1993].

For real-world application it also makes a difference if the design is a small 9-transistor op-amp or a real product like the 30-element LM 709 or the uA 741, which was already the state-of-the-art in 1969! Major practical commercial problems are reliability, IP issues, user-friendliness, license model, etc.

We have little doubt that a satisfying general analog synthesis solution is *technically* possible, but the decision on *how* to make a business out of it is difficult too. Should it be sold as tool, or should there be also a fee for the output, the generated IP? Or is (only) a service model suited? [Iskander2013] demonstrates a basically technology-independent approach at least for circuit sizing, but having a true non-vendor-specific standard (like in the digital domain) would be clearly helpful, because also in this schematic-centric flow several topics are still critical, like who should be able to make adoptions for different IC technologies and devices, for building blocks used in synthesis, for additional performance tests, new sizing strategies, etc.?

All-in-all, realistically, there is little chance for a real breakthrough "event" in analog synthesis, as the introduction of SPICE *was* for circuit simulation, or as the availability of advanced variation-aware environments *is* (e.g., [McConaghy2013, Zhang2016]). People like great operas, stories and personalities, but actually it is more realistic that no hero or good-talented individual, but a long or mid-term hard working team will enable the next bigger step, plus picking up many ideas and implementations available in the industry (not necessarily the silicon industry).

Wer meines Speeres Spitze fürchtet,   *Whosoever fears the tip of my spear*
der durchschreite das Feuer nie!   *shall never pass through the fire!*

In military, any mistake at the beginning, means dead people at the end. So IC designers are anyway already now in a much more comfortable situation; it is mostly only about time and money. Of course scientists are fearless, and mathematicians even more, so some people will probably pass the wall, the ring of fire, created by the loopy demigod Loge.

> **Problem too big? Or not urgent enough?** Indeed chip design is a 50-year *success* story. Often it is a matter of just doing it, take all your experiences, anticipate the challenges, specify, and execute. If something becomes too difficult, you often have fallback solutions, or just a certain feature will not be implemented. So chip design is often more about dealing with complexity, and following rules, but not so much about optimize, and optimize again. So there *are* frameworks for advanced multidisciplinary optimization, but even in this context construction methods are often working fine, i.e., optimization or even co-optimization (like thermal and electrical) of different parts like chip, packaging, and board is usually quite an exception, related to the most challenging parts. Regarding variations the situation is similar: There are standard techniques like corner and MC analysis, high yield methods, contribution analysis, etc., so the benefit of having "more" is limited for many parts of the design.
>
> In conclusion, there are also good reasons for a big diversity in EDA tools, for advanced techniques and addressing a variety of applications, and actually having *many* highly optimized software solutions, instead of one-fits-all.

Appendix

Web Resources

We present here only a short list, which does not reflect any preferences. Although such list of links can be hardly complete, we feel that indeed a lot of useful technical material is accessible this way, from non-commercial pages, but also from the EDA vendors, and others. Note, that deep links may change from time to time, but in such cases it is usually still to get the data with a search engine.

For math topics Wikipedia is generally a good starting point, here are few examples regarding optimization and kernel density fitting:

BFGS: http://en.wikipedia.org/wiki/BFGS_method
Conjugate Gradient: http://en.wikipedia.org/wiki/Conjugate_gradient_method
Brent-Powell: http://en.wikipedia.org/wiki/Powell%27s_method
Hooke-Jeeves: http://en.wikipedia.org/wiki/Pattern_search_(optimization)
KDE: https://en.wikipedia.org/wiki/Kernel_density_estimation

General Design Resources

General Circuit Design, Design IP, etc.:
https://www.semiwiki.com
https://www.semiwiki.com/forum/content/1673-brief-history-rtl-design.html
https://www.chipestimate.com/
http://www.design-reuse.com

Design in New Technologies, Layout-Awareness, etc.
http://www.techdesignforums.com/practice/technique/lde-layout-dependent-effects-fly
http://www.techdesignforums.com/practice/technique/five-key-challenges-20nm-custom-design/

http://www.analog-eetimes.com/news/electrically-aware-design-can-speed-ic-design-flow/page/0/1
http://www.eetimes.com/document.asp?doc_id=1280068

Silicon Foundries:
[TowerJazz], e.g., for material on constraint http://towerjazz.com/pdf/Cadence TGS.pdf
Taiwan Semiconductor Manufacturing Company: http://www.tsmc.com/
United Microelectronics Corporation: http://www.umc.com
Ams: http://asic.ams.com/
xfab: http://www.xfab.com/
GLOBALFOUNDRIES: www.globalfoundries.com

Synopsis:
http://http://www.synopsys.com
http://www.synopsys.com/Community/SNUG

Silvaco:
[SilvacoVM] http://www.silvaco.com

Cadence Design Systems:
https://www.cadence.com
https://community.cadence.com
https://www.cadence.com/content/cadence-www/global/en_US/home/services/cadence-academic-network.html

Material Directly Related to Custom Design:
[Dennison2010] Ian C. Dennison, M. Baker, B. Arsintescu, D. J. O'Riordan, System and method enabling circuit topology recognition with auto-interactive constraint application and smart checking, US Patent 7735036 B2, published 2010.
[Cadence2014]
https://community.cadence.com/cadence_blogs_8/b/cic/archive/2014/04/02/mismatch-contribution-analysis-in-ade-gxl
[CadenceVVO]
https://www.cadence.com/content/dam/cadence-www/global/en_US/documents/tools/custom-ic-analog-rf-design/virtuoso-variation-option-ds.pdf

[CadenceAdvNodes]
https://www.cadence.com/content/dam/cadence-www/global/en_US/documents/
tools/custom-ic-analog-rf-design/monte-carlo-analysis-at-advanced-nodes-wp.pdf
https://www.cadence.com/content/dam/cadence-www/global/en_US/documents/
tools/custom-ic-analog-rf-design/virtuoso-plan-based-analog-verification-wp.pdf
[CadenceVerif]
https://www.cadence.com/content/dam/cadence-www/global/en_US/documents/
tools/custom-ic-analog-rf-design/virtuoso-plan-based-analog-verification-wp.pdf
[Liu2015] Hongzhou Liu, Wangyang Zhang, Device mismatch contribution
computation with nonlinear effects, US Patent 8954910 B1, published 2015.
[Liu2013] Hongzhou Liu, Hui Zhang, Statistical corner extraction using
worst-case distance, US Patent 8589852 B1, published 2013.
[CadenceWCC]
https://community.cadence.com/cadence_blogs_8/b/cic/archive/2014/02/24/what-s-
the-worst-that-could-happen

MunEDA:
https://www.muneda.com
https://www.muneda.com/User-Group-Meetings

Solido:
http://www.solidodesign.com

Mentor Graphics:
https://www.mentor.com
https://www.mentor.com/products/ic_nanometer_design/blog

Intento Design:
http://www.intento-design.com

Thalia Design Automation:
http://www.thalia-da.com/

Miscellaneous, Universities, Organizations, etc.:
[Iastate]
https://wikis.ece.iastate.edu/vlsi/index.php/MonteCarlo_Simulations_using_
ADE_XL
[VeronA] https://www.edacentrum.de/projekte/VeronA

[SI2] http://projects.si2.org/?page=621
[Deepchip2014] http://www.deepchip.com/items/0541-07.html

Math Resources

General Math:
R Archive: https://cran.r-project.org/
Online Statistics Education: http://onlinestatbook.com/2/estimation/confidence.html
http://www.wolframalpha.com/input/?i=3d+plot
http://www.mathopenref.com/quadraticexplorer.html
http://www.math.uri.edu/~bkaskosz/flashmo/contours/combo.html

General Statistics and Confidence Intervals:
http://ion.chem.usu.edu/~sbialkow/Classes/3600/Overheads/Stat%20Narrative/statistical.html
http://statpages.org/confint.html
http://www.danielsoper.com/statcalc3/calc.aspx?id=85
https://www.coursehero.com/sitemap/schools/2623-University-of-Toronto-Toronto/departments/451922-STATS/

Eigenvalue Analysis and Correlation:
http://www.arndt-bruenner.de/mathe/scripts/engl_eigenwert2.htm
http://www.visiondummy.com/2014/04/geometric-interpretation-covariance-matrix/

Sampling Techniques and Misc:
Nice articles and downloads related to sampling methods like Poisson sampling are available e.g., under http://www.coderhaus.com/?p=11, at http://devmag.org.za/2009/05/03/poisson-disk-sampling and https://www.jasondavies.com/poisson-disc

A lecture on sampling & rendering (Computer Graphics, CMU 15-462/15-662, Spring 2016): https://www.cs.cmu.edu/~15462

Here you can find links and resources related to advanced sampling methods: http://marcoagd.usuarios.rdc.puc-rio.br/quasi_mc.html
https://spacefillingdesigns.nl

Here is a Matlab tutorial on Gaussian process modeling, suitable for performance modeling, worst-case searches, global optimization, etc. https://www.mathworks.com/help/stats/gaussian-process-regression-models.html

Optimization:
Evolutionary Multiobjective Optimization: http://delta.cs.cinvestav.mx/~ccoello
/EMOO/

Universities with Excellent Online Material:
University of Florida, Prof. Dr. Nam-Ho Kim: http://www2.mae.ufl.edu/nkim/
Courant Institute of Mathematical Sciences: https://www.cims.nyu.edu/
MIT Engineering Systems Division: https://esd.mit.edu/about.html

Miscellaneous Math-Related Webpages:
JMP (popular general statistical software): http://www.jmp.com
Vose Software (risk analysis): http://vosesoftware.com
Mathworks/Matlab (popular math software): http://www.mathworks.com
Sandia Dakota (a Design and Analysis Kit for Terascale Applications):
https://software.sandia.gov/trac/dakota

Bibliography

To give the reader a better overview, we sorted the references into different
categories.

Note: You can find further (non-printed) material under "Resources from EDA
and other Vendors" and "Useful Web-Based Tools and Links".

Scientific Journals and Magazines

The most up-to-date material and past journal articles you can find, e.g., in
these magazines:

- IEEE Solid-State Circuits Magazine
- IEEE Journals Solid-State Circuits
- IEEE Transactions on Circuits and Systems
- IEEE Transactions on Computer-Aided Design of Integrated Circuits
  and Systems
- IEEE Transactions on Electron Devices

Numerics, Statistics, Sampling, Modeling

General:

- W. H. Press (1992). *Numerical Recipes in C: The Art of Scientific
  Computing*. Cambridge: Cambridge University Press.

- R. E. Walpole et al. (2012). *Probability & Statistics for Engineers & Scientists*, 9th Edn, Upper Saddle River, NJ: Prentice Hall.
- Available at: http://ion.chem.usu.edu/~sbialkow/Classes/3600/Over heads/Stat%20Narrative/statistical.html+Wikipedia!
- H. Schmid and A. Huber (2014). The 3sigma fallacy. *IEEE Solid-State Circuits Mag.* 6, 52–58.
- R. Hoekstra et al. (2014). Robust misinterpretation of confidence intervals. *Psychon. Bull. Rev.* 21, 1157–1164.
- E. T. Jaynes (2003). *Probability Theory the Logic of Science*. Cambridge: Cambridge University Press.
- R. Cotton (2011). *Learning R*. Sebastopol, CA:O'Reilly.
- C. D. Lin and B. Tang (2015). *Handbook of Design and Analysis of Experiments*. Boca Raton, FL: CRC Press.

C_{PK}, Yield Analysis and Normality Tests:

- S. Kotz and N. Johnson (1993). *Process Capability Indices*. Abingdon: Taylor & Francis,.
- C.-C. Wu and H.-L. Kuo (2004). Sample size determination for the estimate of process capability indices. *Inform. Manag. Sci.* 15, 1–12.
- S. V. Shinde (2014). "Intel-22nm squelch yield analysis and optimization," in *Proceedings of the International MultiConference of Engineers and Computer Scientists,* Vol. 2. Hong Kong A. Agresti (2003). Dealing with discreteness: making 'exact' confidence intervals for proportions, differences of proportions, and odds ratios more exact. *Stat. Methods Med. Res.* 12, 3–21.
- S. Weber, T. Ressurreicao and C. Duarte (2016). Yield Prediction with a New Generalized Process Capability Index Applicable to Non-Normal Data. *IEEE Trans. Comput. Aided Design Integ. Circuits Syst.* 35:931f.
- A. MacDonald et al. (2011). A flexible extreme value mixture model. *Comput. Stat. Data Anal.* 55, 2137–2157.
- L. Jäntschi and S. D. Bolboaca (2009). Distribution Fitting: Pearson-Fisher, Kolmogorov-Smirnov, Anderson-Darling, Wilks-Shapiro, Cramer-von-Misses and Jarque-Bera statistics. *Bull. Univ. Agric. Sci. Vet. Med. Cluj Napoca.* 66, 691–697.

Sampling Methods:

- J. Jaffari, and M. Anis (2011). On efficient LHS-based yield analysis of analog circuits. *IEEE Trans. Comput. Aided Design Integr. Circuits Syst.* 30, 159–163.

- A. Singhee and R. A. Rutenbar (2010). Why Quasi-Monte Carlo is Better Than Monte Carlo or Latin hypercube sampling for statistical circuit analysis. *IEEE Trans. Comput. Aided Design Integr. Circuits Syst.* 29, 1763–1776.
- L. Kocsis and W. J. Whiten (1997). Computational investigations of low-discrepancy sequences. *ACM Trans. Math. Softw.* 23, 266–294.
- H. Faure and Chr. Lemieux (2008). *Generalized Halton Sequences in 2008: A Comparative Study.* New York, NY: ACM.
- M. Sobol, D. Asotsky, A. Kreinin and S. Kucherenko (2011). Construction and comparison of high-dimensional sobol' generators. *Wilmott*, 56, 64–79.
- R. Cools, F. Y. Kuo and D. Nuyens (2006). Constructing embedded lattice rules for multivariate integration. *SIAM J. Sci. Comput.*, 28, 2162–2188.
- F.-M. De Rainville, C. Gagné, O. Teytaud and D. Laurendeau (2012). Evolutionary optimization of low-discrepancy sequences. *ACM Trans. Model. Comput. Simul.* 22:25.
- J. Matousek (1998). On the L2-discrepancy for anchored boxes. *J. complex.* 14, 527–556.
- V. R. Joseph and Y. Hung (2008). Orthogonal-maxmin latin hypercube designs. *Stat. Sin.* 18, 171–186.
- E. Thiémard (2001). An algorithm to compute bounds for the star discrepancy. *J. Complex.* 17, 850–880.
- T. Ebert, T. Fischer, J. Belz, T. O. Heinz, G. Kampmann and O. Nelles (2015). "Extended deterministic local search algorithm for maximin latin hypercube designs," in *Proceedings of the IEEE Symposium Series on Computational Intelligence*, Cape Town.
- F. A. C. Viana, G. Venter and V. Balabanov (2010). An algorithm for fast optimal latin hypercube design of experiments. *Int. J. Num. Methods Eng.* 82, 135–156.
- V. D. Husslage, D. Hertog and M. Maximin (2007). Latin hypercube designs in two dimensions. *Operat. Res.* 55, 158–169.
- A. Pilleboue et al. (2015). *Variance Analysis for Monte Carlo Integration.* New York, NY: ACM, 32.
- V. R. Joseph and Y. Hung (2008). Orthogonal-maximin latin hypercube designs. *Stat. Sin.* 18, 171–186.
- A. MacCalman (2012). *Experimental Space-Filling Designs For Complicated Simulation Outpts.* Ph.D. thesis, Naval Postgraduate School, Monterey, CA.

- S. Laine and T. Aila (2006). A Weighted error metric and optimization method for antialiasing patterns. *Comput. Graphics Forum* 25, 83–94.
- L. P. Werner and G. Müller (2012). Design of computer experiments: space filling and beyond. *Stat. Comput.* 22, 681–701.
- J. Dick and F. Pillichshammer (2014). *Digital Nets and Sequences: Discrepancy Theory and Quasi-Monte Carlo.* Cambridge: Cambridge University Press.

Multi-Dimensional Modeling:

- K.-W Pieper and E. Gondro (2008). Statistical modeling with backward propagation of variance and covariance equation. *Paper Presented at the Infineon Technologies CDNLive*, Munich.
- M. Neamtu (2001). *What is the Natural Generalization of Univariate Splines to Higher Dimensions?* Nashville, TN: Vanderbilt University, 355–392.
- D. R. Jones (2001). A taxonomy of global optimization methods based on response surfaces, *J. Glob. Optim.* 21, 345–383.
- A. P. Erikson and K. Astrom (2012). *On the Bijectivity of Thin-Plate Splines, in Analysis for Science, Engineering and Beyond.* Berlin: Springer.
- I. Guerra-Gomez, T. McConaghy and E. Tlelo-Cuautle (2015). "Study of regression methodologies on analog circuit design," in *Proceedings of the Test Symposium 16th Latin-American*, Puerto Vallarta.
- W. Daems, G. Gielen and W. Sansen (2002). "An efficient optimization-based technique to generate posynomial performance models for analog integrated circuits," in *Proceedings of the Design Automation Conference,* New Orleans, LA.
- X. Li and H. Liu (2008). "Statistical regression for efficient high-dimensional modeling of analog and mixed-signal performance variations," in *Proceedings of the Design Automation Conference*, Anaheim, CA.
- A. Lange, C. Sohrmann, R. Jancke, J. Haase, B. Cheng, A. Asenov., et al. (2016). Multivariate modeling of variability supporting non-gaussian and correlated parameters. *IEEE Trans. Comput. Aided Des. Integr. Circ. Syst.* 35, 197–210.
- J. Tropp and A. Gilbert (2007). Signal recovery from random measurements via orthogonal matching pursuit. *IEEE Trans. Informat. Theory* 53, 4655–4666.

- U. Groemping (2006). Relative importance for linear regression in R: the package relaimpo. *J. Stat. Softw.* 17:1.
- W. Zhang, T. Chen, M. Ting and X. Li (2011). Toward efficient large-scale performance modeling of integrated circuits via multi-mode/multi-corner sparse regression. *Des. Autom. Conf.* 2011, 897–902.
- F. Wang, P. Cachecho, W. Zhang, S. Sun, X. Li, R. Kanj, C. Gu (2013). "Bayesian model fusion: large-scale performance modeling of analog and mixed-signal circuits by reusing early-stage data," in *Proceedings of the 50th Annual Design Automation Conference to be Published in IEEE Transactions on Computer-Aided Design of Integrated Circuits and Systems*, Pittsburgh, PA.
- Liu et al. (2014). Computing Device Mismatch Variations Contributions. US 8,813,009 B1.
- H. Moon, T. S. Santner and A. Dean (2010–2015). Two-stage sensitivity-based group screening in computer experiments. *Technometrics* 54, 376–387.
- D. C. Woods and S. M. Lewis (2015). *Design of Experiments for Screening*. Southampton: Southampton Statistical Sciences Research Institute University of Southampton.

Advanced Yield and Worst-Case Analysis:

- T. McConaghy (2013). *Variation-Aware Design of Custom Integrated Circuits: A Hands-on Field Guide*. Berlin: Springer.
- T. H. Fischer, T. Nirschl, B. Lemaitre and D. Schmitt-Landsiedel (2006). Modelling of the parametric yield in decananometer SRAM-Arrays. *Adv. Radio Sci.* 4, 281–285.
- L. Zhang, H. Liu and S. Lewis (2016). *White Paper: Accelerating Monte Carlo Analysis at Advanced Nodes*. San Jose, CA: Cadence Design Systems, Inc.
- A. Singhee, R. A. Rutenbar (2007). "Statistical blockade: a novel method for very fast monte carlo simulation of rare circuit events, and its application," in *Design, Automation, and Test in Europe*, eds R. Lauwereins and J. Madsen (Springer: Amsterdam), 235–251.
- C.-C. Kuo et al. (2012). "Efficient trimmed-sample monte carlo methodology and yield-aware design flow for analog circuits," in *Proceedings of the Design Automation Conference*, San Francisco, CA.
- R. V. Joshi, R. N. Kanj, S. R. Nassif and C. J. Radens (2012). *Statistical Design with Importance Sampling Reuse*, US patent 20120046929 A1.

- S. Sun, X. Li, H. Liu, K. Luo, and B. Gu (2015). Fast statistical analysis of rare circuit failure events via scaled-sigma sampling for high-dimensional variation space. *IEEE Trans. Comput. Aided Des. Integr. Circuits Syst.* 34, 1096–1109.
- S. Sun and X. Li (2014). "Fast statistical analysis of rare circuit failure events via subset simulation in high-dimensional variation space," in *Proceedings of the 2014 IEEE/ACM International Conference on Computer-Aided Design* (IEEE Press: Piscataway, NJ.
- H. Zhang, T. H. Chen, M.-Y. Ting and X. Li (2009). "Efficient design-specific worst-case corner extraction for integrated circuits, in *Proceedings of the 46th Annual Design Automation Conference.* New York, NY: ACM.
- J. B. Kuang (2012). Designing Robust Logic and Array Circuits in the Presence of Variability, StatDes Glasgow.
- S. Jallepalli et al. (2016). Employing scaled sigma sampling for efficient estimation of rare event probabilities in the absence of input domain mapping. *IEEE Trans. Comput. Aided Des. Integr. Circuits Syst.* 35, 943–956.
- C. Gu and J. Roychowdhury (2008). "An efficient, fully nonlinear, variability-aware non-monte-carlo yield estimation procedure with applications to SRAM cells and ring oscillators," in *Proceedings of the 2008 Asia and South Pacific Design Automation Conference*, (Los Alamitos, CA: IEEE Computer Society Press).
- D. Roos (2011). "Multi-domain adaptive surrogate models for reliability analysis," in *Proceedings of the 9th International Probabilistic Workshop*, Braunschweig.
- J. Wang, A. Singhee, R. A. Rutenbar and B. H. Calhoun (2010). Two fast methods for estimating the minimum standby supply voltage for large SRAMs. *IEEE Trans. Comput. Aided Des. Integr. Circuits Syst.* 29, 1908–1920.
- S. Sun and X. Li (2015). *Fast Statistical Analysis of Rare Circuit Failure Events via Baysesian Scaled-Sigma Sampling for High-Dimensional Variation Space.* Piscataway, NJ: IEEE.
- X. Pan and H. Graeb (2012). Reliability optimization of analog integrated circuits considering the trade-off between lifetime and area. *Microelectron. Reliab.* 52, 1559–1564.
- S. Weber and I. Syranidis (2015). Workshop: Variation-Aware Design for RF Engineers, 41st European Solid-State Circuits Conference,

September 18, 2015. (https://www.joanneum.at/fileadmin/downloads/esscirc-essderc/05_Workshop_Weber_Variation-Aware-Design_klein.pdf)

- D. White (2013). A new methodology to address the growing productivity gap in analog design, IEEE/ACM International Conference on Computer-Aided Design (ICCAD).
- Á. Bűrme and H. Habal (2015). Computing Worst-Case Performance and Yield of Analog Integrated Circuits by Means of Mesh Adaptive Direct Search, Journal of Microelectronics, Electronic Components and Materials, Vol. 45, No. 2, 160–170.

Optimization

Basics and Benchmarks:

- M. J. Box (1966). A comparison of several current optimization methods. *Comput. J.* 9, 67–77.
- J. W. Bandler (1987). An automatic decomposition approach to optimization of large microwave systems. *Microw. Theory Techniques* 35, 1231–1239.
- N. Hansen et al. (2010). "Comparing results of 31 algorithms from the black-box optimization benchmarking BBOB-2009," in *Proceedings of the 12th Annual Conference Companion on Genetic and Evolutionary Computation.* (New York, NY: ACM).

Local Optimization:

- D. Agnew (1978). Efficient use of the hessian matrix for circuit optimization. *Circuit Syst.* 25, 600–608.
- McKeown, An Introduction to Unconstrained Optimization, ISBN-10: 0750300256.

Advanced Topics:

- A. Somani et al. (2007). An evolutionary algorithm-based approach to automated design of analog and RF circuits using adaptive normalized cost functions. *Trans. Evol. Comput.* 11, 336–353.
- A. Skajaa (2010). *Limited Memory BFGS for Nonsmooth Optimization.* Master's thesis, New York University, New York, NY.
- M. Pehl, T. Massier, H. Graeb and U. Schlichtmann (2008). A Random and Pseudo-Gradient Approach for Analog Circuit Sizing with Non-Uniformly Discretized Parameters. in *Proceedings of the IEEE*

International Conference on Computer Design (ICCD) (San Francisco, CA: IEEE), 188–193.

- T. Massier, H. Graeb, Senior Member, IEEE and U. Schlicht-mann (2008). The sizing rules method for CMOS and bipolar analog integrated circuit synthesis, *IEEE Trans. Comput. Aided Des. Integr. Circuits Syst.* 27, 2209–2222.

- A. S. O. Hassan, et al. (2014). Yield optimization via trust regions.

- R. Schwencker, F. Schenkel, M. Pronath and H. Graeb (2002). "Analog circuit sizing using adaptive worst-case parameter sets," in *Proceedings of the 2002 Design, Automation and Test in Europe Conference and Exhibition*, Paris.

- B. D. Smedt and G. Gielen (2003). "HOLMES: Capturing the yield–optimized design space boundaries of analog and RF integrated circuits," in *Proceedings of the Design, Automation and Test in Europe Conference and Exhibition*, Grenoble.

- S. Ali, R. Wilcock, P. Wilson and A. Brown (2008). "Yield model characterization for analog integrated, circuit using pareto-optimal surface," in *Proceedings of the IEEE International Conference on Electronics, Circuits, and Systems*, (Piscataway, NJ: IEEE).

- S. Li, M. Tan, I. W. Tsang, and J. T. Y. Kwok (2011). A hybrid PSO-BFGS strategy for global optimization of multimodal functions. *IEEE Trans. Syst. Man Cybern.* 41, 1003–1014.

- J. Zou (2009). *Hierarchical Optimization of Large-Scale Analog/Mixed-Signal Circuits Based-on Pareto-Optimal Fronts*. Ph.D. thesis, Technische Universität München, München.

- S. Shan and G. G. Wang (2011). Turning black box into white function. *J. Mech. Des.* 133.

- G. Gary Wang (2003). Adaptive response surface method using inherited latin hypercube design points. *J. Mech. Des.* 125, 210–220.

- B. Liu, Y. He, P. Reynaert, G. Gielen (2011). "Global optimization of integrated transformers for high frequency microwave circuits using a gaussian process based surrogate model," in *Proceedings of the Design Automation and Test in Europe Conference and Exhibition*, Grenoble, 1–6.

- M. Moustapha et al. (2016). Quantile-based optimization under uncertainties using adaptive Kriging surrogate models. *Structural and Multidiscipl. Optim.* 54, 1403–1421.

- R. K. Brayton, S. W. Director, G. D. Hachtel (1980). Yield Maximization and Worst-Case Design with Arbitrary Statistical Distributions. *IEEE Trans. Circuits Syst.* 27, 756–764.

- A. A. Javid, A. Seifi (2007). The use of stochastic analytic center for yield maximization of systems with general distributions of component values. *Appl. Math. Model.* 31, 832–842.
- A. C. Kammara, L. Palanichamy and A. König (2016). Multi-objective optimization and visualization for analog design automation, *Complex Intell. Syst.* pp. 251–267.

Circuit Design and Layout

- W. M. C. Sansen (2007). *Analog Design Essentials.* New York, NY: Springer.
- T. C. Carusone, D. Johns, and K. Martin (2012). *Analog Integrated Circuit Design.* Hoboken, NJ: John Wiley.
- R. J. Baker (2010). *CMOS Circuit Design, Layout, and Simulation.* Hoboken, NJ: Wiley.
- R. A. Pease (1991). *Troubleshooting Analog Circuits.* Oxford: Butterworth-Heinemann.
- B. Murmann (2015). *The Race for the Extra Decibel.* IEEE Solid-State Circuits Magazine.
- H. Tuinhout, N. Wils, and P. Andricciola (2010). Parametric mismatch characterization for mixed-signal technologies. *IEEE J. Solid State Circuits* 45, 1687–1696.
- R. A. Hastings (2006). *The Art of Analog Layout.* Upper Saddle River, NJ: Pearson Prentice Hall.
- D. Binkley (2008). *Tradeoffs and Optimization in Analog CMOS Design.* Hoboken, NJ: Wiley.
- W. Sansen (2015). Minimum power in analog amplifying blocks. *IEEE Solid State Magazine.* 7, 83–89.
- J. He, S. Zhan, D. Chen, and R. L. Geiger (2009). Analyses of static and dynamic random offset voltages in dynamic comparators. *IEEE Trans. Circuits Syst.* 56, 911–919.
- B. Yu, X. Xu, S. Roy, Y. Lin, J. Ou, and D. Z. Pan (2016). Design for manufacturability and reliability in extreme-scaling VLSI. *Sci. China Inf. Sci.* 59:061406 doi: 10.1007/s11432-016-5560-6
- A. H. Osman (2016). *A GmC-Based Continuous-Time Delta-Sigma Modulator for Neural Readout Applications.* Master thesis, Albert-Ludwigs-Universität Freiburg, Freiburg.
- P. Fronts, J. C. Allen, D. Arceo, and P. Hansen (2008). Optimal lossy matching. *IEEE Trans. Circuits Syst.* 55, 497–501.

- W. Wu, et al. (2011). "Physical-based threshold voltage and mobility models including shallow trench isolation stress effect on nMOSFETs", 2011. *IEEE Trans. Nanotechnol.* 10, 875–880.
- M. Strasser, M. Eick, H. Graeb, U. Schlichtmann, and F. M. Johannes (2008). "Deterministic analog circuit placement using hierarchically bounded enumeration and enhanced shape functions," *Proceedings of the 2008 IEEE/ACM International Conference on Computer-Aided Design*, San Jose.
- F. Lemery (2013). 20nm Automated Full Custom Layout Design Flow, STMicroelectronics, CDNLive Munich, 2013.
- T. Kobayashi, K. Nogami, T. Shirotori, and Y. Fujimoto (1992). "A current-mode latch sense amplifier and a static power saving input buffer for low-power architecture," *Proceedings of the VLSI Circuits Symposium Digest of Technical Papers*, 28–29.

Design Flow, Technology, Synthesis, and Tools

- H. Graeb (2012). "ITRS2011 analog EDA challenges and approaches," in *Proceedings of the Design, Automation Test in Europe Conference Exhibition (DATE'2012)*, Dresden, 1150–1155.
- W. K. Chen (2009). *ComputerAided Design and DesignAutomation.* Boca Raton, FL: CRC Press.
- J. Haase et al. (2011). *Process Variations and Probabilistic Integrated Circuit Design.* New York, NY: Springer.
- J. Chen, M. Henrie, M. F. Mar, and M. Nizic (2012). *Mixed-Signal Methodology Guide.* San Jose, CA: Cadence Design Systems.
- M. Dietrich, and J. Haase (2011). *Process Variations and Probabilistic Integrated Circuit Design.* New York, NY: Springer.
- M. Orshansky, S. Nassif, and D. Boning (2007). *Design for Manufacturability and Statistical Design: A Constructive Approach.* New York, NY: Springer.
- J. Crossley, A. Puggelli, H.-P. Le, B. Yang, R. Nancollas, K. Jung, et al. (2013). "BAG: a designer-oriented integrated framework for the development of AMS circuit generators," *Proceedings of the IEEE/ACM International Conference on Computer-Aided Design (ICCAD'2013)*, (Rome: IEEE), 74–81.
- R. Iskander, M.-M. Louërat, and A. Kaiser (2013). Hierarchical sizing and biasing of analog firm intellectual properties. *Integr. VLSI J.* 46, 172–188.

- H. E. Graeb (1997). *Analog Design Centering and Sizing*. Dordrecht: Springer.
- R. Spence and R. S. Soin (1997). *Tolerance Design of Electronic Circuits*. London: Imperial College Press.
- A. Singhee and R. A. Rutenbar (2010). *Extreme Statistics in Nanoscale Memory Design*. Berlin: Springer Science & Business Media.
- T. McConaghy, P. Palmers, M. Steyaert, G. G. E. Gielen (2011). Trustworthy genetic programming-based synthesis of analog circuit topologies using hierarchical domain-specific building blocks. *IEEE Trans. Evol. Comput.* 15, 557–570.
- H. Graeb, F. Balasa, R. Castro-Lopez, Y.-W. Chang, F. V. Fernandez, P.-H. et al. (2009). "Analog layout synthesis – recent advances in topological approaches," in *Proceedings of the Design, Automation and Test in Europe, DATE 2009*, Nice.
- M. Eick, M. Strasser, K. Lu, U. Schlichtmann, and H. E. Graeb (2011). Comprehensive generation of hierarchical placement rules for analog integrated circuits. *IEEE Trans. Comput. Aided Des. Integr. Circuits Syst.*, 30, 180–193.
- V. Boos and J. Nowak (2011). "A tool framework for technology migration of analog circuits," in *Proceedings of ANALOG'11*, Erlangen.
- G. Venkatakrishnan and N. K. Kadali (2014). "'Dump what you need' – A Coverage Methodology to accelerate SoC Verification," in *Proceedings of the 2014 15th International Microprocessor Test and Verification Workshop*, (Washington, DC: IEEE Computer Society).
- J. R. R. A. Martins and A. B. Lambey (2013). Multidisciplinary design optimization: a survey of architectures, *AIAA J.* 51, 2049–2075.
- J. H. Jacobs, L. F. P. Etman, F. van Keulen, and J. E. Rooda (2004). Framework for sequential approximate optimization. *Struct. Multidiscip. Optim.* 27, 384–400.
- E. S. J. Martens and G. Gielen (2008). *High-Level Modeling and Synthesis of Analog Integrated Systems*, New York, NY: Springer.
- DAC02 Panel discussion (2002) Proceedings of the Design Automation Conference, New Orleans, LA.
- U. Choudhury and A. Sangiovanni-Vincentelli (1993). Automatic generation of parasitic constraints for performance-constrained physical design of analog circuits. *Trans. CAD* 12, 208–224.
- A.-C. Klevbrink (2005). Evaluation of Aptivia and a Place and Route tool, Linköpings tekniska högskola, Sweden, Linköping, 26th August 2005.

Software to the Book

For educational purpose, we add two free RealTime applications to complement our book. For instance, in a book, we can usually explain something, present a formula, and give an example, maybe some numbers, as table or graph. However, often you would need other example parameters or the formula might include a function which is not part of your pocket calculator. For such case and also to explain Monte-Carlo and optimization in general, you can start the apps on your Windows PC and get a kick-start to both topics. This way you can also get support for the questions and answers we present in several chapters—interactively and in real time.

Experience is helpful in analog design! Doing simulations can also extend experiences, but unfortunately both statistical analysis and optimizations can be very time-consuming. Therefore, often designers are just happy to run a 100-point MC analysis on a bigger testbench, inspect the results quickly, but often they are not aware on how large the statistical variations really are! Here it would be extremely helpful a run, e.g., 50 further MC analyses with 100 points, each with different seed values, and just to compare the different MC results quickly. Unfortunately, in real-world designs and design systems, this is usually impractical because often already a single 100-point MC analysis may take some hours; often you get no support at all for calculation across many different MC results!

Actually, each of your circuits is a kind of "distribution generator," or even a distribution inventor! And you can collect much experience by running different circuits and looking to different performances. Also the inspection of different just "mathematically" defined distributions (like log-normal, Student-t, Weibull, Cauchy, Gaussian mixes), e.g., with different number of MC points or different distribution parameters (like the degree of freedom as parameter of the Student-t distribution), is helpful and makes some fun. So supporting this kind of learning-by-doing is one key feature of our auxiliary software to the book! For instance, we explained that estimates for yield, mean, and sigma do not depend on design complexity, and this means you can transfer your experience on how many number of MC points you need one-to-one to any real design setup!

The 2nd RealTime app addresses optimization, and here we have a similar problem. In addition, it is sometimes hard for a designer to understand what an optimizer is really doing or to get a feeling how long an optimization will take. So we created an app executing some optimizations using the famous BFGS quasi-Newton algorithm, most of the built-in examples are related to RF design, and we optimize for up to eight variables. We also included some classical optimization problems like on a pure quadratic function or on Rosenbrook's "banana" function. You can execute the optimization, e.g., for a matching network, and inspect many optimization details to really see how the optimizer works, like how quickly the gradient becomes smaller, or which variables get optimized first or how different starting point impact the final solution—and much more. Also global optimization is demonstrated on the famous problem of the traveling salesman. Here simulated annealing is used and the user can even start a Monte-Carlo analysis on top of the optimization. This makes sense, because the result of such a stochastic optimization depends a little bit on chance. For global optimization, there is almost no guarantee to find the true optimum!

Disclaimer: Both apps run on Windows PCs and are tested under Windows XP, Windows 7, and Windows 8. You should have admin permissions and also read—write permissions for the working directory. The programs include algorithms very similar to the one found in many benchmarks, being integrated to EDA custom IC design environments, math packages, etc. They are created for educational purpose, and we give no warranty on the results.

The newest program versions are available under http://www.riverpublishers.com. Download the zip-file and unzip in your target directory. No need for an installation program.

Statistical Program RealTime MC

The program file name is MC.EXE, and you can start it via double-click from the Windows Explorer. A detailed documentation with tutorial is available as separate PDF file.

Optimization Program RealTime Match

The program file name is ANPASS.EXE and you can start it via double-click from the Windows Explorer. A detailed documentation with tutorial is available as separate PDF file.

Index

About the Authors

Stephan Weber (weberconnect@freenet.de) studied electrical engineering at the Technical University of Berlin. In 1990, he moved to the Hahn-Meitner Institute for Nuclear Physics. The focus here was SPICE and behavioral modeling of discrete semiconductors, integrated circuits, etc. After receiving his Ph.D. on semiconductor modeling, he moved to Siemens Technologies in 1995, where he was designing RF receivers, power amps, etc., in bipolar, SiGe, CMOS, and BiCMOS technologies for high-volume commercial applications. He holds over 20 patents on analog/RF circuits and system topologies. Now, he is a principal application engineer in the Munich Cadence Custom IC team, with focus on simulators and analog design enviroments. He also wrote many articles for European design magazines (*Elektronik, Elrad, Elektor*) and is the creator of the interactive EE encyclopedia *Elekta* (published with Franzis and Noble) in 1998. He is now working in the EDA industry, supporting and developing custom IC design software; and having fun with analog circuit design.

Cândido Duarte (candidoduarte@fe.up.pt) received the *Licenciatura* and Ph.D. degrees in electrical and computer engineering from the Faculty of Engineering of the University of Porto (FEUP), Portugal. Since 2009, he has been with the ECE Department at FEUP as a lecturer in courses of electronics, sensors, and instrumentation. He is also a researcher at INESC TEC (formerly INESC Porto). His scientific interests include wireless transceiver architectures, low-power mixed-signal design, and CMOS circuits for mobile communication systems.